Repl.

Economic Botany

McGRAW-HILL PUBLICATIONS IN THE BOTANICAL SCIENCES

Edmund W. Sinnott, *Consulting Editor*

ARNOLD   An Introduction to Paleobotany
CURTIS AND CLARK   An Introduction to Plant Physiology
EAMES   Morphology of the Angiosperms
EAMES   Morphology of Vascular Plants: Lower Groups
EAMES AND MACDANIELS   An Introduction to Plant Anatomy
HAUPT   An Introduction to Botany
HAUPT   Laboratory Manual of Elementary Botany
HAUPT   Plant Morphology
HILL   Economic Botany
HILL, OVERHOLTS, POPP, AND GROVE   Botany
JOHANSEN   Plant Microtechnique
KRAMER   Plant and Soil Water Relationships
KRAMER AND KOZLOWSKI   Physiology of Trees
LILLY AND BARNETT   Physiology of the Fungi
MAHESHWARI   An Introduction to the Embryology of the Angiosperms
MILLER   Plant Physiology
POOL   Flowers and Flowering Plants
SHARP   Fundamentals of Cytology
SINNOTT   Plant Morphogenesis
SINNOTT, DUNN, AND DOBZHANSKY   Principles of Genetics
SINNOTT AND WILSON   Botany: Principles and Problems
SMITH   Cryptogamic Botany
    Vol. I.   Algae and Fungi
    Vol. II.   Bryophytes and Pteridophytes
SMITH   The Fresh-water Algae of the United States
SWINGLE   Textbook of Systematic Botany
WEAVER AND CLEMENTS   Plant Ecology

There are also the related series of McGraw-Hill Publications in the Zoological Sciences, of which E. J. Boell is Consulting Editor, and in the Agricultural Sciences, of which R. A. Brink is Consulting Editor.

ECONOMIC BOTANY

*A Textbook of Useful Plants
and Plant Products*

Albert F. Hill
Research Fellow in Economic Botany
Harvard University

SECOND EDITION

New York   Toronto   London
McGRAW-HILL BOOK COMPANY, INC.
1952

ECONOMIC BOTANY

Copyright, 1937, 1952, by the McGraw-Hill Book Company, Inc. Printed in the United States of America. All rights reserved. This book, or parts thereof, may not be reproduced in any form without permission of the publishers.

8 9 10 11 12 13 14 15 - MP - 1 0 9 8 7

28789

THE MAPLE PRESS COMPANY, YORK, PA.

PREFACE TO THE SECOND EDITION

The impact of a second World War has served to bring home more forcibly than ever the dependence of our modern civilization on plants and plant products. Once the normal sources of supply were cut off and the exigencies of the war brought about an increased demand, many fibers, oils, fats, insecticides, and drugs, as well as rubber, were quickly classified as strategic materials. An adequate food supply, both for domestic use and for export to less fortunate countries, became of paramount importance.

Greater production both of food and of the strategic materials was imperative. In the case of domestic plants this involved merely an increase in acreage and the utilization of improved methods of cultivation and harvesting. Where foreign plants were concerned, attention was usually first directed to the possible introduction and establishment of these species in the Western Hemisphere. In cases where this was impracticable, it was necessary to secure satisfactory substitutes. Many formerly little-known species suddenly became of great importance.

The past few years have also seen many advances in the field of medicine and the utilization of drugs obtained from plants that were virtually unknown a decade or so ago. The result of all this effort has been a definite change in the agricultural and forest practices of the various countries, particularly those of this hemisphere. Some of these changes will be permanent, others more transitory in nature.

In the present revision an attempt has been made to take cognizance of these matters and to evaluate them as to their possible future significance. In consequence the subject matter has been completely revised and several chapters have been entirely rewritten. Some 140 additional species are discussed and new illustrations have been added.

In the present unsettled state of world affairs statistics of production in the several countries and of imports into the United States vary so from year to year that they are of little value. Consequently, it has seemed advisable to reduce such statistics to a minimum.

The list of species includes all new species discussed in the text, and the nomenclature has been carefully checked and brought up to date. In order not to be too unwieldy the bibliography is restricted to works that have been published since 1936.

The author wishes to express his appreciation to those of his friends and colleagues who have made suggestions and critical comments or otherwise furnished valuable assistance and advice during the progress of the work. He is particularly indebted to Professor Paul C. Mangelsdorf for his interest and friendly cooperation.

The author is also greatly indebted to Dr. Walter H. Hodge of the U.S. Department of Agriculture and Dr. Richard E. Schultes of the Botanical Museum of Harvard University for the use of many of their original photographs and to the Massachusetts Horticultural Society, the McGraw-Hill Book Company, and the Botanical Museum for permission to reproduce several illustrations which have appeared in their publications.

ALBERT F. HILL

CAMBRIDGE, MASS.
  December, 1951

PREFACE TO THE FIRST EDITION

For some years past there has been an ever-increasing feeling among educators that the average college courses in elementary science have fallen far short of meeting the needs of the average student. For the most part such courses have been conducted on the supposition that their sole purpose was to lay the foundation for further advanced work in their particular field. For the man who knows what he wants, this is essential. Many students, however, fall into other categories. Some take a first course because it is required; others to see whether or not they might become seriously interested in a subject; and still others out of idle curiosity or some less tangible reason. In such cases an elementary course should be so constituted as to be interesting and profitable to the extent of adding to the student's general fund of knowledge even if he does not continue in the field. In other words the course should have more of a cultural than a purely technical value. As Gager states it in the preface to his "General Botany," "A subject has cultural value in proportion to the number of human contacts it gives the pupil, the extent to which it broadens his views and extends his interests and sympathies."

The field of applied science, dealing with the practical or economic aspects of a subject, lends itself much better to such treatment than does the field of pure science. This is particularly true of botany. From earliest time plants have been intimately bound up with human existence. Not only have they played an important part in the everyday life of mankind, but they have had a profound influence on the course of history and civilization. A knowledge of the industrial, medicinal, and edible plants cannot fail to broaden one's outlook.

Even though the value of including a considerable amount of economic material in a beginning course in botany may be recognized, the limitations of time or various curriculum requirements usually render such a procedure impracticable. It should be possible, however, to offer at least a half-year course devoted to economic plants as a supplement to the usual first year's work. Such a course would appeal to students in chemistry, economics, and other fields, as well as to those interested particularly in plant science. Moreover, such a course in economic botany ought to be valuable to the science itself. Botany, more than any other science, has suffered from a lack of interest and appreciation on the part of the average person. Any attempt to educate the layman as to the importance of plants cannot fail to be productive of some beneficial results, and may help in establishing botany on a par with chemistry, physics, geology, and zoology in the eyes of the world.

The present book is the outgrowth of several years' experience in presenting a one-semester course dealing with economic plants. The material utilized is of necessity limited, for the whole field of economic botany is too vast a subject, and only the surface can be scratched. An attempt has been made to include the most important plants of America and other parts of the world insofar as they enter into international commerce. It has not seemed advisable to give the detailed morphology of the various species discussed, or to consider too fully their agricultural and commercial aspects. Such information can be obtained from supplementary readings which should be an integral part of the work of the course. A list of 89 important reference works is appended, and the instructor will find numerous articles available in current magazines, government bulletins, and similar sources. For the benefit of anyone interested in the taxonomic phases of the subject, a systematic list of the species discussed is appended.

Although intended primarily as a textbook, this work should have an appeal to the ordinary reader, since material of too technical a nature has been avoided as far as possible.

The author wishes at this time to express his sincere appreciation of all the assistance that has been granted him during the preparation of the book. He desires especially to thank Professor Oakes Ames, who has read the entire manuscript, for his constant interest and valuable suggestions; Professor Samuel J. Record, F. Tracy Hubbard, and Horace N. Lee, who have criticized various sections; and all others who have in any way contributed with advice and comments. Special thanks are due the staff of the Botanical Museum of Harvard University for their courtesy in placing the facilities of the museum at his disposal and for their friendly cooperation in many ways.

The author is also deeply indebted to many institutions and individuals who have contributed photographs for use as illustrations. In this connection his thanks are due the Bureau of Plant Industry and the Forest Service of the United States Department of Agriculture; the Botanical Museum, the Arnold Arboretum, and the Gray Herbarium of Harvard University; the Massachusetts State College; the University of Maine; the University of Minnesota; the Connecticut Agricultural Experiment Station at New Haven; Breck and Company; the United Fruit Company; E. L. Patch and Company; the United States Beet Sugar Association; the Minute Tapioca Company; and the following individuals: Professor S. J. Record, Professor H. W. Youngken, Professor W. H. Weston, Professor D. H. Linder, Dr. F. M. Dearborn, and Mr. R. E. Schultes.

The author further wishes to express his indebtedness to Ginn and

Company, the McGraw-Hill Book Company, Inc., P. Blakiston's Son & Company, The Macmillan Company, World Book Company, and the editors of the *American Journal of Pharmacy* for permission to reproduce various figures which have appeared in their publications, due credit for which is given in each instance.

ALBERT F. HILL

HARVARD UNIVERSITY
April, 1937

CONTENTS

Preface to the Second Edition v
Preface to the First Edition vii

INTRODUCTION

I. The Importance and Nature of Plant Products 1

INDUSTRIAL PLANTS AND PLANT PRODUCTS

II. Fibers and Fiber Plants 18
III. Forest Products: Wood and Cork 52
IV. Forest Resources 84
V. Tanning and Dye Materials 118
VI. Rubber and Other Latex Products 135
VII. Gums and Resins 151
VIII. Essential Oils 175
IX. Fatty Oils and Waxes 191
X. Sugars, Starches, and Cellulose Products 210

DRUG PLANTS AND DRUGS

XI. Medicinal Plants 242
XII. Fumitories and Masticatories 268

FOOD PLANTS

XIII. The History and Nature of Food Plants 286
XIV. The Major Cereals 296
XV. The Minor Cereals and Small Grains 319
XVI. Legumes and Nuts 335
XVII. Vegetables 359
XVIII. Fruits of Temperate Regions 385
XIX. Tropical Fruits 406

Food Adjuncts

XX. Spices and Other Flavoring Materials 436
XXI. Beverage Plants and Beverages 468

Appendix

Systematic List of Species Discussed 495
Bibliography 521
List of Visual Materials 525
Index 529

CHAPTER I

THE IMPORTANCE AND NATURE OF PLANT PRODUCTS

THE IMPORTANCE OF PLANTS AND PLANT PRODUCTS TO MANKIND

The average man is likely to consider himself as a being apart from the rest of the organic world, enabled by reason of his superior intellect to lead a self-sufficient and independent existence. He loses sight of the fact, or is ignorant of it, that he is absolutely dependent on other organisms for his very life, and his material happiness as well. His superior intelligence has made him more dependent rather than less so. Although various animal and mineral products contribute to his welfare, it is the plant kingdom that is most essential to man's well-being.

Man's dependence on plants for the essentials of his existence has been of paramount importance in his life since the human race began. Primitive man probably had few needs other than food and a little shelter. Civilization, however, has brought with it an ever-increasing complexity, and has increased man's requirements to an amazing degree. The man of today is no longer content merely to exist, with food and shelter as his only wants. He desires other commodities as well, and raw materials that can be converted into the many useful articles and products which contribute to his enjoyment of life, and which incidentally increase his debt to plants.

The three great necessities of life—food, clothing, and shelter—and a host of other useful products are supplied in great part by plants. An adequate food supply is, and always has been, man's most outstanding need. In the last analysis all his food comes from plants. To be sure he may eat the flesh of animals, but these lower animals are just as dependent on plants as man himself, and they are equally unable to manufacture any of their food from raw materials. Clothing and shelter, the other prime necessities of life, are derived in great part from plant fibers and from wood. Wood is one of the most useful plant commodities in the world today, and it played an even greater role in the past. Aside from its use as a structural material, wood is valuable as a source of paper, rayon, various chemicals, and fuel. Other types of fuel, such as coal and petroleum, make available for man the energy stored up by plants that lived and died ages ago. Drugs, used to cure disease and relieve suffering, are to a great extent plant products. Industry is dependent on plants for many of its raw materials. Cork; tanning materials and dyestuffs; the oils, resins, and gums used in making paints, varnishes, soap, and perfumes; and rubber, one of the most outstanding materials of modern civilization, are but a few of the valuable products obtained from plants.

Aside from their value as sources of food, drugs, and many of the raw materials of industrialism, plants are important to man in many other ways. The role of colorless plants in the economy of nature; the part that bacteria play in disease and many industries; and the effects of forests and other types of natural vegetation in controlling floods and erosion are but a few examples. The aesthetic value of plants has no small influence on man's enjoyment of life, as evidenced by the host of garden enthusiasts and flower lovers.

The production and distribution of plant products have a profound influence on the economic and social life of the nations of the world, affecting both domestic conditions and international relations, and even changing the course of history. It will not be possible within the limits of the present volume to consider the many aspects involved and their fundamental bearing on human affairs and activities. A few examples, however, may be permitted by way of illustration.

The maintenance of an adequate supply of food and raw materials for the use of industry is essential to the existence, as well as the prosperity, of any nation. Few countries are independent in this respect, with the result that foreign trade, with its many ramifications and consequences, plays a necessary and important part in the life of the world. When the population of a country is small, the problems involved are not very great. Most of the civilized nations, however, not only have a large population, but one that is entirely out of proportion to the country's ability to produce the necessities of life. This tendency to overpopulation in excess of the maximum possible production of food and raw materials is responsible for many of the difficulties and problems that harass the modern world, especially in the case of nations with a restricted land area. The necessity for finding an outlet for their excess population, which all too often is steadily increasing, and the desirability of adding to their domestic supply of commodities have been responsible, in great part, for the policies of aggression that many such countries have pursued in recent years. The story of Japan in Korea and Manchuria and of Italy in Ethiopia are cases in point.

In recent years various economic problems concerned with agriculture have become increasingly important in the United States, and in other countries as well. These have served to bring home more clearly than ever before the intimate relationship between plants and human welfare.

One of the most serious of these agricultural problems is concerned with overproduction, a condition that has frequently arisen in the history of agriculture. Whenever a large supply of any commodity is available for the market, it usually results in lower prices, which often fall below the figure at which a profit can be realized. A particularly serious case of overproduction had developed in the United States in 1929 when the failure of foreign markets and the low buying power at home combined to cause the piling up of a huge surplus of agricultural products. The lowering of prices which followed created such a great discrepancy between the cost of production and the prices received for the products that the farmers were threatened with wholesale bankruptcy and the welfare of the entire nation was impaired. The efforts of the government to deal with this problem through subsidies, crop reduction, crop adjustments, and other means are familiar to all. It has been estimated that there are still several million more acres under cultivation than are necessary to supply all the demands for farm products, both at home and abroad. If this is the case, an obvious method of combating further overproduction would be to remove some of these unnecessary acres from cultivation.

Another agricultural problem concerns the proper utilization of the land, and this is related to characteristics inherent in the plants themselves. The successful pursuit of agriculture in any area depends on the presence of certain environmental factors that are necessary for the particular crop concerned. Each species differs in its soil, moisture, temperature, and other requirements. Satisfactory growth and development can take place only if all these factors are present in proper amounts. This fact has often been ignored and agriculture has been carried on in regions utterly unsuited for crop production, particularly on a commercial scale, with consistently unsatisfactory yields and low financial return as the inevitable result. To remedy this situation, the retirement of these submarginal lands, as they are called, from agriculture has been advocated. This would make possible the utilization of the areas for forests, grazing, wild-life conservation, and human recreation, and at the same time would contribute to crop reduction. The resettlement of some of the farming population, which accompanies the abandonment of agriculture in such areas, obviously has a profound effect on human activities.

Still other agricultural problems are physical, rather than economic, in nature, and are concerned more with productivity than production. The practice of farming necessarily brings about the destruction of the natural vegetation, which has a protective function; this induces conditions that result in the deterioration of the soil. This deterioration may consist of the exhaustion of the mineral nutrients, which is not a serious matter since it can be compensated for by the use of fertilizers, or it may comprise the permanent loss of soil through erosion.

Erosion is caused primarily by the action of water and wind. In the case of water, two types of erosion are produced—sheet erosion and gully erosion. In the former a thin sheet of soil is gradually removed from slightly sloping fields. The process is hardly noticeable and, although widespread, it is not very destructive. Gully erosion, on the other hand, is brought about by the concentrated runoff of water and, where conditions of slope and soil are favorable, results in the formation of deeper and deeper gullies, which eventually render the area unfit for agriculture for all time. Several million acres in the Southern states have been made worthless as the result of this type of erosion. If it is allowed to continue unchecked, its results may be so serious that human life is rendered impossible and barren deserts are the outcome. This has been the case in many parts of China.

Wind erosion is always more or less active on loose and sandy soil, and it is greatly increased as the result of cultivation and overgrazing, which tend to deplete the moisture-containing humus and pulverize the soil. The growing of cereals, which require constant cultivation, is especially likely to bring about conditions that favor both wind erosion and water erosion. The serious situation that developed some years ago in the semiarid regions of the Great Plains is a case in point. Even though the district was unsuited to the purpose, extensive areas of the natural grassland vegetation were plowed up and planted to cereals. The breaking up of the soil and the unusual drought that occurred over a period of several years combined to make conditions exceedingly favorable for wind erosion. This was responsible for the great dust storms that prevailed in the area and brought widespread destruction in their wake, not only wearing away the soil in some places, but depositing the eroded material on fertile ground elsewhere, thus rendering countless additional acres unfit for agriculture, and even for human habitation for many years. It is essential that some sort of soil conservation be put into practice before it is too late. The policies involved in soil conservation include the preservation of soil fertility, the prevention of erosion, the promotion of better land utilization, the stabilization of eroded areas, and various types of crop adjustments.

Plants have been and still are responsible for many of the social ills that beset mankind. In times past the exploitation of workers in various fields of activity concerned with plants has had serious consequences. As examples may be cited slavery, which went hand in hand with the production of cotton in the southern United States; the cruel treatment of the native rubber workers in the Belgian Congo, which shocked the entire civilized world in years past; and more recently the plight of rubber collectors in Brazil.

At the present time the problem of the migratory farm laborer, the share cropper, and the working conditions of farm labor in general are much in evidence.

Perhaps the chief social problem for which plants are responsible is the narcotic drug habit and the illicit trade that has grown up around it. This constitutes one of the most serious aspects of our modern civilization.

The comments made in the foregoing pages, inadequate though they may be, may perhaps serve to give some idea of the many ways in which plants and plant products affect the welfare of mankind.

THE NATURE OF PLANT PRODUCTS

Before one can fully appreciate the importance of plants, however, some knowledge of their structure and activities is desirable. For plants do not manufacture fibers, gums, resins, starch, sugar, and the countless other materials of use to man from any altruistic motive. Each and every one of these products plays a definite role in the life of the plant itself. Some of them contribute directly to the welfare and maintenance of the plant, while others represent waste products of its various activities.

PROTOPLASM AND ITS ACTIVITIES

The living substance in plants, as in animals, is *protoplasm,* and it is this protoplasm which exhibits the various characteristics that distinguish living matter from nonliving. Protoplasm, for example, has a definite chemical composition peculiar to itself and unlike anything in the inorganic world. It is even more distinctive in its behavior. As a result of the normal activities of the organism the existing protoplasm is continually being used up or worn out. This destructive process is compensated for by a constructive phase in which new protoplasm is built up from raw materials. This dual process of waste and repair is constantly going on during the life of the organism, and constitutes its metabolism. Other manifestations of life which protoplasm possesses are its ability to grow, to reproduce, and to respond to stimuli. Finally protoplasm does not occur in a hit-or-miss manner, but has a definite organization. Every plant and animal consists of one or more infinitesimal units known as *cells* (Fig. 1). These little unit masses of protoplasm are the foundation stones of both the structure of the organism and its functional activities. In one-celled organisms all the vital processes are carried on in the single cell. In the higher plants and animals, where the number of cells is well-nigh countless, a division of labor occurs. Some cells will carry on one activity, while others will be adapted for different functions.

Cells that are utilized for some particular function are likely to be similar in structure and appearance, and are usually grouped together to constitute what is known as a *tissue.* The plant body comprises many such tissues, each of which carries on some special work. The *organs* of the plant, such as roots, stems, and leaves, are aggregations of tissues so situated that the particular function involved can be carried on to the best advantage.

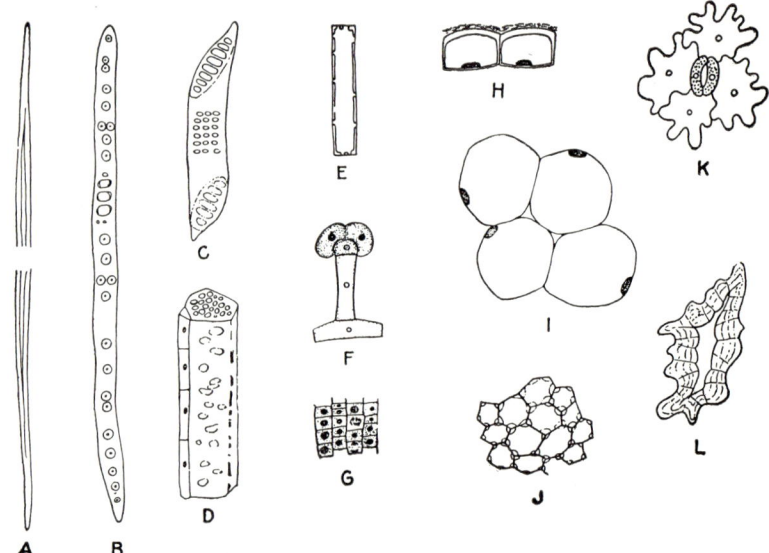

FIG. 1. Various types of cells. *A*, a wood fiber; *B*, a tracheid; *C*, a vessel cell; *D*, a sieve tube with its row of companion cells; *E*, a parenchyma cell from the wood; *F*, a glandular hair, consisting of several cells; *G*, a group of cells from a growing region; *H*, two epidermal cells in section; *I*, four thin-walled parenchyma cells from a storage region; *J*, a group of collenchyma cells; *K*, a stoma with its two guard cells, seen in face view; four adjacent epidermal cells are also shown; *L*, a very thick walled "stone cell" in sclerenchyma. (*Reproduced from Sinnott, Botany; Principles and Problems, McGraw-Hill Book Company, Inc.*)

PHOTOSYNTHESIS

The most significant of these functions in the life of the plant, and for that matter in the life of the whole organic world, is *photosynthesis*. This is the manufacture of food directly from raw materials of the inorganic world. With the exception of a few bacteria, green plants are the only living things that can actually make food. Animals and colorless plants, which do not have this ability, are in the last analysis absolutely dependent on green plants for their existence. Photosynthesis is carried on, in higher plants, chiefly in the leaves. Using the energy of sunlight, which is put to work through the agency of chlorophyll, the green coloring matter, carbon dioxide and water are combined to produce glucose (grape sugar), with oxygen as a by-product.

The grape sugar formed in photosynthesis is transported to every cell of the plant, and within the cells is used as a source of energy, or is further transformed by various physical and chemical processes into all the substances that play a part in the structure and life of the plant. In other words, the utilization of the photosynthetic sugar constitutes the plant's metabolism. This utilization takes several different forms, but there are five main processes involved: (1) the formation of the cell walls, which constitute the plant skeleton; (2) the manufacture of new protoplasm; (3) the elaboration of various food materials for immediate use or for storage as reserve foods; (4) the production of various secretions and excretions; and (5) the release of energy through the breaking down of the sugar as the result of respiration. We shall consider briefly the various substances formed during the four constructive processes involved in metabolism, indicating their importance to the plant and their usefulness to man.

THE PLANT SKELETON

The vast majority of plant cells are enclosed by a protecting and limiting structure known as the *cell wall*. These walls afford strength and rigidity to the organism, serving as a sort of skeleton. The walls are always composed of *cellulose*, either alone, or in combination with other substances. Cellulose is a nonliving material elaborated by the plant from grape sugar. Chemically it is a highly complex carbohydrate with the formula $(C_6H_{10}O_5)_n$. The cell walls, like the cells they enclose, are exceedingly variable in size and appearance. Certain types of cells have walls that are very much thickened, and these *sclerenchyma cells*, as they are called, are the most useful for supporting purposes. As the plant body increases in size, more and more support is required and various sclerenchyma tissues are formed, consisting chiefly of *fibers*. Fibers are long pointed cells with very thick walls and correspondingly small cavities. They tend to interlace and are capable of contracting and stretching. Some fibers, such as the hairs on cotton seeds, have walls that are almost pure cellulose. In other cases, as in the bast fibers found in the bark of many plants, some lignin is present. In the shorter wood fibers (Fig. 2) the walls are almost completely lignified. The presence of lignin greatly increases the strength of a wall without impairing its ability to conduct water. In the parts of a plant where a protective covering is necessary the normal cellulose walls may be infiltrated with waterproofing substances, such as cutin, suberin, or mucilage, all of which, and lignin as well, are manufactured by the plant. In a few instances inorganic materials, silica, for example, may be present in cell walls.

The same properties that make cell walls useful to plants are in many cases responsible for their usefulness to man. Wood, with its lignified walls, has manifold uses wherever a rigid but easily worked material is

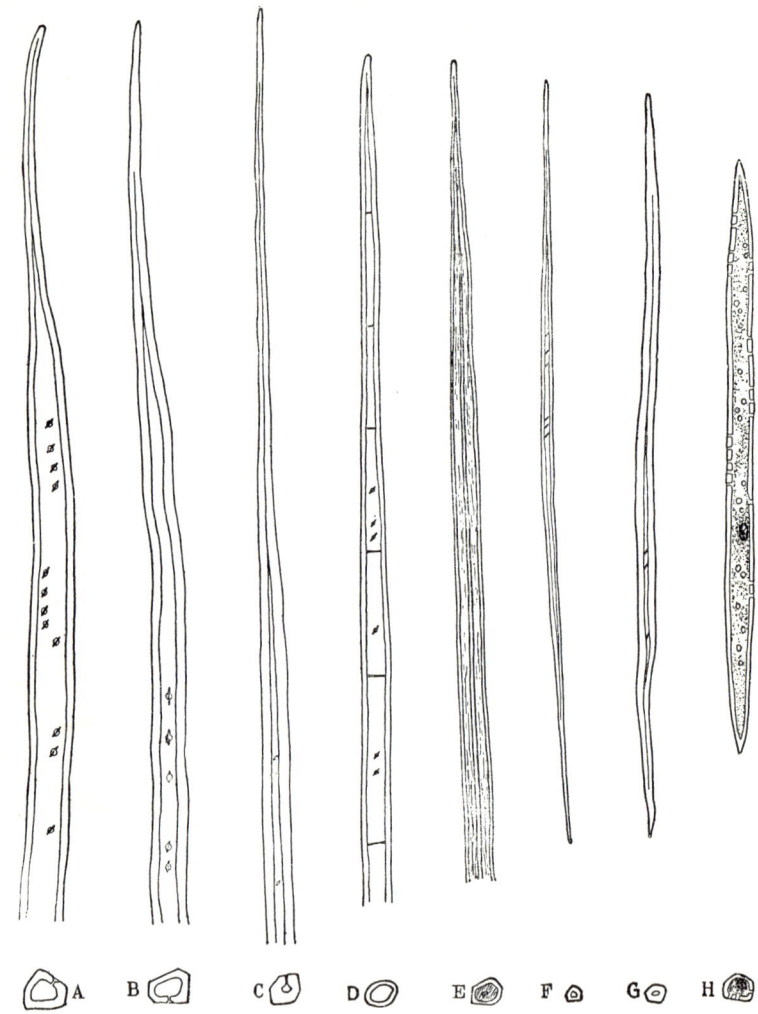

Fig. 2. Wood fibers. A, from *Pyrus Malus;* B, from *Liriodendron Tulipifera;* C, from *Quercus alba;* D, from *Swietenia Mahogani;* E, from *Quercus rubra;* F, from *Carya ovata;* G, from *Guaiacum sanctum;* H, from *Sassafras variifolium.* (Reproduced from Eames, Introduction to Plant Anatomy, McGraw-Hill Book Company, Inc.)

desirable. The more elastic fibers are the basis of the textile industry and, together with wood, constitute the chief raw materials of the paper industry. Cell walls that contain suberin furnish the cork of commerce. Walls that are nearly pure cellulose are utilized in the manufacture of synthetic fibers, explosives, cellophane, and many other industrial products. Since cellulose and its derivatives are highly combustible, all types of cell walls can be used as fuel. Not only is this true of present-day plants, but those of bygone ages as well. Coal is nothing more than the walls of plants which flourished during the Carboniferous Period of the earth's history and which have gradually lost their gaseous elements. A gradual succession of fuels, showing a progressive loss of hydrogen and oxygen, can be traced from cellulose to lignin, peat, soft coal, and hard coal.

LIVING PROTOPLASM

A considerable part of the sugar manufactured in photosynthesis is used directly in the formation of new protoplasm, to replace any that has broken down, and to provide for the growth of the individual. Protoplasm is a highly complex substance, and its chemical nature is but poorly understood, even though only familiar elements are involved. Among the substances that it contains are simple sugars and more highly elaborated carbohydrates; fats in various stages of synthesis; a large amount of protein material, derived in part from grape sugar and in part from nitrates absorbed from the soil; salts of various inorganic elements, such as iron, phosphorus, magnesium, sulphur, calcium, and potassium; and vitamins, enzymes, and other secretions. Living protoplasm is naturally of but little use to man, except as he may utilize fresh plant tissues for food. Our present custom of cooking most of our food greatly alters its original nature. It is not at all unlikely that primitive man, who used raw food, derived a greater benefit, owing to the presence of vitamins and the other protoplasmic constituents in an unimpaired condition.

RESERVE FOOD

Plants usually elaborate a much larger amount of food than can be used immediately for building up the plant body, or as a source of energy. This surplus is stored up in highly modified cells in special locations as a reserve supply to be utilized later for growth and other activities. Underground stems, roots, buds, and seeds are the chief storage organs. Three main types of food materials are manufactured by plants, and all three may occur as reserve food. These classes of foods include carbohydrates, fats, and proteins.

Carbohydrates

Carbohydrates are the simplest of the foodstuffs. They are compounds of carbon, hydrogen, and oxygen, in the proportion of two parts of hydrogen to one of oxygen. The principal carbohydrates are sugar, starch, and the various celluloses.

Sugar. The grape sugar that is manufactured by the plant in photosynthesis is almost universally present in plant cells. This basic material of metabolism, known also as glucose, has the formula $C_6H_{12}O_6$. It is sometimes stored up in large amounts, as in the stems of maize. Fruit sugar, or fructose, another product of photosynthesis, has the same formula, but slightly different properties. It is less common in plants, except in fruits.

The higher, more complex sugars are built up from these simple sugars. The most important of the higher sugars is cane sugar, or sucrose. This substance, which has the formula $C_{12}H_{22}O_{11}$, is accumulated in great quantities in sugar cane and sugar beets, and to a lesser degree in many other plants. All the sugars are soluble in water and so are readily available for use by the plant. They are highly nutritious and constitute a valuable food for the lower animals and man. Man utilizes these sugars, not only as they occur in plant tissues, but by extracting and purifying them as well.

Starch. The starches are insoluble compounds of a more complex nature with the formula $(C_6H_{10}O_5)_n$. Like the sugars, they are derived from grape sugar, and indeed constitute the first visible product of photosynthesis. Starch is the commonest type of reserve food in green plants and is of the greatest importance in their metabolism. Owing to its insoluble nature, however, starch must be digested, *i.e.*, made soluble, before it can be utilized. This is accomplished through the aid of enzymes that are present in the cells. Starch is stored in large thin-walled cells in the form of distinctive grains (Fig. 3). Man is very dependent on starch, which without question constitutes his most important plant food and plays a part in the industrial world as well.

Cellulose. Cellulose is the highest type of carbohydrate. We have already noted its presence in cell walls and discussed its function in that connection. It has little, if any, use as a reserve food, although there is some evidence that certain bacteria can make use of it.

Reserve Cellulose. These substances resemble cellulose physically, but differ in their chemical properties. They include hemicelluloses, pectins, gums, and mucilages. Some of these compounds have a dual role, aiding in the support of the cell walls and serving as reserve food as well. The hemicelluloses may gradually change into pectins, and then into gums.

Hemicellulose. These substances are often found as extra layers of cell walls, especially in the seeds of tropical plants, such as the date and ivory-nut palm. They are readily digested by plants, but only slightly so by man, and consequently have no food value. They are, however, of some use in the industries.

Pectins. Pectins or fruit jellies occur in most plant cells, particularly in fruits and vegetables. They are readily soluble in water and can be used as food by both plants and animals. Pectins also increase the water-holding capacity of cells. The middle lamella, the cementing material that holds cell walls together, consists of compounds of pectin. Pectins solidify after they have been extracted from the plant, and man takes advantage of this property in the preparation of jellies and jams.

Gums. Gums are derived by the breaking down of cellulose or other carbohydrate compounds, and consist of an organic acid in combination with inorganic salts. They may be secreted naturally in the tissues or may arise as the result of wounding. Gums aid in keeping water in the

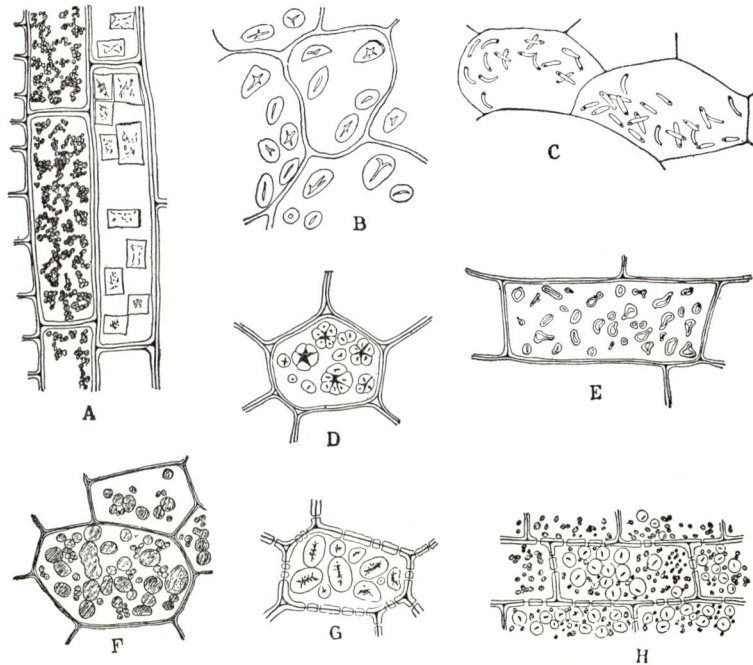

FIG. 3. Starch grains and tannin. Tannin: *A*, in phloem parenchyma of *Pinus* (also crystals); *F*, in pith cells of *Fragaria*; *H*, in ray cells of wood of *Pyrus Malus* (also starch grains). Starch grains: *B*, in pith cells of *Alsophila*; *C*, in outer pericarp of *Musa*; *D*, in cotyledon of *Pisum*; *E*, in ray cell of phloem of *Ailanthus*; *G*, in cotyledon of *Phaseolus*. (*Reproduced from Eames, Introduction to Plant Anatomy, McGraw-Hill Book Company, Inc.*)

plant, and also serve as a reserve food. Man uses them in the industries, in medicine, and as food.

Mucilages. Mucilaginous substances, closely related to gums, are widely distributed in the plant world. When moistened with water they do not dissolve, but form a slimy mass. They are secreted in hairs, sacs, or canals. Their function is varied and they may serve as reserve food, as an aid in checking the loss of water or too rapid diffusion, as a mechanism for water storage, and as a means for facilitating seed dispersal. Mucilage is often associated with cellulose in cell walls. Its chief use to man is in medicine.

Fats

Fats, like carbohydrates, are compounds of carbon, hydrogen, and oxygen, but with a very small amount of oxygen. For this reason they are often referred to as hydrocarbons. The formula for triolein, a typical fat, indicates their chemical nature, $C_{57}H_{104}O_6$. Fats are derived from carbohydrates by two processes: (1) the production of fatty acids, (2) the formation of glycerin. These two products unite to form the fats, which are either liquid or solid in nature. In the former state fats are usually spoken of as oils, or fatty oils, and occur in the form of small globules. Fats are present in small amounts in all living protoplasm, but are stored up as reserve food chiefly in seeds and fruits. They are insoluble and have to be digested before they can be utilized. Because of their high energy content they are a valuable food for both plants and animals. Fats also are important in medicine and in industry.

Proteins

Proteins constitute the third type of reserve food. They are likewise derived, at least in part, from carbohydrates through the formation of

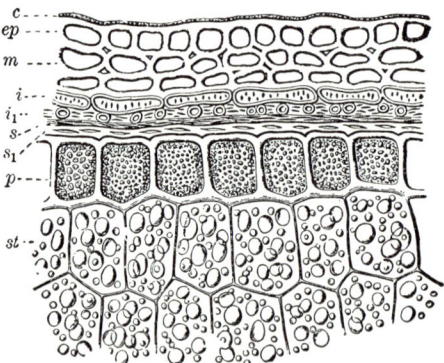

FIG. 4. Section through exterior part of a grain of wheat. *c*, cuticle; *ep*, epidermis; *m*, middle layer of hull; *i*, i_1, layers of hull next to seed coats; *s*, s_1, seed coats; *p*, protein layer with aleurone grains; *st*, cells of endosperm with starch grains. (*After Tschirch, reproduced by permission from Bergen and Davis, Principles of Botany, Ginn and Company.*)

amino acids. These latter simple compounds are then combined with nitrates from the soil, and other substances, to form the highly complex protein molecule. The outstanding characteristic of proteins is their high nitrogen content. Sulphur is also present, and often phosphorus. A typical protein, gliadin, which occurs in wheat, has the formula $C_{736}H_{1161}N_{184}O_{208}S_3$. Although proteins are the chief constituent of protoplasm, they are stored up for the most part only in seeds, where they occur as solid granules, known as aleurone grains (Fig. 4). Hundreds of proteins have been isolated from plant tissues. After proteins have been changed to a soluble form, they constitute an important food for both plants and animals. They are particularly valuable as muscle and nerve builders, rather than as sources of energy, and as such are an essential part of man's diet. Proteins are never extracted from plant tissues for food purposes. They have few industrial uses.

SECRETIONS AND EXCRETIONS

The various secretions and excretions represent different types of substances that are manufactured by plants; they are very diverse in chemical nature and in function. Some are secreted in special cells or tissues (Fig. 5) for a definite purpose, while others have no apparent use and are merely by-products of metabolism. In many cases, however, these materials are of great value to man, and among them are found some of the most valuable plant products. The most important groups include the essential oils, pigments, tannins, resins, latex, waxes, alkaloids, glucosides, organic acids, enzymes, vitamins, and hormones.

Essential Oils

The essential or volatile oils differ from fatty oils in being highly volatile and aromatic. They are formed in glands or special cells. Their function is apparently to attract insects necessary for pollination by means of their pleasing odors, or to repel hostile insects and animals by their acrid taste. They may have some antiseptic and bactericidal action. Man uses these aromatic oils in the preparation of perfumes and soap and in various other industries, as well as in medicine and as food adjuncts.

Pigments

All the coloring materials that occur in plants are manufactured by the plant itself. These pigments are diverse chemically and functionally. The most important is chlorophyll. This exceedingly complex substance, with its associated pigments xanthophyll and carotin, is one of the essential factors in photosynthesis. Other colors are of value only as a means of attracting various insect and other animal agencies of pollination and dispersal, while some are only incidental by-products of the plant's activity. In cases where the pigments are stable, they can be extracted and used as dyes. Formerly natural plant dyestuffs were of great importance in many industries.

Tannins

Tannins are bitter, astringent substances secreted in the bark, wood, or other parts of many plants (Fig. 3). Their function is not fully understood. They aid in the healing of wounds and the prevention of decay and may play a part in the formation of cork and pigments; they also serve as a protection against enemies. Tannins have certain peculiar properties that render them invaluable in certain industries. They have the power of reacting with proteins, such as the gelatin in

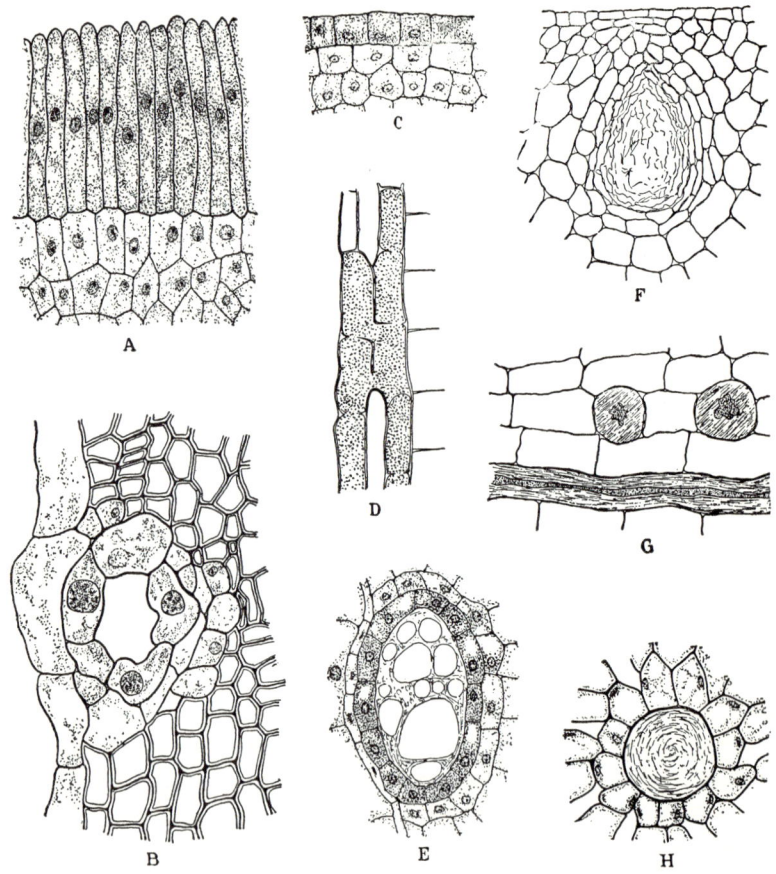

FIG. 5. Secretory tissue. *A*, nectary of *Euphorbia pulcherrima*; *B*, resin canal of *Pinus Strobus*; *C*, floral nectary of *Pyrus Malus*; *D*, latex vessel of *Tragopogon*; *E*, oil canal of *Angelica atropurpurea*; *F*, oil cavity of *Citrus sinensis*; *G*, latex cells of *Euphorbia splendens*; *H*, secretory cell of *Liriodendron*. (*Reproduced from Eames, Introduction to Plant Anatomy, McGraw-Hill Book Company, Inc.*)

animal skins, to produce a hard, firm substance. Consequently they are much used in the tanning of leather. They are also able to react with iron salts to produce a black color. This makes them valuable in the dye industry and the manufacture of inks. Tannins are also useful in medicine because of their astringent properties.

Resins

Resins are complex substances probably derived from carbohydrates. They are secreted in glands or canals and are often associated with essential oils and gums. They are formed either naturally or as a result of injury to the tissues. Resins are insoluble in water and so render any surface impervious to moisture. For this reason they play an important role in the paint and varnish industries. In the economy of the plant resins may serve as a waterproofing medium or aid in resisting decay through their antiseptic action. Resinous substances are also used in medicine.

Latex

Many plants secrete a milky or colored juice which is known as latex. This mixture of resins, gums, hydrocarbons, food, and other substances is formed in special cells, or vessels, usually in the bark or leaves. The significance of latex in the economy of the plant is not known. Man, however, obtains rubber, chewing gum, and other valuable products from this material.

Waxes

The surface of leaves and fruits often has a covering of wax secreted by the plant as a protection against excessive loss of water. This wax is similar to fat in its composition. Waxes are of some slight economic value.

Alkaloids

Alkaloids are vegetable bases containing nitrogen, and they are generally thought to be decomposition products of proteins. They are secreted in special cells or tubes and occur in many different families. Little is known of their biological significance. They may afford protection against enemies because of their bitter taste. Alkaloids are odorless compounds with a marked physiological effect on animals. Consequently they are of the utmost importance in medicine and constitute some of the most valuable drugs. On the other hand, they include some of the most powerful plant poisons and narcotics. Caffeine and theobromine, although they are actually examples of the closely related purine bases, are usually classed as alkaloids and will be so considered in this work.

Glucosides

Glucosides are similar to alkaloids in their properties, but they are derived from carbohydrates rather than proteins. They probably have a protective function as they are usually formed in the bark. It is also thought that they may serve to regulate the acidity and alkalinity of plant cells. Glucosides are useful to man as drugs.

Organic Acids

Organic acids are widely distributed in plants, especially in fruits and vegetables. They may occur in a free state, as salts of calcium, potassium, or sodium, or in combinations with alcohols. These fruit acids are probably attractive to animals and so aid in bringing about dispersal of fruits and seeds. They also play a part in metabolism and growth, and in this respect are as important for man as for plants.

Enzymes

Enzymes are universally present in all living organisms, animals as well as plants. There are many kinds, but they usually occur in such small amounts that it is difficult to extract and analyze them. Their function is to act as catalysts. They bring about all the chemical changes that occur in living matter, without actually entering into the reaction themselves. Perhaps the most important reaction with which they are concerned is digestion, the process by which insoluble substances are broken down into soluble ones and so made available for transportation to all parts of the organism and ultimate utilization. Enzymes are colloidal, and probably protein in nature, and specific in their action. They are concerned not only with oxidation and other destructive phases of metabolism but with the constructive phases as well. They aid in photosynthesis and in the formation of fats and proteins, and are present in every living cell of the plant.

Vitamins

Vitamins are substances which have been discovered comparatively recently and about which little is known. They seem, however, to be absolutely essential for the well-being of both plants and animals. They are formed by plants, and although animals may store them up they are incapable of producing them. Vitamins occur in such minute amounts that it is difficult to determine their exact nature, and only a few have been isolated. They are necessary for normal metabolism, growth, development, and reproduction, and, in fact, seem to control most of the constructive phases of metabolism. Vitamins also are necessary for the prevention of various diseases. Fruits, green vegetables, and seeds are important sources of one or another of the vitamins that have been isolated. Seaweeds are particularly valuable for they contain nearly all the known vitamins.

Hormones

Hormones are substances produced in one part of an organism and then transferred to other parts where, even though present in minute amounts, they may influence some specific physiological process. Investigations in recent years have shown, for example, that one important function of plant hormones is to regulate various growth phenomena such as tropisms, cell elongation, and cell enlargement. Hormones also play a major part in the production of roots and flowers, the formation of fruit, and other activities.

The remainder of this book comprises a discussion of various features concerned with the more important economic plants and plant products that are utilized by man as sources of food, drugs, and the raw materials of industrialism.

The industrial plants, which are perhaps less familiar to the average person, will be considered first, even though they may be less essential to man than the food plants. They include fibers, wood, cork, tanning and dye materials, gums, resins, essential oils, fatty oils, waxes, rubber and other latex products, sugar, starch, and cellulose products. Drugs, together with tobacco and the various narcotics, will be treated secondly. Finally we shall consider the food plants and food adjuncts, which include spices and beverages.

CHAPTER II

FIBERS AND FIBER PLANTS

Plants that yield fibers have without question been second only to food plants in their usefulness to man and their influence on the advancement of civilization. Primitive man in his attempt to obtain the three great necessities of life—food, shelter, and clothing—early turned to plants. Although animal products were available, he needed some form of clothing that was lighter and cooler than skins and hides. For his snares, bowstrings, nets, and the like he needed some form of cordage that was easier to procure than animal sinews and strips of hide. Moreover, some other type of covering for his crude shelters was desirable. All these needs were admirably met by the tough, flexible strands that occurred in the stems, leaves, and roots of many plants.

Almost from the outset plant fibers have had a more extensive use than wool, silk, and other animal fibers. As civilization advanced and man's needs multiplied, the use of these vegetable fibers increased greatly until at the present time they are of enormous importance in our daily life. It is difficult to estimate the number of species of fiber plants, but a conservative figure would be well over two thousand. More than a thousand species of American plants have yielded fibers. Seven hundred and fifty occur in the Philippine Islands alone. Fibers of commercial importance, however, are relatively few, the greater number comprising native species used locally by primitive peoples in all parts of the world.

It is a remarkable fact that the most prominent fibers of the present day are of great antiquity. The cultivation of flax, for example, goes back to the Stone Age of Europe, as evidenced by the remains of the Swiss Lake Dwellers. Ancient Egypt was famous for its fine linen. Cotton was the ancient national textile of India, and was used by all the aboriginal nations of the New World as well. Ramie or China grass has been grown in the Orient from time immemorial.

ECONOMIC CLASSIFICATION OF FIBERS

It is possible to classify fibers in six groups, based on their utilization, as follows:

Textile Fibers. The most important use of fibers at the present time is in connection with the textile industry, which is concerned with the manufacture of fabrics, netting, and cordage. In making fabrics and netting, flexible fibers are twisted together into thread or yarn and then woven, spun, knitted, or otherwise utilized. Fabrics include cloth for wearing apparel, domestic use, awnings, sails, etc., and also coarser materials such as gunny and burlap. The fabric fibers are all of commercial importance. Netting fibers, which are used for lace, hammocks, and all forms of nets, include many of the commercial fabric fibers and a host of native fibers as well. Both commercial and native fibers are used for cordage. For this purpose the individual fibers are twisted together rather than woven. Twine, binder twine, fish lines, rope, hawsers, and cables are among the many kinds of cordage.

Brush Fibers. These are tough and stiff fibers, or even twigs and small stems, which are utilized in the manufacture of brushes and brooms.

Plaiting and Rough Weaving Fibers. Plaits are flat, pliable, fibrous strands which are interlaced to make straw hats, sandals, baskets, chair seats, and the like. More elastic strands are roughly woven together for mattings and the thatched roofs of houses, while supple twigs or woody fibers are used for baskets, chairs, and other forms of wickerwork.

Filling Fibers. These fibers are used in upholstery and for stuffing mattresses, cushions, etc.; for caulking the seams in vessels and in casks and barrels; as stiffening in plaster; and as packing materials.

Natural Fabrics. These are usually tree basts which are extracted from the bark in layers or sheets and pounded into rough substitutes for cloth or lace.

Papermaking Fibers. Papermaking fibers include wood fibers, textile fibers utilized in either the raw or manufactured state, and many other kinds.

It is obvious that any one plant cannot be restricted absolutely to any single group in this economic classification, since the same fiber may be used for different purposes and since the plant may yield more than one kind of fiber. Consequently the several species to be discussed will be considered in the group in which they are of the greatest importance.

STRUCTURE AND OCCURRENCE OF FIBERS

Although put to so many different uses, and perhaps differing in texture, strength, chemical composition, and place of origin, with few exceptions fibers are alike in that they are sclerenchyma cells and serve as part of the plant skeleton. For the most part they are long cells with thick walls, correspondingly small cavities, and usually pointed ends. The walls often contain lignin as well as cellulose. Fibers may occur singly or in small groups, but they are more likely to form sheets of tissue with the individual cells overlapping and interlocking.

Fibers may be found in almost any part of the plant—stems, leaves, roots, fruits, and even seeds. The four chief types, classified according to their origin, include bast fibers; wood fibers; sclerenchyma cells associated with the vascular-bundle strands in leaves; and surface fibers, which are hairlike outgrowths on the seeds of various plants. The use of the term "bast fiber" is open to criticism on the ground that it gives no indication as to the particular tissue or region in which the fibers occur. From a morphological viewpoint it would be preferable to designate those fibers which occur in the outer parts of a stem as cortical fibers, pericyclic fibers, or phloem fibers. However, "bast" is a term of such long standing and so firmly established in commerce that it will be used in the present discussion.

Fibers of economic importance are furnished by many different families of plants, particularly those which occur in the tropics. Among the more important may be mentioned the *Gramineae, Palmaceae, Musaceae, Liliaceae, Amaryllidaceae, Urticaceae, Malvaceae, Linaceae, Bombacaceae, Leguminosae, Moraceae, Tiliaceae,* and *Bromeliaceae.*

TEXTILE FIBERS

Textile fibers have certain special requirements. They must be long and must have a high tensile strength, together with cohesiveness and pliability. They must possess a fine, uniform, lustrous staple and must be durable and readily available. A comparatively small number of the multitude of fibers possess these characteristics, and so can be considered as of commercial importance. The chief textile fibers are included in three classes: surface fibers, soft fibers, and hard fibers. The two last groups are often referred to as long fibers.

The surface or short fibers comprise the so-called cottons. The soft fibers are the bast fibers which are obtained chiefly from the pericycle or secondary phloem of the stems of dicotyledons. These bast fibers are capable of subdivision into exceedingly fine flexible strands, and are used for the best grades of fabrics and cordage. They include flax, hemp, jute, and ramie.

The hard or mixed fibers are structural elements found chiefly in the leaves of many tropical monocotyledons, although they may occur in stems and fruits. They are used for the coarser textile products. Sisal, henequen, abacá, the agaves, coconut, and pineapple are examples of plants that yield hard fibers.

SURFACE FIBERS

Cotton

Cotton (Fig. 6) is the world's greatest industrial crop, the chief fiber plant, and one of the oldest as well as the cheapest. It was known to the ancient world long before written records were made. References to it are to be found in the works of the Greek and Roman writers. Cotton has been in use in India since 1800 B.C., and from 1500 B.C. to A.D. 1500 that country was the center of the industry. The Hindus were the first people to weave cloth. Cotton was introduced into Europe by the Mohammedans, and the word itself is derived from the Arabic "qutn." Apparently the plant had several origins for Columbus found it in cultivation in the West Indies, and it was known to the Peruvians and Mexi-

FIG. 6. A mature cotton plant. (*Reproduced from U.S D.A. Educational Chart, Index No. 3, 11, revised, The Cotton Plant.*)

cans long before the days of the Spanish conquerors. Cotton was first grown in the United States soon after the first settlements were made. The first cotton mill, however, did not commence operations until 1787.

Production of Cotton. In 1948 the world produced an estimated total of 29,285,000 bales of cotton, a bale usually comprising 500 lb. gross weight. The United States led in production with 14,868,000 bales, followed by India and Pakistan with 2,812,000, the Soviet Union with 2,800,000, China and Manchuria with 2,115,000, Egypt with 1,836,000, and Brazil with 1,465,000 bales. Brazil has had a spectacular rise as a cotton-producing country, the output increasing from an average of 455,000 bales in 1932–1933 to a maximum of 2,675,000 bales in 1944, grown chiefly in the state of Sao Paulo. For many years prior to 1934 the United States produced at least one-half of the world's output of cotton. Since that time normal cotton production has been curtailed as a result of governmental regulations. However, an average of about

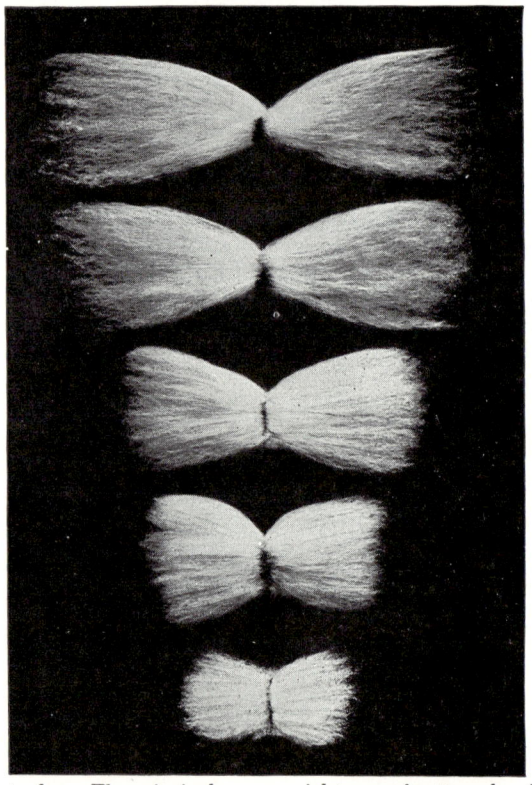

Fig. 7. Cotton staples. The principal commercial types of cotton showing the variation in length of staples. From top to bottom: sea-island, Egyptian, upland long-staple, upland short-staple, Asiatic. (*Reproduced from U.S.D.A. Yearbook Separate* 877.)

25,000,000 acres of cotton has been grown in the United States during the last 15 years, nearly one-third of which was in Texas. The United States, India, Egypt, and Brazil are the chief exporting countries, while Great Britain, Japan, and Germany lead in the importation of cotton.

Characteristics of Cotton. Cotton is obtained from several species of the genus *Gossypium*. The fine fibrous hairs that occur on the seeds constitute the raw material of the industry. These hairs, which are flattened, twisted, and tubular, compose the lint, floss, or *staple*, the length (Fig. 7) and other qualities of which vary in different varieties. The cotton plant is naturally a perennial shrub or small tree, but under cultivation it is treated as an annual. It branches freely and grows to a height of 4 to 8 ft. Cotton thrives best in sandy soil in damp, humid regions that are near water. This type of environment is especially characteristic of the southern United States and the river valleys of India and Egypt. Cotton matures in five or six months and is ready for picking soon after ripening.

Kinds of Cotton. The hundreds of varieties of cultivated cotton have been developed from wild ancestors or produced by breeding during the long period of cultivation. These varieties differ in fiber character as well as other morphological and cultural features. Cotton is a very difficult group to classify, and few authorities agree as to the exact number of species. The cultivated cottons of commercial importance, however, are usually referred to one or another of four species: *Gossypium barbadense* and *G. hirsutum*, which are New World species, and *G. arboreum* and *G. herbaceum*, natives of the Old World.

1. *Gossypium barbadense.* The native home of this species was probably tropical South America. In this cotton the flowers are bright yellow with purple spots. The fruit, or boll as it is called, has three valves, and the seeds are fuzzy only at the ends. Two distinct types of cotton belong here:

SEA-ISLAND COTTON. This cotton has never been found growing wild as it was already in cultivation at the time of Columbus. Its fine, strong, light cream-colored fibers are more regular in the number and uniformity of the twists and have a silkier appearance than those of other cottons. These features are exceedingly valuable and sea-island cotton was formerly in great demand for the finest textiles, lace, yarns, and spool cotton, and also for mercerizing. Sea-island cotton was brought to the United States from the West Indies in 1785. The finest types were developed on the islands off the South Carolina coast and the adjacent mainland. Here staples 2 in. or more in length, surpassing all the others in strength and firmness, were produced. Another form of sea-island cotton is grown along the coast in Georgia and Florida and in the West Indies and South America. This has a staple from $1\frac{1}{2}$ to $1\frac{3}{4}$ in. in length. The maximum yield of sea-island cotton was only 110,000 bales, but this was compensated for by the greater value of the fiber. For some years the growing of sea-island cotton almost entirely ceased owing to the ravages of the boll weevil.

EGYPTIAN COTTON. Egyptian cotton is grown chiefly in the Nile basin of Egypt, where it was introduced from Central America. The plant is quite similar in appearance to sea-island cotton and is probably of hybrid origin. The staple, however, is brown in color and somewhat shorter, measuring from $1\frac{3}{8}$ to $1\frac{3}{4}$ in. in length. Because of its length, strength, and firmness this cotton is used for thread, underwear, hosiery, tire fabrics, and fine dress goods. Egyptian cotton was brought to the United States in 1902 as an experimental crop, and 10 years later it was recommended to farmers in the semiarid regions that are irrigated. A considerable quantity is now grown in California, New Mexico, and Arizona. Repeated selection and breeding have resulted in the development of new and better strains, of which Pima is one of the best. A large amount of

FIG. 8. Leaves, flowers, and fruit of upland cotton (*Gossypium hirsutum*). (*Reproduced from U.S.D.A. Yearbook*, 1904.)

Egyptian cotton is still imported, amounting at times to one-tenth of Egypt's entire crop.

2. *Gossypium hirsutum.* This native American species was grown by the pre-Columbian civilizations. Commonly called upland cotton (Fig. 8), it is the easiest and cheapest kind to grow and constitutes the greater part of the cultivated cotton of the world. The flowers are white or light yellow and unspotted. The bolls are four- or five-valved, and the seeds are fuzzy all over. Upland cotton thrives under a variety of conditions, but prefers a sandy soil with plenty of moisture during the growing and fruiting season and dryness during the time of boll opening and harvest,

together with a temperature range of 60 to 90°F. The northern limit of economic growth is 37°N.L. The great Cotton Belt of the southern United States grows upland cotton almost entirely and produces two-thirds of the world's output of this type of cotton. The fibers are white with a considerable range in staple length, from 5/8 to 1 3/8 in. There are over 1200 named varieties, many of which have been developed as a result of breeding experiments. Typical *Gossypium hirsutum* probably originated in Guatemala or southern Mexico and spread northward to its present limits in the southern United States. A well-marked variety, often recognized as a distinct species, occurs in the West Indies and along the dry coastal areas of South America as far as Ecuador and Brazil. Another variety is found in Central America, northward along the Gulf of Mexico to Florida and the Bahamas, and in the northern West Indies.

3. *Gossypium arboreum.* This species, the perennial tree cotton of India, Arabia, and Africa, was probably the first to be used commercially. At the present time it is but little grown save in India. As in all Asiatic cottons, the staple is coarse and very short, only 3/8 to 3/4 in. in length, but it is strong.

4. *Gossypium herbaceum.* This fourth species of cotton is not found in America, but is the chief cotton of Asia. It has been grown in India from time immemorial and is also extensively cultivated for home consumption in Iran, China, Japan, and elsewhere. It is utilized for low-quality fabrics, carpets, and blankets and is especially suitable for blending with wool.

In addition to these cultivated species several wild species of *Gossypium* are still to be found in some tropical and subtropical countries.

The Cotton Industry. For a long time cotton was a very expensive product for it was difficult to remove the fibers from the seed. The invention of the cotton gin by Eli Whitney in 1793, however, remedied this situation and was responsible for revolutionizing the industry, which rapidly assumed proportions of first importance in both America and England. Cotton today is probably the most important article in the commerce of the world. The economics of this single crop has a profound effect on both the producing and the consuming nations. The steady decrease in the value of the cotton crop in the United States, due to adverse world conditions, has had such a serious effect on the life of the Cotton Belt farmers that government control has been resorted to in recent years in an attempt to improve the situation.

Several operations are necessary in order to prepare the raw cotton fiber, as it comes from the field, for use in the textile industry. In brief these operations are as follows: ginning in either a saw-tooth or a roller gin; baling; transporting to the mills; picking, a process in which a machine removes any foreign matter and delivers the cotton in a uniform layer; lapping, an operation whereby three layers are combined into one; carding, combing, and drawing, during which the short fibers are extracted and the others are straightened and evenly distributed; and finally twisting the fibers into thread.

The Uses of Cotton. The chief use of cotton, either alone or in combination with other fibers, is in the manufacture of textiles of all types, which are too numerous to mention. It is an important constituent of rubber-tire fabrics, and unspun cotton is extensively used for stuffing purposes. Mercerized cotton is made by treating the fibers with caustic soda, thereby imparting a high luster and silky appearance. Absorbent cotton consists of fibers which have been thoroughly cleaned and from which the oily covering layer has been removed. It is almost pure cellulose and constitutes one of the basic raw materials of the various cellulose industries to be discussed later.

One of the most noteworthy advances in the cotton industry has been the utilization of what were formerly treated as waste products. At one time the cotton seed, together with its fuzzy covering of short hairs, or linters, was discarded as valueless. Today, however, all parts of the plant are conserved and yield products that are worth several million dollars annually. The *stalks* contain a fiber that can be used in paper making or for fuel, and the *roots* possess a crude drug. The seeds are of the greatest importance and every portion is utilized. The *linters* furnish wadding; stuffing for pads, cushions, pillows, mattresses, etc.; absorbent cotton; low-grade yarn for twine, ropes, and carpets; and cellulose. The *hulls* are used for stock feed; as fertilizer; for lining oil wells to prevent the caving in of the sides; as a source of xylose, a sugar that can be converted into alcohol, or various explosives and industrial solvents; and for many other purposes. The *kernels* yield one of the most important fatty oils, cottonseed oil, which will be discussed later, and an oil cake and meal which are used for fertilizer, stock feed, flour, and as a dyestuff.

Soft or Bast Fibers

Flax

Flax, once the most valuable and useful of the fibers, is now second to cotton, and possibly jute, in importance. It is much superior to cotton in quality and yields a finer fabric. The native home of flax is not known for the plant has been under cultivation from prehistoric time. It was used by the Swiss Lake Dwellers, the oldest people in Europe who have left remains of their civilization. It was well known to the Hebrews and is frequently mentioned in the Bible. The Egyptians wore linen and used it for their mummy cloths, and carved pictures of the flax plant on their tombs. Long prior to the Christian era the Greeks imported flax.

Five thousand years would be a moderate estimate of the time during which the plant has been under cultivation.

Flax belongs to the genus *Linum*, which contains several wild species of no economic importance, as well as *L. usitatissimum*, the source of the commercial fiber. The flax plant is an annual herb with blue or white flowers and small leaves, growing to a height of from 1 to 4 ft. The fibers are formed in the pericycle, and consist of very tough, stringy strands from 1 to 3 ft. in length, which are aggregates of many long pointed cells with very thick cellulose walls.

Flax grows best in soil that is rich in moisture and organic matter. It is primarily a crop of temperate regions, although it is grown in other

FIG. 9. Flax (*Linum usitatissimum*), showing the straw spread for dew retting. (*Reproduced from U.S.D.A. Farmers' Bulletin* 669, *Fiber Flax*.)

parts of the world to some extent. The preparation of flax is much more laborious than that of cotton, and so is much more costly. The crop is usually harvested by hand, and the stems are broken by a process known as rippling. Ordinarily the fibers are then rotted out by submerging the stems in water or by exposing them to dew (Fig. 9). During this process, which is called retting, an enzyme dissolves the calcium pectate of the middle lamella, which holds the cells together, and so frees the fibers. After retting, the straw is dried and cleaned and the fibers are completely separated from the other tissues of the stem by means of an operation known as scutching. Finally the shorter fibers, which constitute the tow, are separated from the longer fibers, which alone can be used in spinning. This is done by hand or by a hackling machine.

Flax fibers are remarkable for their great tensile strength, length of staple, fineness, and durability. They are used in the manufacture of linen cloth and thread, canvas, duck, carpets, the strongest twine, the best fish and seine lines, cigarette paper, the finest writing paper, and insulating materials. Fibers from the stalks of flax grown for seed are too harsh and brittle for spinning, but are utilized for cigarette paper.

The Northern European countries are the chief producers of fiber flax. The Soviet Union far outstrips the rest of the world, producing 70 per cent of the total output, while Poland, the Baltic states, Belgium, Holland, and France are also important. The finest flax is grown in northern Belgium. Flax was introduced into the United States by the Pilgrims, and until 1900 almost every farmer grew enough for his own use. It was a particularly good crop with which to reclaim native soil, and for a long time its cultivation was confined to the frontier areas of the country. At the present time fiber flax is grown in the United States as a commercial crop only in Oregon. However, in regions where there is little rain, flax is grown for its seed, which is used in medicine and as the source of linseed oil. This valuable industrial fatty oil will be discussed later.

Hemp

There is often considerable confusion as to the real identity of a given commercial hemp fiber, because the term "hemp" is applied rather loosely to include quite a number of very different plants and fibers. The true hemp is *Cannabis sativa*. This plant is a native of Central and Western Asia, but it is extensively cultivated at the present time in both temperate and tropical regions. It often occurs as a troublesome weed.

The hemp plant (Fig. 10) is a stout, bushy, branching annual varying from 5 to 15 ft. in height. It is a dioecious species with hollow stems and palmate leaves. The best grade of fiber is obtained from the male plants. For its best development hemp requires a mild humid climate and a rich loamy soil with an abundance of humus. Calcareous soils are particularly well adapted to hemp culture.

Hemp fiber is a white bast fiber which develops in the pericycle. It is valuable because of its length, which varies from 3 to 15 ft., its strength, and great durability. It lacks, however, the flexibility and elasticity of flax, because it is somewhat lignified. The yield of hemp is large, an acre producing 2 or 3 tons of stems, 25 per cent of which is fibrous material. The plants are harvested by hand or machine and are shocked and dried. The fibers are separated from the rest of the bark by retting, either in dew or in water. They are then broken, scutched, and hackled by hand. Hemp must be harvested when the male flowers are fully out; otherwise the fibers are too weak or too brittle to be of value.

Hemp is a very old crop and has been grown in China for centuries. It was introduced into Europe about 1500 B.C., and that continent is the center of the industry today, the U.S.S.R, Italy, and Poland furnishing

75 per cent of the world output. Hemp has been grown in the United States since colonial days and at one time constituted a considerable industry in Kentucky and later in Wisconsin and other Middle Western states. Since 1930 it has normally been but little grown, as it is an expensive crop; in most regions it has been gradually replaced by other coarse fibers. Wartime shortages made resumption of its cultivation imperative for a time, and a government-supported program produced 120,000,000 lb. in 1944. There is little possibility, however, that hemp will become a permanent crop in the United States. A small amount of hemp fiber is produced in Chile.

FIG. 10. Hemp (*Cannabis sativa*). A field about two months after planting. Hemp is the source of a textile fiber, a drying oil, and a narcotic drug.

Hemp is used for ropes, twine, carpets, sailcloth, yacht cordage, binder twine, sacks, bags, and webbing. Hemp waste and the woody fibers of the stem are sometimes used in making paper. The finer grades of hemp can be woven into a cloth that looks like coarse linen. The short fibers, or *tow*, and ravelings constitute *oakum*. This is used for caulking the seams between the planks used in shipbuilding; in cooperage; and as a packing for pumps, engines, etc. In tropical regions hemp is grown for its seed, and also for a drug that is obtained from the flowering tops and leaves. The seeds contain an oil that is useful in the soap and paint industries as a substitute for linseed oil. The drug, known as hashish, is a resinous substance that contains several powerful alkaloids. Its use and the important part it has played in the economic and social life of many countries will be discussed later.

Jute

Jute is probably used more extensively than any other fiber except cotton, although it is much less valuable than either cotton or flax. Jute is a bast fiber obtained from the secondary phloem of two species of *Corchorus*, an Asiatic genus. The best fiber comes from *C. capsularis*, a species with round pods which is grown in lowland areas subject to

FIG. 11. Jute (*Corchorus capsularis*) grown for fiber production in India.

inundation. The plant is a tall, slender, half-shrubby annual (Fig. 11) with yellow flowers, growing to a height of 8 to 10 ft. It requires a warm climate and a rich, loamy, alluvial soil. Fiber from *C. olitorius*, an upland species with long pods, is but little inferior, and the two are not differentiated for commercial purposes.

The crop is harvested within three or four months after planting, while the flowers are still in bloom. The stems are retted in pools or tanks for a few days to rot out the softer gummy tissues, and the jute, or *gunny*, strands are then loosened by whipping the stems on the surface of the water. The very long pale-yellow fibers, from 6 to 10 ft. in length, are quite stiff, as they are considerably lignified; they have a silky luster. They are very abundant, but are not particularly strong, and they tend to deteriorate rapidly when exposed to moisture, to which they are exceedingly susceptible. In spite of these disadvantages the fact that jute is cheap and easily spun makes it valuable. Practically every civilized country imports some form of jute or jute products.

Although probably a native of Malaya or Ceylon, jute is now almost entirely an Indian crop. Several million acres are cultivated in the valleys of the Ganges and Brahmaputra rivers. As a result of war shortages, Brazil began a large-scale cultivation of jute in the Amazon basin. So successful has this been that some fiber is actually being exported. In the United States jute has been grown experimentally in the Gulf States with good success. However, the absence of an adequate fiber-cleaning machine makes the cost of its production in this country prohibitive.

Jute is used chiefly for rough weaving, and the familiar burlap bags, gunny, wool, and potato sacks, and the covers for cotton bales, are made from it. The fiber is also used for making twine, carpets, curtains, and coarse cloth. Short fibers and pieces from the lower ends of the stalks constitute jute butts, which are used to some extent in papermaking. India not only grows most of the jute, but it is the largest manufacturer and exporter of jute products.

Ramie

Ramie (*Boehmeria nivea*) is a perennial-rooted, herbaceous, or shrubby plant (Fig. 12) without branches when under cultivation. The slender stalks reach a height of from 3 to 6 ft. and bear heart-shaped leaves that are green above and whitish beneath. Ramie is of Asiatic origin and was known to the Chinese at a very early period. It is grown extensively at the present time in China, Japan, Formosa, Indo-China, Borneo, Java, and India in fertile well-drained soil. Experimental cultivation in the southern United States and California has been highly publicized in recent years. The yield is low, but this is compensated for by the fact that several crops may be grown each season.

The fine fibers are obtained from the bast, and are very long, strong, and durable. They are also beautiful with a high degree of luster, and would be very desirable for textile purposes were it not for difficulties encountered during the extraction and cleaning of the fibers. The stems are first immersed in water. Then the bark is peeled off, and the outer portions and green tissue are scraped off by hand or are removed by boiling or mechanical means. The fibers that are left are heavily coated with gum and require further laborious treatment before they can be utilized. They constitute the China grass, or "filasse," which is extensively used in the manufacture of grass cloth and other dress goods in Asia. Ramie is used in Europe for underwear, portieres, upholstery, thread, and paper. Although it is the strongest fiber known, being three times as strong as hemp, ramie is not very generally used because the treatment necessary to remove the fibers is so expensive. The fibers also lack flexibility and are too smooth for good cohesion. A process has recently been developed by which a fiber is obtained which has met all the tests and requirements of the textile manufacturers, and for which a

FIG. 12. Ramie (*Boehmeria nivea*), showing plants eight weeks old.

multitude of uses is predicted. A ramie industry might well be developed in the United States, but cheap labor would be essential for its success.

A variety of ramie, *Boehmeria nivea* var. *tenacissima*, is sometimes differentiated as **rhea**. This plant, a native of Malaya, resembles ramie except that the leaves are green on both sides. For commercial purposes rhea fiber is included under ramie.

Sunn Hemp

Sunn, sun, or san hemp (*Crotalaria juncea*) is an important Asiatic fiber plant. It is not known in the wild state for it has been cultivated for centuries. In fact, it is the earliest fiber to be mentioned in Sanskrit writings. The plant is a shrubby annual legume from 6 to 12 ft. in height, with bright yellow flowers. It is extensively grown in India where over 500,000 acres are planted every rainy season, chiefly in the region around Madras. Sunn hemp is also cultivated in Ceylon and elsewhere in Southern Asia. The bast yields a fiber that is stronger than jute, lighter in color, and more enduring. It is used for cordage, oakum, sacks, nets, and coarse canvas. The United States imports a considerable quantity for cigarette and tissue paper and coarse twines.

Other Soft Fibers

Nearly all the members of the *Malvaceae* yield bast fibers that can be used for textile purposes. Some of these are of considerable commercial importance. Among them may be mentioned:

China Jute or Indian Mallow (*Abutilon Theophrasti*). This annual plant yields a strong, coarse, grayish-white, lustrous fiber with the same general characteristics and uses as jute. It is extensively grown in China, and has been introduced into the United States where it thrives well. In many places it has become a troublesome weed. The fibers have considerable tensile strength, take dyes readily, and are much used in China in rugmaking. They are also utilized in papermaking.

Kenaf (*Hibiscus cannabinus*). This tall herb yields a fiber which has borne some 129 names, among them Deccan, Ambari or Gambo hemp, Java jute, and Mesta fiber. It has long been of value in the Old World tropics as a substitute for hemp and jute in the manufacture of coarse canvas, gunny cloth, cordage, matting, fishing nets, etc. Kenaf has been exploited commercially in India, Java, Iran, Nigeria, Natal, and Egypt. More recently it has been introduced into Europe and the Western Hemisphere. In 1935, the U.S.S.R. had 32,500 acres in production. Since 1941 kenaf has been subjected to an extensive research program in Cuba, El Salvador, and the United States. The plant is adapted to a wide range of climate and soil conditions. It is harvested shortly after the flowers come into bloom. The fibers, which are 5 to 10 ft. long, are usually extracted by retting, although decorticating machines may be used. Kenaf seed yields up to 20 per cent of an oil which is edible after it has been refined.

Roselle or Rama (*Hibiscus Sabdariffa*). This species is cultivated in India, Southeastern Asia, and the islands of the southwest Pacific as a substitute for jute and also for its edible fruit. The light-brown fibers are soft, silky, and lustrous. Roselle is adapted to any well-drained fertile soil where there is a 20-in. rainfall. It grows so rapidly that it can be harvested 90 days after planting. Retting is accomplished in 10 to 12 days, and the fibers are easily slipped from the bark by hand. Recently introduced into Cuba and Central America, it has shown considerable promise as a quick source of a soft fiber.

The red fleshy calices and involucels surrounding the young fruits are very acid and serve as a sour relish. The juice is also used for flavoring and in making jams, jellies, and wine.

Aramina or Cadillo (*Urena lobata*). This species occurs as a weed in all tropical countries. It furnishes a yellowish-white fiber that is more lasting than jute and is used as a substitute in many industries. It is grown commercially in Cuba, Madagascar, the Belgian Congo, Nigeria, and Brazil, where its chief use is in making coffee sacks.

Other malvaceous species yielding fibers of minor importance include **okra** (*Hibiscus esculentus*), **majagua** (*H. tiliaceus*), and several species of the genus *Sida*. *S. acuta* in particular is now being cultivated to a considerable extent in Mexico as a jute substitute. It is an extremely easy plant to harvest and prepare, and the fibers are twice as strong as jute.

The American Indians used the bast fibers of many plants for their bowstrings, nets, and similar purposes. The Colorado River hemp (*Sesbania exaltata*) was extensively used by the western tribes, while the Indian hemp (*Apocynum cannabinum*) and milkweed (*Asclepias syriaca*) yielded the fibers of greatest importance to the eastern Indians.

Hard or Structural Fibers

Abacá or Manila Hemp

Abacá or Manila hemp is the world's premier cordage material. It is obtained from several species of wild plantain or banana. *Musa textilis* (Fig. 13), the principal source, resembles the true banana, but has narrower, more tufted leaves and inedible fruits. The plant consists of a clump of 12 to 30 sheathing leafstalks 10 to 20 ft. high with a crown of spreading leafblades 3 to 6 ft. in length. The fiber is obtained from the outer portion of the leafstalks. The mature stalks are cut off at the roots and split open lengthwise. The pulp and the fiber strands are removed, and the latter are washed and dried. Formerly the strands were removed by hand, a slow and laborious process; more recently decorticators have been used.

The individual fibers (Fig. 14) are 6 to 12 ft. in length, lustrous, and vary in color from white to light ocher. They are light, stiff, elastic, and exceptionally strong, durable, and resistant to both fresh and salt water. For this reason the chief use of abacá is in the manufacture of high-grade cordage, especially marine cables. It is also used for binder twine, bagging, papier-mâché, strong tissue paper, wrapping paper, and Manila paper for sacks. Japan imports a large amount for making the strong paper used for the movable partitions in the houses. The individual fibers cannot be spun, but strands of fibers are utilized in making the lustrous cloth known as *sinamay*.

Although *M. textilis* is found from India to the Philippines, it is only in the latter country, and to some extent in Sumatra and Borneo, that it has been of commercial importance. Abacá was known and used by the

FIBERS AND FIBER PLANTS

Fig. 13. A plantation of abacá or Manila hemp (*Musa textilis*) in the Philippine Islands.

Fig. 14. Abacá fiber in bundles as it is taken from the plantation to the market in the Philippine Islands.

natives for centuries before the arrival of the first European explorers early in the sixteenth century. The first shipment was made to the United States in 1818. From then until 1918 it was the principal export of the Philippines, amounting to some 300,000,000 lb. annually. More recently, sugar and sometimes copra have exceeded it in importance.

Manila hemp requires a warm climate, fertile soil, shade, good drainage, abundant moisture, and an elevation below 3000 ft. It is propagated by rootstalks or suckers. The crop is grown in small fields or on large plantations and matures in 18 to 36 months.

After several failures, abacá was finally successfully introduced into the Western Hemisphere in 1925 in Panama, but little interest was aroused. In 1942, however, the serious shortages of Manila hemp constituted a real threat to the war effort, and the United States government financed a planting project in several Central American countries. Machines for cleaning the leaves were devised and installed, and soon some 26,000 acres were under cultivation which produced 3,000,000 lb. of fiber. The cessation of war found the abacá industry soundly established in Costa Rica. The environment is exceedingly favorable, and the greater proximity to markets, the larger plantations, and the machine methods in use tend to offset the higher costs of production.

Agave Fibers

Agave fibers rank next to cotton in importance in America, their trade value amounting in good years to over $35,000,000. The agaves are stemless perennials with basal rosettes of erect fleshy leaves. These leaves contain the fibers, which are removed by hand or machine. The numerous species are of rather local occurrence. They are very drought resistant and are adapted to dry sterile soils. Several kinds are of commercial importance. Owing to the fact that the trade names for the different agave fibers tend to intergrade, considerable confusion exists as to the identity of the species concerned. The chief types include:

Henequen or Mexican Sisal (*Agave fourcroydes*). This native Mexican species was long used by the Aztecs. At the present time it is grown chiefly in Yucatan and Cuba (Fig. 15). The leaves bear spines, which make them difficult to handle. The fiber, which is scraped out from the leaf tissue, is light straw colored. It is hard, wiry, and elastic, measuring from 2 to 5 ft. in length. Henequen is used chiefly for binder twine, lariats, and similar products. It is not suitable for marine or hoisting cables as it is too heavy and too weak.

A closely allied species, *Agave Letonae*, is the source of the considerable amount of henequen grown in El Salvador.

Sisal (*Agave sisalana*). Sisal is very similar in appearance to henequen but lacks the spines on the leaves. A native species of Mexico and Central America, it is now cultivated in Hawaii, the East and West Indies, and many sections of Africa, especially the British possessions. The plant is exceedingly drought resistant and will grow where all other species fail. Little or no cultivation is necessary. The coarse, stiff, light-yellow to white fibers are removed from the leaves by hand or by means of a "raspador." They are cleaned, dried, and packed in 600-lb. bales for shipment. The United States uses a large amount of sisal for twine, ropes, and cords, most of the supply coming from Mexico, Haiti, and British East Africa.

Istle. Several fibers have been used in Mexico from prehistoric times under the names istle, ixtle, or Tampico fiber. The three most important

FIG. 15. Henequen (*Agave fourcroydes*). A field near Victoria, Tamaulipas, Mexico, from which two crops of leaves have been cut. The plants are six years old.

kinds, and the only ones exported, are jaumave istle from *Agave Funkiana*, tula istle from *A. Lecheguilla*, and palma istle from *Samuela carnerosana* and various species of *Yucca*. The fibers are obtained from immature leaves of wild plants, and, although shorter than those of sisal and henequen, they are very strong and durable. Because of their stiffness and harshness they were formerly used for brushes, but now they serve as cheap substitutes for sisal and abacá in making bagging, twine, and rope. Ore sacks made of istle have been known to last 10 years.

Maguey. The Manila maguey or cantala (*Agave Cantala*) is a Mexican species which was early introduced into India and Southeastern Asia. It is grown on a commercial scale in the Philippine Islands, Java, and elsewhere as a substitute for sisal.

Mexican maguey is obtained from several different species of *Agave*. These latter fibers are of little economic importance. The plants, however, are of considerable interest, for their juice is used in making the Mexican beverages, pulque and mescal.

Mauritius Hemp

Mauritius hemp is obtained from the leaves of the green aloe (*Furcraea gigantea*). This species is a native of tropical America, but is now widely distributed throughout the tropics of both hemispheres, where its fiber is used for domestic purposes. It is grown on a commercial scale in the islands of Mauritius, Madagascar, and St. Helena, and in South Africa, India, Venezuela, and Brazil, where it is known as *piteira*. The plant resembles an agave in habit, but has larger, less rigid leaves and the longest known peduncle or flower stalk. This "pole," as it is called, reaches a height of 20 to 40 ft. The fibers are exceedingly long, 4 to 7 ft., and they are white, soft, very flexible, and elastic. They are weaker than sisal. They are used alone or in mixture for making hammocks, bags, coarse twine, and other small cordage.

Several other species of *Furcraea* in tropical America yield fibers of considerable local importance. These include the *fique* of Colombia (*F. macrophylla*); the *cabuya* of Costa Rica, Panama, and northern Colombia (*F. Cabuya*); and the *pitre* of the West Indies (*F. hexapetala*), sometimes called Cuban hemp.

New Zealand Hemp

This fiber, also known as New Zealand flax, comes from the leaves of the irislike *Phormium tenax* (Fig. 16). This plant is a native of the swampy regions of New Zealand, but it is now found throughout the tropics and in temperate regions as well. It has been introduced into the United States chiefly as an ornamental plant. For a time it was grown in California for its fiber, but the climate was too unfavorable for it to become established on a commercial scale. The fibers are very long, 3 to 7 ft. in length, and have a high luster. They are softer and more flexible than abacá and are used chiefly for towlines, twine and other forms of cordage, and mattings, and to some extent for cloth.

Bowstring Hemp

Numerous species of the genus *Sansevieria* occur as wild plants in various parts of tropical Africa and Asia. These bowstring hemps are herbaceous perennials with basal rosettes of swordlike leaves arising from a creeping rootstalk. The leaves yield a strong white elastic fiber, which has long been used by the native peoples for mats, hammocks, bowstrings, and other types of crude cordage. Wild plants are usually utilized, but some species are cultivated. The Hindus have grown bowstring hemp for a long time. The fibers are removed by hand or by mechanical means. Important species include *S. thyrsiflora* of tropical Africa, grown also in Jamaica and Central America; *S. Roxburghiana* of India; and *S. zeylanica*, cultivated in Ceylon and many other tropical countries. A few species

FIG. 16. New Zealand hemp (*Phormium tenax*). This plant is cultivated in California for the fiber. It is also grown for ornamental purposes.

have been introduced into the United States, prominent among which is the Florida bowstring hemp (*S. longifolia*).

Coir

Coir is the term applied to the short, coarse, rough fibers (Fig. 17) which make up the greater part of the husk of the fruits of the coconut palm (*Cocos nucifera*). Coir is the only prominent fiber that is obtained from fruits. Unripe coconuts are soaked in salt water for several months to loosen the fibers. They are then beaten to separate the fibers, which are then washed and dried. The uses of coir are varied. In tropical Asia and the South Seas it is the source of sennit braid, which is used for hawsers, cables, and small cordage. Coconut fibers are superior to all others for this purpose for they are very light and elastic, and exceedingly resistant to water. Coir is also used for bristles for brushes, doormats, floor coverings, sacks, coarse textiles, upholstery, stuffing for the bearings of railroad cars, and as a substitute for oakum. Ceylon is the center of the commercial production of coir. In Puerto Rico, coir is being prepared for use in horticulture as a peat substitute.

FIG. 17. *Above*, coir, the coarse fibers from the husk of the coconut (*Cocos nucifera*); *below*, sennit, the rope that is made from it. Sennit is the most universally used type of cordage in the South Seas. (*Courtesy of the Botanical Museum of Harvard University.*)

Pineapple

The leaves of the pineapple (*Ananas comosus*), the source of the familiar tropical fruit, furnish fibers of great strength and fine qualities. They are shiny white, very durable and flexible, and are not injured by water. When grown for the fiber, pineapples are planted closer together and develop longer leaves. To be of value the fibers must be taken from leaves that have not attained their maximum growth. Usually two-year-old leaves are cut and the fibers scraped out by hand, a delicate and expensive process. After being dried and combed out, the fibers are tied end to end and can then be woven. In the Philippine Islands piña cloth, one of the most delicate and costly of fabrics, is made from these fibers. Formosa and China also utilize pineapple fiber in making strong fabrics.

Pita Floja

The pita floja plant (*Aechmea magdalenae*) is a pineapplelike species found in dry alluvial soils at low altitudes from southern Mexico to Ecuador. The long leaves furnish a fiber of superior quality known as pita floja or pita. Pita fibers are the basis of one of the most ancient and most important native industries in Oaxaca and are also used in Honduras, Costa Rica, Panama, and Colombia. The fibers are 5 to 8 ft. in length, white or light cream colored, lustrous, finer and more flexible than other hard fibers, and with a high tensile strength. Because they are very resistant to salt water, pita fibers are used for fishlines, nets, etc. They are also frequently utilized for sewing leather. Because of its peculiar characteristics, its abundance and rapid growth, pita floja is one of the most promising fibers not at present in general commercial use.

Caroá

Caroá fiber has recently come into prominence as a substitute for jute. It is obtained from *Neoglaziovia variegata*, a bromeliaceous species of the dry, hot, arid areas of northeastern Brazil. The leaves are mechanically decorticated and yield a soft, white, flexible, elastic fiber three times as strong as jute. Caroá fiber is used for textiles, rugs, sacks, light cordage, twine, and paper.

BRUSH FIBERS

An important use of vegetable fibers is in the manufacture of brushes, brooms, and whisks. Such fibers must be very strong, stiff, and elastic, with a high degree of flexibility. In some cases whole twigs, fine stems, or roots are utilized; in others the fibers are obtained from leafstalks. Among the more important brush fibers may be mentioned:

Piassava

Several species of palms growing in tropical America and Africa are the source of the brush fibers that are known commercially as piassava, piassaba, or bass. These trees have leafstalks or leaf sheaths which yield the stiff, coarse, brown or black fibers used in making brushes for street sweeping.

West African piassava is obtained from the wine palm (*Raphia vinifera*), an exceedingly abundant species in the tidal bayous and creeks of Liberia and other countries on the west coast of Africa. The leafstalks are retted and the bundles beaten out. The long fibers are used for mats as well as brushes. A wine is made from the sap of this palm.

Brazilian piassava comes from two species of palm found everywhere in the lowlands of the Amazon and Orinoco regions. *Attalea funifera* is the source of Bahia piassava. In this species the stiff, wiry, brown fibers are almost bristlelike. They are removed from the swollen bases of the leafstalks with an ax. Their chief use is for the brushes of street-sweeping machines and scrubbing brushes, as the fibers are very durable and retain their resiliency even when wet. Para piassava fibers are produced on the margins of the leaf petioles of *Leopoldinia Piassaba*. They are used for brushes and brooms, and also for ropes, hats, and baskets by the natives.

Several other coarse fibers, such as palmyra and kittul fiber, are classed as piassava in the trade. **Palmyra fiber** is obtained from the Palmyra palm (*Borassus flabellifer*) of the East Indies. This species is one of the most useful of the palms, all parts of the plant being used for some purpose or other. The fibers are made up into rope, twine, paper, and machine brushes. **Kittul fiber** is finer, softer, and more pliable. It comes from the leaf sheaths of the toddy palm (*Caryota urens*) of Ceylon and the East Indies. The black bristles are made into ropes of great strength and durability or into soft brushes, and also serve as substitutes for horsehair and oakum.

FIG. 18. Broomcorn plants (*Sorghum vulgare* var. *technicum*), showing the comparative height of the varieties. From left to right: standard, dwarf standard, and dwarf.

The cabbage palm (*Sabal Palmetto*) of the coastal areas of the southeastern United States yields a valuable fiber known as **palmetto fiber.** The best material is obtained from young leafstalks still in the bud. Coarser fibers come from mature leaves or the bases of the old leafstalks surrounding the bud. For over 50 years a Florida industry has processed this fiber for use as a substitute for palmyra in brushes, especially those which must remain stiff in hot water or caustics. Up to 1,000,000 lb. are produced annually. Palmetto fibers are a red-tan color and 8 to 20 in. in length. The bud of this palm is edible, and the roots contain tannin.

Broomcorn

Broomcorn is a variety of sorghum (*Sorghum vulgare* var. *technicum*). It differs from the other sorghums in having a panicle with long straight branches. This inflorescence, or seed head, is the "brush," the valuable part of the plant. Standard and dwarf forms are recognized (Fig. 18). The brush of the former is stronger and is used for carpet brooms; that of the latter types is used for whisk brooms. In the United States broomcorn is grown chiefly in the Mississippi valley. It is harvested before the flowering season is over by cutting the stems a few inches below the head. The heads are then sorted, threshed, and dried.

FIG. 19. Broomroot or zacaton (*Muhlenbergia macroura*). The fibrous roots of this plant are used in the manufacture of brushes.

The wiry culms of *Spartina spartinae*, a native grass of the southern Coastal Plain from Florida to Mexico, are often used as a cheaper substitute for, or in combination with, broomcorn. Many brooms consist of as much as 50 per cent spartina, surrounded by broomcorn.

Broomroot

Broomroot or zacaton (*Muhlenbergia macroura*) is extensively used in the manufacture of the cheaper brushes. The plant is a grass, found from Texas to Central America, particularly in the mountainous regions of Mexico. It is a perennial species with tufted wiry culms and coarse roots. As the name indicates, the roots are the part utilized (Fig. 19). These are dug up at all seasons of the year by the peons. After washing, cleaning, and drying, the roots are cut from the tops, graded according to quality, length, and color, and baled for shipment.

PLAITING AND ROUGH WEAVING FIBERS

Only a few of the materials used for plaited or coarsely woven articles are of commercial importance, the greater number being used in native manufactures. The raw materials comprise the stems of reeds, rushes, grasses, willows, bamboo, rattan, and many other plants, as well as the leaves and roots. These materials are used entire or split. They are woven or twisted together in the simplest manner and made up into hats, sandals, mats and matting, screens, chair seats, baskets, and similar articles.

Hat Fibers

In many parts of the Orient and in Europe, wheat, rice, barley, and rye are grown for the purpose of making braids or straw plaits for hats. The plants are grown close together so they will have few leaves, and they are harvested before maturity. The stems are split lengthwise before plaiting. The Leghorn and Tuscan hats of Italy are among the best known of the straw hats.

Panama hats are made from the leaves of the toquilla (*Carludovica palmata*), a stemless, palmlike plant which grows wild in humid forests from southern Mexico to Peru. It is cultivated in Ecuador and parts of Colombia. The Panama hat industry is concentrated in Ecuador where over 4,000,000 hats have been made in a year; about 1,500,000 hats are exported annually. Young leaves are collected while they are still folded in the bud and treated with hot water. The coarse veins are removed, and the plaits are separated and split lengthwise into slender strips, which are dried slowly and bleached. They gradually become inrolled, forming fine cylindrical strands known as *jipijapa*. The hats are woven by hand from these strands. About six leaves are required to make one hat. The best Panama hats are characterized by their uniformity and fineness of texture, their strength, durability, elasticity, and resistance to water. The so-called Puerto Rican hats are made from the leaves of the hat palm (*Sabal causiarum*).

Mats and Matting

Commercial mattings are made in many of the Eastern countries from various sedges, rushes, and grasses. Usually the stalks or leaves of these plants are used alone, but they may be combined with a warp of cotton or hemp. Among the principal species utilized may be mentioned the

Chinese mat grass (*Cyperus tegetiformis*) (Fig. 20) and the Japanese mat rush (*Juncus effusus*). The United States imports several million dollars' worth of these products annually.

Native mats are made from a great variety of plants. One of the most outstanding sources, particularly in Southeastern Asia and Oceania, is the screw pine (*Pandanus tectorius*). *P. utilis* is also important. The

FIG. 20. Chinese mat grass (*Cyperus tegetiformis*), showing a piece of matting and the raw material from which it was woven.

leaves of these shrubby plants are also used for sugar bags, cordage, hats, and thatching.

Baskets

The manufacture of baskets from fibers or fibrous materials is an industry belonging to both civilized and native peoples. The various species utilized are too numerous to mention. Roots, stems, leaves, and even woody splints are used. Commercial baskets are usually made from rushes, cereal straw, osiers or willows, and ash or white oak splints. Sweet-grass baskets, made from *Hierochloë odorata*, a common species in swamps along the coast and the Great Lakes, are a familiar sight in eastern North America. Another important source of basket material is the raffia palm (*Raffia pedunculata*), a tree indigenous to Madagascar. Strips of the lower epidermis of the leaves of this palm constitute the raffia of commerce. This fiber is so soft and silklike that it can be woven. Its chief use is as a tie material in nurseries and gardens.

Wickerwork

Hampers, baby carriages, chair seats, and even chairs and other light articles of furniture may be made from willows. Rattan and bamboo are also extensively used for these purposes.

Rattan is obtained from several species of climbing palms (*Calamus* spp.) found in the hot humid forests of the East Indies and other parts of

FIG. 21. Guadua (*Guadua angustifolia*), the most important native bamboo of the northern Andes. (*Photo by Walter H. Hodge.*)

tropical Asia. The stems of these plants are very long, strong, flexible, and uniform in size. They are used, either entire or as splits, throughout Asia for baskets, furniture, canes, and a great variety of other purposes. A considerable quantity of rattan is exported to Europe and the United States.

Bamboos occur in most tropical countries, but they are especially common in the monsoon region of Eastern Asia. They are the largest of the grasses, with more or less woody stems which sometimes reach 1 ft. in diameter and 100 ft. or more in height. The species are very numerous and belong to *Arundinaria, Bambusa, Dendrocalamus, Gigantochloa, Phyllostachys*, and other closely allied genera. The native uses of the bamboo are legion for all parts of the plant are of value. This is particularly true of the stems which are extensively used for all types of construction. Bamboo is shipped to other countries for use in the manufacture of furniture, fishing rods, implements of various kinds, and many other objects. Bamboo splits are utilized for baskets and brushes.

With a few exceptions bamboos have been but little utilized in the Western Hemisphere. *Guadua angustifolia* (Fig. 21) a species with very strong culms, has been a favorite material in Ecuador for furniture and house construction. In recent years, however, more interest has been aroused. It has been found that bamboos thrive in wet soil and are easily propagated by seed or cuttings. In Puerto Rico the production of bamboo culms for furniture and fishing rods has become a profitable business. In 1947, the output was 20,000 lin. ft., obtained chiefly from *Bambusa Tulda*. Early attempts to establish a bamboo industry in Central America were hampered by the fact that most of the common native species were very susceptible to insect attack. Several resistant species have now been developed, and bamboo growing shows promise for the future.

FILLING FIBERS

A considerable number of plant fibers are used for stuffing pillows, cushions, mattresses, furniture, and similar articles. These filling fibers, as they are usually called, are also utilized in caulking the seams of vessels, in the manufacture of staff for building purposes, as stiffening for plaster, as packing for bulkheads and machine bearings, and for the protection of delicate objects during transportation. Surface fibers are commonly used for stuffing purposes, for their staples in general are too short to be spun readily and so are valueless for textile manufacturing. Bast fibers are too expensive, and hard fibers are likely to be too stiff and coarse. The silk cottons constitute the most important source of stuffing materials.

Kapok

Kapok is without question the outstanding silk cotton and the most valuable of all the stuffing materials. Its use is steadily increasing. Kapok is the floss produced in the pods of the kapok tree (*Ceiba pentandra*). This species, originally confined to the American tropics, is now found in Asia and Africa as well. It is an irregular tree (Fig. 22) from 50 to 100 ft. in height, with a buttressed base and a weird habit of growth. It grows very rapidly and begins to bear when only 15 ft. high. A mature tree produces over 600 pods and from 6 to 10 lb. of the cottony fibers.

Fig. 22. A kapok or silk cotton tree (*Ceiba pentandra*). Note the buttresses and the peculiar habit of growth.

Fig. 23. Kapok pods and a mass of the downy fibers produced in the pods. This is the kapok of commerce, the most valuable of the stuffing materials.

The pods are clipped from the branches and usually opened by hand. The floss is removed, and the seeds are separated out by centrifugal force. The floss (Fig. 23) is from ½ to 1½ in. long and whitish, yellowish, or brownish in color. It is very light, fluffy, and elastic and makes an ideal stuffing material for pillows and mattresses. The fibers have a very low specific gravity; they are five times more buoyant than cork and are impervious to water. For this reason kapok is valuable as a filling for life preservers, life cushions, portable pontoons, and similar articles. Its low thermal conductivity and its high ability to absorb sound make kapok an excellent material for insulating small refrigerators and for soundproofing rooms. It is also used for the linings of sleeping bags, gloves for handling dry ice, and in the tropics, surgical dressings. The United States imports a large amount of kapok chiefly from Java, the Philippines, and Ceylon and some from Mexico. During the Second World War, new sources of supply were developed in Brazil and Ecuador. Kapok seeds contain 45 per cent of a fatty oil which is extracted and used for soap and food purposes.

Kapok Substitutes

Numerous other plants have seed hairs or floss which can be used as a substitute for kapok. The red silk cotton or simal (*Salmalia malabarica*), a very large and ornamental tree, supplies a reddish floss known as Indian kapok which has been an important stuffing material in India for centuries. The white silk cotton (*Cochlospermum religiosum*) yields a kapoklike fiber of considerable importance. It is a handsome tree native to India and now widely cultivated in the tropics. This species is one of the sources of kadaya gum.

The madar (*Calotropis gigantea*) and the closely related akund (*C. procera*), shrubs indigenous to Southern Asia and Africa, produce a silk cotton of some importance. These species are now being cultivated in South America and the West Indies. Although much inferior to kapok, this product is frequently used in mixtures with kapok.

In Mexico, the pochotes (*Ceiba aesculifolia, C. acuminata,* and related species) yield a silk cotton fully equal to kapok in buoyancy and resiliency. The South American palo borracho (*Chorisia insignis*) and samohu (*C. speciosa*) also yield large quantities of a glossy, white silk cotton with kapoklike properties.

Several of the milkweeds, all of which have silky hairs on the seeds, are used as a source of stuffing materials. Milkweed floss, one of the lightest materials, is exceedingly buoyant and is a perfect insulator. It was exploited during the Second World War as a substitute for kapok. The pods contain an oil and a wax which are of potential importance, and some species yield textile fibers. In the United States, *Asclepias syriaca* and *A. incarnata* produce the best and most abundant floss; in tropical America, *A. curassavica* is valuable.

The heads of the common cattails (*Typha latifolia* and *T. angustifolia*) yield a floss with a high buoyancy and good insulating properties.

Other Filling Materials

Countless other plants and fibers are useful as filling materials. These include the straw of cereals and other grasses, corn husks, and two products of considerable commercial importance, Spanish moss and *crin végétal*.

Spanish moss (*Tillandsia usneoides*), a conspicuous epiphyte of the Southern states, is an excellent substitute for horsehair after it has been processed. The plant is pulled from the trees with rakes or hooks, or it is collected from the ground or water. It is then cured (fermented) in order to rot off the gray outer covering and ginned to remove any impurities. The prepared fiber is brown or black, lustrous, and very resilient. Some 10,000,000 lb. are produced annually in the United States for use in upholstery and for automobile and car cushions.

Crin végétal is the term applied to the shredded and twisted fibers from the leaves of the dwarf fan palm (*Chamaerops humilis*) of Northern Africa and the Mediterranean region. The United States imports several thousand tons annually.

NATURAL FABRICS

Certain tree basts with tough interlacing fibers can be extracted from the bark in layers or sheets, which can then be pounded into rough substitutes for cloth.

Perhaps the best known of these bark cloths is **tapa cloth,** which constituted the chief clothing of the natives of Polynesia and parts of Eastern Asia until half a century ago. This material is obtained from the bark of the paper mulberry (*Broussonetia papyrifera*). Strips of bark are peeled off the trunk, and the outer coating is scraped off with a shell. After they have been soaked in water and cleaned, these strips are placed on a log of hard wood and pounded with a mallet. The individual strips are united by overlapping the edges and beating them together. Depending on its thinness the finished product varies in appearance from a muslinlike material to a leathery one. Tapa cloth is often dyed.

Similar bark cloths have been manufactured from different sources by the native peoples of many other parts of the world. The South American Indians used the tauary (*Couratari Tauari*) and other species of the same genus. In Mozambique a wild fig (*Ficus Nekbudu*) is used as the source of the native mutshu cloth. A common bark cloth of Ceylon and Malaya is obtained from the upas tree (*Antiaris toxicaria*), which is also the source of an important arrow poison.

The so-called **lace bark** is the product of *Lagetta lintearia*, a small Jamaican tree. The inner bark is removed in sheets and can be stretched into a lacelike material with pentagonal meshes. This is suitable for a number of textile and ornamental uses.

Cuba bast is obtained from *Hibiscus elatus*, a small bushy tree of the West Indies. The inner bark is removed in long ribbon-like strips, which are much used in millinery and for tying cigars.

MISCELLANEOUS FIBERS

The vegetable sponges (*Luffa cylindrica* and *L. acutangula*) (Fig. 24) yield a unique fibrous material. These climbing cucumbers of the tropics bear edible fruits that contain a lacy network of stiff curled fibers. This skeletal material is extracted by retting in water. After cleaning, it is used for making hats, for washing and scouring machinery, in certain

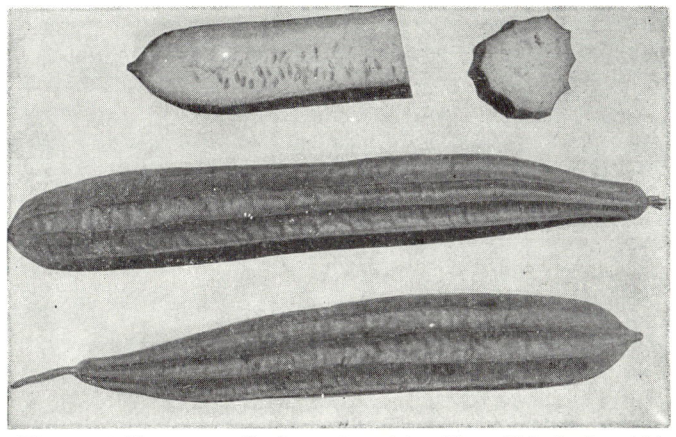

FIG. 24. The vegetable sponge (*Luffa acutangula*). The edible fruits yield a unique fibrous material. (*Courtesy of the Botanical Museum of Harvard University.*)

types of oil filters, and as a substitute for bath sponges. Japan formerly exported large quantities of this material.

PAPERMAKING FIBERS

An important use of fibers is in the manufacture of paper. However, since it is the cellulose present in the fiber, rather than the fiber itself, which is utilized, discussion of this subject will be deferred until Chap. X.

ARTIFICIAL FIBERS

The artificial fibers used in the textile industry are for the most part organic in nature; synthetic glass fibers are an exception. The organic materials utilized are cellulose; proteins, both animal and vegetable; and synthetic resins, such as nylon, which is made from soft coal, water, and air. The cellulose fibers will be considered with other cellulose products in Chap. X. The protein fibers are at present of only minor importance. Although they have many of the general properties of wool, their low wet strength is a detriment. Soybeans, peanuts, and corn are the chief plant sources of protein fibers.

CHAPTER III

FOREST PRODUCTS: WOOD AND CORK

FOREST PRODUCTS

The products of the forest have been of service to mankind from the very beginnings of his history. The most familiar, and the most important, of these products is wood, the manifold uses of which in all types of construction, as a fuel, and as a raw material of the paper and rayon industries are well known to everyone. Wood, however, is by no means the only useful material obtained from trees. Other products include cork, rubber, many of our tanning materials and dyestuffs, resins, gums, oils, drugs, and even sugar, starch, and various chemicals. Moreover the seeds and fruits of many trees often serve as food for man or beast.

Not only are these forest products of value to man, but the forests themselves have many utilitarian features. They help to regulate climate and temperature. They aid in the conservation of the water supply and in flood control by preventing the runoff of water. Their deep roots hold the soil firmly in place and so check erosion. Again they may act as shelter belts against drying winds. In addition forests afford a range for livestock, a shelter for wild life, and offer many recreational aspects for man, the importance of which is just beginning to be recognized.

Our discussion of forest products, as such, will be limited to wood and cork, which will be treated in the present chapter. The other useful materials obtained from trees will be considered in later chapters together with similar economic products from other sources.

THE IMPORTANCE OF WOOD

From the earliest time clothing, shelter, and food have been the three great necessities of mankind. We have already discussed in Chap. II the importance of fiber plants as a source of shelter and of clothing. Wood has been fully as important, and has contributed its share to the advancement of civilization. Primitive man not only used wood in the construction of his rude shelters, but was able, even with the crude stone implements at his disposal, to fashion dugout canoes, implements, and utensils of various kinds. At a later period when metal tools became available the uses of wood increased greatly. When we stop to consider that from the dawn of history to the middle of the last century all ships were made of wood, the influence that this material had on the course of history is obvious. If no wood suitable for seagoing vessels had existed, the great voyages of exploration, the colonization of the newly discovered lands, and many other events that have led to the development of the world as we know it today would have been impossible.

Today wood is the most widely used commodity outside of food and clothing, and for many purposes it has no satisfactory substitute. It is one of the most versatile of the raw materials of industry. It will probably continue to be indispensable in spite of the competition of the various metals, for it is the only raw material in its field that can be renewed. Wood has many other advantages over the metals. It is the most readily available of all raw materials, and with proper care the supply should be inexhaustible. It is cheaper, lighter, more easily worked with tools, and more easily fastened together; wooden structures can be readily altered, moved, or rebuilt. Wood is very strong for its weight, and in its combination of strength, elasticity, and toughness it has no rival. It is a poor conductor of heat, electricity, and moisture and does not rust or crystallize. Its beautiful figure makes it preeminent as a material for fine construction. It is also possible to use wood in the form of thin sheets or veneers. It must not be assumed from the above statements that wood is superior to metals in every respect, for it does have certain very definite disadvantages. In addition to the more familiar uses of wood, modern developments in the field of wood technology have made it possible to convert wood into such diverse products as textile fibers, paper, soap, cattle feed, lubricants, and motor fuel (Fig. 119).

THE STRUCTURE OF WOOD

Wood is a secondary tissue produced chiefly in the stems of gymnosperms and dicotyledons as the result of the activity of a growing layer, the *cambium*. This tissue is responsible for the growth of stems in thickness through the formation each year of new layers of both wood and bark. The nature of this process and the details of wood anatomy are too familiar to students of botany to need much amplification here.

Nature of the Wood Elements. Wood is a heterogeneous tissue composed of several different types of cells, some of which have the function of mechanical support and others that of conduction. In the softwoods, as the woods of gymnosperm origin are commonly called, both these functions are normally carried on in cells known as *tracheids*. In the hardwoods, which comprise woods of angiosperm origin, on the other hand, a division of labor usually occurs. Mechanical support is furnished by the several types of *wood fibers*, which make up a large part of the woody tissue, while conduction of water is carried on in tubular cell fusions known as *vessels*. Tracheids also may occasionally be present.

A third function of the wood elements, the distribution and storage of carbohydrate food, is carried on in the thin-walled *parenchyma cells*. These are the only components of wood which can be said to be alive and which contain protoplasm. They may be wood parenchyma cells, which have a vertical arrangement in the stem, or ray parenchyma cells, which are horizontally arranged.

Diagnostic Features of Woods. The arrangement of the various types of cells in different species affords valuable diagnostic characteristics for the identification of woods, besides making possible the distinctive features that are of commercial importance. Among the gross morphological characteristics that are of value in distinguishing between woods may be mentioned pores, early wood and late wood, growth rings, rays, sapwood and heartwood, and grain and figure.

Porous and Nonporous Woods. The presence or absence, and the nature and arrangement (as seen in cross section) of *pores*, which are cross sections of vessels, serve as a ready means of classifying woods (Fig. 25). Coniferous woods, in which vessels and consequently pores are typically absent, are classed as *nonporous* woods. The porous woods comprise the hardwoods, and two distinct types are recognizable. In some woods, like ash, elm, hickory, and oak, the pores seem to be arranged in concentric circles, the outer and inner portions of which differ with regard to the number and size of the pores. This condition is due to the fact that large vessels are formed at the beginning of each season and smaller ones later on. Such woods are termed *ring-porous*. In other woods, like beech, birch, maple, and walnut, the pores are all small and approximately of the same size and they are scattered uniformly through the wood. These are the *diffuse-porous* woods.

Early Wood and Late Wood. In temperate regions new wood is formed each year during a limited growing season and definite growth layers result, which usually show two distinct areas within each layer. In the spring when growth is resumed, the first wood to be formed contains many large and relatively thin-walled cells, in response to the greater need for conducting elements. This is the early wood, or spring wood, as it has been called. As the season progresses a denser type of wood is laid down with smaller, thicker walled cells, the so-called late wood, or summer wood. As a result of this there is a sharp transition between the cells produced at the end of any one growing season and those formed at the beginning of the succeeding one. This gives rise to what appear in cross section as concentric rings, known as growth rings. The growth ring of a single year is called an annual ring, and the number of these annual rings gives an indication of the age of the tree. In the tropics where growth may be continuous throughout the year, growth zones may occur, but

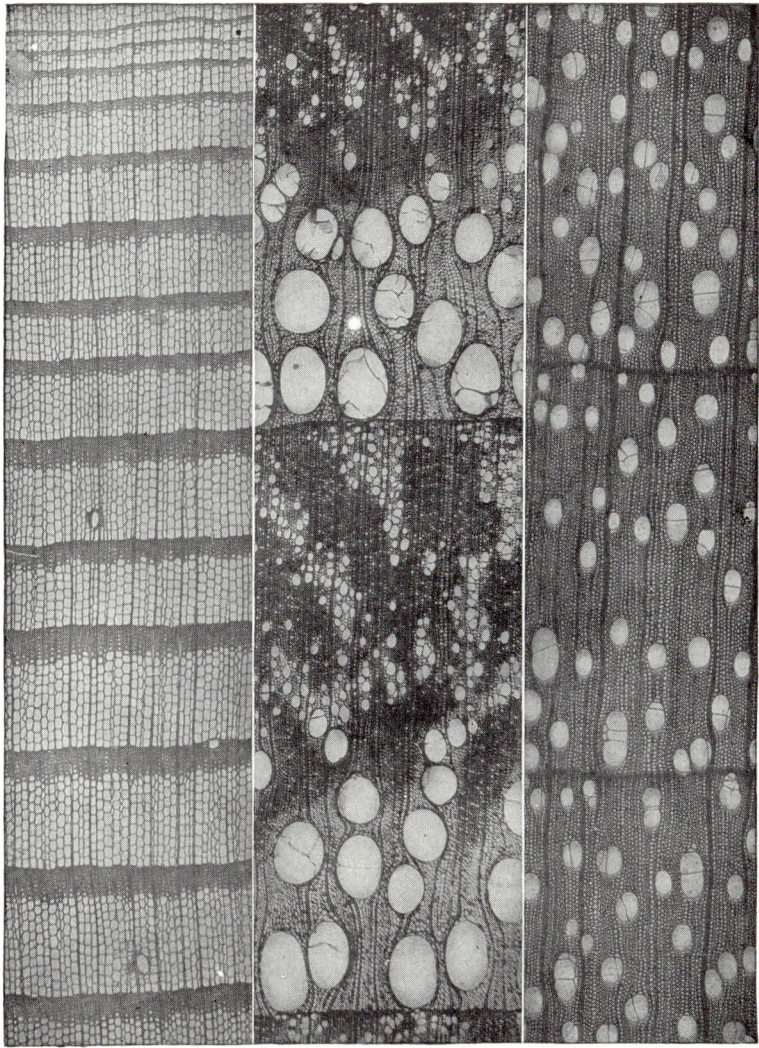

Fig. 25. *Left*, a nonporous wood, western larch (*Larix occidentalis*); *center*, a ring-porous wood, chestnut (*Castanea dentata*); *right*, a diffuse-porous wood yellow birch (*Betula lutea*). (*Reproduced from Holtman, Wood Construction, McGraw-Hill Book Company, Inc.*)

they are due to changes in climate or other causes rather than to definite growth periods as is the case in temperate woods.

Rays. The rays are thin sheets or ribbons consisting chiefly of parenchyma cells that are oriented at right angles to the main axis of the stem. They vary greatly in width, height, and arrangement. Although visible in a cross section as lines radiating from the center of the stem, rays are most conspicuous in radial sections, where they frequently contribute valuable diagnostic features in the variety of their form and arrangement.

Sapwood and Heartwood. Although at first all the cells in wood are physiologically active, sooner or later many of them lose this property and become mere skeletons, serving only to give strength to the tree. Eventually two distinct areas develop—a light-colored outer region of varying width, the sapwood; and a darker inner region, the heartwood. Only the cells in the sapwood are physiologically active, and even in this region only the parenchyma cells can be said to be alive. The older, dead cells of the heartwood often become highly colored and very resistant to decay, owing to the deposition in them of various gums, resins, or other waste materials. Such heartwood is usually capable of a high polish, and

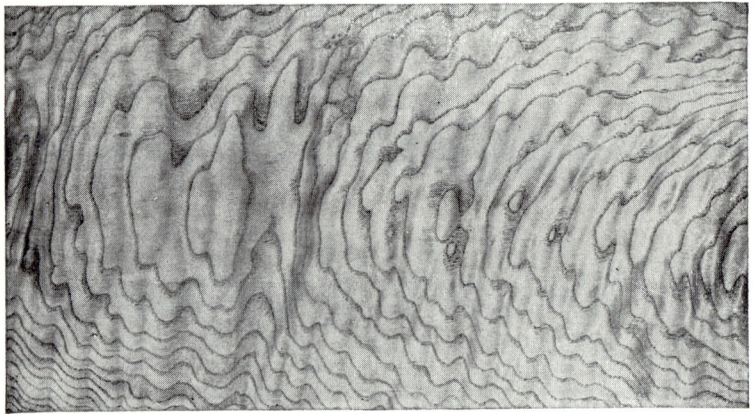

FIG. 26. Wavy grain in white ash (*Fraxinus americana*). The light areas are late wood, the wavy dark lines early wood. (*Courtesy of S. J. Record.*)

is in great demand in the cabinetmaking, furniture, and other high-grade woodworking industries. Although heartwood is usually quite distinct from sapwood in appearance, durability, and other properties, it may not always be so clearly differentiated.

Grain and Figure. Grain, figure, and also texture are terms that are likely to be used in a more or less confusing manner. Strictly speaking, *texture* refers to the relative size and quality of the various wood elements, while *grain* refers to their structural arrangement. *Figure*, on the other hand, applies to the design or pattern that appears on the surface of lumber, and may be due to the kind of grain, the presence of coloring matters that have penetrated the tissues, or both.

Only a few of the many kinds of grain can be mentioned here. In straight-grained wood the various elements occur parallel to the main axis of the stem. Where they are spirally twisted about the axis they constitute spiral-grained wood; and when the longitudinal course of the elements is slightly undulating, wavy grain (Fig. 26) results. Curly grain is due to various irregularities of growth and a diversity of causes.

The attractive and distinctive figures which wood often shows and which make it so valuable for decorative purposes are due chiefly to the different types of grain in combination with the rays, rings, sapwood, heartwood, and the many other cell arrangements. It is possible to accentuate one or another of these variations by using different methods of cutting the wood. In quartersawing, for example, wood is cut parallel to the rays and across the rings, while in plain sawing the wood is cut at right angles to the rays and tangent to the rings. Both quartersawed and plain-sawed wood have their own desirable qualities. Occasionally

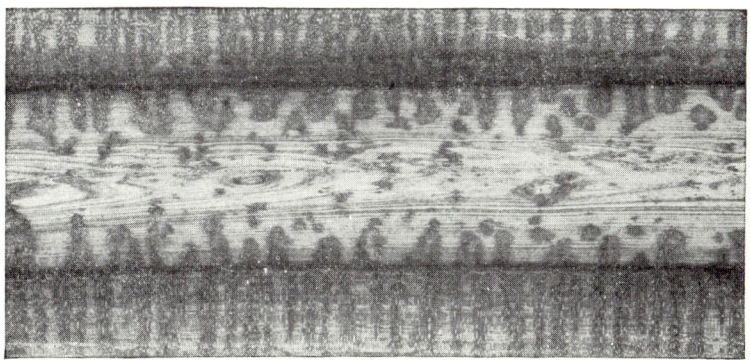

FIG. 27. Figure in snakewood (*Piratinera guianensis*). The streaks of coloring matter that have penetrated the tissues suggest the markings on a snake skin. (*Courtesy of S. J. Record.*)

figures in wood are produced by masses of coloring matter which has penetrated the tissues and which may occur in zones or streaks. In the case of snakewood (Fig. 27) streaks on the tangential surface suggest the markings on the skin of a snake.

THE MECHANICAL PROPERTIES OF WOOD

Wood possesses certain mechanical properties which, either alone or in combination, determine its usefulness and suitability for various purposes. These characteristics differ in different species and even in individual trees. Many tests have been devised to determine the exact nature of the properties that any specific wood may possess, for this information is essential for the consumer.

The mechanical properties of wood are those properties which enable it to resist various external forces that tend to change its shape and size and produce deformations. These external forces induce internal resisting forces, known as stresses, in the wood. When the latter exceed the force of cohesion among the wood elements, some sort of failure occurs. Among the more important mechanical properties may be mentioned strength, stiffness, toughness, hardness, and cleavability.

Strength. Although strength is sometimes used to mean all the mechanical properties of wood, it should more properly be restricted to the ability to resist certain definite forces, such as crushing, pulling, and shearing. Moreover the word should always be modified to indicate the specific type of resistance involved.

Crushing Strength. Crushing or compression strength is the resistance offered to forces that tend to crush wood. These forces may be applied endwise, and so parallel to the grain, as in a column; or sidewise, where they are at right angles to the grain, as in the case of railroad ties. The maximum crushing strength is endwise.

Tensile Strength. Tensile strength is the resistance to forces that tend to pull wood apart, and these also may be applied either parallel to or at right angles to the grain. The maximum tensile strength occurs when the force is parallel to the grain, and it is two to four times greater than the crushing strength.

Shearing Strength. Shearing strength is resistance to those forces which tend to make the fibers slide past one another. These forces may be applied parallel to the grain, at right angles to it, or obliquely. Wood is more resistant to a perpendicular shear.

Cross-breaking Strength. The cross-breaking, or bending, strength of wood is usually applied to beams or other pieces of timber that are supported at both ends and loaded between these points. The strength involved is the resistance to forces that cause the beam to break, and all the above-mentioned forces are involved. The upper part of the beam is under compression, the lower under tension, and shearing also plays a part. Because of its greater tensile strength, a beam fails first by compression, or buckling, on the upper side; then, as the load increases, the tensile strength gradually fails until the beam snaps across on the underside.

The strength of wood is the most important factor in determining the value of any species for structural purposes. It is a very variable property, however, and is influenced by the density of the wood, the moisture content, the presence of defects, and many other factors. The relationship between density and strength is particularly close, so much so that density is considered to be the most satisfactory criterion of strength. Longleaf pine, larch, hickory, sugar maple, and white oak are among the strongest native woods.

Stiffness. Stiffness is the measure of the ability of wood to resist forces that tend to change its shape—in other words, its capacity to withstand deformation under a bending strain. This is often contrasted with flexibility, which is the ability to bend without breaking and which involves toughness and pliability.

Toughness. Toughness is used in several senses, but it is usually considered to mean the ability of wood to absorb a large amount of energy, and so resist repeated, sudden, sharp blows or shocks. It is really a combination of other properties. A tough wood is hard to split, and, although it may rupture, it does not break readily.

Hardness. Hardness is the measure of the power of wood to resist indentations, and also abrasion and wear. It varies with the density and determines the ease with which wood can be sawed or cut.

Cleavability. Cleavability is an expression of the ease with which wood can be split. It is a particularly desirable quality in the case of firewood, and equally undesirable when wood must be able to hold nails or screws. Wood tends to split more easily along the rays and when it is straight-grained.

FACTORS THAT INFLUENCE THE MECHANICAL PROPERTIES OF WOOD

Although there are many factors that influence the strength, hardness, and other properties of wood to a greater or less extent, space will permit the mention of only a few of the more important ones, such as density, moisture, and a few of the defects.

Density. Differences in density—the ratio of the mass of wood substance to its bulk—are among the chief causes of variations in the mechanical properties of wood. It has already been pointed out that, because of the close relationship between the two properties, the density of any wood may be considered the best indication of its strength. Density is usually expressed in terms of *specific gravity*, which is relative density. Whereas density is merely the weight of a unit volume of any substance, and is expressed as weight per cubic foot, specific gravity is the ratio between the weight of the substance concerned and that of an equal amount of pure water.

The density of wood substance is practically the same in all species, and its specific gravity has been estimated as about 1.55. This means that wood is 1.55 times heavier than water, which explains the fact that a piece of wood sinks as soon as the air in the cavities has been replaced by water. The variations in density, however, which are to be noted in different species, individuals, and even parts of the same individual, are due to differences in the actual amount of wood substance present—in other words, to the amount of cell-wall material as compared with cell cavities. Woods with thicker walls and smaller cavities are denser and heavier, while the more porous woods with larger cavities and thinner walls are lighter and less dense.

The specific gravity of wood may be determined in several ways, but usually the weight is calculated from oven-dry material, and the volume is measured with the wood in any desired condition. However, since the moisture content of green, partly seasoned, and seasoned wood is quite different, and since this difference materially affects the specific gravity, it is obvious that figures for specific gravity mean little unless the condition of the wood under which the calculations were made is stated.

Since the amount of actual wood substance has such an important bearing on the density of a piece of wood, it might be assumed that heavier woods are necessarily stronger. This is not always the case for the presence of gums, resins, or other infiltrated substances and also the amount of water may affect the weight without altering the strength.

Oak, hickory, persimmon, and osage orange are among the heaviest native woods. It is often stated that tropical woods are heavier than temperate woods. That this is not the case is evidenced by the fact that both lignum vitae, the heaviest commercial wood, and balsa, the lightest, are products of the tropics.

Moisture. Wood always contains more or less water, the amount in green wood varying in different species and under different conditions from as low as 40 per cent to as high as 100 per cent or more of the dry weight. The water in wood occurs in the cell cavities or in the cell walls, in which case it is known as *hygroscopic water*. The amount of hygroscopic water necessary to saturate the walls is called the *fiber-saturation point*, and constitutes from 20 to 35 per cent of the dry weight. The variation in the amount of water that may be present in wood is due to a number of causes, and is made possible by a characteristic property inherent in the wood. This property, which is known as *hygroscopicity*, is the ability to absorb or give off water under different conditions, with an accompanying swelling or shrinking as the case may be.

As already stated, the moisture content of wood has an important bearing on its weight, density, and often on its strength. If the amount of water present is above the fiber-saturation point, the weight is increased, but the strength is not altered. If, however, the amount of water is brought below the fiber-saturation point through evaporation, then the strength and other mechanical properties as well are changed. This means that it is only the loss of hygroscopic water which is responsible for the increase in strength that accompanies seasoning, as the drying out of wood is called. This loss of hygroscopic water causes the wood to shrink, owing to changes that take place in the cells. As the water passes off, the walls contract, the cells become more closely compacted, and the fibers become stronger and stiffer. This tendency of wood to shrink as it dries is one of the great drawbacks to its use. The amount of shrinkage varies greatly under different conditions, and it is likely to occur unequally and unevenly. Wood, for example, shrinks very little lengthwise, and only about half as much radially as tangentially. As the result of uneven shrinkage, warping, checks, shakes, and other defects may develop, which tend to counteract any increase in strength. In spite of these shortcomings, however, dried or seasoned wood is for the most part stronger, harder, stiffer, and more durable than unseasoned wood. Its desirability is great enough so that artificial methods of seasoning are utilized in an attempt to control the process as far as possible.

Artificial Seasoning. The two chief types of artificial seasoning are air seasoning and kiln drying. In *air seasoning*, the moisture is removed by exposure to air without resorting to artificial heat. It is carried out in the open until the wood ceases to lose weight. The final moisture content varies from 12 to 30 per cent, depending on the species, the duration of the process, and the weather conditions. The principal objects of air seasoning are to reduce the weight, the amount of shrinkage, and possible defects; to render the wood less subject to decay; to increase its strength and combustibility; and to prepare it for painting, preservative treatment, and kiln drying.

In *kiln drying*, heat is applied to wood in an enclosed space. Either unseasoned or seasoned wood can be utilized. The moisture is removed more rapidly and more completely, the moisture content of the finished product varying from 4 to 12 per cent. If green lumber is kiln dried, checks, warping, and defects due to fungi or insects may often be prevented.

Defects. The mechanical properties of wood are frequently affected by the presence of various types of defects, which may be of major or minor importance. These defects are due to a considerable diversity of causes. Some of them may be normal characteristics, but ones which limit the usefulness of wood. For example, wood is dimensionally unstable; it swells, warps, and checks with changes in humidity and temperature. Its strength is unidirectional in that it is strong with the grain and weak across the grain. These disadvantages may be overcome in part by the use of plywood, wood alloys, or reconstructed wood.

As noted above, certain defects may develop during the seasoning of wood; others, such as knots and cross grain, may be inherent in the wood structure; still others may be due to external agencies. Among the latter may be mentioned insects, fungi, marine borers, birds, parasitic seed plants, frost, lightning, and fire. Defects due to insects and fungi are the most important.

Insect Damage. The damage to wood resulting from insect injuries is far greater than is usually realized, amounting to about $9,000,000,000 annually. All sorts of wood from standing timber to lumber and wood products may be attacked. Holes produced by wood-boring insects constitute the chief type of injury. Insects are most destructive when in the larval stage. In recent years the ravages of termites, or "white ants," have been increasingly serious.

Decay. Decay in wood is caused by fungi. Four conditions are essential for the development of these lower plants. Unless a favorable temperature, sufficient moisture, at least a small amount of oxygen, and an adequate food supply are available, fungous decay does not occur. The food supply is furnished by the cellulose and lignin in the cell walls, and is made available by enzymes, which are secreted by the fungi. The so-called brown rots remove the cellulose, leaving behind a brittle brown mass of lignin compounds; the white rots utilize the lignin and leave the white cellulose behind. Still other fungi are able to utilize both cellulose and lignin.

FIG. 28. A creosoting plant in the Holy Cross National Forest, Colorado. (*Photo by U.S. Forest Service.*)

Woods vary greatly in their natural resistance to decay. This property is known as *durability*, and is so important that it alone may determine the ultimate use of wood, especially in the case of posts, poles, mine timbers, etc., which are exposed to moisture. Sapwood decays more readily than heartwood, for the latter often contains resins, gums, tannins, and other substances that resist fungous activity. Seasoned wood is also less apt to decay because of the lower moisture content. Among native woods that are naturally very durable may be mentioned the cedars, redwood, cypress, osage orange, and locust; the least durable woods include basswood and balsam fir.

Since the greatest possible durability of wood and wood products is highly desirable, it is frequently the practice to render the wood more immune to decay by treating it with preservatives that are poisonous to fungi. This process of *wood preservation* has developed into a considerable industry in recent years (Fig. 28). Various chemicals are used as preservative agents, chiefly creosote, a coal-tar product; and zinc chloride. The methods used are brushing or spraying the surface, dipping in open tanks, and various pressure processes that make possible a deeper penetration of the liquids.

THE USES OF WOOD

The uses of wood are so many and so varied that it will be impossible to discuss them at length within the limits of the present volume. It will also be impracticable to cite definite figures regarding the amount of wood and wood products utilized, since conditions vary so much from year to year. Anyone desirous of obtaining such figures may do so from government reports and other sources. For purposes of comparison, however, some such data are desirable, and consequently reference will be made to a few of the figures published in Zon and Sparhawk's "Forest Resources of the World." According to these authors the total annual consumption of wood in the world amounts to 56,000,000,000 cu. ft. Of this amount 28,000,000,000 cu. ft. (50 per cent) is used in North America, 17,000,000,000 cu. ft. in Europe, 8,000,000,000 cu. ft. in Asia, 2,500,-000,000 cu. ft. in South America, and the remainder in Africa, Australia, and Oceania. The United States uses about 24,000,000,000 cu. ft. The per capita consumption in North America is 188 cu. ft., a figure five times greater than that for any other continent. For the United States it is 228 cu. ft. This is significant when compared with Great Britain's 15 cu. ft., France's 26 cu. ft., and Germany's 27 cu. ft. It must be stated, however, that in other countries where the forests are still abundant the per capita consumption is even greater than in the United States.

In view of the fact that a great deal more wood is cut in the United States each year than is replaced by normal growth, it is obvious that the present rate of consumption cannot continue indefinitely. This does not mean necessarily that all the forests are doomed to extinction. There is every reason to believe that under proper management and with adequate methods of conservation in the future, together with an integrated forest industry, there is sufficient potential forest area in the United States to supply all our needs. Our policy should be to grow more and waste less.

It has been estimated that during the next five years the United States will use 14,593,000,000 cu. ft. of wood and wood products. The lumber industry will need 8,670,000,000 cu. ft.: 1,836,000,000 cu. ft. will be used for fuel, 1,660,000,000 for pulp, 600,000,000 for posts, poles, and piling, and the balance for veneers, cross ties, cooperage, mine timbers, shingles, distillation, and other such purposes. This estimate does not take into account the potential and highly desirable utilization of large quantities of wood and wood waste as the source of the many products made available by modern wood technology.

Fuel

Fuel is one of the great necessities of modern life and it is indispensable, both in the home and in industry, as a source of heat and power. Any material that burns readily in air can be utilized, and a great variety of plant products are so used. The most important of these are wood, peat, and coal, which represent different stages in the carbonization of the original plant tissue.

Wood. Even though much less wood is used for fuel at the present time than formerly, the total consumption is probably greater than for any other purpose. It is difficult to obtain reliable figures regarding the amount used, for much of the supply is obtained from small wood lots and other local sources. It must amount to 40,000,000 to 50,000,000 cords. Farms and rural communities account for 90 per cent of the total, with the Southern and Western states using more than the rest of the country. The extensive use of wood for fuel does not represent a proportional drain on the forest resources, for much waste material is used, as well as wood unfit for other purposes.

Wood makes an excellent fuel since it is 99 per cent combustible when perfectly dry, and so leaves only a small amount of ash. It is also a flaming fuel and well adapted for heating large surfaces. The value of any particular wood for heating purposes depends on the amount of moisture present; thus seasoned wood is obviously better than green wood. Hardwoods have the greatest fuel value, especially such woods as hickory, oak, beech, birch, maple, and ash. In the South longleaf pine is used, while in the West favorite species include Douglas fir, western yellow pine, western hemlock, and western larch.

Peat. Peat consists of deposits of vegetable matter which have accumulated in bogs and swamps and slowly decomposed, becoming somewhat carbonized and compacted (Fig. 29). The various plant tissues can still be made out. The process of peat formation is constantly going on. Peat is a valuable fuel in countries where wood is scarce. It is more bulky to handle, and leaves from five to fifteen times as much ash.

At the lowest depths of some peat bogs a soft brown coal, known as **lignite,** may be found. This also has the original plant structure still visible.

Coal. Coal comprises the fossilized remains of plants that lived in former geological periods. The original plant tissue has been more fully decomposed and converted into carbon. Coal is much harder and more compact than peat or lignite, and has a greater heating power. It also yields a larger amount of smoke and ash. Anthracite or hard coals are the oldest and contain about 95 per cent carbon. Bituminous or soft coals are of more recent origin and consequently are less completely carbonized; they tend to soften and fuse at temperatures below the combustion point. Cannel coal consists of fossilized spores. It is very compact and oily and burns with a candlelike flame. Unlike other coals it does not soil the hands.

Coal is a cheap source of power and heat, and also of many valuable chemical products. Among the latter, which are obtained by destructive

FIG. 29. Section of a peat deposit several feet in depth. (*Reproduced from Weaver and Clements, Plant Ecology, McGraw-Hill Book Company, Inc.*)

distillation, may be mentioned oils, such as benzol and naphtha; coal gas, which is used for fuel and illuminating purposes; ammonia; coal tar, the source of dyes, antiseptics, and countless other materials; and coke.

Coke bears the same relationship to coal that charcoal does to wood. It is obtained by the smothered combustion of coal in piles or special ovens, usually as a by-product of the illuminating-gas industry. It is nearly pure carbon and burns without smoke or flame. Coke is an excellent fuel, and is also used in metallurgy.

Petroleum. Although no trace of the original structure remains, it is generally supposed that petroleum had an organic origin and was formed under pressure from the minute floating plant and animal life of former shallow seas. Crude petroleum has many uses, but the substances derived from it by fractional distillation are even more important. Among these products may be mentioned gasoline, kerosene, petroleum jelly, and paraffin.

Lumber

Wood, in the form of lumber, has been used for building purposes and other types of construction from colonial days. This demand for lumber led to the establishment of a lumber industry at an early date. The first sawmill was started in Maine in 1631. From this beginning there has gradually developed a huge industry. The total output is around 40,000,000,000 board feet annually, a figure larger than that of any other nation.

Man was quick to take advantage of the grosser physical characteristics of wood. A tree which has been felled and trimmed is a ready-made structural component with compressive strength and stiffness and can be utilized directly for posts, poles, mine props, and rough construction. By proper sawing in the direction of the grain, however, these characteristics can be carried over into the various shapes of lumber, which, in addition to being light in weight compared with their strength, are easily shaped to dimensions and assembled and have good insulating powers.

The term "lumber," as generally used, means wood that has been prepared to some extent for future use. The larger sizes of lumber, intended for heavy construction, are often referred to as "timber." This term, however, is also frequently used for the forest trees themselves. The standard unit of measuring lumber is the board foot, which is the equivalent of a piece of wood 1 in. thick, 12 in. wide, and 1 ft. long.

The Lumber Industry

During the long history of the lumber industry in the United States there have been many changes in the location of this industry, in the species utilized, and in the actual woods operations.

The center of the industry has always been in a region where large stands of virgin timber were available. Until 1830 Maine was the principal lumber-producing state, and for the next 40 years New York and Pennsylvania led. By 1870 the center had shifted to the Lake States, with first Michigan and later Wisconsin assuming the lead. From 1910 on, the Southern states became the leaders, with southern pine replacing the northern hardwoods and white pine. More recently the center has moved to the Pacific Northwest and is utilizing the immense stands of Douglas fir and other softwoods. At the present time the Southern states as a group still produce the most lumber, although Washington and Oregon are the leading individual states. The Northwestern states must eventually assume the leadership, for over half the remaining timber supply is located in this region.

Since the beginning of the lumber industry some 150 native species have been utilized. The softwoods, however, have furnished about 78 per cent of all the lumber cut. For many years the eastern white pine was the outstanding timber tree, and indeed was one of the most valuable trees in the whole world. The demand for it was so great that the supply soon became greatly diminished and it was replaced in importance by other species. Oak and hemlock have also had their heyday. At the present time the most important trees are southern yellow pine, Douglas fir, western yellow pine, oak, hemlock, white pine, and red gum, each of which yields more than 1,000,000,000 board feet of lumber. The first two of these, however, produce twice as much as all the others together. Many woods that at one time were considered worthless are of commercial importance today. Examples include red gum, sycamore, beech, and tupelo.

In the early days the original forests were so abundant that the timber supply was thought to be inexhaustible. As a result the lumbering operations were usually very wasteful. At times every bit of wood would be stripped from an area, leaving only great quantities of slash as an ever-present fire hazard and in other ways rendering the region unfit for future forests. Or all the finest specimens would be cut, leaving only inferior trees, so that little or no reproduction was possible. The inferior quality and quantity of our eastern forests bear witness to these wasteful methods. In recent years, however, there have been many improvements in lumbering operations looking toward an elimination of waste and more complete and efficient utilization of the present supply, together with conservation and maintenance of a supply for the future.

The Utilization of Lumber

Normally about 8 per cent of the lumber cut each year is exported; 32 per cent is used for structural timbers and rough lumber for construction; 33 per cent goes to the planing mills; and 27 per cent is used in other woodworking industries.

Export Lumber. The exporting of lumber has been an important item in American foreign trade since the early days of New England, and it increased in amount until the First World War. Great Britain and the West Indies have been good customers, and in recent years there has been a considerable export business with Mexico, several of the South American countries, Australia, New Zealand, Germany, Italy, China, and France.

Structural Timbers. Structural timbers are the large sizes of lumber that come from the sawmills, and are used for bridges, buildings, and other types of heavy construction. They include beams, girders, stringers, posts, joists, rafters, planks, caps, sills, and boards for sheathing, roofing, and flooring. Before wooden vessels were replaced to such an extent by steel vessels, immense quantities of heavy timbers were used in shipbuilding (Fig. 30). Structural timbers are obtained chiefly from the softwoods because large sizes are more readily available. The timbers are usually sawed from the heart of the tree, and, even though defects may be present, the bulk is so great that the strength is not impaired. Strength is the chief requirement of a good structural timber, particularly the resistance to stresses that can be estimated. Numerous timber-testing experiments have made possible the estimation of the working stresses with remarkable accuracy. Soundness, durability, and ease of working are other desirable characteristics. The principal species used in the United States for structural timbers are southern yellow pine,

FIG. 30. Structural timbers used in shipbuilding. (*Courtesy of S. J. Record.*)

Douglas fir, hemlock, eastern white pine, western yellow pine, spruce, larch, and redwood.

Planing-mill Products. Planing mills, which are often associated with sawmills, use a large amount of lumber representing some 60 different species. Their products are sometimes classified as "factory lumber," which is lumber that has been recut to smaller dimensions and reworked. The chief products of the planing mills are sashes, doors (Fig. 31), blinds, window frames, and interior finish. The best wood for millwork has a straight grain and soft uniform texture. It must not shrink or swell, and it should be easy to work and capable of taking paints and varnish. The cheaper products are made from white pine, yellow pine, Douglas fir, and other softwoods. For the more expensive products, oak, birch, red gum, walnut, maple, mahogany, and other hardwoods with a fine figure are utilized. Veneers are now being used extensively for door panels.

Interior finish includes baseboards, cornices, columns, grills, mantels, balusters, stairwork, posts, scrollwork, porch work, and trimming. Almost all the native and imported woods that are decorative and have good wearing qualities are used for these purposes.

Hardwood flooring has been a comparatively recent development of the planing mills, and is widely used. Only the most durable woods, which also have attractive figures, are used. These are kiln dried and thoroughly seasoned. The woods chiefly utilized are maple, beech, birch, oak, tupelo, and yellow pine.

FIG. 31. Sanding, the last process in the manufacture of doors, an important product of planing mills. (*Reproduced by permission from Atwood, United States among the Nations, Ginn and Company.*)

Other Woodworking Industries. Woodworking industries other than planing mills use about 27 per cent of the lumber cut. These industries are so numerous and their requirements and products so varied that it will not be possible to discuss them. The majority use a relatively small amount of wood, less than 1 per cent. A few, however, are more important. These include the box and crate, railroad car, furniture, vehicle, agricultural implement, and woodenware industries.

Boxes and Crates. The manufacture of boxes, crates, baskets, and other containers for the transportation of canned goods, farm products, and a host of other articles is steadily increasing. The industry uses about 15 per cent of the lumber cut, as well as a large amount of veneers and timber in bolt form. For box-making purposes wood should be light, strong, and easy to work, with good nail-holding power and a surface than can be printed upon. Lower grades of softwood lumber and the softer hardwoods are used. The chief species utilized are southern pine, western yellow pine, white pine, red gum, hemlock, and spruce.

Furniture and Fixtures. The manufacture of furniture requires wood of special strength, hardness, and durability. It must not shrink or warp, and should be ornamental and capable of a high polish. Oak, maple, birch, walnut, and red gum are the most important native species, although tulip, chestnut, beech, elm, and basswood are also used. A large number of imported woods that have attractive color or figure are utilized. Altogether some 60 species are used, with a consumption of nearly 1,500,-000,000 board feet. Veneers are increasing in importance, for they can be used to cover cheaper and less attractive woods.

Railroad Cars. About 1,000,000,000 board feet of lumber are used each year in the building of the various kinds of railroad cars. At least 40 different species of wood are used, although yellow pine and Douglas fir furnish the greater part. Oak, maple, hemlock, cypress, and several ornamental woods are also important. The sills, brake beams, posts, etc.; the sides and roofing; and the interior finish require different kinds of woods.

Vehicles. The manufacture of vehicles has always been an important industry. The use of wood increased in magnitude in the early days of the automobile industry, but has fallen off since metal has so extensively replaced wood in that industry. The annual consumption of wood in vehicle construction is still about 1,000,000,000 board feet. Maple, red gum, and oak supply about half the amount. The different parts of vehicles have their own special requirements. The bodies must be light and stiff, while the running parts require toughness and elasticity. Hickory, elm, ash, oak, locust, and birch are generally employed for wheels. Large quantities of softwoods are used by automobile manufacturers for crating purposes.

Other Industries. The following list of products of other woodworking industries, arranged approximately in order of importance, will serve to give an idea of the extensive uses of wood in this country: agricultural implements, caskets and coffins, refrigerators and kitchen cabinets, ship and boat building, matches, woodenware and novelties, musical instruments, tanks and silos, signs and supplies, professional and scientific instruments, electrical machinery and apparatus, machine construction, toys, laundry appliances, handles, supplies for dairymen, poultrymen and beekeepers, tobacco boxes, patterns and flasks, sporting and athletic goods, boot and shoe findings, shade and map rollers, brooms and carpet sweepers, picture frames and moldings, motion-picture and theatrical scenery, brushes, plumber's woodwork, shuttles, spools and bobbins, trunks and valises, sewing machines, pumps, wood pipe and conduits.

airplanes, toothpicks, printing materials, playground equipment, dowels, clocks, paving materials, saddles and harness, gates and fencing, butcher blocks and skewers, bungs and faucets, firearms, scales, elevators, whips, canes and umbrellas, tobacco pipes, and artificial limbs.

Posts

It is difficult to estimate the annual consumption of wood for fence posts since such a large number are cut and used locally. The total, however, must be at least 600,000,000. Fence posts are used chiefly on farms and along roads and railroad right of ways. The old rail fences have almost disappeared. Posts are usually cut 7 ft. in length and from 4 to 6 in. in thickness. They are used in the round or are split. Strength, light weight, and durability in the soil and weather are the main requirements. A great variety of woods are utilized, principally redwood, cedar, chestnut, oak, tamarack, ash, black locust, osage orange, and cypress. In order to increase their durability, fence posts are usually treated with preservatives.

Mine Timbers

A large amount of wood, both round and sawed, is used in mines for shafts and for the various types of supporting structures (Fig. 32), such as

Fig. 32. Mine timbers. A stretch of gangway in a Pennsylvania mine. (*Photo by U.S. Forest Service.*)

props, caps, and collars. Strength and durability are particularly important for safety is a prime requisite and the wood is used in an environment especially favorable for the agencies of decay. Almost any kind of wood is likely to be used, however, for most mine timbers are only for temporary use; frequent replacements are necessary as the result of breakage, wear, and decay. Oak, pine, chestnut, tamarack, maple, beech, hemlock, and Douglas fir are among the more important species. Very little mine timber enters the trade, for most of the supply is cut locally from whatever wood is available.

Poles and Piling

From 3,000,000 to 5,000,000 poles are used each year for telephone, telegraph, electricity, and power-transmission lines. About the same number of piles are used for harbor work of various kinds and bridge, trestle, and similar construction.

Poles are used in the round, and, as they are likely to decay at the ground level, only woods with a durable sapwood are used. Even these woods are treated with preservatives. Other requirements are strength, light weight, and accessibility. The shape is also important. Cylindrical, slightly tapering poles with large butts are the most desirable. Piles, in addition to the above requirements, must be able to withstand heavy top loads and must be capable of being driven.

Southern pines, cedars, especially the western red and northern white cedar, Douglas fir, and lodgepole pine are used to the greatest extent. Less favored species include the chestnut, oak, cypress, redwood, larch, elm, and locust.

Cooperage

The cooperage industry, which comprises the manufacture of wooden containers bound together with hoops of wood, is a very old industry, dating back to Biblical and Roman times. Since 1907, however, production has been in steady decline owing to the competition of other types of containers. There are two principal classes of products: *slack cooperage*, for holding dry substances; and *tight cooperage*, for holding liquids.

Slack Cooperage. This type of cooperage includes a great variety of tubs, barrels, casks, pails, buckets, kegs, churns, and other containers. They are used to transport fish, meat, tobacco, vegetables, fruit, flour, sugar, cement, crockery, glassware, and a host of other materials. A single barrel usually consists of 15 staves, one set of heading, and six hoops. These various parts may be manufactured in different factories and even in different regions. There are many grades of slack cooperage ranging from flour and sugar barrels, with tongued and grooved staves, to the loose-fitting and cheaply constructed cement barrels.

Wood for use in slack cooperage must be cheap, light, easy to work, elastic, and free from warping and twisting. Otherwise any sort of wood is permissible. Limbs, tops, defective logs, and other forms of waste lumber can be utilized. Veneers are being used to a considerable extent. Staves and heads are usually made from red gum, pine, elm, beech, maple, oak, Douglas fir, ash, and a variety of other woods. Wooden hoops are made chiefly from elm, although hickory, oak, gum, and hackberry are occasionally used. Metal hoops made from wire or beaded steel, however, are now utilized to a great extent.

Tight Cooperage. Barrels and kegs that are to be used as containers for beer, wine, oil, and other liquids are more carefully constructed. Woods which will impart no odor or taste to the liquids and which are impermeable are essential. White oak has long been the outstanding

FIG. 33. White oak staves in Alabama ready for shipment. (*Photo by U.S. Forest Service.*)

wood used (Fig. 33), especially when the liquids are to remain in the barrels for a long time. Red gum, red oak, white ash, yellow birch, sugar maple, and Douglas fir are also utilized. The finished product is usually treated on the inside with paraffin to insure that no leakage will occur. Hoops are made chiefly of strap steel.

Heavy Cooperage. Large-sized tanks and vats constitute heavy cooperage. They are made with staves and heads of white oak, Douglas fir, cypress, or redwood, bound together with metal straps.

RAILROAD TIES

From 1880 to 1910 there was a great demand for railroad ties, including cross ties, switch ties, and bridge ties, the number increasing from 35,000,000 in 1880 to 142,000,000 in 1910. This was the period of expansion for both steam and electric roads. More recently there has been a reduction in the consumption of ties, due in part to a lessened demand and in part to the longer life of the ties as a result of preservative treatment. The number of ties used each year will always be considerable, probably around 40,000,000, considering that there are about a billion ties in use and that the average life of an untreated tie is only five or six years.

Ties are usually hand hewn from a seasoned wood but may be sawed. Strength, durability in the soil, and the ability to resist impact, crushing, and spike pulling are essential. Treated ties are fully as serviceable as those made from naturally durable woods. Seventy per cent of the ties are made from oak, southern pine, red gum, and Douglas fir. Other species used include cypress, cedar, chestnut, beech, maple, tamarack, and hemlock. Douglas fir ties are often sawed.

Veneers

Veneers are thin slices or sheets of wood of uniform thickness. Although they can be cut as thin as $1/110$ in., the commercial product is usually about $1/20$ in. thick; $3/8$ in. is the maximum. Veneers do not represent a new development in the utilization of wood, for they were known to the Romans, Egyptians, and other early civilizations. Their use in this country was retarded for many years because wood was so abundant and so cheap. In recent years, however, a large demand for veneers has arisen, due partly to their own inherent qualities, partly to the necessity of conserving the supply of valuable timber, and partly to the fact that wood can be used which otherwise would be wasted.

In making veneers, logs or pieces of wood are peeled and boiled and then cut with a knife. Although several processes are used, 90 per cent of veneers are cut by the *rotary process* (Fig. 34). This involves turning a log on a lathe against a stationary knife, with the formation of a continuous sheet of veneer. This process is quite wasteful as a core from 6 to 10 in. in diameter is left. This core, however, can be used for fuel, boxes, crates, excelsior, and various industries. The figure in this type of veneer is not very striking, since the sheet is cut parallel to the annual rings. In the *slicing process* logs are quartersawed and consequently show a more attractive figure. The logs are first quartered and then sliced with a stationary knife, thus yielding separate sheets. This is the least wasteful process and is used for many of the imported woods. In the *sawing process* quartered logs are cut with a circular saw. Although the resulting veneers are quite thick and although the process is a rather wasteful one, the most valuable woods are sawed because the fibers are but little torn and the material can be more readily worked up. Freshly cut veneers are usually wet and must be thoroughly dried before adhesives can be applied.

At first veneers were used only in connection with furniture and cabinetwork to cover up inferior woods, and they were made from mahogany, walnut, and other woods with a beautiful color, grain, and figure. Now many species are used and the veneers are utilized in the manufacture of boxes, baskets, cooperage, door panels, trunks, mirrors, musical instruments, and many other articles. Veneers make possible maximum strength with minimum weight. Any wood is suitable which comes in large sizes, has a symmetrical grain and figure and few defects, and is inexpensive. A large variety of both domestic and foreign woods are used in the manufacture of veneers, but over one-half of the total output in the United States is made from red gum or Douglas fir. Three kinds of hardwood veneers are manufactured. *Face veneers* are sliced or sawed from carefully selected logs and are used for only the finest work.

FIG. 34. Producing rotary-cut veneer. (*Courtesy Wood Mosaic Co., Louisville, Kentucky. Reproduced from Panshin, Harrar, Baker, and Proctor, Forest Products, McGraw-Hill Book Company, Inc.*)

Normally some 80 or more kinds are available. Chief among native woods are black walnut, quartered red and white oak, red gum, and sugar maple. *Commercial veneers* are rotary-cut to order and are used for plywood, concealed parts of furniture, etc. Beech, birch, maple, basswood, tulip, cottonwood, tupelo, sycamore, and oak are all utilized. *Container veneers*, the cheapest kind, are made from any inexpensive wood and are used for boxes, barrels, crates, and similar containers. Softwood veneers, either rotary-cut or sliced, are manufactured chiefly on the Pacific Coast. Eighty per cent of them are made from Douglas fir, but Sitka spruce, western yellow pine, and Port Orford cedar are also used. Softwood veneers are utilized almost entirely for structural plywood or interior paneling.

Plywood

A recent development in overcoming some of the natural defects in wood is the manufacture of the so-called "built-up" stock or plywood. This involves gluing together from three to nine thin veneers. The grain of each successive layer is at an angle to the next, so the strength is redistributed and the dimensional instability of one layer is compensated for and reduced by the others. Consequently the finished product is very strong and stable and much less likely to warp or twist than ordinary wood. Nails and screws may be driven close to the edge with no danger of splitting the plywood.

The manufacture of a five-ply plywood panel consists of preparing the face, back, cross band, and core sheets; applying the adhesive; pressing the glued stock into a panel; and drying and finishing the product. The softer woods are generally utilized, as they can be glued more easily. One simple type of plywood has a $3/8$-in. core of poplar with $1/10$-in. birch veneers on each side. Various animal and plant adhesives are used as well as large quantities of synthetic resins. Modifications of the ordinary process result in molded or curved plywood and in wood alloys.

Plywood is now a fully recognized engineering material with its own peculiar properties. It has extensive uses in the home for doors, walls, flooring, partitions, shelves, cabinets, furniture, and interior trim. Enormous quantities are also used in making concrete forms, prefabricated houses, airplanes, boats, railroad cars, and the bodies of trailers and station wagons.

Wood Alloys

In the manufacture of wood alloys or densified wood, such as compreg, impreg, and uralloy, from plywood the natural wood structure is impregnated with synthetic resins and bonded under high pressure. The resins establish strong physical and chemical bonds with the wood fibers and create a material with new properties, which is also very hard, strong, stable, and resistant to decay.

Reconstructed Wood

Reconstructed wood is characterized by a reorganization of the fibers whereby they are taken out of their original unidirectional grain and rearranged in multidirectional patterns. This may be accomplished by physical or chemical means. Since the use of chemicals involves breaking down the wood into cellulose and lignin, the discussion of a chemically reconstructed wood, such as paper, will be reserved until Chap. X.

Masonite is an example of a wood which has been reconstructed by physical means. Wood chips are subjected to high pressure in a steam vessel and are then exploded with the abrupt release of the pressure. This tears the fibers apart and also reactivates the lignin, a natural plastic, which fixes and binds the fibers in their new orientation. At first this new material was used as an insulating fiberboard, as it was an excellent insulator against heat, sound, and electricity. Later it was found that the application of heat and pressure converted the boards into a homogeneous grainless synthetic board with extreme hardness and resistance to water.

Shingles and Shakes

Shingles are thin pieces of wood used to protect the roofs and sides of buildings from the weather. Until recently they were made by hand by

Fig. 35. A shingle mill in the Olympic Forest, Washington. (*Photo by U.S. Forest Service.*)

riving bolts of wood and shaving off the shingles with a drawing knife. Handmade shingles were probably among the first of the wood products manufactured by man. Shingles are now chiefly machine made (Fig. 35), and some are sawed. Shingle wood must be durable, light in weight, easy to split, and able to hold nails without loosening, and it must not warp. Straight, even-grained woods are the best. The durability of shingles is increased by treating them with preservatives. The most outstanding shingle wood is western red cedar. Northern white and southern white cedars, redwood, and cypress are also used. When available, virgin white pine is an excellent material. The output of shingles reached its peak in 1909 when 15,000,000,000 were produced. Since that time the industry has suffered from the many substitutes.

Shakes are split shingles and naturally are much thicker. They were important in colonial and pioneer days, but now are but little used save for special architectural effects and in the more remote regions such as the southern Appalachians. In the West shakes are still manufactured on the Pacific Coast and in the northern Rocky Mountain area from western red cedar, sugar pine, and redwood.

Excelsior

Excelsior consists of thin curled strands or shreds of wood and is made by placing wood on frames and pressing it against rapidly moving knives or steel teeth. This material was first known as *wood fiber*, and "excelsior" was later introduced as a trade name. Excelsior is light and elastic and makes an excellent material for packing and shipping glassware and other breakable articles. It is resilient and free from dust and dirt, and so can be used for stuffing mattresses and upholstery. A very fine grade of excelsior, known as *wood wool*, is used in filters and in the manufacture of rugs and matting.

The production and use of excelsior are now on the decline. Almost any wood that is light in weight and color, odorless, and free from resins or gums can be used. Basswood, which is very soft and springy, is the best material. Cottonwood and the other poplars and yellow pine are used to a great extent. Other woods include buckeye, tulip, birch, spruce, maple, white pine, and red gum. Wood waste from the other industries is also used. It is estimated that a cord of wood yields about 1500 lb. of excelsior.

Sawdust and Shavings

The principal use of sawdust is for fuel, usually in the form of briquettes. It is also increasingly valuable for many industrial purposes. Often it is separated by species and graded by size. It serves as a bedding for stables and kennels, as a floor covering to absorb moisture, for cleaning, drying, and polishing metal, as a packing medium, as a soil conditioner and an insulating material. It plays a part in the making of leather and the conditioning of fur and is an ingredient of composition flooring, artificial wood, wallboard, abrasives, floor-sweeping compounds, and many other products.

Shavings find their greatest use as fuel and as a packing material, although they are utilized in many other ways. White pine shavings are most desirable.

Wood Flour

Wood flour, which consists of finely ground sawdust, shavings, and other forms of wood waste, is of increasing importance in the manufacture of linoleum, plastics, nitroglycerin explosives, veneer bonds, composition flooring, insulating brick, and many other products. It is used as a filler, an absorbent, or a mild abrasive. Only light-colored woods with a low resin content can be used. White pine furnishes about 75 per cent of the total amount. Formerly imported, domestic production is now sufficient to meet the demand.

Charcoal and Wood Distillation

Charcoal

The heating of wood in order to convert it into carbon or charcoal is a practice as old as civilization. It was probably the first chemical process discovered by man. Charcoal is still a valuable fuel, since it has twice the heating power of wood and burns without flame or smoke. It is

Fig. 36. Cleaning the dust in a charcoal pit in Pennsylvania. (*Courtesy of S. J. Record.*)

extensively used in many European and Asiatic countries, particularly where forests are abundant, and it is the chief domestic fuel in most tropical countries. Charcoal is also used in medicine, as a reducing element in the iron and steel industry, and in the manufacture of chemicals, gunpowder, explosives, and certain cosmetics. It is the best material for absorbing impurities and bad odors from both water and the atmosphere. During the First and Second World Wars it was extensively used in gas masks. About 7,500 tons are used annually in curing shade-grown tobacco. The best yields of charcoal are obtained from the denser hardwoods, such as beech, maple, birch, oak, and hickory. Willow charcoal is especially desirable for explosives. The conversion of wood into charcoal was formerly carried on in open-air pits (Fig. 36) by a process of partial combustion. This method was extremely wasteful, as all the volatile material contained in the wood was lost. Beehive kilns and portable ovens have also been used. Today charcoal burning has been almost entirely replaced by wood distillation, and the valuable gases and other by-products are recovered as well as the charcoal.

Wood Distillation

The process of wood distillation is also ancient and was known to the Egyptians. At the present time it is important, not only in rendering

Fig. 37. Wood distillation retorts in Wilmington, N.C. (*Photo by U.S. Forest Service.*)

available the volatile wood elements, but as a factor in forest conservation. One of the chief sources of wood for destructive distillation is the waste left by lumbering operations, sawmills, and planing mills. There are two distinct types of wood distillation.

Hardwood Distillation. This process utilizes the denser and heavier hardwoods and has been carried on in the United States for over 100 years. The wood is heated in large oven retorts (Fig. 37). The immediate products are charcoal, pyroligneous acid, which condenses from some of the gases given off, tar and oil, and noncondensable wood gases. The tar and oils are allowed to settle out, and the pyroligneous acid is passed through a series of stills where more tar and oils are removed. Eventually slaked lime is added, and in a final distillation wood alcohol (methanol) passes off and acetate of lime is left as a residue. A recent modification of the process results in the recovery of acetic acid directly from the pyroligneous acid. The average yield per cord is 900 to 1080 lb. of charcoal, 180 to 220 lb. of acetate of lime or 103 to 125 lb. of acetic acid, 9.5 to 11 gal. of wood alcohol, 22 to 25 gal. of wood tar and oils, and 7,000 to 11,500 cu. ft. of wood gas.

Acetate of lime is used chiefly in the manufacture of acetic acid, which has wide applications in the textile, paint, leather, film, and plastics industries, and acetone, which is extensively used as a solvent. Wood alcohol also finds its greatest use as a solvent, especially in the paint and varnish industry; it is the source of a great variety of chemical products as well, such as aniline dyes and formaldehyde, and has a multitude of uses as a fuel, illuminant, denaturant, and ingredient of medical, chemical, and other industrial preparations. The wood tar and oils are used as a fuel or as the source of many industrial oils. Wood gas also serves as a fuel and may be converted into a substitute for gasoline in internal-combustion engines.

Softwood Distillation. This is a more recent development and utilizes resinous woods, chiefly southern yellow pine. The products are charcoal, wood turpentine, oils, tar, and wood gas. Softwood pyroligneous acid contains only small amounts of wood alcohol and acetic acid, and no attempts are made to recover them. The wood turpentine is used in the manufacture of paints, varnishes, and synthetic camphor. The tar and oils also have a large number of industrial uses; one important product is creosote.

Other Uses of Wood

The use of wood as a raw material for the paper and textile industries and as a source of tanning and dye materials, food, alcohol, and other products is discussed elsewhere.

CORK

Cork is a forest product of great antiquity, its use dating back at least as far as Greek and Roman times. It is obtained commercially for the most part from the cork oak (*Quercus Suber*), a tree native to the Mediterranean region. This is a species from 20 to 60 ft. in height and 4 ft. in diameter, with a short trunk and densely spreading crown. The evergreen leaves resemble those of holly, but are spongy and velvety. The acorns serve as a food for swine. The cork oak is found from the Atlantic to Asia Minor, and is especially abundant in Portugal, Spain, Algeria, Tunisia, southern France, Morocco, Italy, and Corsica. About 3,764,000 acres of cork forests occur, over a million of which are found in Algeria. The tree thrives best on rocky siliceous soil on the lower slopes of the mountains.

Cork or "corkwood" consists of the outer bark of the tree. This can be removed without injury to the tree, and, provided the inner bark is not damaged, it is renewed, new layers being formed each year. The operation consists of making vertical and horizontal cuts with hatchets or saws, and then prying off large pieces of the bark (Fig. 38). Great care is taken not to injure the inner bark, which would prevent the formation of new bark and might even endanger the life of the tree. The rich dark-red color of the exposed areas is one of the characteristic sights in a cork forest. The stripping is usually carried on in midsummer when weather conditions are most favorable.

Fig. 38. A cork gatherer in Gibraltar, Spain. (*Photo by U.S. Forest Service.*)

The bark of both the trunk and larger branches is usually utilized, although in some countries the cutting area is restricted to the first 6 ft. of the trunk. Cork is first removed when the trees are about 20 years old. This first yield, which is known as virgin cork, is very rough and coarse and of little value. Subsequent strippings occur every nine years. The second yield is better, but the best quality of cork is not obtained until the third cutting and thereafter. The trees live for 100 to 500 years, and have an average yield of 40 to 500 lb. per tree. The best grade of cork consists of inch-thick layers obtained from young vigorous trees.

After stripping, the pieces of cork are dried for several days and weighed and then shipped to some central point for further treatment. They are first boiled in large copper vats. This removes the sap and tannic acid, increases the volume and elasticity, and flattens the pieces. It also loosens the outermost layer, which is scraped off by hand. The rough edges are then trimmed off, and the flat pieces are sorted and baled.

Cork possesses many properties that make it very valuable in industry. In spite of its bulk it is very light and exceedingly buoyant, owing to the fact that it is composed entirely of dead watertight cells. It can be readily compressed and is very resilient. Even after 10 years of use cork stoppers will recover 75 per cent of the original volume. It is durable, a low conductor of heat, and is resistant to the passage of moisture and liquids. It also absorbs sounds and vibration and has certain frictional properties.

Uses of Cork

Cork is used for many purposes and for the manufacture of a great variety of products. In some cases the natural cork is utilized; in others composition cork, made of coarse or finely ground pieces treated with adhesives and molded.

Articles made from natural cork include stoppers, which are cut by hand or machine; hats and helmets for use in tropical countries; tips for cigarettes; carburetor floats; handles for golf clubs, penholders, fishing rods, and other articles; mooring buoys, floats, life preservers, life jackets, and surf balls; baseball centers; decoys; mats; tiles; and many other products. Corkboard, made by heating natural cork, is used as an insulating material for houses, cold-storage plants, and refrigerators. It serves as a means of improving the acoustics of rooms and rendering them soundproof. It is very resilient and consequently is a valuable material for machinery isolation.

Composition cork is used for the lining of crown caps, the metal tops for sealing bottles; gaskets; toes, counters, and innersoles for shoes; polishing wheels; friction rolls; and several types of floor covering. *Linoleum* is made from cork or wood flour, linseed oil, resins, such as rosin or kauri gum, pigments, and burlap. The oil is boiled and allowed to solidify by dripping on pieces of cloth. The solidified oil is ground up and melted with the resins. This mixture is cooled and hardened, and after several days of "curing" it is mixed with the cork, which has been ground as fine as dust, and with the dry color pigments. It is then pressed into burlap cloth with hydraulic presses. The linoleum is then seasoned in ovens and finished by giving it a protective surface of nitrocellulose lacquer. *Linotiles* are individual tiles made from ground cork and linseed oil, but much thicker and denser than linoleum.

Over 300,000,000 lb. of cork are produced annually, chiefly in Spain, Portugal, and Algeria. The United States imports from 50 to 60 per cent of the output. Cork is such an essential material in our national economy, both in peace and war, that attempts have been made to develop at least an emergency supply which would not be subject to wartime restrictions. Experiments have been carried on in California since 1940. Some 5000 cork oak trees have been located, acorns have been collected and distributed, and 200,000 seedlings have been set out. This domestic cork is of good quality and regrows rapidly and satisfactorily after stripping. There is every reason to believe that a flourishing cork industry may be developed in California and the Southwestern states.

CHAPTER IV
FOREST RESOURCES

DISTRIBUTION OF FOREST LANDS

Climatic conditions, in some parts at least, of all the continents are favorable for the development of forests. Although our knowledge concerning the exact nature and distribution of these forests is still incomplete, it is possible to estimate roughly the amount of forest land in the world as about 7,500,000,000 acres, or 22½ per cent of the earth's surface. These forest lands are distributed on the various continents as follows: Asia, 28 per cent; South America, 28 per cent; North America, 19.3 per cent; Africa, 10.6 per cent; Europe, 10.3 per cent; Australia and Oceania, 3.8 per cent. In many countries forests were originally much more abundant than they are at the present time.

The softwoods or conifers occupy 35.4 per cent of the total forest area, occurring in pure stands or mixed with hardwoods. They are especially characteristic of the colder areas, about 95 per cent occurring in the north temperate zone. When they are found in warmer regions, conifers are restricted to the higher altitudes. In the case of the hardwoods, a distinction is usually made between temperate and tropical hardwoods. The former occupy 16 per cent and the latter 35.4 per cent of the total forest area. As in the case of the conifers, most of the temperate hardwoods (89 per cent) are found in the north temperate zone. It is of interest to note that three-quarters of the world's population lives in this area, and consumes nearly 90 per cent of all the wood used. The fact that both the softwoods necessary for general construction and the hardwoods have been readily available throughout this area has been of great economic importance. It has been only recently, as a result of the depletion of the native forests, that attention has been turned to the almost untouched tropical forests, save as a source of ornamental woods, dyewoods, and similar products.

FORESTS OF NORTH AMERICA

Forests occupy about 26.8 per cent of the land area of North America, with conifers comprising 72.4 per cent, temperate hardwoods 20.1 per cent, and tropical hardwoods 7.5 per cent of the forests. The northern part of the continent—Alaska, Canada, Newfoundland—is predominantly coniferous, with 93 per cent of the softwoods and 7 per cent of temperate hardwoods. The United States has 62 per cent conifers and 38 per cent temperate hardwoods; Mexico has 47 per cent conifers, 34 per cent temperate hardwoods, and 19 per cent tropical hardwoods. In Central America tropical hardwoods account for 75 per cent and conifers 25 per cent of the total. The West Indies have an even higher percentage of tropical hardwoods (87 per cent), with 13 per cent of conifers.

In the United States at the present time there are about 469,000,000 acres of woodland, comprising 24.7 per cent of the total land area. Of this amount about five-sixths is capable of producing lumber on a commercial scale. The present forest area is only a little over half that of the original forests that existed at the time of the first English settlements three centuries ago.

Forest Areas of North America

The forests of North America are found in at least six well-defined areas (Fig. 39): (1) the northern coniferous forest; (2) the eastern deciduous forest; (3) the southeastern coniferous forest; (4) the Rocky Mountain forests; (5) the Pacific Coast forests; and (6) the tropical and subtropical forests.

Northern Coniferous Forest. This great evergreen forest extends across the continent from Newfoundland and Labrador to the lower Hudson Bay region and Alaska, south of the treeless arctic tundra. The chief trees are white spruce, black spruce, balsam fir, and larch, with some paper birch, aspen, and balsam poplar. From Nova Scotia and northern New England to Minnesota, and southward along the summits of the Appalachians, there exists a transitional region between the coniferous forest and the eastern deciduous forest, with species of both these areas intermingled. Prominent trees of this so-called "northern hardwood region" include the white pine, red spruce, white cedar, beech, sugar maple, hemlock, yellow birch, and locally the red pine and jack pine.

Eastern Deciduous Forest. This forest, one of the oldest on the continent, covers most of the eastern and central part of the United States. It attains its best development in the lower Ohio valley and on the slopes of the southern Appalachians. It extends as far north as Ontario and southern Quebec. The more important trees are the oaks, hickories, chestnut, tulip, black walnut, basswood, ash, and elm. Toward its northern limits, as we have just seen, the beech and maple become prominent, intermingled with various conifers. Similarly toward its southern and southwestern limits the oaks and hickories often occur with many of the pines that characterize the southeastern coniferous forest. Westward the deciduous forest becomes gradually restricted to the river valleys of the prairie region.

Southeastern Coniferous Forest. This forest is found along the sandy Atlantic coastal plain from Virginia to Texas. The various pines, principally longleaf, shortleaf, loblolly, and slash pines, occur on the

FIG. 39. The forest formations of North America. North of the northern evergreen forest is the tundra formation. On the unshaded areas south of it are the prairie, plains, and desert formations. (*Reproduced by permission from Transeau, General Botany, World Book Company, Yonkers-on-Hudson, N.Y.*)

uplands; in lower ground are found red gum, tupelo, live oak, and magnolia. Southern white cedar and cypress occur in the swamps.

Rocky Mountain Forest. Like all the forest areas of Western America, the Rocky Mountain forest is made up almost entirely of coniferous species. The area extends from northern British Columbia southward across the United States and Mexico, and even into Central America. Naturally in so large an area there are many differences in the nature of the forest due to latitude and altitude. Western yellow pine is the most characteristic species. Others include the lodgepole pine, Douglas fir, white fir, and western larch, with Engelmann spruce and alpine fir at the higher elevations. In northern Idaho and adjacent Montana a forest occurs which is quite similar to some of those on the Pacific Coast, with western red cedar, western hemlock, and western white pine as the chief species.

Pacific Coast Forests. Within the Pacific Coast region there are several distinct forest areas. Along the coast from Alaska south to Washington, Sitka spruce is the chief species. In southern British Columbia, the Puget Sound region, and eastern Oregon and Washington as far east as the summits of the Cascade Range there is situated one of the most magnificent conifer forests in the world. Few, if any others, can surpass it in density of the stand and the size of the trees, which reach a height of 200 to 250 ft. and a diameter of 8 to 15 ft. The mild winters, due to the nearness of the ocean, and the very great precipitation, about 100 in., are in part responsible for the development of this forest. Douglas fir is the dominating species, and with it are associated western hemlock, western cedar, Sitka spruce, and several firs. This region is the location of the largest lumbering operations in the country at the present time.

From Oregon south to San Francisco Bay the forests along the Coast Range are dominated by the redwoods, the tallest of all the conifers. This region has been exploited for many years and much of the timber has been cut and the land given over to agriculture.

East of the Cascade Range the forests tend to merge with those of the Rocky Mountains. The principal trees are western yellow pine, Douglas fir, western larch, lodgepole pine, and locally western white pine, Engelmann spruce, and alpine fir. This forest also extends southward along the Sierra Nevadas in California. Here western yellow pine, sugar pine, incense cedar, Douglas fir, and white fir are common with red fir at the higher altitudes. The Big Trees are found locally along the lower slopes of the Sierras.

Tropical Forest. Subtropical conditions are found in southern Florida, but none of the trees are of commercial importance. True tropical forests are found in the Mexican lowlands, on the eastern slopes of Central America, and in the West Indies. Most of the original forest in the latter two regions has been destroyed as a result of migratory agriculture and has been replaced by the dense, almost impenetrable tropical jungle. The more important tropical species of North America will be considered together with those of South America.

The Principal Woods of Temperate North America

In North America there occur over 500 species of woody plants, exclusive of tropical species. About a hundred of these are of commercial importance. The most prominent of these will be considered briefly.

SOFTWOODS

Cedar

The cedars of commerce all have a fragrant, light, soft wood which is even grained and resistant to decay. The chief use of the wood is in millwork and the manufacture of instruments, woodenware, caskets, boats, and laundry appliances. Commercial cedar comprises seven different species:

The Port Orford cedar (*Chamaecyparis Lawsoniana*), found in Oregon and California, is the hardest of all the cedars and has other exceptionally good technical properties. It yields a strong, durable, heavy, stiff timber which takes a good polish. The wood is used for boats, airplanes, furniture, millwork, matches, floors, interior finish, and posts.

The Alaska cedar (*Chamaecyparis nootkatensis*) ranges from Alaska to Oregon along the Pacific Coast. The light, stiff, hard, strong wood is easy to work and durable. It is used for boats, shingles, fences, and interior finish.

The southern white cedar (*Chamaecyparis thyoides*) is a smaller tree found in the Atlantic and Gulf Coast states. The wood is used for boats, shingles, poles, posts, ties, and cooperage. In colonial days it was used for houses, but it is too light to support the weight of second stories.

The incense cedar (*Libocedrus decurrens*) of California, western Nevada, and Oregon has a close-grained reddish wood of considerable value on the Pacific Coast. It is used for building purposes, posts and piling, and to a large extent for lead pencils.

The eastern red cedar (*Juniperus virginiana*) is one of the most characteristic trees of the eastern United States. The wood is very durable and is especially resistant to weather conditions. It is used for poles, fence posts, crosstrees, ties, etc. The sapwood is white, and the heartwood rich red and very fragrant owing to the presence of an essential oil. The wood is soft, with a fine, even grain, and it can be whittled easily, so, until it became scarce, it constituted the standard wood for lead pencils. Chests, cigar boxes, pails, panels, veneers, and interior finish are also made from it.

The northern white cedar or arbor vitae (*Thuja occidentalis*), a common tree of New England, the Lake States, and adjacent Canada, is one of the lightest of woods. It is soft and easily worked and is utilized for woodenware, canoe and boat building, shingles, fence posts, railroad ties, poles, tanks, and silos.

The western red cedar (*Thuja plicata*) of the northern Rocky Mountains and Pacific Northwest is the largest of the cedars, reaching a height of 150 ft. and a girth of 30 ft. The brownish-red, close, even-grained wood is very soft, but extremely durable. Two-thirds of the output is used for shingles. Other uses include fences, poles, interior finish, cabinetwork, and cooperage. The coast Indians used this species for their totem poles and war canoes, and also made ropes and textiles from the bast fibers of the inner bark.

Cypress

The bald cypress (*Taxodium distichum*), a characteristic tree of swamps along the Atlantic Coast from Delaware to Texas and in the lower

FIG. 40. A stand of Douglas fir (*Pseudotsuga taxifolia*) in the Siskiyou National Forest, Oregon. (*Photo by U.S. Forest Service.*)

Mississippi valley, is one of the strongest and heaviest of the softwoods. The tree reaches a height of from 80 to 140 ft. and a girth of from 5 to 12 ft. Conspicuous features are the "knees," conical outgrowths from the roots for purposes of aeration. Unlike most conifers, the cypress sheds its leaves. The wood is a rich red color with a distinct grain. It is soft and coarse and works well. Its chief utilization is in millwork for cabinet and interior-finish purposes. It is also used for shingles, ties, posts, sides of greenhouses, tanks and cisterns, and other structures exposed to decay, for the wood is very durable and long lived. Boxes, boats, and cooperage are other products.

Douglas Fir

The Douglas fir (*Pseudotsuga taxifolia*) is one of the two most important American woods at the present time, and more of it is available than any other species. The tree covers a wide range of territory in the northwestern United States and adjacent Canada, attaining its best development on the Pacific Coast from central California to British Columbia. It is very large, reaching a height of 200 ft. or more and a diameter at the base of 8 to 10 ft. (Fig. 40). The lowest branches are high up, so the trees are frequently used for masts, spars, and flagpoles. The size of the trunk also makes possible timbers of any length and size. The wood is resinous with a close, even, well-marked grain, and is of medium weight, strength, stiffness, and toughness. It is very durable and, when well seasoned, does not warp. A great deal of the Douglas fir cut is used in heavy construction. Other uses include piles, ties, paving blocks, veneers, airplanes, floors, and millwork. The bark is now being processed for use as a cork substitute in making plastics, adhesives, and explosives. It also has insecticidal properties.

Fir

The true firs are of comparatively little commercial importance. The wood is very light, soft, and brittle, and finds its greatest use in the manufacture of boxes and crates. Fir is also used in millwork and as a source of pulpwood. The most important species are the balsam fir (*Abies balsamea*) of the northern coniferous forest region and four western species, the lowland white fir (*A. grandis*), the noble fir (*A. procera*), the red fir (*A. magnifica*), and the white fir (*A. concolor*).

Hemlock

The eastern hemlock (*Tsuga canadensis*), one of the most characteristic trees (Fig. 41) of the northeastern transition forest, furnishes a cheap coarse lumber used chiefly for framing timbers, scantling, sheathing, laths, rafters, and other types of rough construction. Other uses include pulp, ties, boxes, and plank walks. The wood is coarse grained and splintery, but is very strong, tough, stiff, and easily worked. The bark has long been used as a tanning material.

The western hemlocks (*Tsuga heterophylla* and *T. Mertensiana*) are larger trees and furnish a superior wood, which is heavier, stronger, stiffer, and more suitable for heavy construction. Other uses are similar to those of the eastern hemlock. Both are rich in tannin. These trees are the most important sources of pulpwood in the Northwest.

FIG. 41. A pure stand of hemlock (*Tsuga canadensis*) in the Hearts Content area, Pennsylvania. (*Photo by U.S. Forest Service.*)

Larch

The eastern larch or tamarack (*Larix laricina*) and the western larch (*L. occidentalis*) furnish the larch wood of commerce. The former species is found in the northeastern United States and across Canada, while the latter, a larger tree, grows chiefly in Montana, Idaho, and Washington. Larch is one of the heaviest, strongest, and toughest of the softwoods and finds its chief use in furnishing heavy timbers for general building construction. Because of its durability larch is used for fences, posts, poles, ties, and paving blocks. The naturally curving lower parts of the trunk furnish ideal material for boat "knees," ribs, and other forms of ship timber. Larch is also used for planing-mill products, tanks, and boxes.

Pine

Pine has always been one of the most important of the commercial woods and today constitutes about one-half of the total lumber supply. The wood is obtained from several different species, belonging to either the soft-pine or the hard-pine group. Although 28 species are of some value, only eight are outstanding. The *soft or white pines* have a straight-grained soft wood of mellow and uniform consistency that is comparatively free from resin and easy to work. It is used for rough carpentry work, cabinetwork, patterns, cooperage, toys, crates, and boxes. The *hard or yellow pines* have a resinous, heavy, hard, strong, and durable wood, which finds its greatest use in buildings, bridges, ships, and other types of heavy construction. Because of its durability yellow pine is much used for floors, stairs, planks, and beams.

FIG. 42. A white pine (*Pinus Strobus*), showing the characteristic habit. (*Photo by E. H. Wilson; courtesy of the Arnold Arboretum.*)

White Pines. The northern white pine (*Pinus Strobus*) (Fig. 42) is one of the most valuable timber trees of the world. It was formerly used more than any other species, both for domestic consumption and for the export trade. So much of this wood has been cut that today the white pine has fallen to sixth place in importance. The tree is found in the northeastern United States and adjacent Canada, and along the Appalachian Mountains to Georgia. The best stands today are found in Michigan, Wisconsin, and Minnesota. It is a conspicuous member of the forest flora in this region, often attaining a height of from 100 to 200 ft. and a diameter of 3 to 9 ft. The wood is very light and easy to work as it is one of the softest of timbers. It is also very durable. Houses constructed in colonial days are still in good condition. The sapwood is white and the heartwood a pinkish brown, with a fine, even grain and lustrous surface. At the present time white pine is used chiefly for doors, window sashes, pattern making, cabinetwork, boxes, and matches. Several other species with a similar wood and uses are classified as white pine by the lumber trade. These include the western white pine (*P.*

FIG. 43. The bark, leaves, and fruit of the southern yellow or longleaf pine (*Pinus australis*). (*Photo by U.S. Forest Service.*)

monticola), the sugar pine (*P. Lambertiana*) one of the largest and most valuable timber trees of the West, and the lodgepole pine (*P. contorta*).

Yellow Pines. These are a much more heterogeneous group and, although often classed together, each species has a distinctive wood. The southern yellow or longleaf pine (*Pinus australis*) (Fig. 43) is one of the two most important timber trees of the United States at the present time. It is found in the Southeastern states from North Carolina to Texas. The wood has a fine, smooth, compact grain and is the heaviest, hardest, strongest, stiffest, and toughest of the softwoods. It is very durable and capable of bearing great weights. Its chief use, consequently, is for beams, joists, and other timbers for heavy construction, and for wharves, bridges, ships, cars, and railroad ties. Some yellow pine is used for millwork and boxes. The wood is very resinous and the tree is the chief source of the naval-stores industry to be discussed later. It is also an important pulpwood. Associated with the yellow pine in the southern forests, and often classified with it, are the shortleaf pine (*P. echinata*), the slash pine (*P. caribaea*), and the loblolly pine (*P. Taeda*).

Other important hard pines are the western yellow pine (*Pinus ponderosa*) and the red or Norway pine (*P. resinosa*) of the Eastern states.

Redwood

The redwood (*Sequoia sempervirens*), restricted to the Coast Range of northern California and adjacent Oregon, is one of the largest trees in the

FIG. 44. *Sequoia gigantea*. The General Sherman Tree, shown here, is believed to be the largest and possibly the oldest tree in the world. (*Photo by U.S. Forest Service*.)

world, reaching a height of from 200 to 300 ft. and a diameter of from 8 to 22 ft. Its only rival is the famous Big Tree of California (*S. gigantea*) (Fig. 44) whose dimensions and age, the latter estimated at 3600 years, are equaled by no other living organism. The wood of the Big Tree is of little commercial value, so these giants of the forest have been spared destruction. The redwood, however, has been extensively exploited in recent years, and only the strenuous efforts of conservationists and nature lovers have made possible the setting aside of a few stands, in which this magnificent species will be protected for all time.

The wood is fine and straight-grained, strong, light, and very soft. The sapwood is pale in color, while the heartwood is a rich dull red. The cinnamon-brown bark, which is often 1 ft. in thickness, is very striking. Redwood does not warp or shrink readily and is very durable, particularly after seasoning. It is used for general construction purposes, shingles and siding, tanks, coffins, silos, posts, water pipes, ties, furniture, cabinetwork, and interior finish. The fibrous bark is used as an insulating and stuffing material, and it yields a textile fiber for use with wool. The fine bark dust is a good soil conditioner.

Spruce

The several species of spruce have a light, soft, compact, straight-grained wood which is stiff, strong, easy to work, and comparatively free from resin. Spruce has increased in importance as white pine has become scarce. The chief uses are for pulpwood, light construction, boxes, millwork, and cooperage, and as Christmas trees. The wood is resonant and so is much used for sounding boards of pianos and the bodies of violins and similar instruments. The principal species is the white spruce (*Picea glauca*), one of the most characteristic trees of the great coniferous forest that stretches from the northeastern United States to Labrador and across the continent to Alaska. Other eastern species are the red spruce (*P. rubens*) and the black spruce (*P. mariana*), both of which have a more restricted distribution. The most important western species is the Sitka or tideland spruce (*P. sitchensis*). This large tree, which reaches a height of from 200 to 300 ft., grows along the coast from Alaska to northern California. In addition to the uses already mentioned for spruce, Sitka spruce is used for boats, oars, and other products that require a light, strong, and elastic wood. It was formerly used extensively in airplane construction. Engelmann spruce (*P. Engelmannii*), found in the Rocky Mountain and Cascade Range region from Canada to Arizona and New Mexico, is also of importance.

HARDWOODS

Ash

Ash wood is strong, elastic, tough, hard, stiff, and light in weight. It is light reddish brown in color, easy to split, and hard to nail. It is often beautifully figured and is capable of a high polish. The characteristics of ash adapt it to a wide range of uses other than structural. Among the articles made from ash may be mentioned handles, oars, bats, tennis rackets, rods, cues, clothespins, toys, barrels, and baskets. It is also used for carriages, cars, boats, farm implements, furniture, cooperage, refrigerators, and interior finish. The most important species is the white ash (*Fraxinus americana*), a tree of the eastern deciduous forest. Other prominent species include the red ash (*F. pennsylvanica*) and its variety, the green ash (var. *lanceolata*); the blue ash (*F. quadrangulata*); the black ash (*F. nigra*)—all eastern species; and the Oregon ash (*F. oregona*) of the West coast.

FIG. 45. An American beech (*Fagus grandifolia*) growing in the open in Tennessee. This tree is 110 ft. high and has a diameter at breast height of 50 in. (*Photo by U.S. Forest Service.*)

Basswood

The basswood or linden (*Tilia americana*) is a tree primarily of the eastern deciduous forest. It is a large species reaching a height of 80 ft. The wood is light-colored and straight-grained with a smooth uniform texture. It is the lightest, softest, weakest, and least tough of the more important hardwoods. However, because of its color, even grain, and extreme ease of working it is widely used. Its uses include boxes and crates, millwork, pattern making, woodenware and novelties, furniture, trunks, Venetian blinds, picture frames, carriage bodies, beehives, plywood, cooperage, pulp, charcoal, and excelsior.

Beech

The beech (*Fagus grandifolia*) (Fig. 45) is one of the three most characteristic trees of the northeastern transition forest. The fine-grained pinkish-brown wood (Fig. 46) is moderately hard, strong, and heavy, and has a wide range of usefulness. Beech is extensively used for boxes and crates because it does not impart any taste or odor. Flooring, interior finish, furniture and fixtures, tool handles, woodenware, laundry appliances, clothespins, wagon stock, shoe lasts, and ties are among the other products made from beech. The wood is also used for fuel, charcoal, and wood-distillation purposes.

Birch

Birch wood (Fig. 46) is hard, heavy, strong, and tough, with a fine wavy grain that is often beautifully figured and capable of a high polish. It is often stained and finished to imitate cherry or mahogany. The yellow birch (*Betula lutea*), which ranges from New England and the Lake States to Georgia, and the black birch (*B. lenta*), which has a more restricted distribution, furnish most of the wood used for furniture, doors, window frames, floors, and other forms of millwork. Other articles made from birch include: handles, clothespins, shoe pegs and lasts, wheel hubs, woodenware, boxes and baskets, dowels, yokes, veneers, and spools.

The white or paper birch (*Betula papyrifera*) is a more northern tree ranging from the northeastern United States across Canada. The wood, which is very strong and elastic with a fine uniform texture, is used chiefly for plywood, spools, toothpicks, boxes, handles, dowels, bobbins, and shoe lasts and pegs, and in turnery. The bark, which peels off in characteristic layers, finds a use in the manufacture of canoes and fancy articles of various kinds. Birch is also a good wood for fuel and distillation.

Cherry

The wild black cherry (*Prunus serotina*) is the only one of the several species of the genus which has a wood of commercial importance. This tree occurs in the deciduous forest area from Ontario to Florida and from the Dakotas to Texas. It is especially abundant in the southern Appalachians. The wood is hard, with a fine, straight, close grain. It varies in color from a light to a dark red depending on the age, and is often stained before use. Because of its beautiful grain and color and the ease with which it can be worked, cherry is especially desirable for furniture and interior finish and cabinetwork. At the present time it is but little used, as the supply has greatly diminished. Minor uses include the bases of scientific instruments, printer's supplies, pattern making, and turnery.

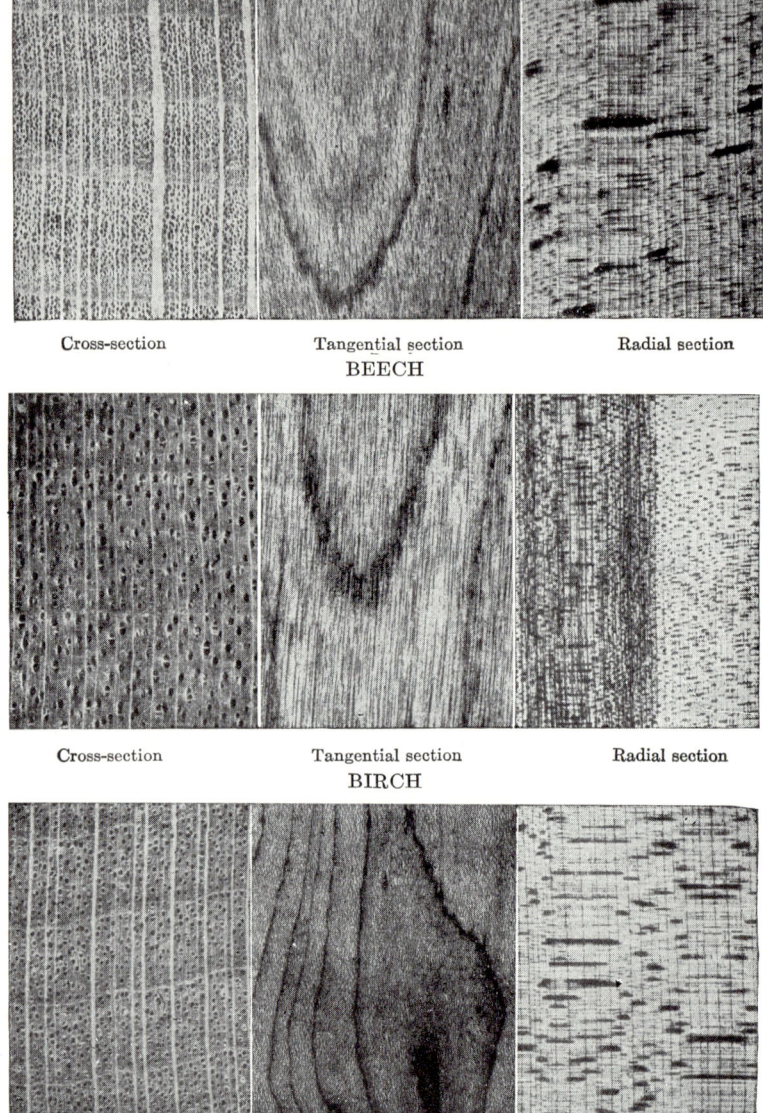

FIG. 46. Cross, tangential, and radial sections of the wood of the beech (*Fagus grandifolia*), birch (*Betula* sp.), and sugar maple (*Acer saccharum*). (*Reproduced from Holtman, Wood Construction, McGraw-Hill Book Company, Inc.*)

The bark has medicinal properties, and the ripe fruits are sometimes used for flavoring liqueurs.

Chestnut

Until its almost complete destruction by the chestnut-blight disease, the chestnut (*Castanea dentata*) was one of the most important timber trees of the eastern United States. At one time it was a conspicuous member, in both size and abundance, of the deciduous forests from Maine to Tennessee, attaining its best development in the southern Appalachians. The brown, soft, light open-grained wood is very durable and easy to work. Chestnut is used for millwork, furniture, caskets, musical instruments, boxes, woodenware, veneers, and plywood cores. Because of its durability it is an excellent wood for posts, poles, piling, ties, fence rails, cooperage, and shingles. It is also a source of pulpwood and a tanning material.

Elm

Several species of elm, especially the rock elm (*Ulmus Thomasi*) and the familiar, ornamental white elm (*U. americana*), yield a valuable wood with a beautiful grain. The wood of the rock elm is tough, strong, hard, heavy, elastic, pliable, and durable. Next to hickory it is the most important source of hubs, spokes, fellies, and rims of wheels. It is also used for agricultural implements and tool handles, butcher blocks, veneers, and cooperage, and in the manufacture of furniture, refrigerators, musical instruments, baskets, and woodenware. The white elm (Fig. 47) has a lighter, softer, and weaker wood, which, however, is tough and fibrous and is used for much the same purposes.

Hickory

The hickories are found in the eastern deciduous forest area from Ontario to Minnesota, Florida, and Mexico. Although the wood of all the species is used to some extent, the most important sources are the shagbark hickory (*Carya ovata*), the mockernut (*C. tomentosa*), and the pignut (*C. glabra*). Hickory is one of the toughest, hardest, heaviest, and strongest of woods and is used where both strength and the ability to withstand shocks are required. It is a coarse straight-grained wood. The sapwood is preferable to the heartwood. Hickory is used chiefly for spokes, fellies, axles, and other parts of wagons; also for ax, pick, and hammer handles; baseball bats; agricultural implements; shafts of golf clubs; pump rods; and cooperage. It is the standard for fuel wood and for smoking meat.

Locust

The black locust (*Robinia pseudoacacia*), another tree of the eastern deciduous forest region, also yields an exceedingly heavy, strong, hard, durable, and elastic wood. Locust has a coarse, open, crooked, compact grain and a smooth, satiny surface. Its chief use is for insulator pins and brackets. It is also commonly used for tree nails, boat ribs, fence posts, ties, sills, wagon hubs, and mine timbers. The wood is of more importance in England, where it is used for furniture and shipbuilding.

FIG. 47. A typical American or white elm (*Ulmus americana*), showing its characteristic graceful habit   (*Reproduced by permission from Foster, Elementary Woodworking, Ginn and Company.*)

The honey locust (*Gleditsia triacanthos*), with similar properties and uses, is of minor importance.

Maple

Maple is one of the more important woods. The chief source of the commercial wood known as hard maple is the rock or sugar maple (*Acer saccharum*). This tree, a conspicuous species of the eastern deciduous and northeastern transition forests, ranges from southeastern Canada to Georgia. Two other eastern species, the silver maple (*A. saccharinum*) and the red maple (*A. rubrum*), and the Oregon maple (*A. macrophyllum*), found in Washington and Oregon, furnish soft maple wood which is of less importance. Hard maple wood is heavy, tough, compact, strong, and very hard. It is light brown in color with a dense even grain and fine texture (Fig. 46). It is susceptible of a fine polish, and is often beautifully grained and figured, as in the case of curly and bird's-eye maple. These latter characteristics make it one of the best woods for furniture, veneers, flooring, bowling alleys, and interior finish. It is also used for novelties, shoe lasts, rulers, tool handles, inlays, panels, keels of vessels, pianos, violins and other musical instruments, bowls, cooperage, charcoal, fuel, and wood-distillation products. The sap of the maple furnishes us with maple syrup and maple sugar.

Oak

Oak is the most important of the hardwoods and stands in fourth place among all the woods. Not only is the timber important, but the oaks are

FIG. 48. A famous white oak (*Quercus alba*), the Avery oak in Dedham, Mass. It is estimated that this tree was full grown in 1636. (*Photo by E. H. Wilson; courtesy of the Arnold Arboretum.*)

the largest and finest of the hardwood trees of the forest. Oak wood is hard, tough, durable, resilient, and elastic. Its great strength and ability to resist heavy strains render it valuable for shipbuilding and all types of heavy construction. It is also capable of a high polish and is unrivaled for decorative work, especially in the form of quartered oak. Over fifty species of oak occur in the United States, some twenty of which are of commercial importance. These belong either to the white oak, or to the red oak group. It is difficult to distinguish between the wood of the individual species and they are known in the trade as either white, or red, oak.

White oak lumber is harder, stronger, and more durable. It is used for bridge and building timbers, piling, ties, parts of machinery, agricultural implements, furniture, flooring, cabinetwork, interior finish, and cooperage. In the latter connection it is of interest to note that oak barrels are the only satisfactory containers for beer, wine, and alcoholic spirits. The most important species in this group include the white oak (*Quercus alba*) (Fig. 48), the bur oak (*Q. macrocarpa*), the post oak (*Q. stellata*), the chestnut oak (*Q. montana*), the swamp chestnut oak (*Q. Prinus*), the Oregon oak (*Q. Garryana*), and the swamp white oak (*Q. bicolor*).

Red oak lumber is softer, more porous, more open-grained, and less durable. It has practically the same uses, but is less highly esteemed. The principal species include the red oak (*Quercus borealis*), the black oak (*Q. velutina*), the scarlet oak (*Q. coccinea*), the pin oak (*Q. palustris*), the turkey oak (*Q. laevis*), the willow oak (*Q. phellos*), the Texas red oak (*Q. texana*), and the shingle oak (*Q. imbricaria*).

The **live oak** (*Quercus virginiana*), a quite distinct, evergreen species, has the hardest, strongest, and toughest wood of all the oaks. It is used in the construction of wagons, ships, and farm tools. Only a comparatively small amount is available.

Osage Orange

The osage orange (*Maclura pomifera*) is a small tree, native to the Gulf States, but cultivated elsewhere. Its wood is the heaviest, toughest, hardest, and most durable of all the native hardwoods. Only a little is available and this is used chiefly for fellies, insulator pins, tree nails, posts, stakes, and woodenware. The Indians used this wood for bows. It is also the source of a dyestuff.

Poplar

The true poplars of the United States are known commercially under several different names. The most important tree is the cottonwood (*Populus deltoides*) of the Central and Eastern states. It has a soft, light, even-grained fibrous wood, which is easily worked. Its chief uses are as a substitute for basswood in the manufacture of boxes and excelsior, and as a pulpwood. It is also used for millwork, woodenware, and plywood. The balsam poplar (*P. balsamifera*), several other large poplars, and the much smaller aspens (*P. tremuloides* and *P. grandidentata*) have similar uses. Aspen wood is also used in furniture, in cooperage, and for matches and pulp for making book paper.

Red Gum

The red gum or sweet gum (*Liquidambar Styraciflua*) (Fig. 49) has greatly increased in importance as a commercial wood during the past few decades. The tree ranges from Connecticut to the mountains of Central America, attaining its best development in the Southeastern states. The wood is light and soft, but tough and resilient. It is reddish brown in color, with a fine, straight, close grain, and it polishes well.

FIG. 49. A large sweet gum (*Liquidambar Styraciflua*) in the De Soto National Forest, Mississippi. (*Photo by U.S. Forest Service.*)

Red gum is often stained to imitate cherry, walnut, or mahogany. It is extensively used for veneers and also for furniture, cabinetwork, interior finish, fancy boxes, and cooperage. Red gum wood is known in England under the name of satin walnut. The tree is the source of storax, a medicinal product.

Sycamore

The sycamore (*Platanus occidentalis*), with its characteristic bark which peels off in large patches, is a familiar tree throughout the eastern deciduous forest region. It is the largest hardwood tree in the United States. The wood is hard, tough, strong, and very durable. It is light reddish brown in color with a close uneven grain. When quartersawed, a mottled figure with lustrous rays is obtained. This is often marketed as *lacewood*. Sycamore is extensively used for tobacco boxes and other containers as it does not impart any odor. It is also used for furniture, millwork, interior decoration, butcher blocks, yokes, boxes, crates, woodenware, cooperage, brushes, and plywood.

FIG. 50. An old-growth tulip poplar (*Liriodendron Tulipifera*) in Virginia. (*Photo by U.S. Forest Service.*)

Tulip

The tulip tree (*Liriodendron Tulipifera*) is one of the largest trees (Fig. 50) of the eastern deciduous forest, reaching a height of from 125 to 250 ft. and a diameter of 6 to 14 ft. The wood, which is known commercially as yellow poplar or whitewood, is of great importance. It is soft, light, and easily worked, with a fine straight grain. It is also stiff and durable, although not very strong. It is used for millwork, boxes, furniture, carriage bodies, musical instruments, woodenware, toys, boats, light construction, and veneers.

Tupelo

Two species are known commercially as tupelo, black gum, or sour gum. *Nyssa sylvatica* is found from Maine to Michigan and south to Florida and Texas, while *N. aquatica* is restricted to the swamps of the Southern states. The wood is pale yellow with a dense, fine, twisted, and interwoven grain. It is soft, light, tough, stiff, and resistant to wear. Tupelo has increased in importance in recent years. It is used for flooring, tobacco boxes, wheel hubs, woodenware, veneers, ties, handles, yokes, pulp, rollers, and piling.

Walnut and Butternut

The black walnut (*Juglans nigra*) has always been one of the most valuable native woods. It is a large tree (Fig. 51) of the deciduous forest region, ranging from Massachusetts to Florida and from Minnesota to Texas. The wood is moderately hard, tough, and strong and is easily worked. It is very durable. The color varies from a rich dark brown to a purplish black. The wood has a fine even grain and a good figure and is capable of taking a high polish. Black walnut has been extensively exploited and consequently is scarce; it is so high priced that it is even sold by the pound. A new supply is being assured through cultivation on farms. Since the seventeenth century it has been the chief wood used for gun stocks. Other important uses include furniture, cabinetwork, millwork, musical instruments, airplane propellers, sewing machines, and veneers. Before it became so valuable the wood was used locally for fences, barns, and light construction.

Fig. 51. A fine specimen of the black walnut (*Juglans nigra*). (*Courtesy of the Massachusetts Horticultural Society.*)

The butternut (*Juglans cinerea*), which extends from New Brunswick to Minnesota and south to Georgia and Arkansas, has a wood of somewhat similar nature and uses. The figure is like that of black walnut, so it makes a good substitute for the former. The wood, however, lacks the color and is not so strong. It is used chiefly for furniture, boxes and crates, excelsior, millwork, and woodenware. Sugar is sometimes obtained from the sap of the butternut and a dye from the green husks that surround the fruit.

Minor Species

While all the other woody plants serve some useful purpose in industry, or are valuable, at least for fuel, it will be possible to mention only a few of them.

Apple (*Pyrus Malus*). Apple wood, usually obtained from old orchards, is very strong, hard, and compact, with a uniform close grain. It is used principally for tool handles, pipes, knobs, mallet heads, rulers, canes, and turnery.

Red Alder (*Alnus rubra*). This, the largest of the alders, is of commercial importance in Oregon and Washington west of the Cascade Range. The wood, which has a fine even grain, uniform texture, and a reddish-brown color, works and polishes well and gives a good imitation of mahogany and black walnut. The factories in the states where the tree is native use red alder more than all other woods combined, chiefly for furniture, millwork, handles, and novelties.

Blue Beech (*Carpinus caroliniana*). This small tree of Eastern North America has a heavy, strong, and very stiff wood. No other wood has been found which is as suitable for levers. It is also used for tool handles and for charcoal.

Buckeye (*Aesculus octandra*). This tree of the Middle West furnishes a soft, light, easily worked wood used for boxes, excelsior, millwork, piano keys, furniture, trunks, and artificial limbs.

Cucumber (*Magnolia acuminata*). This tree, the largest and most abundant of the magnolias, is particularly characteristic of the southern portion of the deciduous forest area. Its soft, light, durable wood is used for millwork, woodenware, boxes, excelsior, and cheap furniture. It is often sold as yellow poplar.

Dogwood (*Cornus florida*). This small tree, found throughout the eastern United States, reaches its best development in the South. The wood is very hard and heavy, with a fine, lustrous, close grain. Dogwood is used primarily for shuttles for cotton mills, as it is very resistant to wear. Other uses include mauls, wedges, bobbins, golf-club heads, engraver's blocks, and cogs. Only the sapwood is used.

Catalpa (*Catalpa speciosa*). This small tree, a native of the lower Ohio valley, is extensively planted throughout the Middle West. The wood is very durable and is much used for ties and fence posts.

Coffee Tree (*Gymnocladus dioica*). This species of the eastern deciduous forest has a strong and durable wood used for furniture, interior finish, sills, bridges, posts, ties, and fuel.

Hackberry (*Celtis occidentalis*). This eastern tree has a tough, strong, heavy, and moderately hard wood, used chiefly in millwork and for boxes, woodenware, vehicles, furniture, and cooperage.

Holly (*Ilex opaca*). This characteristic tree of the Southern coastal states and the lower Mississippi and Ohio valleys is best known by its leaves and fruit, which are almost universally used for Christmas decorations. The tough, close-grained, whitish wood is used for inlays, brushes, woodenware, and fancy articles, and is often stained to imitate ebony.

Hornbeam (*Ostrya virginiana*). The wood of the hornbeam, which occurs in Eastern North America, is one of the hardest, toughest, and strongest known, but is available only in small amounts. It is used for handles, carriage parts, levers, and fence posts.

Persimmon (*Diospyros virginiana*). The sapwood of this southeastern species is very heavy, tough, hard, strong, elastic, and resistant to wear. It is used chiefly for shuttles, and also for boot and shoe findings, for golf-club heads and other sporting goods, and in turnery.

Sassafras (*Sassafras albidum*). This small tree of the eastern United States is especially abundant in the Southern and Gulf States. Small amounts of the fragrant, durable, soft, and light wood are available. It is used for millwork, furniture, cooperage, fence rails and posts, and boxes.

Willow (*Salix nigra*). This is the only one of the numerous willows which is of commercial importance. It reaches its best development on the flood plains in the Mississippi and Ohio valleys. The smooth wood is soft, light, tough, and fairly strong. It is used for boxes and crates, plywood cores, excelsior, bats, boats, water wheels, and charcoal. The long, slender, pliable young shoots are used for wicker baskets and furniture.

FORESTS OF SOUTH AMERICA

The forests of South America cover 44 per cent of the land area. They consist principally of tropical hardwoods (89.3 per cent), with some temperate hardwoods (5.5 per cent) and conifers (5.2 per cent). Two types of tropical hardwoods occur. The most abundant is the dense humid rain forest which characterizes the Guianas and the great Amazon and Orinoco River basins and which also occurs along the eastern coast of Brazil. This forest is noteworthy for the great number of species and for the size and frequency of the individual trees. It is estimated that there are over 2500 different tree species in the Amazon forest alone. This forest, which is often said to be the most extensive body of solid forest in the world, has but little vegetation on the forest floor, owing to the density of the canopy. Epiphytes and lianas, however, are frequent and characteristic. In drier parts of Brazil and Argentina an open deciduous type of tropical forest occurs. Mixed forests of conifers and temperate hardwoods occur along the northern Andes, and again in the southern Cordilleras. Conifer forests, consisting chiefly of Paraná pine (*Araucaria angustifolia*) cover large areas in southern Brazil and Argentina, while similar areas, with *A. araucana* the dominant species, occur in Chile. In Paraguay and Argentina there are extensive areas of open forests composed chiefly of quebracho (*Schinopsis Lorentzii* and *S. Balansae*), the most important source of tanwood.

Important Woods of Tropical America

The tropical forests of Mexico, Central America, the West Indies, and South America have long been utilized as the source of most of the high-grade cabinet and furniture woods of commerce, and to some extent for timber for general construction. However, these forests have been but little exploited as a source of ordinary lumber up to the present time, and they constitute a vast reservoir for future needs. The number of available species is enormous. Brazil alone is said to have over 3000 woody species, including some 50 of the most valuable cabinet woods. Because of their importance in the United States, a few of these tropical woods will be discussed.

FIG. 52. Logs of balsa wood (*Ochroma pyramidale*) awaiting shipment. (*Courtesy of S. J. Record.*)

Balsa (*Ochroma pyramidale*). This soft and pithy wood is the lightest of commercial timbers, weighing only 10 lb. per cu. ft. The tree is a native of tropical forests from southern Mexico to northern Peru. Ecuador furnishes 99 per cent of the world's supply (Fig. 52). Balsa is obtained chiefly from wild trees, although there are some plantations. It grows faster than any other jungle tree save the papaya and reaches a height of 12 ft. within a year. It can be cut when only two years of age, but usually trees six to nine years old are selected.

Because of its light weight and its buoyancy, due to the presence of air in the cell cavities, it is used for life preservers, buoys, swimming belts, rafts, floats, pontoons, sea sleds, surfboards, toys, and motion-picture sets and in airplane construction. Balsa possesses good insulating properties and is used to line refrigerators, autotruck bodies, and the holds of ships. It is also a sound deadener and so is utilized for ceilings and partitions and under heavy machinery to prevent the transmission of vibrations. The seed hairs are used as a stuffing material.

Boxwood. The Venezuelan boxwood or zapatero (*Gossypiospermum praecox*) is much used as a substitute for the European boxwood in the manufacture of engraver's blocks, rulers and other scientific instruments, shuttles, spools, bobbins, musical instruments, inlays, and veneers. The supply comes chiefly from Venezuela.

FIG. 53. A Spanish cedar (*Cedrela odorata*) in woodlands near Cienfuegos, Cuba. (*Photo by Walter H. Hodge.*)

Cedar. The Spanish or cigar-box cedar (*Cedrela odorata*) (Fig. 53) is widely distributed throughout tropical America, both as a native and as an introduced species. It is the most important timber tree for domestic use in tropical America. It has a reddish-brown aromatic heartwood with a straight coarse grain. In the United States the wood is used chiefly for cigar boxes, as it is insect repellent. It is also used for closets, chests, shingles, and as a substitute for mahogany.

Cocus Wood. This wood (*Brya Ebenus*), also known as American ebony or granadillo, is used for knife handles, musical instruments, principally flutes and clarinets, turnery, inlays, and cabinetwork. The wood is very hard and durable and polishes well. It comes chiefly from the West Indies.

Cocobolo. Cocobolo, obtained from *Dalbergia retusa*, a tree found from Mexico to Panama, is one of the showiest and most strikingly colored of the exotic woods. The heartwood is orange to orange-red in color, streaked with jet black. The wood is very hard, tough, and strong. It is used for instruments, knife and umbrella handles, steering wheels, inlays, lacquer, and turnery.

FIG. 54. Felling a greenheart tree (*Ocotea Rodioei*) in British Guiana. (*Photo by C. D. Mell; courtesy of S. J. Record.*)

Crabwood. The crabwood (*Carapa guianensis*) attains its best development in the Guianas, although it occurs in other parts of South America and in the West Indies. The very strong and hard rich-brown wood is used as a substitute for mahogany.

Greenheart. The greenheart (*Ocotea Rodioei*), a native of British Guiana (Fig. 54), furnishes structural timbers of great value. The wood is one of the strongest, and is very hard, heavy, tough, resistant to decay and insect injury, and elastic; it is much used, particularly in Europe, for bridges, piles, wharves, paving blocks, and shipbuilding. Other uses include shafts, spokes, and fishing rods. The wood receives its name from its peculiar greenish-brown color.

Lancewood. Two West Indian species, the lancewood (*Oxandra lanceolata*) and the degame or lemonwood (*Calycophyllum candidissimum*), furnish the lancewood of commerce. This wood is yellowish, with a fine close grain, and is very tough, strong, and elastic. It is used for fish poles, spars, shafts, ramrods, whips, cues, bows, lances, and turnery.

Letterwood. The letterwood or snakewood (*Piratinera guianensis*) of British Guiana yields a very heavy ornamental wood. It is brown in color with peculiar black markings (Fig. 27) and a close, fine, lustrous grain. It is used for canes, umbrellas, drumsticks, violin bows, veneers, and inlays.

Lignum Vitae. This important wood (Fig. 55) is obtained chiefly from two species, *Guaiacum officinale* and *G. sanctum*, found in southern Mexico, Central America, and the West Indies. Lignum vitae is one of the hardest of the commercial woods and is naturally very tough, strong, and resistant. It is dark brown, streaked with black, with a very fine, intricately woven grain and contains a resin which acts as a natural lubricant and preservative. The most important use is for the bearings or bushing blocks for steamship propeller shafts. Other uses include bowling balls and other sporting goods, pulley blocks and conveyors, instruments, and furniture. The wood was once thought to possess extraordinary remedial powers for many of man's worst diseases; hence the name lignum vitae, or wood of life. The gum resin, guaiacum, used in medicine, occurs in the form of tears excreted from the living tree.

Locust. The West Indian locust (*Hymenaea Courbaril*) is one of the most important timber trees of tropical America. The hard tough wood

FIG. 55. Shipping logs of lignum vitae (*Guaiacum officinale*) in Haiti. (*Photo by C. D. Mell; courtesy of S. J. Record.*)

is used for general carpentry work, furniture, shipbuilding, and cabinetwork. This tree is the source of South American copal.

Mahogany. Mahogany is the most important export wood in tropical America and is the source of the world's most valuable timber and premier cabinet wood. It was used for woodwork as early as A.D. 1514. The cathedral of Santo Domingo, built in 1550, contains some beautiful mahogany carvings still in excellent condition. The early Spanish explorers utilized the timber for shipbuilding, and it was used for building construction in England as early as 1680. Its use as a furniture wood dates from the beginning of the eighteenth century, with the years 1750 to 1825 marking the golden age of the great master craftsmen, Chippendale, Adam, Hepplewhite, and Sheraton.

There are three species of mahogany, only two of which are of commercial importance. *Swietenia Mahogani*, the West Indian or Spanish mahogany (Fig. 56), found in the Florida Keys and the West Indies, was the first known and the first to be exploited. It has been introduced into Central America and other tropical countries. *Swietenia macrophylla* occurs from Yucatan to northeastern Colombia and Venezuela and along most of the southern affluents of the Amazon in Brazil, Peru, and Bolivia. Although several other species have been segregated, for practical purposes all the native mahogany cut in North and South America may be considered as belonging to this species.

FIG. 56. A large mahogany tree (*Swietenia Mahogani*) in a tropical jungle in British Honduras. (*Photo by U.S. Forest Service.*)

The mahogany is an ornamental evergreen tree reaching a height of 40 to 50 ft. with large buttresses at the base. The trees occur scattered through rich moist forests about one to the acre. The wood is reddish brown in color with a crooked grain. It is very heavy, strong, hard enough to resist indentations, yet easy to work, and it polishes and glues well. It is used for furniture, fixtures, musical instruments, millwork, cars, ships and boats, caskets, airplanes, foundry patterns, veneers, and plywood. Mahogany has many substitutes and imitations. There is a sufficient supply of wild trees to last for many years if economically used. Plantations are also proving successful, especially in Honduras.

Majagua. The majagua or blue mahoe (*Hibiscus elatus*) of the West Indies has already been referred to as the source of Cuba bast. The hard and slightly aromatic wood has a lustrous, richly variegated, open grain. It is used for cabinetwork, carriages, gun stocks, fishing rods, and ship's knees.

Mora. The mora (*Mora excelsa*), of British Guiana, Venezuela, and Trinidad yields a brown wood, which is very hard, tough, and even more durable than teak. It has been used chiefly in Europe for shipbuilding, platforms, ties, and all types of heavy construction.

Prima Vera. This cream-colored wood, sometimes called white mahogany, is obtained from *Cybistax Donnell-Smithii* of southern Mexico, northern Honduras, Salvador, and Guatemala. It is used for furniture and fixtures, millwork, ships, boats, cars, and veneers. The tree is very ornamental with a wealth of yellow flowers appearing before the leaves.

Purpleheart. *Peltogyne paniculata*, the purpleheart of Brazil, the Guianas, and Trinidad, has a very hard, tough, strong, durable brown wood, which turns a purple color on exposure. It is used for heavy construction, furniture, billiard tables, cues, fishing rods, inlays, and turnery.

Rosewood. Of several different rosewoods, the Brazilian rosewood (*Dalbergia nigra*) is the best known. The dark purple, almost black, wood is often striped and has a coarse, dense, even grain. Rosewood is one of the finest cabinet woods and is also used for scientific instruments, furniture, pianos, cars, sporting and athletic goods, handles, and brushes.

Satinwood. The West Indian satinwood (*Zanthoxylum flavum*) has long been a favorite furniture wood. It was especially esteemed in England by such builders as Sheraton, Adam, and Hepplewhite. The creamy or golden-yellow wood is smooth and lustrous and slightly oily. It has a very close, dense, even grain. Other uses include millwork, musical instruments, caskets, brushes, cabinetwork, inlays, and veneers.

FORESTS OF EUROPE

In Europe forests occupy about 31.3 per cent of the land area. Seventy-four per cent of these forests are classed as coniferous, 24 per cent as temperate hardwoods, and 1.6 per cent as mixed forests. The coniferous woodlands are especially characteristic of the northern portion of the continent, while the hardwood and mixed forests are found in Southern and Western Europe. The original forest cover of Europe has been severely depleted as the result of long utilization and the necessity of clearing the land for agriculture and other purposes. In Great Britain only 5 per cent of the original forest is left, and in France, Spain, Belgium, Italy, and Greece from 10 to 20 per cent. Sweden and Finland still have over half their forests available and, together with the U.S.S.R., are the most heavily forested regions in Europe.

The most striking feature of the European forests to an American is the presence of the same genera that are found in the New World, although these are usually represented by different species. The principal conifers are the Scotch pine (*Pinus sylvestris*) and Norway spruce (*Picea Abies*), which furnish the woods known as yellow deal and white deal respectively. Other conifers include the cluster pine (*Pinus Pinaster*), the stone pine (*P. Pinea*), the silver fir (*Abies alba*), larch (*Larix decidua*), and yew (*Taxus baccata*). The American white pine and Douglas fir are extensively planted.

The most important hardwoods are the oaks, chief among which are *Quercus Cerris*, *Q. Robur*, and *Q. petraea*. Other prominent hardwoods include the black alder (*Alnus glutinosa*), ash (*Fraxinus excelsior*), beech (*Fagus sylvatica*), birches (*Betula pendula* and *B. pubescens*), cherry (*Prunus Cerasus*), chestnut (*Castanea sativa*), elm (*Ulmus procera*), hazel (*Corylus Avellana*), holly (*Ilex Aquifolium*), hornbeam (*Carpinus Betulus*), lime (*Tilia cordata*), maple (*Acer pseudoplatanus*), plane (*Platanus orientalis*), rowan (*Sorbus Aucuparia*), blackthorn or sloe (*Prunus spinosa*), and willow (*Salix alba*).

A few European species are imported into the United States. These include the boxwood, walnut, briar, and olive.

The so-called Turkish boxwood (*Buxus sempervirens*), of Southern Europe, Asia Minor, and Northern Africa, has been used for so many centuries that the supply has almost disappeared. The wood is very hard, with a fine, dense, uniform grain, and a smooth, somewhat lustrous texture. It is used for blocks for wood engraving, rulers and other instruments, shuttles, bobbins, firearms, whips, canes, umbrellas, inlays, and turnery.

The English walnut (*Juglans regia*), known commercially as the Circassian walnut, occurs from the Black Sea region across Asia Minor and Persia to northern India. Its hard wood is beautifully figured and takes a high polish. Material obtained from burls is especially desirable. It is used for furniture and fixtures, millwork, musical instruments, firearms, cabinetwork, turnery, and veneers.

Olive wood from *Olea europaea* is imported for brushes, canes, and turnery; and briar root (*Erica arborea*) for tobacco pipes.

FORESTS OF ASIA

In Asia forests cover 21.6 per cent of the land area. They are especially abundant in the northern and eastern parts of the continent and very sparse in the southwest. As in other localities, much of the original forest area has been destroyed. This is especially true of China, where centuries of cultivation have not only destroyed the forests, but much of the arable land as well, as a result of too extensive cultivation and subsequent erosion.

Conifers comprise 42.4 per cent of the forest area. They are characteristic of most of Siberia and also occur in the Himalayas and the mountains of Asia Minor, China, and Japan. Temperate hardwoods make up 27.3 per cent of the forests. These hardwoods, and mixed forests as well, are found in the southern provinces of the U.S.S.R., Afghanistan, Iran, Asia Minor, China, and Japan. In the case of both the conifers and the temperate hardwoods, European species are found in the western part of the continent, giving way to more distinctive Asiatic species of the same genera in the east. Pine, spruce, fir, juniper, cedar, maple, ash, basswood, poplar, alder, birch, walnut, oak, larch, and yew are all represented. The United States imports some oak from Siberia, and maple from Japan.

Tropical hardwoods make up 30.3 per cent of the total forest area, and are found south of the Himalayas. In many countries they comprise 100 per cent of the woody species. This is true of Ceylon, Thailand, the East Indies, and Malaya. As in other tropical forests elsewhere, the number of species is very great. India is estimated to have 2000 and Japan 1000 different species. Many of these Asiatic forests are dominated by teak, while others are composed chiefly of members of the *Dipterocarpaceae* and *Leguminosae*. Seventy-five per cent of the Philippine forest trees belong to the former family. A considerable number of these Asiatic woods enter into foreign trade, and a few are imported into the United States. These include ebony, padouk, satinwood, and teak.

Ebony. Several species furnish a wood known as ebony, the most important of which is the Macassar ebony (*Diospyros Ebenum*), which is found from India and Ceylon to the East Indies. Ebony is a black wood with brown stripes. It is very hard and heavy, and has a fine grain. It takes a high polish and has long been a famous cabinet wood. Other uses include whips, canes, umbrellas, piano keys, sporting and athletic goods, handles, inlays, veneers, and turnery.

Padouk. This wood, also known as Burmese rosewood, obtained from *Pterocarpus indicus*, is very ornamental. It is red in color with black stripes and is very lustrous. It is hard and durable and polishes well. Padouk is used for furniture, fine cabinetwork, millwork, car construction, turnery, and veneers. It has been introduced into Honduras.

Satinwood. The East Indian satinwood (*Chloroxylon Swietenia*) has an extremely hard yellowish or dark-brown heartwood, which is sometimes mottled. It has a satiny luster and a fine, dense, even grain. Satinwood is used for furniture, cabinetwork, veneers, and brushes.

Teak. Teak (*Tectona grandis*) (Fig. 57), native to Southeastern Asia and Malaya, is one of the most durable of woods and one of the most important commercial timbers of the tropics. It is hard and does not warp, split, or crack, and so makes a valuable timber for general construction. It is also very resistant to decay and termites. The wood is yellowish brown in color and greasy to the touch. It is extensively used in ship and boat building, and for furniture, cabinetwork, millwork, piles, railway cars, flooring, and greenhouses. Introduced experimentally into Panama and Honduras, teak has shown phenomenal growth, in one instance attaining a height of 64 ft. and a diameter of 18 in. in 16 years.

Other important woods of tropical Asia include the acle or pyinkado (*Xylia xylocarpa*), deodar (*Cedrus deodara*), sal (*Shorea robusta*), and sissoo (*Dalbergia Sissoo*), all used for general construction, and the fol-

FIG. 57. Logs of teak (*Tectona grandis*) in Burma. Elephants are often used in the lumbering operations. (*Photo by U.S. Forest Service.*)

lowing cabinet woods: Moulmein cedar (*Cedrela Toona*), laurelwood (*Calophyllum inophyllum*), rosewood (*Dalbergia latifolia*), and sandalwood (*Santalum album*).

FORESTS OF AFRICA

In Africa forests cover only 10.7 per cent of the land area. Tropical hardwoods predominate, comprising 96.9 per cent of the forests, with temperate hardwoods accounting for 2.2 per cent, and conifers only 0.9 per cent. Conifers are restricted to the Mediterranean region, the high mountains of Central and Eastern Africa, and South Africa. Two types of tropical forest occur. The most extensive is the dense rain forest which covers much of equatorial Africa, particularly the West African coast and the Congo basin. This region has a uniformly distributed rainfall amounting to 60 to 160 in. Important elements of this forest are the mangrove swamps along the coast. An open parklike forest occurs in regions where the rainfall amounts to 30 to 40 in. Large areas of this type of forest occur in the northeastern and southern parts of Africa, particularly in Angola and Rhodesia. There are many important woods that resemble the better known woods of tropical America and Asia, but they have been but little exploited as yet. The United States imports only one wood, the African mahogany (*Khaya senegalensis*), a West Coast species. This is a beautifully figured and richly colored wood, which is much used as a substitute for true mahogany.

FORESTS OF AUSTRALIA AND OCEANIA

Forests cover only 5.8 per cent of the land area in Australia, while in New Zealand 25.7 per cent of the land is forested, and in Oceania 71 per cent. The most heavily forested regions are New Guinea with 80 per cent, Samoa with 70 per cent, and Tasmania with 64.7 per cent. In Australia 4 per cent of the forests are coniferous, 11 per cent are temperate hardwoods, and the remainder are tropical hardwoods. Conifers occur in New South Wales, Queensland, and Tasmania, and temperate hardwoods only in Tasmania. Tropical hardwoods occur in all the states except Tasmania. These forests rival those of Pacific North America in size and density. The species are different from those found in other parts of the world. Most of them are either eucalypts or acacias. There are over 70 commercial species of *Eucalyptus*, the most important of which are the karri (*E. diversicolor*) and jarrah (*E. marginata*) of Western Australia. The former reaches a height of 300 ft., with a clear length of 180 ft. Other valuable trees are the blackwood (*Acacia melanoxylon*) and silky oak (*Grevillea robusta*). The principal coniferous species is *Araucaria Cunninghamii*.

In New Zealand 68 per cent of the forests are coniferous and the remainder temperate hardwoods. The chief species is the huge kauri pine (*Agathis australis*), one of the largest timber-producing trees in the world. The wood is strong and durable and very free from knots. It is exported in large quantities. Kauri is also the source of an important resin. Other valuable species include the white pine (*Podocarpus dacrydioides*), totara (*Podocarpus Totara*), red pine (*Dacrydium cupressinum*), and several beeches.

The forests of Oceania are entirely composed of tropical hardwoods, and none of them are of importance, except locally.

CHAPTER V

TANNING AND DYE MATERIALS

TANNINS

Tannins are organic compounds, chiefly glucosidal in nature, which have an acid reaction and are very astringent. Their biological function is problematical. They may be concerned with the formation of cork or pigments, or with the protection of the plant. Tannins are of interest economically because of their ability to unite with certain types of proteins, such as those in animal skins, to form a strong, flexible, resistant, insoluble substance known as leather. Because of this property of "tanning" hides, tannin-containing materials are in great demand. Tannins also react with salts of iron to form dark-blue or greenish-black compounds, the basis of our common inks. Because of their astringent nature they are useful in medicine. Tanning materials are often utilized in oil drilling to reduce the viscosity of the drill without reducing the specific gravity.

Although nearly all plants contain some tannin, only a few species have a sufficient amount to be of commercial importance. Tannins are found in the cell sap (Fig. 3) or in other definite areas in bark, wood, leaves, roots, fruits, and galls. In most cases these structures are of little value otherwise, so the extraction of tannin is usually incidental to other industries.

THE TANNING INDUSTRY

Tanning is a very old industry, dating back possibly 5000 years. The Chinese tanned leather over 3000 years ago. The Romans used oak bark for tanning skins. In more recent times the American Indians utilized several native plants in curing their buffalo hides. The first tannery in the United States was established in Virginia in 1630. Twenty years later there were over 50 in New England alone. The industry developed in this region because it was here that the chief raw material, hemlock bark, was most abundant and cheapest. By 1816 the business was worth over $200,000,000, and Boston had become one of the chief leather markets of the world. Fifty years later the industry began to shift westward and southward, as the hemlock became scarce, and many tanneries were established in Pennsylvania. Here oak was used extensively. Still later other tanneries sprang up in the Southern states, using chestnut as a raw material. As the native species became scarce, other sources began to be utilized, including western trees, and foreign products as well. The development in recent years of concentrated extracts with a high tannin content has greatly increased the number of available sources. At first each company made its own supply, but now extracts can be imported, and many foreign species have been developed as raw materials of first importance. The world's supply of tanning materials is very abundant and much of it is as yet untouched. The United States uses about one-half of the world's tannin output, but supplies only a small part, chiefly hemlock, oak, chestnut, sumac, and canaigre.

Sources of Tanning Materials

Nearly all the sources of tannin occur in the wild state, very few being cultivated. The more important raw materials will be discussed according to their morphological origin.

Barks

Hemlock. From the earliest time the hemlock (*Tsuga canadensis*) has been the chief domestic source of tannin in the United States. Its continued utilization for this purpose has contributed greatly to the gradual elimination of this species as an important member of our forests. Wasteful methods were employed, and often the trees were felled and allowed to rot after the bark was removed. Hemlock bark (Fig. 58) contains 8 to 14 per cent tannin. It is used for sheepskins and for sole and other heavy leathers, either alone or in combination with oak. Extracts are now available with a tannin content as high as 28 to 30 per cent. In recent years attention has been directed to the western hemlock (*T. heterophylla*), which seems destined to be one of the chief native sources of the future.

Oak. Oak bark has been used extensively as a source of tannin, and several species are available. The most important American species is the chestnut oak (*Quercus montana*). This tree is very abundant in the Appalachian region and many tanneries have been established there. The tannin content is 6 to 11 per cent. The extract, with a 26 to 30 per cent content, is widely used for heavy leathers. The black oak (*Q. velutina*), with about the same amount of tannin, yields the extract quercitron. This is much used although it dyes the leather a curious yellow. The California tanbark oak (*Lithocarpus densiflora*) has been utilized since 1850. This species has a yield of tannin up to 29 per cent and is of increasing importance. Red oak (*Q. borealis*) and white oak (*Q. alba*) are used somewhat, but they contain only a small amount of tannin. European species of oak are much used in England and on the Continent.

Mangrove. Mangrove bark has become very important in recent years both in this country and abroad. It gives great promise for the

Fig. 58. Hemlock bark ready for use, stacked at a Maine tannery. (*Photo by U.S Forest Service.*)

Fig. 59. A mangrove (*Rhizophora Mangle*) swamp in the Solomon Islands.

future as the supply is extensive and virtually untouched. Although all mangroves contain tannin, the red mangrove (*Rhizophora Mangle*) is the chief source. This tree is very abundant in the tropical swamps of both hemispheres in tidal areas (Fig. 59). The present commercial supply comes from East Africa, the East Indies, Central and South America, where it is a steadily growing industry. Mangrove bark is very hard and heavy, and contains 22 to 33 per cent tannin. The leaves may also be used. The extract is the cheapest of the tanning materials. It is rarely used alone as it gives a bad color to the leather.

Wattle. Wattle bark is an important source of tannin. It is used chiefly in Great Britain, but is of steadily increasing prominence in this country. It is the product of several species of acacias, chiefly *Acacia decurrens* and its varieties *dealbata* and *mollis*, and *A. pycnantha*. These small trees are natives of Australia, but are now cultivated extensively in South Africa, East Africa, Ceylon, Brazil, and elsewhere. The bark, which may have a tannin content as high as 40 to 50 per cent, is removed from the trees when they are from five to fifteen years of age. It is dried and ground to a powder. An extract is also available. Wattle bark yields a solid, very firm, faintly pink leather, much used for soles. Experiments have shown that wattle trees thrive in California, and a profitable domestic industry may well be established there. Wattle wood is a good pulpwood and can be used for mine timbers, posts, and crossties. A gum is also produced.

Other Barks. Many other barks have been used as a source of tannin, but the content is too low for them to be of much commercial value. In Europe the larch (*Larix decidua*), Norway spruce (*Picea Abies*), and several birches and willows are used. Birch bark is much favored in the U.S.S.R., and the fragrance of Russia leather is due to the presence of an essential oil in the bark. Willow bark furnishes a light-colored, soft, pliable leather much used for gloves. Several tropical barks are of minor importance. Mallet bark, obtained from *Eucalyptus occidentalis* of Western Australia, has a tannin content of 35 to 50 per cent. The demand for this bark in the past has been so great that it has led to the depletion of the supply and now only a little is available. Avaram bark, from *Cassia auriculata*, is the most important tanbark of India. The babul (*Acacia nilotica*), another Indian species, has tannin-containing bark, and pods as well, both of which are exported to Europe for making extracts. Tanekaha bark from *Phyllocladus trichomanoides*, a New Zealand tree, is used for glove leather as it contains an orange-yellow dye in addition to the tannin.

Woods

Chestnut. The wood of the chestnut (*Castanea dentata*) was made available when the extracting process was developed. The tannin is extracted at high temperatures from chips of wood. The resulting solution is cleared, filtered, and evaporated, and eventually has a content of 30 to 40 per cent tannin. Chestnut is used for all heavy leathers. It constitutes about half of the native tanning materials. In Europe the wood of *C. sativa* is utilized.

Quebracho. Quebracho wood is the world's most important source of tannin at the present time. It constitutes over 65 per cent of the tanning materials imported by the United States, and 30 per cent of the amount used. Quebracho is the heartwood of two large South American trees, *Schinopsis Lorentzii* (Fig. 60) and *S. Balansae*. The word "quebracho" means "ax breaker." The wood is one of the hardest known, with a specific gravity of 1.30 to 1.40. It was originally used for railroad ties. The first shipment of tannin was made in 1897. Argentina and Paraguay are the chief centers of the industry. Formerly the bark and sapwood

FIG. 60. A quebracho tree (*Schinopsis Lorentzii*) in Argentina. The wood of this species is the most important source of tanning materials. (*Photo by U.S. Forest Service.*)

were removed and then the logs were hauled out, a very laborious process. At the present time, instead of shipping the logs, extracts are made in factories located close to the source of supply. The logs are chipped and cooked with steam in copper extractors until the liquor is very concentrated, with a tannin content of 40 to 60 per cent. Quebracho is one of the quickest acting tans. It is used either alone or in combination for all kinds of leather, especially sole leather to which it imparts extra durable qualities.

Leaves

Sumac. In the United States the dried leaves of three native sumacs, *Rhus glabra, R. typhina,* and *R. copallina*, are an important source of tannin. The leaves are picked in the fall when they begin to turn red, and are dried and ground into a powder. The tannin content is higher, amounting to 10 to 25 per cent, in plants growing in the Southern states.

Sicilian Sumac. The Sicilian sumac (*Rhus coriaria*) is of even greater importance and is extensively used both in the United States and abroad. The leaves have a tannin content of 20 to 35 per cent. It yields a leather of a pale color and soft texture, the best for gloves and bookbinding. Sicilian sumac is one of the few tanning plants to be cultivated. Most of the supply comes from Sicily and southern Italy.

Gambier. Gambier or white cutch is a resinous substance extracted from the leaves and young branches of *Uncaria Gambir*, a climbing shrub of the Malay Peninsula and Indonesia. Under cultivation the plant becomes shrubby. The trees are cropped four times a year and the tannin is extracted from the tissues with boiling water. It crystallizes out as a semisolid whitish substance. A considerable amount of gambier is exported to Europe and the United States, usually in the form of small cubes or larger blocks (Fig. 62). It has a tannin content of 35 to 40 per cent and is used for all kinds of leather. Gambier is also used as a dyestuff, as a masticatory, and in medicine.

Fruits

Myrobalan. Myrobalan nuts are the unripe fruits of two species of Indian trees, *Terminalia chebula* and *T. Bellerica*. These trees have long

Fig. 61. Myrobalan nuts, the unripe fruits of *Terminalia chebula* and a related species, are imported from India for use as a tanning material. (*Photo by S. J. Record.*)

been grown in India for both the fruits and the timber. The nuts (Fig. 61) have a tannin content of 30 to 40 per cent. When used alone they yield a spongy leather of a light-yellow color, but in combination they are more satisfactory. They are used with calf, goat, and sheep skins, and for sole and harness leather.

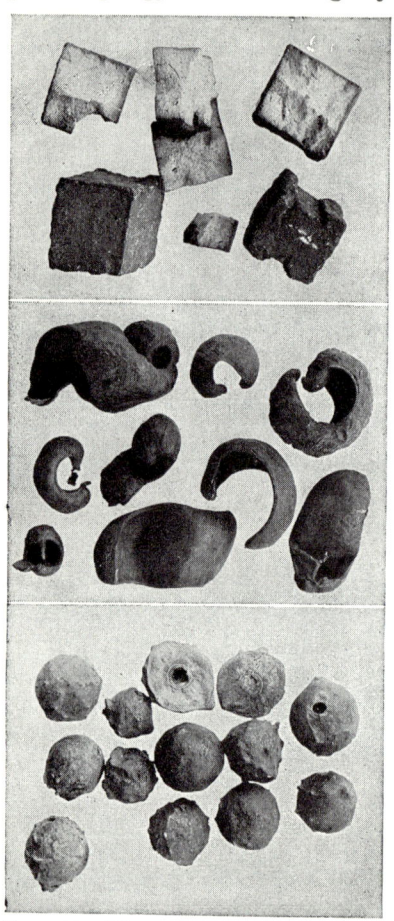

FIG. 62. Three important sources of tannin. *Above*, cubes of gambier or white cutch obtained from *Uncaria Gambir;* center, divi-divi pods, the fruit of *Caesalpinia coriaria;* below, Aleppo or nut galls from *Quercus infectoria.* (Photos by S. J. Record.)

Divi-divi. This material is obtained from the dried twisted seed pods (Fig. 62) of *Caesalpinia coriaria*, a small leguminous tree of the West Indies and South America. The product is very cheap and has a high tannin content, 40 to 50 per cent. In spite of this, only a small amount is used in the United States, although it is very popular in Europe.

Tara. *Caesalpinia spinosa*, a stocky shrub or small tree widely distributed in tropical America, has fruits which are very rich in tannin, 43 to 51 per cent. Tara is cultivated in Peru and also in North Africa. It is used for tanning high-grade leather, as it affects the color only slightly; in making ink; and for a black dye.

Algarobilla. This Chilean species (*Caesalpinia brevifolia*) also has pods with a high tannin content. It is generally used in combination with materials of a lower tannin concentration.

Valonia. The sun-dried acorn cups of the Turkish oak (*Quercus macrolepis*), a tree of Asia Minor and the Greek archipelago, furnish valonia. This material, which has a tannin content of 45 per cent, is used for the finer grades of leather, almost always in combination.

Roots

Canaigre. Canaigre is an American tanning material obtained from the tanner's dock (*Rumex hymenosepalus*), a species native to the southern

United States and Mexico. The plant is extensively cultivated. The roots are sliced and the tannin is extracted. Canaigre extract, with a tannin content of about 30 per cent, yields a bright-orange, firm, heavy leather.

Palmetto. The roots of the palmetto (*Sabal Palmetto*) have been used to a slight extent as a source of tannin, but the content is too low, 10 per cent, to be of much value.

The Manufacture of Ink

Writing inks have been in use from the earliest recorded time. The Egyptians used ink on their papyrus as early as 2500 B.C., and the oldest Chinese writings go back at least to 2600 B.C. At first a carbon ink was used, a combination of charcoal, gums, and varnish. The charcoal was obtained from plant sources, such as charred date stones, or was of animal origin.

There are several types of modern inks, the most important being the carbon and the tannin inks.

Carbon inks differ from all the others in that they are paintlike and remain on the surface of the paper whereas the rest are dyelike and soak into the paper where they combine chemically with the fibers. *Chinese or India ink*, a virtually permanent ink, is made from the carbon black, lampblack, or soot obtained by burning pine wood or a vegetable oil such as tung or sesame, mixed with glue, gum arabic, or some similar sizing material. *Printing ink* contains carbon obtained from natural gas, petroleum, or other materials, in combination with rosin, a drying oil (usually linseed), some chemical drier, and often soap.

Tannin inks, which utilize the property that tannin has of combining with iron salts to give a blue-black color, are the most important kind at the present time. Most of them are gallotannate in nature, the tannin being derived from insect galls which also contain gallic acid. Tannin inks were first described in the eleventh century. Aleppo or nut galls (Fig. 62) have from the outset been the chief source of the tannin. These galls are formed on the twigs of the Aleppo oak (*Quercus infectoria*), as a response to the injuries caused by the egg-laying activities of an insect. The plant is a small shrub, 5 or 6 ft. in height, found throughout the Mediterranean region. The small spherical or pear-shaped galls are produced in great abundance and have an exceedingly high tannin content. In making ink, either the galls or an extract made from them are combined with ferrous sulphate; an agglutinant, such as gum arabic; and a coloring material, such as logwood. Similar galls produced on *Rhus chinensis* and other Chinese and Japanese species of sumac are sometimes used as substitutes for Aleppo galls, although they are much inferior. During the First World War these sumac galls were the only kind available. Other types of oak galls are occasionally utilized. Tannin inks are also made to some extent from other sources of tannin, such as logwood and chestnut. Logwood ink is especially noteworthy because it contains both tannin and a coloring agent as well.

Colored inks are prepared from natural dyestuffs or aniline dyes in combination with alum, water, and a gum. The best grade of red ink utilizes the coloring material in brazilwood.

DYES AND PIGMENTS

Natural dyestuffs and stains, obtained from the roots, bark, leaves, fruit, or wood of plants, have been in use among all nations from earliest time. The cultivation of the plants and the preparation of the dyes have constituted an important industry in many countries. About the middle of the last century the natural products began to be supplanted by the synthetic or aniline dyes obtained from coal-tar products. These latter dyes are brighter, more permanent, cheaper, easier to use and afford a wider range of colors. Their use has gradually led to the abandonment of most of the plant products.

Over 2000 different pigments are secreted by plants. The majority of these are used only locally by primitive peoples, if at all. A comparatively small number, about 150, have been of commercial importance, and of these a mere handful have been able to compete with the artificial colors.

The chief use of dyes has been in connection with the textile industry. Before they can be taken up by the fabrics, however, they must be rendered insoluble so they will not run. This is accomplished by the use of mordants, which are salts of various metals. When fabrics are steeped in a solution containing a weak salt of iron, chromium, aluminum, or tin, a fine layer of the metallic oxide is deposited on the cloth. The dye forms an insoluble compound with this oxide. Dyes are also used for coloring paints, varnishes, leather, ink, paper, wood, furs, food, cosmetics, and medicines.

It would be too tedious a task even to list all the dyes that have been used in the United States at various times. The Indians made use of many native species and the early settlers followed the example of the Indians. Several dyes, not now in use, have been of importance. Butternut bark, for example, was used for dyeing homespun, and later for dyeing the uniforms of the Confederate army. As in other countries, most of the plant pigments have been generally supplanted by aniline dyes. For a while after 1914, when the First World War cut off the supply of synthetic colors, 90 per cent of which were made in Germany, the United States returned somewhat to the use of the natural products. Before long, however, a domestic synthetic dye industry was built up which by 1939 was producing products valued at $83,000,000.

Nearly all colors were available in one or another of the plant pigments. Red dyes were obtained from alkanna, barwood, brazilwood, cudbear, logwood, safflower, sappanwood, and sandalwood; yellow dyes from annatto, fustic, gamboge, henna, osage orange, Persian berries, quercitron, saffron, turmeric, and weld; blues from cudbear, indigo, and woad; greens from chlorophyll and lokao; and brown from cutch. These more important natural coloring materials will be considered according to their morphological nature.

Woods

Logwood. Logwood is one of the oldest dyestuffs and one of the most important. At the present time it is used more than all the others together. The dye is obtained from the heartwood of *Haematoxylon campechianum*, a small, thorny, leguminaceous tree (Fig. 63) with peculiarly corrugated and clustered trunks. Logwood is a native species of Mexico, but it has been introduced throughout the tropics. The United States obtains its supply chiefly from the West Indies and Central America. The trees can be propagated from seed and are now being cultivated to a considerable extent. They are cut when from 10 to 12 years of age and the bark and sapwood are removed. Either the logs or extracts are exported. The dye, of which as much as 40,000 tons has been made annually, has a purplish-red color. It is known as haematoxylin. It is used in its natural state or with a mordant. The presence of a considerable amount of tannin enables logwood extract to react with iron salts to give a black color, which is extensively used for dyeing cotton and woolen goods, leather, furs, and silk. This black is the best and most lasting obtainable. As has been noted, logwood is also used in making some inks. Haematoxylin stain finds its greatest use in histological work. A related species, the brazilette or hypernic (*H. Brasiletto*) furnishes a red dye used for dyeing leather.

FIG. 63. A logwood tree (*Haematoxylon campechianum*) in Haiti, showing the clustered and corrugated trunks. Logwood is an important dyestuff and tanning material. (*Photo by C. D. Mell; courtesy of S. J. Record.*)

Fustic. Fustic is the principal source of natural yellows, browns, and olives, and it ranks next to logwood in importance. It is used for leather and, in combination with logwood, for wool, silk, rayon, and nylon. It is obtained from the heartwood of *Chlorophora tinctoria*, a tree of the dense tropical forests of the West Indies, Central and South America. The light yellowish wood turns a dark yellowish brown on exposure to the air. Fustic is exported in the form of short logs, chips, paste, or powder. This dye is often called *old fustic* to distinguish it from *young fustic*, a one-time substitute obtained from the twigs of *Cotinus Coggygria*.

Cutch. Cutch is a term used for several kinds of raw materials used both as dyes and in tanning. Gambier or white cutch has already been discussed. Black cutch or catechu is the source of the most important brown dye. It is obtained from the heartwood of *Acacia Catechu*, a tree of India and Burma. Small pieces of the wood are boiled in water and the extract is evaporated down to a purplish-black, gummy, semisolid substance, which is molded into blocks for export. The dye is strictly fast and is used for the various brown, fawn, olive, and drab colors, including the familiar khaki. Catechu is also used as a masticatory and in medicine.

Osage Orange. The osage orange (*Maclura pomifera*) is a native tree from southern Missouri to Texas, and is commonly planted elsewhere in the United States for ornamental purposes and for hedges. It is also one of the few native sources of a dyestuff. The bright orange wood yields a dye that is used for orange-yellows and gold and as a base for greens. It was known to the Indians and in recent years has come into prominence as a substitute for fustic and the aniline dyes.

Sappanwood and Brazilwood. These two soluble red wood dyes have had an interesting history. The first red dyewood to be discovered was the heartwood of *Caesalpinia Sappan*, now known as sappanwood. This tree is a native of India and Malaya and cultivated elsewhere in the Asiatic tropics. The wood was introduced into Europe during the Middle Ages and was called "bresel wood." Later, when the Portuguese discovered a similar wood in South America, they naturally gave it the same name "bresel," and this name was also applied to the country in which the tree was found, *i.e.*, Brazil. The source of this New World brazilwood, as it is still called, is *Caesalpinia echinata*, which occurs in tropical America and the West Indies. The heartwood yields a red dye used for cotton and woolen cloth and for red ink. The wood is a valuable material for violin bows. To some extent the bark and pods can be used as a source of the dye. Sappanwood has the same uses as brazilwood. In both cases the color is fugitive and the dyes are not much used today.

Barwood and Camwood. Much confusion exists in regard to the identity of these two insoluble red wood dyes, obtained from several West

African trees. *Baphia nitida* and a related species are probably the source of camwood, while barwood is obtained from *Pterocarpus erinaceus* and *P. Soyauxii*. They yield shades of brown, red, and violet and are used in the United States chiefly for dyeing wool.

Red Sanderswood or Red Sandalwood. *Pterocarpus santalinus*, an East Indian tree, has a hard, fragrant, reddish wood which is the source of an insouble blood-red dye, known as red sanderswood or red sandalwood

Leaves

Indigo. For many years indigo or anil was known as the "king of the dyestuffs." Nothing had been found to equal the permanency and strength of its deep-blue color. Now, however, it has almost entirely been replaced by a synthetic product. It is obtained from the Asiatic *Indigofera tinctoria*, the tropical American *I. suffruticosa*, and several other species of the same genus. The plants are stiff-stemmed, weedy-looking annuals or shrubby perennials. The indigo industry was formerly widespread and was even carried on in the Southern states at one time. Indigo is still cultivated in India, Java, and Natal. Curiously enough, the dye is not present in the plant itself. The leaves contain a soluble colorless glucoside, known as indican, which oxidizes in water to form the insoluble indigo. Fresh plants, collected in the flowering season, are broken up and steeped in water for 12 or more hours. The liquid is constantly stirred to bring about complete oxidation, and the indigo gradually settles out as a bluish sediment, which is made up into small cubes for export. Indigo has been used as a dyestuff in India and other parts of the East from earliest time. It was introduced into Europe in the sixteenth century and soon spread all over the world.

Chlorophyll. Chlorophyll, the source of a green dye, is present in all green plants. It is especially characteristic of the leaves of the higher plants, from which it can readily be extracted with various solvents. Chlorophyll is of increasing importance as a coloring material for foods, soap, and similar products. Its value lies in the fact that it is absolutely harmless and may even serve as a deodorant.

Henna. This orange dye is obtained from the leaves and young shoots of *Lawsonia inermis*, a small tree 6 to 8 ft. in height, native to Egypt, Arabia, Iran, and India. It is widely cultivated in the tropics and subtropics as an ornamental and dye plant. The branches are first cut when the tree is three years old, and thereafter twice annually. The leaves are dried and ground into a paste. Henna is a very fast dye and was formerly used chiefly for fabrics and leather. It is also used in many countries for dyeing the hair, eyebrows, and fingernails, and for other forms of personal adornment.

Woad. One of the first blue dyes to be used in Europe was obtained from the leaves of the woad (*Isatis tinctoria*). This dye was used by the primitive Britons to paint their bodies. The leaves are moistened, slightly fermented, molded into balls, and dried. The plant is still cultivated to some extent.

Weld. Weld is another old-time European dyestuff. It is a deep-yellow dye much used for silks. It is obtained from the leaves of the weld (*Reseda Luteola*), which was formerly cultivated throughout Europe. The plant was introduced into America by the early settlers, and still persists in many localities.

Roots and Tubers

Madder. Madder was formerly one of the most important natural dyestuffs and was widely cultivated in all the Mediterranean region. It

FIG. 64. Turmeric plants (*Curcuma longa*) cultivated in Tela, Honduras. (*Photo by Walter H. Hodge.*)

is still grown in Italy and the Levant. The dye is obtained from the roots of the madder (*Rubia tinctorum*), which occurs as a wild plant in Greece,

Asia Minor, and the Caucasus. An infusion made from three- or four-year-old roots shows a brilliant scarlet color, the familiar "turkey red." The coloring material is a glucoside, alizarin, which was one of the first dyes to be made synthetically.

Alkanna. The roots of *Alkanna tinctoria* yield a red, violet, or crimson dye which is used for coloring oils, pomades, medicines, and wines, and as a stain for histological work.

Turmeric. Turmeric, one of the most important coloring materials of India, is obtained from the tubers of *Curcuma longa* (Fig. 64). The natural dye is orange-red or reddish brown. It is much used to impart a yellow color to cloth and foods, such as curries. Turmeric also serves as a chemical indicator as it changes color depending on the presence of alkalies or acids.

Barks

Quercitron. The ground or crushed bark of the black oak (*Quercus velutina*), a familiar tree of the eastern United States, yields a bright-yellow dye known as quercitron. This is used for leather, cotton, and woolen goods with a tin mordant. Black-oak bark is also a tanning material. An exceptionally strong preparation of quercitron is marketed as flavin.

Lokao. Lokao or Chinese green is one of the few natural green coloring matters. It is a powdery substance made from the bark of two Chinese species of buckthorn, *Rhamnus globosa* and *R. utilis*. It is used chiefly for silks and cotton.

Flowers

Safflower. The safflower (*Carthamus tinctorius*), a very important Asiatic dye plant, is a native of India, but is now widely distributed in most warm countries. It is one of the great tropical crops. Not only are the flowers (Fig. 65) used for a dye, chiefly in coloring food, but the seeds furnish an edible oil and the leaves are used as a salad vegetable. In India over 1,000,000 acres are planted to safflower. The plant is grown to some extent in the United States for the oil. The yellow or orange thistlelike heads are picked in dry weather, dried out, and pressed into cakes. Both a red dye, used for fabrics and rouge, and a yellow dye are obtained. Safflower is grown commercially for the dye in southern France and Bengal.

Saffron. The saffron crocus (*Crocus sativus*) is the source of this old and powerful yellow dye. This plant is a native of Greece and Asia Minor, but is cultivated in many parts of Europe and in India and China as well. The blue or lavender flowers blossom in the fall. The stigmas and tips of the styles are used for the extraction of the dye. These are clipped as soon as the flowers open and are dried naturally or with artificial heat. It takes at least 4000 flowers to furnish an ounce of the dye. The coloring material is readily soluble in water so is not used for fabrics. It is, however, much used for coloring medicines and food to which it imparts a characteristic flavor.

FIG. 65. The leaves and inflorescence of the safflower (*Carthamus tinctorius*) (*Reproduced from U.S.D.A. Circular* 366, *Safflower, A Possible New Oil-seed Crop for the Northern Great Plains and the Far Western States.*)

Fruits

Persian Berries. The dried unripe fruits of a buckthorn, *Rhamnus infectoria*, which grows in Southern Europe, Asia Minor, and Iran, are known as Persian berries. An extract of these berries yields yellow and green dyes of some use in European countries.

Sap Green. Another native European buckthorn, *Rhamnus cathartica*, is the source of sap green. This water-color pigment is obtained from the fruits, which also have medicinal properties. This species has been introduced into the United States.

TANNING AND DYE MATERIALS

Seeds

Annatto. This important coloring material comes from the seeds of *Bixa Orellana*, an evergreen bush or small tree (Fig. 66) native to tropical America. It is cultivated in many other tropical countries. There are few more satisfactory trees as regards growth and behavior. They begin to bear fruit the second year, and average 300 to 600 lb. to a tree. Each spiny pod contains from 30 to 50 seeds, which are surrounded by a scarlet

FIG. 66. *Bixa Orellana.* An important yellow dye, annatto, is obtained from the aril that surrounds the seeds of this tropical American plant.

aril. This aril yields a bright-yellow dye. Either the seeds are exported or the aril is scraped off and made into a paste for shipment. Annatto is nearly tasteless, and so is well adapted for coloring butter, margarine, cheese, and other foodstuffs. It is also used for wool and calico goods, paint, varnish, lacquer, and soap. Many South American Indians use urucú, as it is called, to paint their bodies red.

Gum Resin

Gamboge. This yellow dye is obtained from a gum resin which is exuded from the Siamese gamboge tree (*Garcinia Hanburyi*) and allied species. These trees grow in Ceylon, Thailand, and the East Indies. Incisions are made in the bark and a yellow viscid juice oozes out, which dries on exposure. It is usually collected in hollow bamboos, where it hardens into cylinders. The beautiful yellow dye is soluble in water, alcohol, or oil; it is much used by artists, and also gives a gold tinge to the varnishes used for lacquer and metalwork. Gamboge has a medicinal use as a violent cathartic.

Lichens

Archil and Cudbear. This blue or purple dyestuff, known variously as archil, orchil, orseille, or cudbear, is obtained from several species of lichens from various parts of the world, chiefly *Roccella tinctoria*. It was formerly used for wool and silks and for staining wood and coloring wine. It is now utilized for coloring drugs, sauces, and bitters. The dye is prepared by treating the macerated lichens with ammonia and exposing them to the air. A blue archil liquor is then extracted with water. When this is heated to drive off the ammonia, it changes to red archil. This is evaporated and ground to a fine powder or paste, which is known as cudbear.

Litmus is obtained from the same lichens by a somewhat different process. They are treated with an alkali and allowed to ferment for several days. Lime is then added and the dye is extracted with water. The liquid is evaporated down and mixed with chalk or powdered gypsum, or is applied to paper. Litmus is used as a chemical indicator for acids and alkalies, for its natural purplish color is changed to red by acids and to blue by alkalies.

CHAPTER VI

RUBBER AND OTHER LATEX PRODUCTS

RUBBER

Rubber is obtained from the milky juice, or *latex*, of various erect or climbing woody plants of the tropics or subtropics. The majority of the rubber plants belong to the *Moraceae, Euphorbiaceae,* or *Apocynaceae*. Although well over fifty species are available as sources, only a few have been important commercially and at the present time *Hevea brasiliensis* stands preeminent. Wild trees were formerly the only source of rubber, but now cultivated *Hevea* trees, the so-called plantation rubber, furnish about 98 per cent of the supply. Rubber is the most recent of the major crops of the world. The industry is little more than 100 years old, and cultivation has been carried on only 60 years or so. In view of this, the increase in the production of plantation rubber from 800 long tons in 1900 to 305,000 tons in 1920 and 1,395,000 in 1940 must be considered as one of the greatest triumphs in modern agriculture. This great development of the rubber-growing industry has not been without its drawbacks, however. Overproduction has seriously affected the industry financially in recent years, and many attempts have been made at some sort of regulation. The British and Dutch, in particular, have tried to restrict production and exert other methods of control in Malaya, Java, Sumatra, and other plantation rubber centers within their empires. The recent successful development of synthetic substitutes for rubber, after many years of experimentation, may further tend to jeopardize the natural rubber industry. However, while these substitutes are superior for some purposes, such as the conduction of oil, natural rubber is still preferred for tires, which utilize three-quarters of the rubber output.

Latex occurs in special cells or in a series of special vessels which permeate the bark, leaves, and other soft parts of the tree. Usually only the latex from the lower part of the trunk is of importance commercially. Latex is a gummy white liquid full of minute globules. It is a varying mixture of water, hydrocarbons, resins, oils, proteins, acids, salts, sugar, and caoutchouc, the substance used as the source of rubber. The significance of latex to the plant is obscure. It is of some value in the healing of wounds, and it may serve for protection, nutrition, the transport of materials or as a fluid reservoir.

The properties of rubber have long been known. The primitive

Central American Indians were familiar with it, as were the Incas of Peru. Their word *cauchuc* has been altered to the present caoutchouc. Columbus was the first to report the existence of rubber to Europeans, but it was not introduced into Europe until 1735. The name "rubber" was first applied in 1770 by Priestley, owing to the fact that caoutchouc could be used for removing pencil marks. Mackintosh in 1823 developed a process for waterproofing cloth, but it was not until 1839, when the discovery of the vulcanizing process was made by Goodyear, that rubber really came into its own. From then on the rubber industry developed rapidly, and today rubber has a vast and constantly increasing number of uses and industrial applications. The properties which make it so valuable include its plasticity and elasticity, its resistance to abrasion and to electrical currents, and the fact that it is impermeable to liquids and gases.

The most important rubber-yielding species will be discussed in detail.

Hevea Rubber

The Hevea or Para rubber tree (*Hevea brasiliensis*) normally is the source of from 95 to 98 per cent of the rubber produced throughout the world. This tree is a native of the hot damp forests of the southern affluents of the Amazon River in South America. It has been estimated that there are over 300,000,000 trees in the vast area traversed by the Amazon and the Orinoco. Within this region the optimum conditions for its development are found—a uniform climate with a temperature range from 75 to 90°F. and a rainfall of from 80 to 120 in. The trees may reach an age of at least 200 years and may attain a height of 60 to 140 ft. The leaves are three-lobed, and the flowers are small and inconspicuous. The fruits (Fig. 67) have three seeds, which contain 23 to 32 per cent of a fatty oil. This oil is sometimes extracted and used as a drying oil. The resulting oil cake is rich in proteins and is used as a stock feed. Although rubber trees are natives of swampy floodlands, they grow best on deep, fertile, well-drained upland soils at an elevation of 600 to 1500 ft. above sea level. The latex occurs in a series of vessels in the cortex. Several other species of *Hevea* are tapped, including *Hevea Benthamiana*, which has a high-quality latex, but their production is insignificant compared with that of *Hevea brasiliensis*.

Wild Rubber. In the early days of the industry only wild rubber trees were available as a source of supply. The maximum production was reached in 1910 with an output of 83,000 long tons. Thereafter with the development of plantation rubber, wild rubber production steadily declined to as little as 8,500 tons in 1932. Since that time the output has fluctuated, depending on price and demand. Naturally during the recent wartime shortages wild rubber was again actively exploited.

The methods employed in obtaining and preparing the latex have changed but little over the years. The collection of the latex is in the hands of natives, called *seringuieros*, who are usually in the employ of absentee owners. Each seringuiero is responsible for a single tapping circuit which contains from 35 to 180 trees, about 2 or 3 to the acre. When a new tree is located, it is tapped at a point about 3 ft. above the ground by cutting several short downward 30-degree panellike incisions in the bark with a special knife. The cuts are sufficiently deep to sever the latex vessels but do not extend far enough to injure the cambium. Cups are placed below the incisions to receive the latex, which flows for

Fig. 67. Foliage and fruits of the Para rubber tree (*Hevea brasiliensis*). (*Photo by Walter H. Hodge.*)

several hours. The seringuiero makes the round of his trees each day. Successive tappings consist of entirely new incisions made slightly below the previous ones. The latex is collected from the cups and carried to the camp for coagulation. This is accomplished by coating a pole with latex and suspending it over a fire made of palm nuts or special woods. These yield a dense smoke containing acetic acid, creosote, and tars which coagulates the latex, forming a layer of crude rubber. The process is repeated until balls weighing 125 to 200 lb. are obtained. In some localities paddles are dipped in the latex and held in the smoke. The balls are shipped to mills for processing.

Plantation Rubber in the Eastern Hemisphere. In 1876, Henry Wickham took 70,000 *Hevea* seeds from the Amazon to England. These were grown at Kew, and enough seedlings were obtained to start a rubber industry in the eastern tropics, at first in British Malaya and Ceylon and later in Java, Sumatra, and other areas. There had been earlier attempts at rubber cultivation, but plantation rubber did not become permanently established until the beginning of the twentieth century. In 1910, when Amazon rubber production was at its peak, the East produced only 11,000 tons. Four years later, however, it had outstripped its rival, and by the outbreak of the Second World War it was responsible for 98 per cent of the world's output of 1,500,000 tons, with 9,000,000 acres under cultivation. The phenomenal development of plantation rubber in the East is due to many factors: favorable climate, freedom from disease, low costs, etc., and especially to an extensive research program. Every step in the production of the crop was carefully investigated, and the best methods of planting, cultivation, tapping, and coagulation were determined. A great increase in yield was made possible by a selection and breeding program. When bud grafting was found to be feasible, buds of high-yielding types were grafted on ordinary seedlings. The resulting clones differed widely, and only those with the most desirable characteristics were kept. The use of these "approved clones" began in 1925. By 1945, 10 per cent of the plantations had trees which were the result of this program, and they obtained a yield of 1500 lb. to the acre. The 90 per cent which still had a random mixture of trees had a yield of only 450 to 500 lb. Breeding experiments give much promise for the future. Where controlled crosses of high-yield clones have been made, yields of as much as 2000 lb. per acre have been obtained. Not all plantation rubber is produced on large estates. The acreage devoted to "native rubber," grown in small family gardens, is actually more than that of "estate rubber."

Plantation Rubber in the Western Hemisphere. After 1924, when export restrictions were first imposed on eastern rubber, the United States became seriously interested in establishing plantations in other parts of the world. Firestone selected Liberia, but Ford and Goodyear turned to tropical America, the native home of *Hevea*. In 1928 the 2,500,000-acre Fordlandia was started in Brazil, while Goodyear chose Panama and Costa Rica. All the American ventures failed after a promising start, owing primarily to the ravages of a leaf blight, *Dothidella Ulei*, to which plantation rubber in the Western Hemisphere was particularly susceptible. It soon became obvious that rubber could not succeed as a crop in the American tropics without adequate disease control. A research program was instituted which has had highly satisfactory results. An effective spray, consisting of copper and sulphur fungicides with various spreaders and stickers, has been developed and can be used on seedlings. Yield and resistance tests were made on thousands of individual wild trees, and the most desirable were selected for propagation and nursery stock. Most fortunate was the successful importation of hundreds of approved eastern clones from a Goodyear plantation in the Philippines. When these were budded on resistant stock from the jungles, desirable combinations of characters were obtained. Double budding has also proved very satisfactory. This consists of budding a high-yielding eastern clone onto a native root as soon as the plant is large enough. This is grown in a nursery under spray control until it is 6 to 10 ft. high. Then the top is budded with a disease-resistant strain.

FIG. 68. A plantation of Hevea rubber in lowland Honduras. (*Photo by Walter H. Hodge.*)

The resulting rubber tree has a native root system, a high-yielding eastern clonal trunk, and a disease-resistant crown. Long-range breeding programs are being continued, as well as extensive explorations in the jungle, in the hope of obtaining strains which combine disease resistance and high yield and so will obviate the expensive spraying and double-budding practices.

With the cooperation of the United States and the governments of the Latin-American countries involved, rubber is now being grown successfully in many areas (Fig. 68). Emphasis is being placed on small family plantings, rather than large plantations. The rubber trees are usually set out at intervals of 12 to 24 ft. in rows 20 ft. apart. Other crops are often planted between the rows. Normally from one-third to one-half of the acreage is resting; the balance is tapped on some alternating program. At each successive tapping a thin shaving of bark is sliced off the original panel until it nearly reaches the ground level. One-third, one-half, and full spirals are used.

After the latex has been collected, a little ammonia or some other anticoagulant is often added to keep it liquid until it reaches the mill, where it is concentrated, or made into *sheet rubber*. This is accomplished by cleaning the latex and pouring it into large pans; a coagulant, such as formic acid or acetic acid, is added, and in a few hours a mass of soft rubber results. Blocks of this are washed and run between rollers to form thin sheets. These are smoked and baled for shipment. Other types of crude rubber may be prepared. *Worm rubber* consists of irregular wormlike pieces cut from sheets with a pair of shears. *Crepe rubber* is made by passing washed and bleached coagulated latex through a creping machine which turns out long, thin, perforated strips of rubber. In the preparation of *sprayed rubber*, latex is dropped on whirling disks and little particles of rubber are thrown off. Any moisture quickly evaporates. This kind of rubber is exceedingly pure and clean.

A steadily increasing amount of liquid latex is exported. This requires immediate processing in the field and is feasible only for large plantations.

Castilla Rubber

Castilla or Panama rubber is obtained principally from *Castilla elastica*, a tall native tree of Mexico and Central America. It has been known, under many local names, since 1794 and was the chief source of rubber until 1850. Several other species of *Castilla* may be utilized, among them *C. Ulei* of the Amazon region, the source of caucho rubber. *C. elastica* grows in deep loamy soil on high ground and may reach a height of 150 ft. It requires a temperature above 60°F. and must have shade when young. Today, as in the past, the trees are all too frequently cut down in order to obtain the largest possible yield of latex. This wasteful method might lead to the extermination of the species, and more conservative methods have been recommended. The trees are tapped (Fig. 69) when eight to ten years of age, adult trees yielding up to 50 lb. of latex. This is coagulated with plant juices, alum, and by boiling or exposure to the air. The crude rubber is shipped in flat cakes. Castilla rubber was at one time extensively cultivated in Central America under plantation conditions, but it could not compete successfully with Hevea rubber and most of the plantations were abandoned. The surviving trees were the source of much of the emergency supply of Castilla rubber used during the Second World War.

Ceara Rubber

Ceara or manicoba rubber is obtained from *Manihot Glaziovii*, a small tree (Fig. 70) native to the desert regions of Brazil. It grows well in dry rocky ground, and so can be utilized in areas unsuitable for other types of rubber. The tree grows very rapidly, reaching its maximum height of 30 ft. in a very few years. Ceara rubber is now grown in Ceylon, India, and

FIG. 69. The base of a Panama rubber tree (*Castilla elastica*), showing incisions and methods of collecting the latex.

many other tropical countries. The trees are tapped when four or five years of age and yield a good grade of rubber. The latex is coagulated by exposure to air or smoke. The crude rubber is exported as blocks or flat cakes.

Assam Rubber

Assam rubber, or India rubber as it is more familiarly called, is obtained from *Ficus elastica*, a native tree of northern India and Malaya. The tree requires a hot climate and a large amount of rainfall. It often starts life as an epiphyte. It grows to a great height, developing huge buttresses or

Fig. 70. The Ceara rubber tree (*Manihot Glaziovii*) growing in Ceylon.

Fig. 71. An India rubber tree (*Ficus elastica*) in Colombo, Ceylon, showing the huge buttresses and prop roots.

prop roots (Fig. 71). Our familiar greenhouse plants resemble young wild plants. The roots as well as the stem are tapped. The latex is allowed to drip onto bamboo mats, where it coagulates. A large amount coagulates on the trunk as well. This crude rubber is scraped off, cleaned, and dried. The native methods of tapping the wild trees have been very wasteful, so the tree is now cultivated to some extent. The yield is low and the trees do not mature until about 50 years of age. Assam rubber is of low grade and is of little or no commercial importance at the present time.

Lagos Silk Rubber

Lagos silk rubber comes from *Funtumia elastica*, a large tree of tropical West Africa. It was discovered in 1894, but was immediately exploited by such wasteful methods that it has been nearly exterminated. In 1898 over 6,000,000 lb. were exported. The tree is now being cultivated and an attempt made to reestablish the industry. Plantations of Hevea rubber, however, are the chief source of rubber as far as West Africa is concerned.

Landolphia Rubber

Other former sources of African rubber were several woody climbers belonging to the genus *Landolphia*. The most important of these were *L. Kirkii* on the east coast and *L. Heudelotii* and *L. owariensis* on the west coast. These huge vines, sometimes 6 in. in diameter, were pulled down and cut into small pieces. The latex that exuded was coagulated with plant juices or the heat of the sun, sometimes even on the bodies of the native collectors. The most wasteful methods were employed, for a large immediate yield was all that was desired. The worst feature in connection with this particular rubber industry, however, was the barbarous treatment of the natives. The story of the operations in the Belgian Congo during the reign of Leopold II will always remain one of the blackest pages in history. Although there is still a considerable amount of wild landolphia rubber, there seems to be but little future for it, for the African natives are the most ignorant of all rubber collectors and cannot be taught proper methods of tapping. Furthermore cultivation is impracticable owing to the habit of the plant.

Guayule Rubber

The guayule (*Parthenium argentatum*), a native American species, has been utilized since 1910 as a minor source of rubber. This low semishrubby plant (Fig. 72) grows in the arid regions of Mexico and the southern United States. It was known to the early Indians, who obtained the caoutchouc by chewing the plants. Unlike the other sources of rubber, there is no latex present in the guayule plant, but little granules of caoutchouc are scattered all through the tissues. These are extracted mechanically or by means of solvents. The entire plant is pulled up and chopped up or ground with water. The particles of caoutchouc float to the surface and are removed. They are then dried and pressed into slabs. If the resinous materials are removed by solvents the product is as good as the best Hevea rubber. Guayule rubber is especially good for mixing with synthetic rubber substitutes. Guayule has been adapted to planta-

FIG. 72. A mature guayule plant (*Parthenium argentatum*), the only important source of rubber native to the United States.

tion culture, and it was extensively investigated as part of the Emergency Rubber Program. The yield has been increased, the life cycle shortened, and improved cultural, harvesting, and processing machinery developed. Production costs are still too high for it to compete with Hevea rubber save in an emergency. The importance of guayule for the future lies in the fact that it constitutes a living stockpile of rubber on land which otherwise would be idle.

Dandelion Rubber

The Russian dandelion (*Taraxacum kok-saghyz*), a close relative of the common dandelion, was discovered in 1931 by Russian botanists who were searching for new economic plants. Rubber is present in considerable quantities in latex tubes in the long tap roots. The yield has been increased by selection and breeding, and the plant is grown on some 2,000,000 acres in the Soviet Union. Seeds were brought to the United

States in 1942 and planted experimentally in 42 states. Vigorous plants with greater root weight and rubber content were segregated. Continued research may well bring about a yield of 400 to 500 lb. per acre. The chief advantage of the Russian dandelion is that it is a temperate zone plant and well adapted to the Northern states and Canada, where it can be harvested the year it is planted. Under the most favorable conditions it might be profitable to grow this species as a minor source of rubber. It is cultivated for this purpose in Argentina.

Minor Sources of Rubber

Many other species of plants have been experimented with as a possible source of rubber. Among these may be mentioned intisy (*Euphorbia*

FIG. 73. *Euphorbia Intisy*, a minor source of rubber, growing in the Atkins Garden of Harvard University in Cuba. (*Photo by Walter H. Hodge.*)

Intisy). This leafless shrub (Fig. 73) of the arid regions of Madagascar contains a latex that has had considerable use locally. The rubber coagulates on the surface of the plant in long elastic strands and is of high grade. Attempts have been made to cultivate intisy in the United States, for it is well adapted to the desert conditions of the Southwest.

Cryptostegia grandiflora and *C. madagascariensis*, ornamental woody climbers native to Madagascar but now found in the tropics and subtropics of both hemispheres, were widely heralded during the Second World War as a new source of rubber. These species have been grown for many years in India as the source of the high-grade palay rubber. They are abundant in Mexico, occurring in a wide variety of soils and climates. They seem to be the hardiest and fastest growing of all rubber plants, attaining harvesting age within six months. Apparently they afford considerable promise for the future, but actual attempts in Haiti to grow them on a large scale failed owing to the difficulties of extracting the rubber.

Other rubber-bearing plants of local importance include the mangabeira (*Hancornia speciosa*), a shrub or small tree of Bolivia, Brazil, and Paraguay; a species of *Micrandra* which yields the caura rubber of Venezuela; and various species of *Sapium* in Northern South America. A few native North American plants have a rubber-containing latex. Among those which have attracted some attention may be mentioned the desert milkweed (*Asclepias subulata*); the Indian hemp (*Apocynum cannabinum*); several goldenrods, particularly *Solidago Leavenworthii*, which was extensively investigated by Thomas A. Edison; species of the rabbit brush (*Chrysothamnus*), the source of chrysil rubber; and species of *Cnidoscolus*, from which chilte rubber is obtained.

Production and Use of Rubber

From 1938 to 1940 the world's trade in crude rubber averaged 1,085,420 tons, with 99 per cent coming from Southeastern Asia. British Malaya led in exports with 415,918 tons, followed by Indonesia with 402,745 and Ceylon with 66,421. Other important producing countries were Indo-China, Thailand, Borneo, India, and Burma. The United States is the greatest consumer of rubber, often using as much as the rest of the world combined. Great Britain, France, Italy, the U.S.S.R., and normally Germany and Japan are next in order.

Rubber is one of the most indispensable of plant products and it has a wider range of industrial uses than any other material. Over 75 per cent of the crude rubber is used for tires and inner tubes. Other uses include rubber boots and shoes; mechanical goods such as hose, tubes, and belting; waterproof clothing; druggist's supplies; insulated wire and other electrical goods; toys; machine packing; and cements. Hard rubber, which is prepared by vulcanizing crude rubber with 30 per cent sulphur, has many additional uses and is especially valuable for surgical appliances and telephone and radio parts. A new and promising use is in road construction.

GUTTA-PERCHA

Gutta-percha is a nonelastic rubber obtained from the grayish-white latex of several members of the *Sapotaceae*. The chief source is *Palaquium Gutta*, a tree of Malayan origin which now grows in Borneo,

Sumatra, the Philippine Islands, and other tropical countries as well. The latex is produced in sacs, which occur in the cortex, phloem, pith, and leaves. It is obtained by making incisions from which the milky juice runs out very slowly, or by felling the trees (Fig. 74). This latter is the more usual method. The bark is removed in strips 1 in. in width and 1 ft. apart, and the latex is caught in coconut shells or in palm or plantain leaves. The latex soon coagulates into grayish-yellow masses of a hard substance, which is odorless and heavier than water. This crude product contains several resins and other impurities, and is purified by washing in hot water. The whole mass is boiled and then kneaded

FIG. 74. A gutta-percha tree (*Palaquium Gutta*) felled and ringed. This is a very destructive method of obtaining the latex. (*Photo by the Philippine Bureau of Forestry.*)

into blocks, or it is chopped or sliced up and the pieces are washed, strained and kneaded, and then rolled into thin sheets. The value of gutta-percha depends on the amount of a hydrocarbon, gutta, that is present.

Gutta-percha is hard at ordinary temperatures. It deteriorates very rapidly in the air through oxidation and should be kept under water. It softens at 77°F., can be kneaded at 122°F., and melts at 248°F. Because it is an exceedingly poor conductor of electricity it is much used for insulation. No other material can replace gutta-percha and the similar balata in the construction of submarine cables, which require a substance that is resistant to salt water, pliable, and with just the right amount of rigidity. Other uses include splints, supports, pipes, golf balls, speaking tubes, telephone receivers, waterproofing, and adhesives. It is also utilized for protecting wounds and in dentistry. Although little known outside a limited circle, gutta-percha is indispensable in the world's work. It has been known since 1842.

BALATA

Balata is a nonelastic rubber that is obtained from the latex of *Manilkara bidentata*, formerly known as *Mimusops Balata*, and other species of the genus. *Manilkara bidentata* is a native of Trinidad and South America. It is a magnificent tree which grows to a height of over 100 ft. When mature, its purplish wood is very hard and durable and is much used for ties and building purposes under the name of bully wood or bulletwood. The fruit is edible. The latex is obtained by tapping the trees three times each year. It flows freely and readily coagulates in the air. A tree 3 ft. in circumference will yield from 50 to 100 lb. of dry balata. After coagulation it is cleaned and molded into cakes. Balata contains about 50 per cent of gum. It serves the same purpose in industry as does gutta-percha. It is particularly well adapted for machine beltings as it grips tightly and never stretches. It is also used as a substitute for chicle. Balata has been known since 1859. Unlike gutta-percha, it has never been cultivated and probably will always remain a wild crop, since no successful system of tapping has been devised without fatal injury to the tree.

The Amazon region produces several inferior types of balata, chief of which is abiurana or coquilana obtained from *Ecclinusa Balata*.

JELUTONG

Jelutong is obtained principally from *Dyera costulata* and related species. These Malayan trees have an astonishing flow of latex, greater than all the other latex species together. From 1910 to 1915 this material was exploited as a source of rubber. The latex, however, is full of gums, resins, and other impurities, and it yielded such a poor grade of rubber that it was soon discarded. Jelutong is now used chiefly as a substitute for chicle.

CHICLE

The sapodilla or naseberry (*Achras Zapota*) is a tall evergreen tree, a native of the Yucatan Peninsula. It is cultivated in tropical America and Florida for its edible fruit. The bark contains a latex, 20 to 25 per cent of which consists of a gutta-percha-like gum. This gum is known as chicle, and is the basis of the chewing-gum industry. It is also used in making surgical tape and dental supplies.

The most primitive method of obtaining the chicle is to tap the trunk and then scrape the thickened exudate from the bark. The crude or loaf chicle consists of pink or reddish-brown pieces mixed with 25 to 40 per cent of impurities. In southeastern Mexico and British Honduras, where the industry is carried on most extensively, the native collectors, or *chicleros*, are more careful. Zigzag gashes are cut in the trunk with machetes, up to a height of 30 ft. The latex runs to the base of the tree where it is collected in rubberized bags (Fig. 75), leaves, or even hollows in the earth. This accounts for the grains of sand that are sometimes found in chewing gum. The flow of latex lasts for several hours and the

Fig. 75. Collecting the latex from a sapodilla (*Achras Zapota*). This latex contains chicle, the basis of the chewing-gum industry. (*Photo by H. M. Heyder; courtesy of S. J. Record.*)

yield may be as much as 60 qt. In order to conserve the supply, plantations are now being established. These are not very practicable, however, for the trees can be tapped only every two or three years.

The hardened chicle is boiled, a process that requires considerable skill, for the chicle must be poured off when the moisture content reaches 33 per cent. It is then molded into blocks for shipment. Raw chicle contains resin, gutta, arabin, calcium, sugar, and various soluble salts. It is purified by being broken into small pieces, washed in a strong alkali, neutralized with sodium acid phosphate, washed again, and then dried and powdered. The final product is an amorphous pale-pink powder. This is insoluble in water and forms a very sticky mass when heated. As it ages, it is partially oxidized, turns brown, and becomes very brittle. The final steps in the manufacture of chewing gum involve cleaning, filtering, sterilizing, and compounding with various flavoring materials. As it comes from the final processing, 13 lb. of chicle make about 5000 pieces of chewing gum. A piece of gum usually contains about 15 per cent chicle, the remainder being chicle substitutes, sugar, and flavoring substances. Many attempts have been made to discover other natural or synthetic products that might be used as substitutes for chicle. Inferior latex from other sapodillas, balata, and jelutong are all utilized

FIG. 76. *Couma macrocarpa*, the source of sorva, a chicle substitute, showing the abundant flow of latex. (*Photo by Richard E. Schultes.*)

to some extent. Sorva or leche caspi, obtained from *Couma macrocarpa*, a large tree of the upper Amazon and its tributaries, was extensively exploited during the Second World War. The tree, which has a very abundant latex (Fig. 76), is cut down, ringed, and the latex is collected in cups made from palm leaflets. The latex is coagulated by boiling and is shipped in the form of large blocks. The white wood of this species is used for furniture, and the fruits have a mucilaginous but edible pulp.

The United States is the great chewing-gum nation and uses about the entire output of chicle. The supply is imported chiefly from British Honduras, Guatemala, and Mexico.

CHAPTER VII

GUMS AND RESINS

GUMS

The true gums are formed as the result of the disintegration of internal tissues, for the most part from the decomposition of cellulose, through a process known as gummosis. Gums contain a large amount of sugar and are closely allied to the pectins. They are colloidal in nature and soluble in water, either dissolving completely or swelling, but are insoluble in alcohol and ether. They exude naturally from the stems, or in response to wounding. The commercial gums reach the market in the form of dried exudations. Gums are especially common in plants of dry regions. They find their greatest use as adhesives, and are also used in printing and finishing textiles, as a sizing for paper, in the paint and candy industries, and as drugs. The three most important commercial gums are gum arabic, gum tragacanth, and karaya gum.

Gum Arabic. This is a dried gummy exudation obtained from *Acacia Senegal* and related species, small native trees of arid Northern Africa. They are extensively cultivated in the Sudan. The trees are tapped between February and May, when the fruits are ripe. Transverse incisions are made with a small ax (Fig. 77) and thin strips of the outer bark are torn off. The gum slowly exudes as a viscous liquid, collects in a drop, and hardens. After three to eight weeks these "tears" are collected. They are bleached by the sun, and the impurities are removed before shipping. Gum arabic was used by the Egyptians as early as 2000 B.C. Sudan gum has been an article of commerce since A.D. 100. Several kinds reach the world's markets. Kordofan or hashab gum is exported from the region around Cairo and Port Sudan, while Senegal gum comes from north of the Senegal River. Gum arabic is slowly and completely soluble in cold water and has a high degree of adhesiveness and viscosity. Most of it is used in the textile, mucilage, paste, polish, and confectionery industries, and as a glaze in painting. In medicine it is used as an emulsifying agent and as a demulcent.

Gum Tragacanth. This gum forms as a result of the transformation of the pith and medullary-ray cells into a mucilaginous substance that exudes naturally or after the bark has been punctured or excised. It comes from *Astragalus gummifer* and other species of the genus, thorny shrubs of the arid regions of Western Asia and Southeastern Europe.

The gum is allowed to dry on the bark before it is collected. It reaches the market in one of three forms: tears, which are the dried natural exudation; vermiform gum, which consists of narrow twisted coils or strings; and flakes, which are ribbonlike pieces. Most of the commercial supply comes from Iran and Turkey. Gum tragacanth is used in calico printing and for other industrial purposes. It is one of the oldest drugs and was known three centuries before the Christian era. In modern

FIG. 77. A native tearing away the bark of a gum-arabic tree (*Acacia Senegal*) so that the gum may exude. The gum-arabic industry is carried on in the Sudan and other regions of Northern Africa. (*Reproduced by permission from Toothaker, Commercial Raw Materials, Ginn and Company.*)

medicine it serves as an adhesive agent for pills and troches and for the suspension of insoluble powders.

Karaya Gum. This gum has become very important in recent years as a substitute for tragacanth, and several million pounds are imported annually from India. It is used in the textile, cosmetic, cigar, paste, and ice cream industries. Karaya gum is obtained from *Sterculia urens*, a large tree of central India. Incisions are made into the heartwood; the gum oozes into these and accumulates as large irregular knobs. These are collected, sorted, and graded. The gum enters the trade in the form of tears or a powder. It is known by a great variety of names: karaya, kadaya, katira, kuteera, katilo, kullo, India gum, and Sterculia gum.

Other Gums. Many other plants produce gums of some commercial importance. Gum ghatti, obtained from *Anogeissus latifolia*, a large tree native to India and Ceylon, is used as a substitute for gum arabic. The leaves are used for tanning. *Feronia Limonia* and *Cochlospermum religiosum* in India, Burma, and Java yield gums which are also used in place of gum arabic. The Asiatic *Cycas circinalis* is the source of cycas gum. The carob (*Ceratonia Siliqua*) produces tragasol, a mucilaginous hemicellulose occurring in the pods. In the United States, mesquite gum is obtained from *Prosopis juliflora*, *P. glandulosa*, and other species, while cherry gum comes from various species of *Prunus*.

RESINS

Resins represent oxidation products of various essential oils and are very complex and varied in their chemical composition. The resin is usually secreted in definite cavities or passages. It normally oozes out through the bark and hardens on exposure to the air. Usually tapping is necessary in order to obtain a sufficient amount to be of commercial value. Commercial resins are also often collected from fossil material. Resinous substances may occur alone or in combination with essential oils or gums. Unlike gums, resins are insoluble in water, but dissolve in ether, alcohol, and other solvents. Resin production is widespread in nature, but only a few families are commercially important. These include the *Anacardiaceae*, *Burseraceae*, *Dipterocarpaceae*, *Guttiferae*, *Hamamelidaceae*, *Leguminosae*, *Liliaceae*, *Pinaceae*, *Styracaceae*, and *Umbelliferae*. It is often difficult to trace the exact botanical origin of a resin, especially in the case of fossil and semifossil types.

Resins are probably of some service to the plant by preventing decay. This is due to their high antiseptic qualities. They may also tend to lessen the amount of water lost from the tissues.

Resins possess certain characteristics that render them of great importance in industry. Their ability to harden gradually, as the oil that they contain evaporates, makes possible commercial varnishes. The resins are dissolved in solvents, and surfaces are then painted with the mixture. As the solvents and oils evaporate, a thin waterproof layer of the resin is left. Resinous substances have been utilized for waterproof coatings, and also for decorative coatings, for ages. The ancient Egyptians varnished their mummy cases. The utilization of lacquer in the arts has been practiced in China and Japan for centuries. The Greeks and Romans were familiar with many of the same resinous materials that are in use today, as, for example, amber, mastic, and sandarac. Another property of resins that is of industrial importance is their ability to dissolve in alkalies to form soap. Resins are also used in medicine; for sizing paper; as a stiffening material for mats; in the preparation of sealing wax, incense, and perfumes; and for many other purposes.

The classification of resins is in a very chaotic condition. The same term is often used for very diverse materials. In the trade, resins are often referred to as gums, while such terms as varnish resins, hard resins, spirit varnishes, balsams, gum resins, damar resins, soft resins, and many others are used more or less indiscriminately. The chemical differences between the various groups are much more definite. For our purposes we shall distinguish three groups: the hard resins, oleoresins, and gum resins.

HARD RESINS

The hard resins contain only a little, if any, essential oil. They are usually solid, more or less transparent, brittle substances with no particular odor or taste. They are readily fusible and burn in air with a smoky flame. They are nonvolatile and are very poor conductors of electricity. When friction is applied, they become negatively electrified. The hard resins constitute the best source of varnishes, owing to their low oil content and the readiness with which they dissolve in alcohol. The most important commercial resins, such as the copals and damars, belong to this group. Hard resins are also used in paints, inks, plastics, sizing, adhesives, fireworks, and many other products.

Copals

The copals comprise a considerable group of resins of recent, semifossil, and fossil origin, which are found in many tropical and subtropical countries. The word "copal" is of Mexican origin. Many of the harder copals are known as animés, especially in England. The copals contain almost no oil, and yield a hard elastic varnish, which is much used for outdoor work. Several types are known which are quite diversified in character and source.

East African Copals. *Zanzibar copal* and the closely allied *Madagascar* and *Mozambique copals* are derived from *Trachylobium verrucosum*. Zanzibar copal is the hardest of all resins except amber and is very valuable. The resin exudes naturally from the trunk, branches, and fruit. Most of the commercial supply, however, is obtained from semifossil material, derived from still living trees, and fossil material from trees that no longer exist. This fossil copal, as it is dug from the ground, is covered with a crust of oxidized material. After this has been removed, the copal shows a characteristic surface, known as "goose skin," which consists of large and small excrescences. The interior is clear and transparent and varies from yellow to brownish red in color.

Inhambane copal, of little commercial importance, is obtained from *Copaifera conjugata*, a valuable timber tree of the Southeast African coast.

West African Copals. The West African copals include a considerable number of hard resins, and are usually designated by the name of the region from which they are obtained. The most important of these are the Congo, Angola, Sierra Leone, Accra, and Benin copals.

Congo copal is the most important of the West African copals, as it is very hard, and has been extensively exploited in recent years. It is derived from *Copaifera Demeusii* and *C. mopane*, which are characteristic trees of the Congo basin. Although living trees furnish some of the supply, the greater part is obtained from the ground or from watercourses, and is more or less fossil in nature. The white and red *Angola copals* are derived from the same two species.

Sierra Leone copal is a light-yellow, hard, and brittle resin that is obtained from *Copaifera copallifera* and *C. Salikounda*. Living trees are wounded and the resin exudes and hardens in the form of globular tears. Some fossil material is also obtained. Because of the value of this copal, the trees are protected by the Sierra Leone government.

Accra and *Benin copals* are probably derived from *Daniella Ogea* and related species. These large trees are found in the coastal forests of Liberia, the Gold Coast, and Nigeria. The resinous exudation, known locally as ogea gum, is becoming increasingly important as a varnish resin.

Kauri Copal. Kauri copal or kauri gum is one of the most valuable of the hard resins. It is obtained from the kauri pine (*Agathis australis*), which is New Zealand's largest and most important tree (Fig. 78). The copal is chiefly fossil in nature and is dug up on ridges and in swamps and boggy ground. "Swamp gum" furnishes the bulk of the supply, and ranges in size from pieces 1 or 2 in. in diameter to lumps weighing 100 lb. "Range gum" yields the best grade of kauri. An inferior "bush gum" is obtained by tapping living trees. Kauri is yellow, transparent, and very hard. It is an exceedingly valuable varnish resin, especially for marine and outside work. It is also used in making linoleum. Kauri constitutes one of the chief exports of New Zealand, and the industry is controlled by the government.

Manila Copal. The first shipments of this important copal were made from Manila and the name has persisted, although today 75 per cent of the product is shipped from Indonesia. The source of all the East Indian, Philippine, and Malayan copals, of which there are many different kinds, is *Agathis alba*. This tall conifer, which reaches a height of 200 ft., is a characteristic tree on high ground. The resin exudes naturally, and is also obtained by systematic tapping. Some of the supply is derived from fossil material and consists of large, irregular, angular, milky pieces with a yellowish interior. Hard, semihard, and soft copals are included among the many different kinds in the trade. Pontianak copal, a semifossil type from Borneo, is the hardest variety and is especially popular in the United States. Manila copal varnishes are durable, but they do not adhere strongly to surfaces and they are not very brilliant. They are used chiefly for interior work and enamels. These copals are often erroneously called damars, although they are quite distinct from the true damars.

South American Copals. The South American locust (*Hymenaea Courbaril*), a tall tree of Brazil and other parts of tropical America, is the

FIG. 78. The kauri pine (*Agathis australis*), New Zealand's largest and most important tree, is the source of kauri copal, one of the most valuable of the hard resins.

chief source of the South American copal. The stems, twigs, and even the fruits exude a large amount of resin, which trickles to the ground. The commercial resin is collected from the base of living trees, and former trees as well, and is marketed as Demerara, or Para copal. It is the softest of all copals, and consequently the least valuable.

Damars

Considerable confusion exists in regard to the application of the term "damar." The word is of Malayan origin, and is used by the natives to indicate a torch made of decayed wood and bark, mixed with oil and powdered resin, wrapped in leaves, and bound with strips of rattan. Originally it did not refer to any specific tree or resin. Gradually the word came to be a collective term for a great variety of hard resins of quite different origin, and included even kauri and manila copal. In the commercial trade, however, the term "damar" is practically restricted to resins that are obtained from members of the *Dipterocarpaceae*. A few resins from species of the *Burseraceae* are included. This distinction should be maintained, for the true damars are very different chemically from the various coniferous resins. For example, unlike Manila copal, damar is insoluble in chloral hydrate, but completely soluble in alcohol and turpentine.

The trees that yield damars (Fig. 79) are characteristic of all Southeastern Asia, and are particularly abundant in Malaya and Sumatra. Although all members of the *Dipterocarpaceae* secrete resin, comparatively few species are of commercial importance. These are found chiefly in the genera *Balanocarpus*, *Hopea*, and *Shorea*. Damars are especially important in Malaya and are obtained by tapping the trees. The most important Malayan varieties are Damar Mata Kuching from *Hopea micrantha* and related species, Damar Penak from *Balanocarpus Heimii*, and Damar Temak from *Shorea hypochra*. The principal damars of India are sal damar from *Shorea robusta*, white damar from *Vateria indica*, and black damar from *Canarium strictum*. Damars are also produced in Borneo, Java, Sumatra, Thailand, and Cochin China. The most outstanding commercial variety of damar is the so-called Batavian damar, a product of *Shorea Wiesneri*, a species found in Java and Sumatra.

FIG. 79. A dipterocarpus tree boxed for gathering the resin known commercially as damar. (*Photo by the Philippine Bureau of Forestry.*)

The so-called East India resins are products of the same trees that yield damars. They are older and harder and are often gathered from the ground or from watercourses.

Damars are used chiefly in spirit varnishes and the manufacture of nitrocellulose lacquers. Damar varnishes adhere better than Manila copal varnishes, but are softer and less durable. They are particularly well adapted for varnishing paper because of their luster and light color. They are also used for indoor work and in histology.

Amber

Amber is a fossil resin found chiefly along the shores of the Baltic Sea. The principal source of Baltic amber was the now extinct pine, *Pinus succinifera*, a species that flourished on the shores of a former sea in Eocene time. Amber is an exceedingly hard and brittle substance. It occurs in several forms, the most important of which is succinite. Some of these forms are transparent and others are almost opaque. The color varies from yellow to brown and even black. When rubbed, amber takes on a high polish and becomes negatively electrified. It also gives off a characteristic aromatic odor. Amber has been known for thousands of years. The Swiss Lake Dwellers were familiar with it, and it was

FIG. 80. Oriental amber carvings and a block of unpolished amber. Amber is a fossil resin, secreted chiefly by the now extinct *Pinus succinifera*. (*Courtesy of the Botanical Museum of Harvard University.*)

highly prized by the Greeks and Romans. It has always been used for beads and other ornamental purposes, and is often carved (Fig. 80). Today the chief use of amber is for the mouthpieces of pipes and holders for cigars and cigarettes. The darker grades yield a valuable varnish, but it is too expensive to be used much. Amber is also used to increase the elasticity of rayon fibers and as the source of an essential oil. Scientifically amber is of interest for there are often found imbedded in it the remains of plants, insects, and other objects which existed at the time the fresh resin was exuded from the pines.

Lacquer

Lacquer is a natural varnish that is exuded from various Asiatic trees, and enormous quantities of it are used in oriental countries for ornamental purposes. The principal source is the lacquer tree (*Rhus verniciflua*), a native of China, but long cultivated in Japan. The trees are carefully cultivated and systematically tapped. The exudation is a milky liquid which darkens and thickens rapidly on exposure. It can be kept unchanged, however, for long periods by storing in closed containers. Before use it is filtered. When applied as a varnish, the thin film rapidly hardens in a moist atmosphere, owing in part to oxidation. Lacquer affords a remarkable protection as it is unchanged by acids, alkalies, alcohol, or heat up to 160°F. When pigments are used, they are mixed with the lacquer before drying.

The art of lacquering originated in China centuries before the beginning of the Christian era, and reached its highest development in that country during the Ming dynasty (A.D. 1368–1644). In Japan the first records go back to the fourth century, when lacquer was used for many purposes.

FIG. 81. Gold and cinnabar lacquer, the work of Japanese artists. (*Courtesy of the Botanical Museum of Harvard University.*)

The earliest specimens extant belong to the sixth century. The art reached its height during the seventeenth century, though much fine work was produced as late as the nineteenth century. The Japanese have outstripped their predecessors as regards the excellence of the products (Fig. 81) in all fields except the carving of lacquer. They have been especially skilled in the use of gold as a color. The process of lacquering is a very complicated and tedious one. In some cases from 300 to 400 coats are applied and the whole operation requires several years for completion. The technique was kept a secret for many years.

Burmese lacquer is obtained from *Melanorrhoea usitata*. It dries more slowly than Japanese lacquer, but has been much used in recent years in an attempt to build up a native lacquer industry in Burma.

Natural lacquers are also obtained in Formosa and Indo-China. In the latter country *Rhus succedanea* is the source.

Shellac

Although not strictly a plant product, shellac deserves some consideration. It is prepared from stick-lac, a resinous substance secreted on the twigs of many trees by an insect, *Tachardia lacca*. The lac insect derives its food from the sap of the trees and secretes the resin as a sort of cocoon for the protection of itself and its young. Although some 40 species may serve as hosts for the insect, only seven are important, and these are often cultivated. They include *Butea monosperma, Schleichera oleosa, Zizyphus xylopyrus, Ficus religiosa, Acacia nilotica, Cajanus Cajan*, and *Zizyphus Jujuba*. *Butea monosperma* was used as a host as early as A.D. 250. At first a valuable red dye which was obtained from the insect was the only desired product. Since A.D. 1590 the resinous excretions have been more important. Most of the shellac of commerce is prepared by native workers. India furnishes over 97 per cent of the total output, the remainder coming from Burma, Thailand, and Indo-China. The crude stick-lac is removed from the twigs and soaked to extract the red dye. It is then dried and powdered to a granular consistency. This seed-lac is melted and thin sheets of it are allowed to harden. These are broken up into the semitransparent, brittle, orange-red flakes which constitute shell-lac. If the melted seed-lac is poured out in drops, it hardens into the thick round pieces known as button-lac. Shellac is often bleached.

Shellac has many industrial uses. It can be molded readily and is the most satisfactory material for the manufacture of phonograph records. It is a high-grade insulator and is extensively used in the electrical industry. It is the principal spirit-varnish resin yielding a tough film with a smooth finish, which is capable of a high polish. Shellac varnishes cannot be used for outside work for they are not water resistant. Shellac is also used in making sealing wax, drawing inks, some water colors, and nitrocellulose lacquers; for sizing papers; for stiffening felt hats; and, in India, for numerous ornamental purposes.

Acaroid Resins

The acaroid or grass-tree resins are obtained from Australian trees belonging to the genus *Xanthorrhoea*. These plants are among the few monocotyledons, other than palms, which have an arboreal habit. They consist of a short woody stem, composed of the old leaf bases, surmounted by a tuft of long rushlike leaves. The resin collects around the bases of the old leaves and is removed by beating the stem. The yellow acaroid resins reach the market in the form of elongated or round reddish-brown pieces. They are obtained chiefly from *Xanthorrhoea hastilis*. Red acaroid resins from *X. tateana, X. australis*, and allied species are much more common. They consist of uneven pieces of a brownish color. The acaroid resins are used in making sealing wax, gold size, and spirit varnishes for use in coating metals; as a substitute for rosin in paper sizing and inks; as a mahogany stain; as a source of picric acid; and in medicine.

Sandarac

Sandarac is a soft pale-yellow resin obtained chiefly from *Callitris quadrivalis* (or *Tetraclinis articulata*, as it should be called), a small tree of Northern Africa. Australian species of *Callitris* are also a source of sandarac. The resin is formed between the inner and outer layers of the bark and is excreted in the form of small tears, which quickly become opaque. Sandarac yields a hard, white, rather brittle spirit varnish, especially useful for coating labels, negatives, cardboard, leather, and metals. It was formerly used in medicine and was well known to the older civilizations.

Mastic

Mastic is a very old resin and was known at least 400 years before the Christian era. The most useful variety is Chios mastic, derived from *Pistacia lentiscus*, a small tree of the Mediterranean region. Although the resin exudes naturally, the flow is aided by removing strips of bark. Some of the resin adheres to the trunk in the form of long, ovoid, pale-yellow, brittle tears, while the remainder falls to the ground. Bombay mastic, which consists of large irregular pieces of a dull, milky color, comes from *P. cabulica*. Mastic yields a pale varnish used for coating metals and pictures, both oils and watercolors. It is also used in lithographic work, in perfumery, in medicine, and as a cement for dental work. It is one of the most expensive and high-grade resins.

Dragon's Blood

Dragon's blood includes various deep red substances of a resinous nature. Sumatra dragon's blood is obtained from *Daemonorops Draco*, a climbing rattan palm of Eastern Asia. The dark reddish-brown resin occurs as small granules on the scaly fruits. It is used chiefly in the manufacture of red spirit varnishes for metals and in making zinc line engravings. During the eighteenth century the great Italian violin makers used dragon's blood in their varnishes. Socotra dragon's blood, a resin that exudes from the stem of *Dracaena cinnabari* of Western Asia, is also used to some extent for varnishes, dyes, and stains. Dragon's blood is sometimes obtained from tropical American species of *Dracaena* and other Asiatic species of *Daemonorops*.

Kinos

Kinos, or gum kinos as they are usually called, are derived from several sources. Malabar kino consists of the dried juice of *Pterocarpus Marsupium*, a large Indian tree. The trees are tapped and the juice is boiled down. It reaches the market in the form of small, brownish-red, brittle pieces. West African kino is a red resin from *P. erinaceus*. Bengal kino comes from *Butea monosperma*. Several Australian species of *Eucalyptus* are important sources of gum kino. The principal species is the red gum (*Eucalyptus camaldulensis*). The kino is secreted in cavities between the wood and bark, and oozes out after incisions have been made. In the air the resin hardens into a solid reddish mass. Several tropical American trees, chiefly *Dipteryx odorata* and *Coccoloba uvifera*, also yield kinos. Kinos find their chief use in medicine for throat troubles, and are used to some extent in tanning.

OLEORESINS

The oleoresins contain a considerable amount of essential oils in addition to the resinous materials, and consequently they are more or less liquid in nature. They have a distinct aroma and flavor. Among the oleoresins are included the turpentines, balsams, and elemis. The distinctions between these groups are very slight and there is often a confusion of terms.

TURPENTINES

Turpentines are oleoresins obtained almost exclusively from coniferous trees. They are viscous, honeylike liquids or soft and brittle solids. The resin is secreted and stored in ducts near the cambium layer and exudes naturally as a soft, sticky substance, often called pitch. For commercial use crude turpentine is obtained by tapping the trees. On distillation turpentines yield the essential oil or spirits of turpentine, and rosin, both of which are exceedingly useful products, around which an important industry has been built up. This turpentine or naval-stores industry is one of the oldest of the forest industries. The Trojans and Greeks were familiar with pitch and its uses, and it is mentioned in the Bible. Today the industry is valued at from $35,000,000 to $40,000,000 annually. The United States leads in production with one-half of the total output. Turpentine and rosin are also produced in many European countries, and even in India and Indo-China.

The Turpentine Industry in the United States

The history of the turpentine industry in the United States is closely identified with the economic development of the South. Tar and pitch were among the earliest exports of the country, and the industry was practically the only source of livelihood in the early days of the Carolinas. At first the products were used chiefly in connection with sailing vessels, a fact that gave the name "naval stores" to the industry. This name still holds, although today turpentine and rosin are the products.

The most important source of the crude turpentine used in the naval-stores industry is the longleaf pine (*Pinus australis*). This species (Fig.

82) furnishes 90 per cent of the raw material. The Cuban or slash pine (*P. caribaea*) is used to some extent. The western yellow pine (*P. ponderosa*) and other western species are potential sources. The industry is carried on chiefly in the eight coastal states from North Carolina to Texas, with Georgia the present center of production. The industry is waning at the present time, owing partly to the depletion of the virgin timber supply and partly to the lack of demand for the products.

Attempts are being made to conserve the supply by reforestation and better methods of tapping. In the best year, 1909, over 750,000 bbl. of

FIG. 82. A virgin stand of longleaf pine (*Pinus australis*). This species is the chief source of turpentine and rosin. (*Photo by U.S. Forest Service.*)

turpentine and over 2,500,000 lb. of rosin were produced. Normally about half the product is exported, chiefly to Europe.

Turpentine oozes out from the resin canals after the cambium layer has been exposed by a cut, and at the same time the development of new ducts above the cut is stimulated. In the species used commercially the ducts are exceedingly large, and there is a correspondingly heavy production of the oleoresin. Turpentine stands usually contain from 50 to 200 trees per acre. The original methods used for collecting the resin were very wasteful. Cavities, known as boxes, were cut near the base of the tree during the winter. Later, triangular pieces of bark and wood just above the corners of the box were removed. This "cornering" enabled the resin to flow into the box more freely, but it was injurious to the tree. The resin was removed from the boxes every three weeks.

At the present time several types of cup-and-gutter systems are in use. Basal incisions are made in the trunk at a 20-degree angle, and metal gutters or aprons are slipped into the cut. These guide the resin into a metal cup or some other type of container. A strip of bark and wood, known as the advance streak, is removed just above the gutter. This stimulates the flow of resin and induces the formation of new ducts. At regular intervals a narrow strip is removed above the advance streak. This practice of "chipping" assures a continuous flow of gum. The wounded area resulting from this periodic chipping is called a "face."

Fig. 83. Collecting the resin from a longleaf pine which has been tapped by the cup-and-gutter method.

The oleoresin that collects in the cup is called the "dip." The cups are emptied about every four weeks (Fig. 83), and the contents are transported to stills. Some of the oleoresin solidifies as it runs down the face. This "scrape" is removed at intervals during the season or only at the end. Upon distillation it yields a much smaller amount of turpentine and rosin than does the dip.

The distillation of the turpentine is carried on in copper stills (Fig. 84), the process lasting from 2 to 3 hours. The distillate is collected in barrels, where the *oil of turpentine* rises to the top and is run off for storage. The residue, which is *rosin*, while still hot, is run through a series of screens to remove any impurities, and then into a cooling vat. When cool, it is transferred to slack barrels where it completely hardens within 24 hours. Not all the turpentine is removed during distillation as this yields a better grade of rosin.

Several central processing plants are now in operation. These utilize an efficient method of washing and cleaning the resin before distillation, resulting in a better quality and higher yield of rosin and more economical operation.

The production of an inferior grade of naval stores as the result of the destructive distillation of pine wood has already been discussed. Considerable turpentine, rosin, and pine oils are obtained from old pine stumps and logging waste by a steam and solvent process. Some turpentine, known as sulphate turpentine, and tall oil, a liquid rosin, are recovered as by-products of the sulphate pulp industry.

The oil of turpentine, usually referred to as "turpentine" or "spirits of turpentine," has many uses. It is of major importance in the paint and varnish industry, where it has a thinning action, due to its properties as

FIG. 84. A turpentine distillery. The crude resin yields oil of turpentine and rosin as products.

a solvent. It is used in connection with the printing of cloth, particularly cotton and woolen; as a solvent for rubber and gutta-percha; in medicine; and in the manufacture of many chemicals.

Rosin or colophony, a brittle, friable, faintly aromatic solid, is even more important in industry. It is used in the manufacture of soap, varnishes, paints, oilcloth, linoleum, sealing wax, printer's ink, roofing and floor coverings, adhesives, plastics, rubber, drugs, and various chemicals. It serves as the chief sizing material for paper, and constitutes the brewer's pitch used for lining beer barrels. Rosin oils are utilized as greases, lubricants, and solvents.

The Turpentine Industry in Foreign Countries

The naval-stores industry is highly developed in France, where *Pinus Pinaster*, known in that country as the maritime pine (*P. maritima*), is cultivated and tapped for its turpentine. The products are of the highest quality and are preferred to those produced in the United States. Spain is the third largest producer of naval stores and is becoming of increasing importance. Four species of pine are tapped, *P. Pinaster, P. halepensis, P. nigra,* and *P. Pinea.* Most of the forests are under government control. In Portugal the industry is just becoming prominent and vast areas are as yet untouched. *P. Pinaster* and *P. Pinea* are the sources. Greece is also increasing in importance as a producer of naval stores, utilizing *P. halepensis.* In the U.S.S.R., Poland, and Germany inferior grades of rosin and turpentine are obtained from *P. sylvestris.*

The turpentine industry is now being developed in many parts of Asia, particularly Indo-China, Indonesia, and India. In India several species of pine are available, and also the only non-coniferous tree that is a source of turpentine and rosin. This is the Indian frankincense (*Boswellia serrata*). The great distances and lack of transportation facilities have handicapped the industry in India, but in spite of all obstacles it continues to expand.

Turpentines of Minor Importance

Crude turpentines from various species of conifers are often used in their natural state for purposes other than the production of naval stores. Some of them have been important in industry, in the arts, and especially in medicine for over four centuries. Among the best known may be mentioned:

Canada Balsam. This oleoresin is a true turpentine, rather than a balsam, as the name would indicate. It is secreted by the balsam fir (*Abies balsamea*) of the northern United States and Canada. The resin collects in elongated blisters on the bark and only small amounts are obtainable. It is estimated that a tree will yield from 8 to 10 oz. a year. Collectors use a pot with a spout cut at an angle. This is forced into the blisters and held in place while the balsam drains out. The balsam is a viscid yellowish or greenish substance. Its chief use is as a mounting medium for microscopic work and as a cement for optical lenses. It is very transparent and has a high refractive index, which results in a minimum dispersal of light. The medicinal value of Canada balsam was recognized as early as 1607. It is used as an irritant, stimulant, and antiseptic, and is a component of collodion and many plasters. It is also utilized as a fixative for soap and perfumes.

Oregon balsam, obtained from the Douglas fir, has similar properties and uses.

Spruce Gum. Spruce gum is the natural exudation of the various spruces of the northern United States and Canada, with *Picea rubens* the chief source. It is usually the result of an injury to the sapwood. The thin, clear, bitter, sticky oleoresin is secreted in blisterlike cavities (Fig.

85) in the bark or in longitudinal fissures in the wood. It hardens on exposure to the air and is collected when hard or semisoft. Before the advent of chewing gum about 500,000 lb. were used each year as a masticatory. The gum softens in the mouth and assumes a reddish color. It has a pleasing resinous taste.

Venetian Turpentine. Venetian turpentine is obtained from the European larch (*Larix decidua*), a common tree in the mountains of Central Europe. Unlike all other conifers, the resin ducts are located in the heart of the tree so that holes must be bored in order to obtain the resin. The trees are tapped in the spring. Venetian turpentine has been an

FIG. 85. A red spruce (*Picea rubens*) in Maine, showing the spruce gum oozing from fissures in the bark. (*Photo by V. C. Isola; courtesy of S. J. Record.*)

important product since the middle of the eighteenth century. It is a yellowish or greenish liquid with a characteristic odor and taste. It is used in varnishes, histology, lithographic work, and veterinary medicine.

Other crude turpentines of less importance include Bordeaux turpentine from *Pinus Pinaster*, Strasbourg turpentine from *Abies alba*, and Jura turpentine from *Picea Abies*. In the case of the first two of these European turpentines, the crude exudation is used to some extent, but more often it is strained and filtered through cloth. The residue, known as Burgundy pitch, is a stimulant and counterirritant, and is used in plasters, ointments, and other pharmaceutical preparations.

BALSAMS

Technically balsams are oleoresins that contain benzoic or cinnamic acid and so are highly aromatic. The term "balsam," however, is often applied erroneously to quite different substances, such as Canada balsam, copaiba balsam, etc. The true balsams contain much less oil than the turpentines and are more or less viscous substances. They yield essential oils on distillation. Balsams are used in medicine and as fixatives in the perfume industry.

Balsam of Peru

Balsam of Peru is obtained from *Myroxylon Pereirae* (Fig. 86), a tall tree of Central America which is cultivated in many tropical and subtropical countries. The wood resembles mahogany and is quite valuable.

FIG. 86. Balsam trees (*Myroxylon Pereirae*), the source of balsam of Peru, in dry forests in El Salvador. (*Photo by Walter H. Hodge.*)

The balsam is a dark, reddish-brown, thick, syrupy, viscous liquid, and is a pathological product formed as a result of wounding the tree. The trunks are beaten with a blunt instrument and the injured bark is then charred. It soon falls off naturally or is removed. Sometimes small "windows" are cut in the bark. The balsam exudes from the exposed surface of the wood and is collected on rags. These rags are later boiled to free the balsam, which is purified. Balsam of Peru is used in medicine for healing slow wounds and skin diseases. During the Second World War it was used in field dressing stations for quick applications of a protective covering to the surface of wounds. It is also used in the treatment of coughs, bronchitis, and similar ailments, because of its stimulative and antiseptic effect on the mucous membranes. It is utilized in perfumes as a fixative for the heavier odors and has served as a substitute for vanilla. Its common name is a misnomer, for the tree does not grow in Peru   The United States imports its supply chiefly from San Salvador.

Balsam of Tolu

Balsam of Tolu is a pathological product obtained from *Myroxylon Balsamum*, a tree of Venezuela, Colombia, and Peru. V-shaped incisions are made in the trunk and the balsam slowly exudes and is collected in gourds. It is a brown or yellowish-brown plastic substance with a pleasant aromatic taste and odor. It is used for salves and ointments and as an expectorant and antiseptic in the treatment of coughs, colds, and bronchitis. It is sometimes used to flavor cough syrups. Considerable amounts are used as fixatives in the perfume and soap industries.

Styrax

Two varieties of styrax or storax enter into commerce. The most important type, which is the styrax of antiquity, is known as *Levant styrax*. This is obtained from *Liquidambar orientalis*, a small tree common along the coasts of southwestern Asia Minor. The balsam is a pathological product formed as a result of wound stimulation. The outer bark is bruised and soon the balsam exudes into the inner bark. The outer layers are discarded and the balsam is recovered by boiling the inner layers in sea water. The residual bark is dried and used for fumigating. Styrax is a semiliquid, grayish-brown, sticky, opaque substance with a pronounced aromatic odor. It is used in soaps and cosmetics; as a fixative for the heavy "oriental" type of perfume; in adhesives, lacquers, and incense; as a flavoring for tobacco; and in medicine as a stimulant to the mucous membranes and for the treatment of scabies.

American styrax is obtained from the sweet gum (*Liquidambar Styraciflua*). Although this tree ranges from New England to Mexico and Central America, the commercial supply of styrax comes only from Guatemala, Honduras, and Venezuela. This balsam is a thick, clear, brownish-yellow, semisolid or solid substance and has the same uses as the Levant styrax.

Benzoin

Benzoin is a solid balsam and is often classed as a resin or balsamic resin. It is a pathological product obtained from various species of *Styrax*, found in Southeastern Asia and the East Indies. The balsam oozes out from incisions made in the trunk and branches. Two varieties are known in commerce. *Siam benzoin* comes from *Styrax tonkinense* and *S. benzoides* and occurs as yellowish or brownish pebblelike tears with a milky white center. The tears are hard and brittle at ordinary temperatures and occur separately or adhere together only slightly. *Sumatra benzoin*, from *Styrax Benzoin*, occurs in reddish or grayish-brown blocks or lumps, composed of masses of tears stuck together. Benzoin is exceedingly aromatic with a vanillalike odor. It is used in medicine as a stimulant and expectorant; and in the preparation of heavy sweet perfumes, soap, toilet waters, lotions, tooth powders, incense, and fumigating materials.

OTHER OLEORESINS

There are numerous oleoresins that do not belong to either the turpentines or balsams. Among the more important may be mentioned copaiba, elemi, and Mecca balsam.

Copaiba

Copaiba, known also as copaiba balsam or copaiva, is a natural oleoresin obtained from several species of *Copaifera* native to tropical South America. The copaiba trees are small (Fig. 87) with strong, tough, durable wood, which contains resin ducts of large size. So much oleoresin is secreted and stored in cavities that sometimes the pressure causes the trunk to burst open. Copaiba is obtained by boring holes into the heartwood. The secretion flows out very rapidly. It is a thin, clear, colorless liquid at first, but turns yellow and becomes more viscid with age. It has a peculiar aromatic odor and a persistent bitter taste. Several commercial varieties occur, which differ in the amount of resins and essential oils present as well as in the source. The most important of these are Maracaibo copaiba from Venezuela, obtained from *Copaifera officinalis*, which contains considerable resin and is rather thick; and Para copaiba, a very fluid grade from Brazil, obtained from *C. reticulata*. Copaiba is used in making varnishes, lacquers, and tracing paper, as a fixative in scenting soaps and perfumes, and in photography to emphasize half tones and shadows. In medicine it is used as a disinfectant, laxative, diuretic, and mild stimulant. Long used by the native Indians for its healing properties, copaiba became known to Europeans early in the sixteenth century.

FIG. 87. *Copaifera officinalis*, a source of copaiba balsam. (*Photo by C. D. Mell; courtesy of S. J. Record.*)

There are many substitutes for copaiba, chief among which are gurjun balsam and illurin balsam.

Gurjun balsam is obtained from *Dipterocarpus turbinatus* and related species of India and the East Indies. The thick, opaque, grayish oleoresin is obtained by cutting holes in the trunk.

Fig. 88. *Daniella thurifera*, from which illurin balsam is obtained, growing in the Botanical Garden at Roseau, Dominica, B.W.I. (*Photo by Walter H. Hodge.*)

Illurin balsam or African copaiba, a very fragrant oleoresin with a thick, pungent, pepperlike odor, comes from a West African tree, *Daniella thurifera* (Fig. 88). This species, sometimes known as Sierra Leone frankincense, is characteristic of the drier open forests and savannahs of Upper Guinea. The resin exudes in a copious flow from the base of the trees. A similar product comes from *D. Oliveri*, a species of the denser moist forests. This oleoresin, under the name of wood oil, is one of the chief products of Nigeria.

Elemi

The term "elemi" is used as a collective name for several oleoresins of different origin; it is also used erroneously as a synonym for some of the softer copals. Elemis differ considerably in their characteristics. These oleoresins exude as clear pale liquids, but they tend to harden on exposure. Some remain soft, while others become quite hard.

Manila elemi is the most important and the best known of the elemis. Its source is the pili tree (*Canarium luzonicum*) of the Philippine Islands. The oleoresin is secreted in the bark and oozes from the trunk in fragrant white masses. The natives use elemi for torches and for caulking their boats. Manila elemi is used in lithographic work and the manufacture of inks, adhesives, and cements; in the varnish industry to give toughness and elasticity to the products; in perfumes; and in medicine in plasters and ointments.

Less important varieties include African elemi from *Boswellia Frereana;* Mexican elemi from *Amyris elemifera, A. balsamifera,* and related species; and Brazilian elemi from *Protium heptaphyllum, Bursera gummifera,* and several other trees.

Mecca Balsam

Mecca balsam is a greenish turbid oleoresin with an odor of rosemary. It is obtained from *Commiphora Opobalsamum*, an Arabian species. This material has long been used in incense and for perfumes of an oriental type. It has some medicinal value. The supply of Mecca balsam is limited and consequently it is a rare and costly product.

GUM RESINS

Gum resins, as the name indicates, are mixtures of both true gums and resins and naturally combine the characteristics of both groups. They often contain small amounts of essential oils and traces of coloring matter. Gum resins occur naturally as milky exudations and collect in the form of tears or irregular masses. They are also obtained by injuring or tapping the plants. They are produced for the most part by plants of dry, arid regions, especially species of the *Umbelliferae* and *Burseraceae*. Three of the umbelliferous species, the sources of ammoniacum, asafetida, and galbanum, are very common in Iran and Afghanistan and furnish the characteristic aspects of the vegetation of the plains and steppes in those countries. During the dry season these plains are barren, but shortly after the rainy season sets in these plants send up thick stems from their perennial rootstalks. When fully grown the plants attain a height of 5 or 6 ft., and are so abundant that they form a sort of open forest.

Ammoniacum

Ammoniacum is obtained from *Dorema Ammoniacum*, a tall, stout, naked, hollow-stemmed perennial found in the deserts of Iran, southern Siberia, and other parts of Western Asia. It has a milky juice which exudes from the stem and flowering branches and hardens into tears. Insect injury often causes the exudation to occur. The brownish-yellow tears are hard and brittle and occur singly or in masses. Ammoniacum is used in medicine as a circulatory stimulant, and in perfumery.

Asafetida

The sources of asafetida are *Ferula assafoetida* and allied species, stout perennial herbs of Iran and Afghanistan. The cortex of the thick fleshy roots exudes a milky juice during the rainy season. The crown of the roots is cut off and protected from the sun. The gum resin gradually collects on the surface in the form of tears, or masses of tears of varying colors imbedded in a thick, gummy, grayish or reddish matrix. Asafetida has a powerful and foul odor and a bitter acrid taste, due to sulphur compounds present in the essential oil. In spite of this it has been used throughout the East for flavoring sauces, curries, and other food products, and as a drug. In Europe and America it is used in perfumes and for flavoring only when exceedingly dilute and after certain impurities have been removed. Asafetida has many valuable medicinal properties and is used in the treatment of coughs, asthma, and other nervous afflictions and as an aid to digestion and metabolism.

Galbanum

Galbanum is a gum resin excreted from the lower part of the stems of *Ferula galbaniflua*, another stout herbaceous perennial of Northwestern Asia. It occurs in the form of separate tears or brownish and yellowish-green masses. It has been used for centuries in medicine. Galbanum has a powerful tenacious aromatic odor.

Myrrh

Myrrh is one of the oldest and most valuable of the gum resins. There has been considerable confusion in regard to its source, due to the fact that apparently two forms occur.

Herabol myrrh is derived from *Commiphora Myrrha*, a large shrub or small tree of Abyssinia, Somaliland, and Arabia. The gum resin oozes naturally from the stems or as a result of wounding. The pale-yellow liquid gradually solidifies and becomes brown or even black in color. Herabol myrrh is used in perfumery and for medicinal purposes as a tonic, stimulant, and antiseptic and is often a constituent of mouthwashes and dentifrices.

Bisabol myrrh or sweet myrrh comes from *Commiphora erythraea*, an Arabian species of similar appearance. This is the myrrh of antiquity and has been used for centuries in incense, perfumes, and embalming. Myrrh was an important product in Biblical times, ranking with gold in value. It is still used in perfumes and incense for religious ceremonies. It is one of the constituents of Chinese joss sticks.

Frankincense

Frankincense or olibanum is obtained from *Boswellia Carteri* and related Asiatic and African species. The clear yellow resin exudes from incisions made in the bark and hardens as small yellow grains. Frankincense, like myrrh, has been a valuable material since Biblical times. It is still an indispensable ingredient of incense for religious observances, and is also used in perfumes, because of its excellent fixative properties, face powders, pastilles, and fumigating powders.

Opopanax and Bdellium

These gum resins are much less important. **Opopanax** is derived from two very distinct plants, *Commiphora Kataf* of the *Burseraceae* and *Opopanax Chironium* of the *Umbelliferae*. It is used in perfumery and was formerly of importance in medicine.

Bdellium is a bitter aromatic gum resin obtained from *Commiphora Mukul* of India and *C. africana* of Africa. It is used to some extent in perfumery.

OTHER RESINS

Several other resinous substances, which find no use in industry, will be considered with the medicinal plants. These include aloes, guaiacum, jalap, and podophyllum. Gamboge, a gum resin, has already been discussed under dye plants.

CHAPTER VIII

ESSENTIAL OILS

The essential oils, or volatile oils as they are often called, are found in many different species of plants. These oils are distinguished from fatty oils by the fact that they evaporate or volatilize in contact with the air and possess a pleasant taste and strong aromatic odor. They can be readily removed from plant tissues without any change in composition. Essential oils are very complex in their chemical nature. The two principal groups are the terpenes, which are hydrocarbons, and the oxygenated and sulphuretted oils.

The physiological significance of these oils as far as the plant is concerned is not obvious. They probably represent by-products of metabolism rather than foods. The characteristic flavor and aroma that they impart are probably of advantage in attracting insects and other animals which play a role in pollination or the dispersal of the fruits and seeds. When present in high concentration, these same odors may serve to repel enemies. The oils may also have some antiseptic and bactericidal value. There is some evidence that they may play an even more vital role as hydrogen donors in oxidoreduction reactions, as potential sources of energy, or in affecting transpiration and other physiological processes.

All distinctly aromatic plants contain essential oils. They occur in some 60 families and are particularly characteristic of the *Lauraceae, Myrtaceae, Umbelliferae, Labiatae,* and *Compositae.* The amount of oil varies from an infinitesimal quantity to as much as 1 to 2 per cent. The oils are secreted in internal glands or in hairlike structures. In some instances, as in wintergreen and mustard, the oil is not present in the plant, but develops only as the result of chemical action when the ground-up plant tissue is extracted with water. Almost any organ of a plant may be the source of the oil: flowers (rose), fruits (orange), leaves (mint), bark (cinnamon), root (ginger), wood (cedar), or seeds (cardamom), and many resinous exudations as well.

Essential oils are extracted from the plant tissues in various ways depending on the quantity and stability of the compound. There are three principal methods: distillation, expression, and extraction by solvents.

Distillation. The oldest and simplest type of distillation is boiling in water, but this is now of no practical importance. The more usual method is by steam distillation. Whole or ground material is placed in a still, and live steam is introduced. The oil vaporizes and, together with the steam, passes into a condenser. Upon cooling, the oil, or *essence*, collects on the surface of the water and is removed and filtered. Although oils obtained in this way are often of high quality, distillation cannot always be used. In the case of flowers with delicate odors the heat adversely affects the constituents and the oils must be removed by other means which do not involve the risk of chemical changes and which conserve the natural odor.

Expression. The so-called citrus oils are obtained from the rinds of oranges, lemons, and other citrus fruits by submitting them to various types of pressure.

Extraction by Solvents. Both nonvolatile and volatile solvents can be used in the extraction process. The oldest method involves the use of nonvolatile solvents, usually oils or fats, such as a high-grade lard or suet, which absorb the odors and yield *pomades*. There are two chief types, enfleurage and maceration.

In *enfleurage* the extraction is carried on in a normal temperature. Glass plates are covered with the cold fat. The flowers are placed on it and allowed to remain for several days. The fat dissolves out and eventually absorbs the perfume material. Enfleurage was formerly used for all flowers, but it is now restricted for the most part to jasmine and tuberose, which continue the production of the essential oil even after they have been removed from the stem.

In *maceration* the plant material is digested with hot oil or melted fat, and the flowers are often broken up to aid the process. Rose, violet, orange, and cassie flowers are usually treated in this way. After absorption of the perfumes is completed, the pomades are treated with ethyl alcohol to dissolve out the oil and yield *floral extracts*.

The direct extraction of perfumes by means of volatile solvents is a much more recent process, dating from 1879. Petroleum ether is generally used. After the flowers have been exhausted, the solvent is distilled off, leaving a semisolid residue, the *concrete*, which consists of the oils and insoluble plant waxes. Alcohol is added, the waxes are removed by filtration or by freezing, and the alcohol is then eliminated. The resulting *absolute* is a highly concentrated form of perfume oil. Dried plant material, resinous substances, and even plant juices may also be utilized.

Essential oils have very varied industrial applications. Because of their odor and high volatility they are extensively used in the manufacture of perfumes, sachets, soap, and other toilet preparations. Many are used as flavoring materials or essences for candy and ice cream and in cooking, and for cordials, liqueurs, and nonalcoholic beverages as well. Still others have therapeutic, antiseptic, or bactericidal properties and so are valuable in medicine and dentistry. In fact, nearly all the essential oils have been used at some time in some country for medical purposes, although comparatively few are official at the present time. Some of the oils are used as clearing agents in histological work; as solvents in the paint and varnish industries; as insecticides and deodorants; in the manufacture of various synthetic odors and flavors; and in such widely diversified products as chewing gum, tobacco, shoe polish, library paste, printer's ink, tooth paste, and fish glue.

Although an enormous quantity of volatile oils is utilized in our industries, the growing and distillation of oil-producing plants are practiced only to a slight extent in the United States. This country leads in

FIG. 89. A Japanese mint-oil still in Colorado.

the cultivation of peppermint and spearmint and in the production of the oils (Fig. 89). Wormwood, wormseed, tansy, and dill are also grown commercially to some extent. Recent cultural experiments with lavender and geranium seem to indicate that the commercial production of various perfume oils in the United States is at least possible. A few oils are extracted from native wild plants. These include the oils of turpentine, sassafras, wintergreen, sweet birch, witch hazel, eucalyptus, and pennyroyal. In other cases oils are obtained as by-products of various industries. Among such are lemon, orange, grapefruit, lime, apricot, bitter almond, cedarwood, and hop oils. Most of the essential oils used in the United States, however, are imported.

Any economic classification of essential oil plants is exceedingly difficult, for the various groups intergrade. For example, clove oil, obtained from the familiar spice, is used for flavoring, in perfumery, in medicine, in histology, and as a source of synthetic vanilla. In numerous cases the same oil is used for flavoring and in medicine. Our discussion of spices and other food adjuncts, of flavoring materials, and of medicinal plants will be reserved until later. At the present time we are concerned only with those oil-yielding plants which are used for perfumery and similar purposes or which have other strictly industrial applications.

PERFUMES

It has often been said, as in the case of many other economic plant products, that the history of perfumes is the history of civilization. Certainly perfumes have been in vogue since the earliest recorded times. We know that the Egyptians and ancient Hebrews used them for both personal and religious purposes. They played an important part in the life of the Romans and Greeks, reaching such a high degree of specialization in the case of the latter people that a special perfume was required for each part of the body. Later on we find that Catherine de' Medici knew as much about perfumes as she did about poisons, that in the time of Queen Elizabeth a gift of rare perfumes was a sure way to win the royal favor, while the court of Louis XIV at Versailles had a particular perfume for each day in the year, the preparation of which was superintended by the king himself. In those days perfumes were of considerable hygienic, as well as aesthetic, value, for they acted as true antiseptics and deodorants and masked offensive odors at a time when personal cleanliness was too often overlooked. Today perfumes are still in great demand. The consumption of the natural products is increasing in spite of the many synthetic substitutes that the chemist has put on the market. These latter materials are not so lasting as those obtained directly from the plants. They are used for cheap grades, to fortify the natural products, and for blends. The most valuable perfumes are combinations of several essential oils. Frangipani, for example, contains sandalwood, sage, neroli, orris root, and musk, while one of the formulas for Eau de Cologne, dating from 1709, calls for neroli, rosemary, lemon, and bergamot dissolved in pure alcohol and aged. The expert perfumer must be able to blend the several oils at his command as an orchestra leader combines the various instruments into a perfect whole.

In addition to the aromatic oils, the finer perfumes contain fixatives— substances which are less volatile than the oils and which delay and so equalize evaporation. These may be of animal or plant origin. Musk, ambergris, and civet are frequently utilized for this purpose. Balsams and oleoresins, such as benzoin, styrax, and oak moss; essential oils with a low rate of evaporation like orris, patchouli, clary sage, and sandalwood; and various synthetic materials are also used.

The cultivation of perfume plants is carried on for the most part in countries bordering on the Mediterranean Sea and the Indian Ocean.

Most of the natural perfumes are made in southern France in the region around Grasse and Cannes in the French Riviera. In this area garden flowers are cultivated on a large scale, and from 10,000,000,000 to 12,000,000,000 lb. are normally gathered each year. These include over 5,000,000 lb. of orange blossoms, over 4,000,000 lb. of roses, 440,000 lb. of jasmine, and 330,000 lb. of violets. Large quantities of cassie, tuberoses, jonquils, thyme, lavender, rosemary, and geraniums are grown, and many other fragrant species to a lesser degree. Flowers are also grown for the perfume industry to some extent in Reunion, North Africa, England, and various European and Asiatic countries. When the Second World War cut off supplies, the United States developed substitutes and initiated or increased the cultivation of several essential-oil plants in Central America. Of the 75 essential oils regularly used in perfumery, however, only 8 are normally produced in the Western Hemisphere, and but one of these, oil of petitgrain, is of real importance.

Perfume Oils

Some of the more important essential oils used for perfumes will be considered at this time.

Otto of Roses. This valuable oil, sometimes called attar or ottar of roses or rose oil, has long been known and is still one of the favorite perfumes, either alone or in combination. Bulgaria supplies most of the commercial supply at the present time, utilizing the flowers of the damask rose (*Rosa damascena*) (Fig. 90). Over 12,000 acres on the southern slopes of the Balkans are devoted to the cultivation of this small shrub. The harvest period covers about three weeks during May and June. The flowers are picked in the early morning just as they are opening and are distilled as soon as possible. Until recently the peasant cultivators have utilized their own primitive stills, but now large modern distilleries are in operation. The oil is colorless at first, but gradually turns a yellowish or greenish color. About 20,000 lb. of the flowers are required to make 1 lb. of the essence, which is worth about $200. Very little pure otto reaches the markets, for it is almost always adulterated with geranium or palmarosa oil or geraniol, all of which have a roselike odor. Otto of roses is also manufactured in France, Italy, North Africa, Asia Minor, and India. In France the cabbage rose (*R. centifolia*) is used, and the perfume is obtained by both hot and cold enfleurage, as well as by distillation. Large quantities of rose water are also made. This consists for the most part of the water left after distillation, which still contains some otto. It is sometimes prepared by dissolving a small amount of otto in water.

Geranium. The leaves of several species of *Pelargonium* yield an essential oil on distillation. Geranium oil is widely used as an adulterant

Fig. 90. The damask rose (*Rosa damascena*). The flowers of this plant are the source of otto of roses, one of the most expensive of perfumes. (*Photo by E. H. Wilson; courtesy of the Arnold Arboretum.*)

Fig. 91. Rose geraniums (*Pelargonium odoratissimum*) under experimental cultivation

of, or substitute for, otto of roses in making perfumes and soap. *P. graveolens*, the species most frequently cultivated, is grown chiefly in Algeria and Reunion and to a lesser extent in southern France and Spain. In recent years cultural experiments have been carried on in Florida, Texas, and California with *P. odoratissimum* (Fig. 91), the rose geranium. The plants are easy to propagate from slips and are productive for five or

FIG. 92. The ylang-ylang (*Cananga odorata*). The flowers of this Asiatic species yield one of the most important essential oils used in the perfume industry. A few specimens of this tree, one of which is pictured here, are to be found in Florida.

six years after reaching maturity. They can be grown only where there is freedom from freezing temperatures. A good grade of oil is obtained from the leaves, and the plant may prove a desirable addition to the essential-oil plants that are cultivated in the United States.

Ylang-ylang. This is one of the most valuable and important oils in the perfumer's art, and is present in almost every perfume. The name means "flower of flowers." The ylang-ylang tree (Fig. 92) is an Eastern Asiatic species, *Cananga odorata*. Its yellowish-green bell-shaped flowers have an exceedingly delicate and evanescent fragrance. The oil, often known as cananga oil, is derived by simple distillation or extraction from the petals of fully opened blossoms. The Philippine Islands were once the chief producers of ylang-ylang, but now the French colonies in the Indian Ocean have a virtual monopoly. The tree grows wild and is also cultivated in various parts of Southern Asia and the East Indies. Long known to the native peoples of the East, this oil first reached Europe about 1864. Since that time it has been in great demand even though it is rather expensive.

Cassie or Acacia. The flowers of *Acacia Farnesiana* yield an essential oil almost as valuable as ylang-ylang or otto of roses. The plant is a thorny native shrub of the West Indies, but it occurs spontaneously in many tropical and subtropical countries. It is extensively cultivated in southern France, Algeria, Egypt, Syria, and India as a source of perfume. The oil is removed from the petals by maceration with cocoa butter or coconut oil, or by extraction. It has an odor like violets, and is much used for pomades, sachets, and powders.

Neroli. Oil of neroli, which is extensively used in blends and for mixing with synthetic perfumes, is obtained from orange blossoms. True oil of neroli, or *neroli bigarade*, is distilled from the flowers of the bitter orange (*Citrus Aurantium*), while *neroli Portugal* comes from the sweet orange (*C. sinensis*). Southern France leads in the production of neroli, although a considerable amount is made in the other Mediterranean countries and in the West Indies, especially in Haiti.

Oranges are the source of other essential oils used in perfumes. The leaves and twigs (and formerly the small immature fruits) supply *petitgrain* oil. This is widely used to add a pleasant bouquet to scents, cosmetics, and soap. It is produced chiefly in Paraguay. Both bitter and sweet oranges are utilized and the oil is obtained by distillation. *Oil of orange* is obtained by expressing the ripe peel. Neither of these oils is so valuable as neroli.

Bergamot. This greenish oil is expressed from the rind of the bergamot (*Citrus Aurantium* subsp. *Bergamia*). It has a soft sweet odor and is extensively used in the United States for scenting toilet soaps and in mixed perfumes. It is imported chiefly from Italy and Sicily.

Orris. The rhizomes of *Iris pallida, I. florentina*, and allied species contain an essential oil that has the odor of violets. A tincture of orris root is often used to adulterate pure extract of violets, while the powdered root is the basis of violet powder. The plant is cultivated in Southern Europe, Iran, and northern India. Italian orris root is considered the best. The rhizomes are peeled and dried in the sun, and the odor gradually develops. Orris is used to some extent as a flavoring substance.

Calamus. Calamus root is the sweet and aromatic rhizome of the sweet flag (*Acorus Calamus*), a common plant of marshy ground in

Europe, Asia, and America. In a powdered form calamus is used for sachet and toilet powders, while the distilled oil is used in perfumery. It is also used for medicinal and flavoring purposes. The candied root was at one time a popular confection.

Grass Oils. The grass family is the source of several important essential oils which are extensively used in perfumery. The genus *Cymbopogon*, formerly included in *Andropogon*, is especially rich in perfume plants.

Oil of Citronella. This widely dispersed commodity is distilled from the leaves of *Cymbopogon Nardus* (Fig. 93). Thousands of acres of citronella grass are cultivated in Java and Ceylon. The pale-yellow oil is inexpensive and is much used for cheap soaps and perfumes and as an

FIG. 93—Citronella grass (*Cymbopogon Nardus*), the source of the familiar oil of citronella.

insect repellent. The oil contains 80 to 90 per cent geraniol, and so is important as a substitute for otto of roses. In recent years citronella has been introduced into Central America, and a considerable industry has been developed in Guatemala and Honduras with some 4500 acres under cultivation. The crop is harvested by hand. Cutting stimulates growth, and a new crop is ready in three months. The oil is removed by steam distillation.

Lemon-grass Oil. The leaves of *Cymbopogon citratus* yield on distillation a reddish-yellow oil with a strong odor and taste of lemons. This is due to the unusually high content of citral, 70 to 80 per cent. The oil is used in soaps and in medicine. Citral is extensively used in perfumes, bath salts, cosmetics, and toilet soaps and as a flavoring substance. It is also the source of the aromatic substances known as ionones, which have many uses. One of the ionones is necessary in the synthesis of vitamin A; another is the raw material from which synthetic violet is made.

Lemon grass is common everywhere in the eastern tropics and is cultivated in India, Ceylon, British East Africa, the Belgian Congo, and Madagascar. It has now been successfully introduced as a crop plant in the Western Hemisphere, and large quantities of the oil are exported from Guatemala, Haiti, Brazil, Salvador, and other Latin-American republics. Lemon grass is also grown on some 1500 acres of Florida mucklands.

Palmarosa and Ginger-grass Oils. These nearly identical oils are much used as adulterants of otto of roses, as they contain a large amount of geraniol. *Cymbopogon Martinii* and a variety are cultivated in India as a source of these oils, which are exported in large quantities.

Oil of Vetiver. This oil is obtained from the roots and rhizomes of the khuskhus plant (*Vetiveria zizanioides*), a native of India and Bengal, but now grown throughout the tropics and subtropics. The roots are very sweet scented and are made up into mats, fans, screens, awnings, sunshades, baskets, sachet bags, and pillows. The leaves are odorless. The plant has been introduced into the West Indies and Louisiana, where every French garden has some. It readily escapes from cultivation and has become naturalized in many places. On distillation the roots yield an oil much like citronella which is used for high-grade perfumes, soaps, and in medicine. It is one of the finest fixatives.

Oil of Bay. The leaves of *Pimenta racemosa* on distillation yield this oil, which is used in perfumery and in the preparation of bay rum. The plant is a native of the West Indies, and the bay-rum industry is located in that region. Formerly the leaves were distilled in rum and water, but now the oil is dissolved in alcohol, with which are mixed various aromatic materials. Bay rum has soothing and antiseptic properties.

Lavender. Lavender is a very old perfume and was used by the Romans in their baths. It is still one of the most important scents. The true lavender plant (*Lavandula officinalis*) is a native of Southern Europe, occurring on dry, barren soil. It is a low shrub (Fig. 94) with terminal spikes of very fragrant bluish flowers. Many horticultural forms and hybrids occur. Lavender is grown in southern France at altitudes varying from 1500 to 1800 ft., and large amounts are raised in England. Lavender has a clean odor and the dried flowers are used in sachets and for scenting chests and drawers. The oil is an important constituent of Eau de Cologne and other high-grade perfumes and is also used in soaps, cosmetics, and medicine as a mild stimulant. Lavender water, a mixture of the oil in water and alcohol, is a highly popular toilet article in England.

Spike Lavender. This plant (*Lavandula latifolia*) is coarser and yields an inferior grade of oil. It can be grown at lower altitudes than true lavender and is extensively cultivated in France and Spain. It is used in perfumes and cosmetics and to flavor the meat jellies known as aspic.

Violet. Violet is one of the most popular perfumes and has a sweet and delicate odor. Blue and purple double varieties of *Viola odorata*, a native European species, are grown chiefly in the vicinity of Nice. The oil is extracted by solvents or maceration with hot fats. It occurs in such minute amounts that 15 tons of flowers are required to obtain 1 lb. of oil. Real violet perfume today is rare and expensive, and it has been almost entirely replaced by synthetic products derived from ionone.

Jasmine. Jasmine is one of the most highly esteemed of perfumes and the plant is extensively cultivated in southern France and other Mediterranean countries. The principal source is *Jasminum officinarum* var.

FIG. 94. A field of lavender (*Lavandula officinalis*).

grandiflorum, which is usually grafted on a less desirable variety. The flowers are picked as soon as they are open and the oil is extracted by enfleurage.

Carnation. There are over 2000 horticultural varieties of carnation, all derived from *Dianthus Caryophyllus*, a species of Southern Europe, Northern Africa, and tropical Asia. Forms with the most desirable form, size, and color have the least odor. Less highly cultivated strains have the richest odor and are used for the perfume. The oil is extracted by solvents. Most of the present-day supply is synthetic.

Rosemary. The rosemary (*Rosmarinus officinalis*), a native of the Mediterranean region, has long been a favorite sweet-scented plant and has played an important role in the folklore of many countries. It is one of the least expensive and most refreshing odors. The plant is a small evergreen shrub and is cultivated in Europe and the United States. The oil is obtained by distillation of the leaves and fresh flowering tops or by extraction. It is used in Eau de Cologne, toilet soap, and medicine. The leaves are used in cooking.

Hyacinth. This plant (*Hyacinthus orientalis*) is a native of Western Asia and Asia Minor. It was introduced into Europe during the sixteenth century and was grown as an ornamental plant, particularly in Holland.

FIG. 95. A field of hyacinths (*Hyacinthus orientalis*). These familiar ornamental plants are extensively grown in France for making perfume.

It is a familiar species in the United States (Fig. 95). Hyacinths are grown for perfume in southern France. The odor is heavy, sweet, and somewhat overpowering. Solvents are used to obtain the oil, which is generally utilized greatly diluted.

Oak Moss. Oak moss or *mousse de chêne* is a recent and extremely valuable addition to the raw materials of the perfume industry. It comprises various lichens that grow on the bark of trees. The principal sources are European species of *Ramelina* and *Evernia*, especially *R. calicaris*, *E. furfuracea*, and *E. prunastri*. These lichens contain oleoresinous substances which are extracted by means of solvents. After they have been collected, the lichens are thoroughly dried, as the perfume develops during storage. Oak moss not only has a heavy, penetrating odor and blends well, but it also has a high fixative value. It is an essential element in lavender perfumes and soap and in better grades of cosmetics.

Linaloe or Bois de Rose. There are several sources of linaloe, lignaloe, or bois de rose oil, a very aromatic substance widely used in perfumes, soaps, and cosmetics and for flavoring foods and beverages. Mexican linaloe is distilled from chips of the wood of two Mexican species, *Bursera penicillata* and *B. glabrifolia*. Cayenne linaloe or bois de rose is derived from *Aniba panurensis* of the Guianas, while Brazilian bois de rose is obtained from *A. rosaeodora* var. *amazonia*, a tree of the lower Amazon basin.

Sandalwood. Sandalwood oil is obtained by distillation from the wood of *Santalum album* and allied species. The sandalwood tree grows wild in India and other parts of Southeastern Asia and is cultivated in many other countries. The oil is used throughout the Orient as a perfume and also in medicine. It is an excellent fixative and is much used in blends. The sweet-scented wood is utilized for chests and boxes. The demand for sandalwood has been so great that in many parts of the East the true sandalwood has been practically exterminated. This has been brought about largely by careless methods of lumbering. There are numerous substitutes used in various parts of the world.

Patchouli. Patchouli is obtained from the fleshy leaves and young buds of *Pogostemon Cablin*. The plant is a small shrub that grows wild in Southeastern Asia and is cultivated in China. The leaves are fermented in heaps and are then distilled. The dark-brown oil has a powerful odor, resembling that of sandalwood. It is one of the best fixatives for heavy perfumes. It is also used in soaps, hair tonics, and tobacco. It is responsible for the characteristic odor of cashmere shawls, which are always shipped in patchouli-scented containers.

Champaca. Champaca oil, which constitutes one of the most famous perfumes of India and other oriental countries, is obtained from *Michelia Champaca*, a large handsome tree of the eastern tropics. The conspicuous yellow flowers (Fig. 96) are very fragrant and are much worn by the natives. The oil is obtained from the flowers by maceration or extraction and rivals ylang-ylang in its delicious fragrance.

Several of the gum resins, chiefly frankincense and myrrh, which have already been discussed, have been used in perfumery for thousands of years.

Garden flowers, in addition to those mentioned above, which are extensively cultivated for their perfume include heliotrope (*Heliotropium arborescens*), lily of the valley (*Convallaria majalis*), jonquil (*Narcissus Jonquilla*), mignonette (*Reseda odorata*), narcissus (*Narcissus Tazetta*), clary sage (*Salvia Sclarea*), and tuberose (*Polianthes tuberosa*).

FIG. 96. Flowers and fruit of *Michelia Champaca*, the source of the fragrant champaca oil. (*Photo by Walter H. Hodge.*)

Other sources of essential oils used in perfumes will be discussed later in connection with other topics. Among these are anise, caraway, cassia, cinnamon, clove, lemon, peppermint, thyme, wintergreen, and zedoary.

ESSENTIAL OILS USED IN OTHER INDUSTRIES

Camphor

Camphor is the most important of the essential oils used in industry. Commercial camphor, known as camphor gum, consists of tough, white, translucent masses or granules with a penetrating odor and pungent, aromatic taste. It is solid at ordinary temperatures, thus bearing the same relation to the other essential oils that vegetable fats do to the fatty oils. It volatilizes very slowly.

Camphor is obtained by distillation of the wood of the camphor tree (*Cinnamomum Camphora*). This tree is very tall and striking in appearance (Fig. 97) with shiny, dark, evergreen leaves. It is a native of China, Japan, and Formosa, but has been widely introduced into tropical and subtropical regions elsewhere, chiefly as an ornamental plant. The

Japanese has a virtual monopoly of the camphor industry, since 75 per cent of the product comes from Formosa. The earlier crude methods of obtaining camphor were very destructive and the existence of the trees was threatened. Now only trees 50 years of age or older are used, and every stage in the process is carefully supervised. The wood is reduced to chips or ground to a fine powder and the leaves are also ground up. This material is then distilled with steam for about 3 hours, and the crude camphor crystallizes on the walls of the still. This is removed and must

Fig. 97. A camphor tree (*Cinnamomum Camphora*) in Florida. (*Photo by U.S. Forest Service.*)

be purified before it is ready for market. At the present time 80 per cent of the camphor produced is synthetic, made for the most part from pinene, a turpentine derivative.

The United States is a large consumer of camphor. Because of its importance in this country, the growing of camphor has been experimented with and today camphor is a crop of increasing prominence, especially in Florida. The trees are propagated from seed in nurseries and are later transplanted. Nearly 10,000 acres are now devoted to camphor growing. Twigs and leaves are utilized for distillation rather than the old wood. Clippings can be made from the trees when they are three or four years old, and several times a year thereafter.

The principal use of camphor is in the manufacture of celluloid and the various nitrocellulose compounds. Camphor also has a wide range of medicinal uses, both internally and externally. It is also used in perfumery.

Borneo camphor, which is obtained from *Dryobalanops aromatica* of the East Indies, has been used as a substitute.

Cedarwood Oil

Several of the essential oils have a high refractive index and are valuable as clearing agents in the preparation of permanent microscopic mounts and for use with oil-immersion lenses. The most important of these is cedarwood oil. This inexpensive oil is obtained by steam distillation from the heartwood of the eastern red cedar (*Juniperus virginiana*) and allied southern species. Chips, sawdust, waste from the lead pencil and other industries, old stumps, roots, and even fence rails are utilized. Cedarwood oil is also used in perfumery, soaps, deodorants, liniments, cleaning and polishing preparations, and as an adulterant of expensive sandalwood and geranium oils. It has insecticidal properties and is used as a moth repellent and in fly sprays.

Clove and bergamot oils are also utilized as clearing agents.

Miscellaneous Oils

Essential oils are useful as solvents in the paint and varnish industry. The most important of such oils, oil of turpentine, has already been discussed. Various other oils, chiefly eucalyptus oil from *Eucalyptus dives*, an Australian species, are employed in the flotation process for the separation of minerals from their ores. Still other volatile oils are used in the preparation of cleaning materials and for many other industrial purposes.

CHAPTER IX

FATTY OILS AND WAXES

FATTY OILS

Another type of oil that occurs in plants is the fatty oil. The fatty oils are also called fixed oils because, unlike the essential oils, they do not evaporate or become volatile, and they cannot be distilled without being decomposed. Chemically these vegetable fatty oils are close to animal fats. They consist of glycerin in combination with a fatty acid. The so-called oils are liquid at ordinary temperatures and usually contain oleic acid. The fats, on the other hand, are solid at ordinary temperatures and contain stearic or palmitic acid. The fatty oils are insoluble in water, but soluble in various organic solvents. When fats break down, they yield the fatty acids and glycerin, of which they are composed, and usually develop a rancid odor and taste. When a fat is boiled with an alkali, it decomposes and the fatty acid unites with the alkali to form soap. If potash or lye is used, a soft soap is obtained; if soda is used, a hard soap is the result.

Fatty oils are produced in many families of plants, both tropical and temperate. They are stored up, often in large amounts, in seeds (Fig. 98) and, to a less extent, in fruits, tubers, stems, and other plant organs; they are often associated with proteins. This type of reserve food material is available as a source of energy for the processes involved in the germination of the seed. The fatty oils are bland and lack the strong taste and odor and the antiseptic qualities of the essential oils. Consequently they are available as food for man. These edible oils contain both solid and liquid fats and form indispensable articles of human food. The demand for edible oils has so increased in recent years that various processes have been developed whereby the nonedible oils have been rendered available. This is usually done by hydrogenation, the adding of hydrogen.

The method of extraction of the oils varies in different cases. Usually the seed coats have to be removed, and then the material is reduced to a fine meal. The oils are removed by solvents or by subjecting the meal to screw or hydraulic pressure. This latter method is used primarily for the edible oils. The residue is rich in proteins and is valuable as a fertilizer and as a cattle feed. The pressure causes the cell walls to break and the fats escape. The extracted oils are filtered and may be further purified. The higher grades are edible, and the lower are used in the industries. The increasing demand for these industrial oils since the First World War has led to the improvement of methods of cultivation and preparation, and also to a search for new sources the world over. Fatty oils also have a medicinal value.

Four *classes of vegetable fatty oils* are recognized: (1) drying oils, (2) semidrying oils, (3) nondrying oils, and (4) fats or tallows. The drying

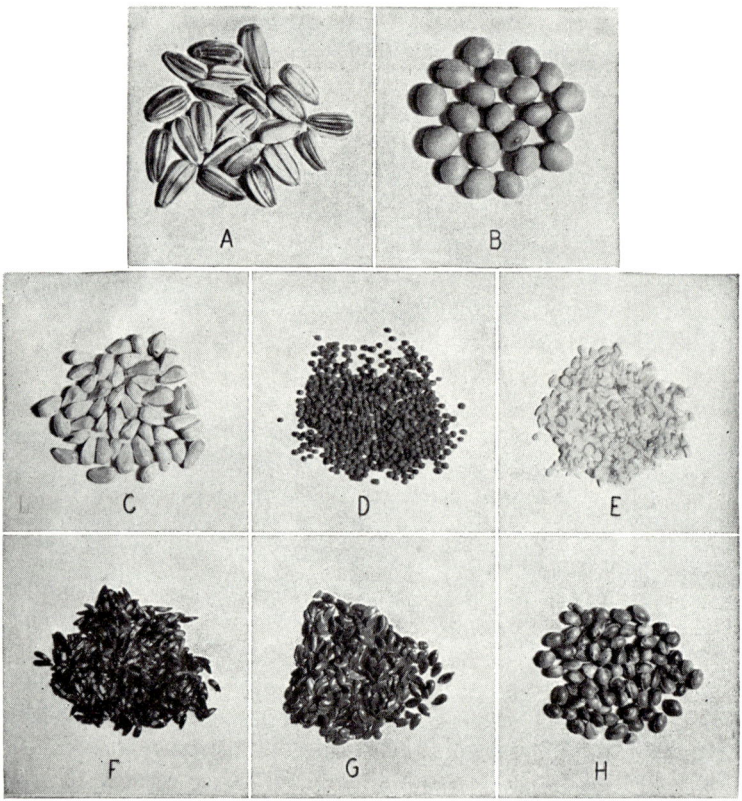

FIG. 98. A group of oil seeds. *A*, sunflower (*Helianthus annuus*); *B*, soybean (*Glycine Max*); *C*, safflower (*Carthamus tinctorius*); *D*, rape (*Brassica Napus*); *E*, sesame (*Sesamum indicum*); *F*, niger (*Guizotia abyssinica*); *G*, flax (*Linum usitatissimum*); *H*, hemp (*Cannabis sativa*).

oils are able to absorb oxygen and on exposure dry into thin elastic films. These oils are of great importance in the paint and varnish industries. The semidrying oils absorb oxygen slowly and only in limited amounts. They form a soft film only after long exposure. Some of these oils are edible; others are used as illuminants or in making soap and candles. The nondrying oils remain liquid at ordinary temperatures and do not form a film. These oils are edible, and can be used for soap and lubricants. The fats are solid or semisolid at ordinary temperatures. They are edible and are also useful in the manufacture of soap and candles. Drying and semidrying oils are of more frequent occurrence in plants of temperate regions, while nondrying oils and fats predominate in tropical species.

Drying Oils

Linseed Oil. The seed (Fig. 98*G*) of the flax plant (*Linum usitatissimum*) has long been the source of one of the most important of the drying oils. The oil content is 32 to 43 per cent. Seeds are collected and stored for several months, the impurities are removed, and the seeds are ground to a fine meal. The oil is usually extracted by pressure with heat or by the use of solvents. Cold-pressed oil is produced in Eastern Europe, where it is used for edible purposes. Linseed oil varies from yellow to brownish in color and has an acrid taste and smell. On oxidation it forms a very tough elastic film. This drying property is increased by heating the raw oil to 125°C., producing the so-called boiled linseed oil. Linseed oil is used chiefly in making paints, varnishes, linoleum, soft soap, and printer's ink. After extraction, the oil cake is used as a cattle feed. Linseed is grown on a commercial scale chiefly in Argentina, where over 6,000,000 acres are devoted to seed-flax cultivation. The U.S.S.R., India, Uruguay, China, Canada, and the United States also produce a considerable quantity. Minnesota, Montana, and the Dakotas are the centers of seed-flax production in this country, the annual yield amounting to about 40,000,000 bu.

Tung Oil. Tung oil, sometimes called chinawood oil, is widely used in the varnish industry as a substitute for linseed oil. It is obtained from the seeds of two Chinese species of *Aleurites*, *A. Fordii*, the tung-oil tree (Fig. 99), a species native to central and western China, and *A. montana*, the mu tree, restricted to southwestern China. The oils from these trees are practically identical in composition and properties, and imported tung oil is frequently a mixture of the two. The Chinese have used tung oil for centuries for waterproofing wood, paper, and fabrics. It is a good preservative and very resistant to weathering, so it is particularly valuable for outside paints. Boatmen prefer it, as it is little affected by water. The United States began the cultivation of *A. Fordii* in 1905, and at the present time tung-oil trees are being grown with great success on some 75,000 acres in the Gulf States, principally in Mississippi and Florida. The trees are handsome and are often planted for ornament. The establishment of a domestic tung-oil industry has been of great importance to southern agriculture. Not only is it a profitable source of income, but it has helped to solve the problem of eroded and other waste land, for tung trees will grow on soils unsuitable for other types of agriculture. The outer husk of the fruit is removed, and the oil is expressed from the seeds by expeller presses. Tung oil is pale yellow to dark brown in color. It dries very rapidly and has preservative and waterproofing qualities. As a consequence its chief use is in varnish and paint manufacture, where it has largely replaced kauri and other hard resins. Large quantities are also used in making linoleum, oilcloth,

FIG. 99. A tung-oil tree (*Aleurites Fordii*) growing in a field of opium poppies. Fengtu Hsien, western Szechuan, China. (*Photo by E. H. Wilson; courtesy of the Arnold Arboretum.*)

brake linings, leather dressings, soap, inks, insulating compounds, and fiberboard. The oil cake is a good fertilizer but cannot be used as a feed.

Soybean Oil. The soybean (*Glycine Max*) (Fig. 100), indigenous to China and a food plant of paramount importance in Eastern Asia, is the source of an exceedingly valuable oil. Midway between linseed and cottonseed oils in its characteristics, it is sometimes classed as a drying and sometimes as a semidrying oil. Soybean oil is extracted from the seeds (Fig. 98, *B*) by expression with hydraulic or expeller presses or by the use of solvents. The oil content of the improved varieties now under cultivation is 19 to 22 per cent. After refining, soybean oil can be used as a salad or cooking oil and for other food purposes. Over 50 different food products are made from it, principally margarine and shortenings. The oil has great versatility, and the list of industrial uses is constantly growing. Among them may be mentioned the manufacture of candles, soap, varnishes, lacquers, paints, greases, linoleum, rubber substitutes, cleaning compounds, insecticides, and disinfectants. The oil cake or meal has a 40 to 48 per cent protein content and is a valuable feed for livestock and a source of soy flour. It is also used in making adhesives, plastics, spreaders, foaming solutions, sizing, fertilizers, a synthetic textile fiber, and many other products.

FIG. 100. A field of soybeans (*Glycine Max*) in Connecticut. (*Courtesy of the Connecticut Agricultural Station in New Haven.*)

Oiticica Oil. The seeds (nuts) of *Licania rigida*, a large evergreen tree of northeastern Brazil, furnish the oiticica oil of commerce. The United States imports a considerable amount for use as a substitute for tung oil. Oiticica oil is extracted by solvents or by hydraulic presses. It is used in the paint and varnish industry; in making linoleum, printing inks, and brake bands; and for improving the elasticity of rubber products. In Brazil the oil has been used for many years as an illuminant and in medicine.

Perilla Oil. Perilla oil is obtained from the seeds of *Perilla frutescens*, an aromatic annual 3 to 5 ft. in height with numerous branches (Fig. 101). The plant, a native of northern India, China, and Japan, is extensively cultivated in the Orient, particularly in Manchuria and Japan. It matures slowly and has to be harvested before it is quite ripe or the seeds fall from the capsules. The oil, which is expressed from roasted and crushed seeds, is edible and has been used for food purposes from earliest time. The industrial uses of perilla oil, however, are much more important. It is used in the manufacture of the famous Japanese oil papers, cheap lacquer, paper umbrellas, waterproof clothes, artificial leather, and printer's ink. The United States imports a large amount for use as a substitute for linseed oil in the paint and varnish industries.

Candlenut Oil. This oil is obtained from the hard-shelled seeds or "nuts" of *Aleurites moluccana*, which is native to Malaya and the Pacific islands and is cultivated elsewhere. It has long been important in the Philippine Islands, where it is known as lumbang oil. It is a good

FIG. 101. Young plants of *Perilla frutescens*, the source of perilla oil.

drying oil and is much used in making paints, varnishes, lacquer, linoleum, and soft soap and as a preservative for the hulls of vessels. The nuts were formerly used in Hawaii and the other Polynesian islands for illumination, hence the name candlenut. The oil cake is poisonous and serviceable only as a fertilizer.

Walnut Oil. Mature and old kernels of the English walnut (*Juglans regia*) yield a drying oil used for white paint, artists' oil paints, printing ink, and soap. Hot-pressed oil is best adapted for these purposes. The fresh oil and cold-pressed oil have a pleasant smell and nutty flavor and are edible. In California waste kernels from the shelling operations are used as a source of the oil.

Niger Seed Oil. This pale yellow oil is obtained from the seeds (Fig. 98, *F*) of *Guizotia abyssinica*, an annual plant native to tropical Africa. It is extensively cultivated in Africa, India, Germany, and the West Indies. The higher grades have a pleasant aromatic odor and are used for food, while the poorer grades are used for soap and as illuminants. This oil is but little used in the United States.

Poppy Oil. An important drying oil is obtained from the seeds of the opium poppy (*Papaver somniferum*). This famous drug plant is grown for its seeds in northern France and Germany and in India. The first (cold) pressing yields a white edible oil, while a second (hot) pressing furnishes a reddish oil used for lamps, soap and, after bleaching, for oil paints.

Safflower Oil. *Carthamus tinctorius*, already mentioned as the source of a dye, is extensively cultivated in Egypt, India, and the Orient and to some extent in the United States for its oil-containing seeds (Fig. 98, *C*). The oil is used for soap, paints, varnishes, as an edible oil, and as an illuminant.

Tall Oil. This so-called oil is not a true oil but is a by-product of the sulphate pulp industry. The waste liquor from pine pulp mills is concentrated by evaporation, and the soap curds are removed and acidified. The crude tall oil which results contains fatty acids, resins, and other materials. It is refined by steam distillation. Tall oil is used in soap manufacture and, after treatment with glycerin, as a drying oil.

Other Drying Oils. Many other species furnish drying oils of some commercial importance. Among them may be mentioned the hemp (*Cannabis sativa*) grown in China, Japan, and European countries for the oil which is extracted from the seeds (Fig. 98, *H*) and used for soap, paints, varnishes and as a lamp oil. The seeds of tobacco (*Nicotiana Tabacum*); *Hevea brasiliensis*, *Manihot Glazovii*, and other sources of rubber; figs; grapes; and raisins all furnish drying oils of some value.

Semidrying Oils

Cottonseed Oil. This is the most important of the semidrying oils and is used as the standard of comparison. The United States is the chief producer, but nearly all the cotton-growing countries contribute to some extent. Over 1,000,000,000 lb. of the oil are expressed annually. The industry has been developed since about 1880, prior to which time cotton seeds were waste products. The seeds are carefully cleaned and freed from impurities and the linters and usually the hulls are removed. The kernels are then crushed and heated and are finally subjected to hydraulic pressure or expeller presses. The oil is pumped into tanks where the impurities settle out. The pure refined oil is of great value as a salad and cooking oil and for making oleomargarine and lard substitutes. The residue is the source of various products that have a wide range of industrial uses. Among these may be mentioned soap, washing powders, oilcloth, artificial leather, insulating materials, roofing tar, putty, glycerin, and nitroglycerin.   Cottonseed meal is important as a foodstuff and fertilizer.

Corn Oil. The kernels of maize or Indian corn contain about 50 per cent of oil present in the embryo. Until recently the embryos were a waste product of the milling industries, but today the production of corn oil is of increasing importance, amounting to about 200,000,000 lb. annually. It can be used for nearly all the purposes to which any oil is put. Refining methods have made possible the utilization of 75 per cent of the oil for edible purposes. The familiar "Mazola" oil can be used for cooking, in bakeries, and for mixing with other oils. The crude oil has many industrial uses, such as the manufacture of rubber substitutes,

Fig. 102. A field of sesame (*Sesamum indicum*) which has been pulled and stacked to allow the pods to dry.   Pangkwangchen, Shensi, China.   (*Photo by F. N Meyer; courtesy of the Arnold Arboretum.*)

soaps, and cheap paints. Like cottonseed oil, it is of little use as a lubricant.

Sesame Oil. This oil, known also as gingelly oil, is the product of the seeds (Fig. 98, *E*) of an annual herb, *Sesamum indicum*. It is the chief oil of India and has been cultivated there and in China from remote times. Today over 3,000,000 acres are devoted to this crop. Its use has spread to other tropical regions and it is now grown in many Asiatic (Fig. 102), African, and Latin-American countries. China produces about one-half of world output, India one-third, and Africa and Latin America the balance. Sesame is the principal vegetable oil in Mexico. Sesame oil was brought to the United States by the slaves, and the Southern negroes grow the plant to this day. The seeds contain about 50 per cent oil, which is easily extracted by cold pressure. The finer grades are tasteless and nearly colorless and are used as a substitute for olive oil in cooking and in medicine. European countries use enormous quantities, as it is a compulsory addition to margarine and other food products. Marseilles imports over 100,000,000 lb. The poorer grades are used for soap, perfumery, and rubber substitutes, and to some extent as lubricants. In India the oil is used for anointing the body, as fuel for lamps, and as food. The oil cake is a good cattle food. Sesame seeds are also used in the confectionery and baking industries.

Sunflower Oil. The seeds (Fig. 98, *A*) of the common sunflower (*Helianthus annuus*) contain 32 to 45 per cent of a light golden-yellow oil equal to olive oil in its medicinal and food value. It is an excellent salad oil and is used in margarines and lard substitutes. The seeds are a good bird and poultry food, the oil cake is excellent for stock, and the whole plant is often grown for ensilage. The oil has semidrying properties which render it useful in the paint, varnish, and soap industries. Probably a native of South America, the sunflower has been cultivated for so long in various parts of the world that its exact place of origin is unknown. The U.S.S.R., Rumania, and Argentina are large producers of sunflower seeds. In recent years sunflowers have become an important crop in the United States and Canada.

Rape and Colza Oils. The seeds of several species of *Brassica*, particularly *B. campestris*, *B. Napus* (the rape) (Fig. 98, *D*), and *B. Rapa*, yield oils with similar characteristics which are classified commercially as rape or colza oils. The oil content is 30 to 45 per cent, and the oil is extracted by expression or solvents. Rape seeds are extensively cultivated in China, Japan, India, and Europe. The crude oil is edible when cold pressed and is much used for greasing loaves of bread before baking. It is also used in lamps, in oiling woolen goods, in the manufacture of soap and rubber substitutes, and for quenching or tempering steel plates. The refined oil, generally referred to as colza oil, is also edible, and it is used as a lubricant for delicate machinery.

Other Semidrying Oils. Among these may be mentioned camelina oil from *Camelina sativa*, grown in many European countries for its seed and used for soap and as an illuminant; croton oil, a powerful drug to be discussed later; and argemone oil from *Argemone mexicana*. The seeds of a great variety of cultivated plants, such as apples, pears, apricots, cherries, peaches, plums, citrus fruits, cereals, tomatoes, canteloupes, watermelons, pumpkins, and black and white mustard, also contain semidrying oils.

Nondrying Oils

Olive Oil. Olive oil, obtained from the fruits of the olive (*Olea europaea*), is the most important of the nondrying oils. The tree is a small evergreen and is cultivated chiefly in the Mediterranean countries and to some extent in Australia, South Africa, South America, Mexico, and the United States. Normally the world production of olive oil has been about 2,000,000,000 lb., with Spain, Italy, Greece, and Portugal the leading countries. The United States produces about 42,700,000 lb., chiefly in California, but has to import nearly fifteen times as much more. The oil is squeezed from the pulp either by hand or mechanically. The finest grades are obtained by the former method. These oils are golden yellow, clear and limpid, odorless, and edible. They are used chiefly as salad and cooking oils, in canning sardines, and in medicine. Inferior grades have a greenish tinge and are used for soapmaking and as lubricants. The poorest grades are obtained by the use of solvents after several pressings. Fully ripe olives give the largest yield. Olive oil is one of the most important food oils, as it will keep for a long time and becomes rancid only when exposed to the air. The oil cake is used for stock feed and as a substitute for humus in soil conditioning.

Peanut Oil. Peanut oil is obtained from the seeds or "nuts" of the common peanut (*Arachis hypogaea*), to be discussed later. The chief producers of peanuts are China, India, the United States, and Africa. The seeds are shelled, cleaned, and crushed, and the oil is expressed by both hydraulic presses and expellers. The filtered and refined oil is edible and is used as a salad oil, for cooking, for packing sardines, in making margarine and shortenings, and as an adulterant for olive oil. Inferior grades are used for soapmaking, lubricants, and illuminants. The oil cake is one of the best stock feeds, as it has a higher protein content than any other similar product. Peanut oil is in great demand in Europe, where it is extracted with solvents as well as by expression. In the United States Spanish peanuts are grown for oil production, as they have a higher oil content.

Castor Oil. This very versatile oil comes from the seeds of *Ricinus communis*, a coarse erect annual herb cultivated in both temperate and tropical regions. In the United States it is a favorite ornamental plant, and it is now being grown for the oil. The seeds, which are very characteristically marked (Fig. 103), contain 35 to 55 per cent of a thick colorless or greenish oil which is obtained by expression or solvent extraction. Formerly the chief use of castor oil was in medicine, where it acts as a purgative; at the present time 99 per cent is utilized in industry in the manufacture of some 25 different products. It is water resistant and so is used for coating fabrics and for protective coverings for airplanes, insulation, food containers, guns, etc. It is an excellent lubricant especially for airplane engines. When hydrated, it is converted into a quick-drying oil extensively used in paints and varnishes. Castor oil is also used in making soap, inks, and plastics; for preserving leather; and as an illuminant. The leaves have insecticidal properties, while the stalks are a source of paper pulp and cellulose. The oil cake or pomace is poisonous but makes an excellent fertilizer. Formerly castor beans were imported chiefly from India. More recently Brazil and Mexico have become important producers and to some extent China and Manchuria.

Other Nondrying Oils. Nondrying oils of minor importance include kapok oil from the seeds of the kapok tree, already mentioned, and used as a substitute for cottonseed oil; tea-seed oil from *Camellia Sasanqua*, a valuable oil in China, Japan, and Assam; and oil of ben from *Moringa oleifera*. Nondrying oils are also obtained from almonds, pecans, filberts, pistachio, and pili nuts and from the flesh of avocados.

FIG. 103. Castor-bean seeds (*Ricinus communis*), the source of castor oil. (Reproduced by permission from Youngken, Textbook of Pharmacognosy, P. Blakiston's Son & Company.)

Vegetable Fats

Coconut Oil. This is one of the most extensively used of the fatty oils. It is obtained from the dried meat of the coconut (*Cocos nucifera*) to be discussed later. This oil is pale yellow or colorless and is solid below 74°F. After the nuts have been harvested, the husks are removed and the nuts split open and dried by either natural or artificial heat. The dried meat, or copra as it is called, is then easily removed. This is ground up, and the oil is expressed by various methods. The cake is sometimes put through hydraulic presses a second time, and still more oil is removed. The yield is about 65 to 70 per cent. Recently, fresh meat has been utilized in the presses and this yields 80 per cent or more. Refined coconut oil is edible and is now extensively used for food products, chiefly margarines. It is particularly well adapted for this purpose as it is solid at ordinary temperatures. It is almost indispensable for making candy bars and similar types of confectionery. Coconut oil has long been used for the best soaps, cosmetics, salves, shaving creams, shampoos, and other toilet preparations. It is the only oil used in marine soaps. It is also useful as an illuminant. The cake is an excellent stock food. About 500,000 tons of coconuts are used annually for the oil. Copra is produced chiefly in Ceylon, the Straits Settlements, India, Polynesia, the Philippine Islands, and the West Indies. Most of the oil is expressed in Europe, the United States, and Japan, although Ceylon and India export large amounts.

Palm Oil. Palm oil is a white vegetable fat, solid at ordinary temperatures, which is obtained from the nuts of the African oil palm (*Elaeis guineensis*). This tree is a native of Western Africa but has spread all through the tropics of both hemispheres and now covers enormous areas. It occurs spontaneously and is also cultivated in Brazil, Haiti, and Honduras, where it is known as dendê. The oil palm is a very productive

FIG. 104. Fruit of the oil palm (*Elaeis guineensis*). Both the fibrous pulp of the fruits and the kernels yield an important vegetable fat. (*Photo by the Philippine Bureau of Forestry.*)

tree. It begins to bear at the age of 5 to 6 years, reaches full bearing at 15, and continues until 60 or 70 years of age. Each tree bears 10 bunches of 200 nuts a year. The fibrous pulp of these fruits (Fig. 104) contains 30 to 70 per cent of fat. The oil is obtained chiefly by crude native methods. A few new methods have been developed making possible a more efficient extraction. It is yellow-orange or brownish red in color. Over 200,000 tons of the oil enter the world trade, coming normally from Sumatra, Java, and the west coast of Africa. Palm oil is used in making soap and in the manufacture of tin plate, terne plate, and cold reduced sheet steel. The refined oil is used in margarine and vegetable shortenings. In parts of Africa it is used as a fuel for Diesel motors.

Palm-kernel Oil. This white and much more valuable oil is also obtained from the African oil palm, but from the kernels rather than from the pulp. It is extensively used in the margarine and candy industries, as it has a pleasant odor and nutty flavor. It is also used for making glycerin, shampoos, soap, and candles. The natives express a little oil for their own use, but most of the kernels are shipped to the oil mills of the United States and Europe, where the oil is extracted by hydraulic presses or by solvents. Over 500,000 tons of the kernels are used annually. Palm-kernel cake is a good cattle feed.

Brazilian Palm Oils. When supplies of copra, coconut oil, and palm oil were cut off during the First World War, attention was directed to the rich resources of Brazil, which included some 500 species of palms, many of them potential sources of oil. During the last 20 years the United States has imported an ever-increasing amount of palm oils from this country. Although not native, both the coconut and the African oil palm occur in great abundance. In addition several native species have become of economic importance. These include the babassu, cohune, licuri, tucum, and murumuru palms, all of which furnish kernel oils.

Babassu Oil. Babassu oil, the most important of the New World palm oils, is obtained from the babassu palms *Orbignya Martiana* and *O. oleifera*, magnificent trees 60 ft. in height with a vase-shaped crown of leaves. From two to eight enormous clusters of fruit are produced, each weighing around 200 lb. and containing from 200 to 600 fruits. The outer portion of the fruit is a dense, tough, fibrous husk. This surrounds a thin mealy mesocarp and the nut, which has a thick exceedingly hard shell. Development of the industry was retarded when only hand labor was available to crack the nuts. Now machines have been developed which exert 10,000 to 25,000 lb. pressure. The nuts contain from two to six kernels with a 63 to 70 per cent oil content. Babassu oil is expressed from the kernels and when refined is used as a substitute for coconut oil in the margarine, baking, confectionery, and soap industries. It is also used in making bulletproof glass, explosives, lubricants, and as a fuel for Diesel engines.

Cohune Oil. The nuts of the cohune palm (*Orbignya Cohune*) (Fig. 105), a native of Central and South America, contain 40 per cent of a firm yellow fat. Over two million acres of this tree occur in British Honduras alone, and the yield is from 1,000 to 2,000 nuts per tree. Like the babassu, the nut is very hard to crack, but now that effective machines have been devised, the oil is becoming of increasing importance.

Licuri Oil. *Syagrus coronata*, the licuri or ouricuri palm, characteristic of dry areas in eastern Brazil, is important commercially as the source of a palm-kernel oil and also of a wax which occurs on the leaves. The fruits resemble miniature coconuts.

Murumuru Oil. The murumuru palm (*Astrocaryum Murumuru*) is the chief source of palm-kernel oil in the state of Para, Brazil, and the oil is now being exported. The tucum palms (*Astrocaryum Tucuma* and *A. vulgare*) of Northern South America yield both a palm kernel and a pulp oil and also a fiber of commercial importance.

Vegetable Fats of Minor Importance. These include:

Cocoa Butter. This white or yellowish fat with a chocolate odor and flavor is expressed from the beans of the cacao or cocoa (*Theobroma*

Fig. 105. Fruiting cluster of a cohune palm (*Orbignya Cohune*) growing in the Atkins Garden of Harvard University in Cuba. (*Photo by Walter H. Hodge.*)

Cacao) during the process of making cocoa. It is firm at ordinary temperatures. Its chief use is in making chocolate; it is also used for cosmetics and in perfumery and medicine.

Carapa Fat. This thick white or yellow oil is obtained from the seeds of several species of the genus *Carapa* and is used for soap. The South American natives use the oil from *C. guianensis* to grease their skins and drive off insects. *C. moluccensis* is a native of East Africa, India, Ceylon, and the Moluccas. Carapa oil is also used as an illuminant.

Shea Butter. The seeds of *Butyrospermum Parkii*, an African tree, furnish shea butter, a greenish-yellow fat with a pleasant odor and taste. The fat is edible, and is also used mixed with, or as a substitute for, cocoa butter in chocolate manufacture. Inferior grades are utilized for soap and candles. A considerable amount is exported to Europe.

Mowra Fat. Two species of the genus *Madhuca*, *M. indica* and *M. longifolia*, are the source of various Indian products, which are known as mowra fat, bassia fat, mahua butter, or illipe butter. The trees grow wild and are also extensively cultivated. The kernels contain 55 to 65 per cent of a soft yellow oil widely used locally for cooking and tallow. Over 66,000,000 lb. are exported to Europe for use as a margarine and chocolate fat and in soap manufacturing. The cake is unfit for food but makes a good fertilizer. *M. butyracea* is the source of a similar product, phulwara butter, used locally.

Borneo Tallow. This fat is a hard yellowish-green brittle solid sometimes known as green butter. It is obtained from *Shorea aptera* and several other species of the same or of an allied genus native to the East

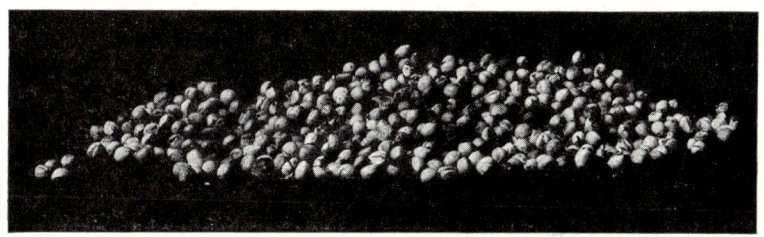

FIG. 106. The seeds of *Sapium sebiferum*, the source of Chinese vegetable tallow. Ochany, western Hupeh, China. (*Photo by E. H. Wilson; courtesy of the Arnold Arboretum.*)

Indies. The kernels, which contain 50 to 70 per cent of fat, are dried and expressed by the inhabitants for their own use or are exported to Europe for soapmaking and as a substitute for cocoa butter.

Chinese Vegetable Tallow. This material occurs as a thick layer of hard white fat on the seeds (Fig. 106) of a Chinese tree, *Sapium sebiferum* which has been introduced into many other countries, including the southern United States. After proper treatment the tallow is used in soap and candle manufacture. The seeds of this tree contain a drying oil that is of some value.

Nutmeg Butter. The seeds of the nutmeg (*Myristica fragrans*) and allied species contain about 40 per cent of a yellow fat with the flavor and consistency of tallow. Nutmegs that are unfit for use in the spice trade are roasted and powdered and the oil is extracted between warm plates. Several varieties of nutmeg butter are on the market, all used for ointments or for candles. Mace yields a similar material.

Other Vegetable Fats. Locally many other vegetable fats are of some importance. Among such may be mentioned *pongam oil* from the seeds of *Pongamia pinnata*, used for illumination and medicine in India and Ceylon; *macassar oil* from the seeds of *Schleichera oleosa*, a soft yellowish-white fat used in India, Ceylon, and the East Indies for cooking purposes, as a hair oil, and for illumination; and *ucuhuba* and *otoba butter* obtained from various species of *Virola* in Northern South America.

Many members of the *Flacourtiaceae* in Asia, Africa, and South America have oily seeds which contain chaulmoogric and usually hydnocarpic acid. These oils have long been used by the natives in treating skin diseases. The most important one, *chaulmoogra oil*, will be discussed under medicinal plants.

WAXES

Waxes are usually found on the epidermis of leaves and fruits, where, because of their impervious character, they serve to prevent too great loss of water through transpiration. Waxes are harder than fats and have a higher melting point. They do not become rancid and are less easily hydrolyzed. In chemical composition waxes are quite similar to fats, but are esters of monohydric alcohols rather than glycerides. Only a few are of commercial importance. Among these may be mentioned:

Carnauba Wax. This is the most important vegetable wax and occurs as an exudation on the leaves of the wax palm (*Copernicia cerifera*), a native of Brazil and other parts of tropical South America (Fig. 107). This slender palm is known locally as the "Tree of Life," since nearly every part of the plant is put to some use. The commercial supply of wax is obtained chiefly from wild trees in northeastern Brazil. Young leaves are carefully selected and gathered before they are fully open and are then dried in the sun for several days until the wax is a flourlike dust. This is removed by threshing and is melted down in clay vessels. It is then strained, cooled, and formed into cakes or broken into small pieces for shipment. Several grades are recognized. The crude product is greenish-gray in color and is very hard with a high melting point. Carnauba wax is used in the manufacture of candles, soap, high-luster varnishes, paints, carbon paper, phonograph records, batteries, sound films, insulation, salves, ointments, and many other products.

A similar wax, obtained from the trunk of *Ceroxylon andicola* of the South American Andes, is often used as a substitute.

Candelilla Wax. This wax is obtained from *Euphorbia antisyphilitica*, a low, light bluish-green desert shrub of Texas, Mexico, and Northern Central America. Its tiny leaves are quickly deciduous. The wax exudes from pores and forms a thin film on the stems. The amount produced increases during the winter, so the plants are collected at that time. Wild plants are the principal source. The wax is extracted by solvents or by boiling. The crude material is white, but the refined product is a light tan color and has a sweetish odor. Candelilla wax is softer, contains more resin, and has a lower melting point than carnauba, so it is not as valuable. It is generally used as an extender wax in mixture with others. However, its qualities may be improved chemically so that it may be used alone.

Myrtle Wax. The berries of the bayberry (*Myrica pensylvanica*) (Fig. 108) and the wax myrtle (*M. cerifera*), both native to the eastern United States, are covered with a thick layer of wax. This material is actually

FIG. 107. The carnauba palm (*Copernicia cerifera*) in Brazil. This palm yields the most important vegetable wax. (*Reproduced from Pan American Bulletin, March,* 1932.)

a fat, rather than a wax. It is removed from the fruits by boiling in water and is used for candle and soap manufacture.

Cauassú Wax. The leaves of the cauassú (*Calathea lutea*), a tall herb of the lower Amazon region, is potentially an important source of a commercial wax. The wax is produced on the underside of the large leaves. Drying in the sun for only 2 to 3 hours is sufficient for the tissue-paper-thin scales to form; they are removed by scraping. New wax-yielding leaves are produced within a year, whereas it takes 8 to 10 years for carnauba leaves to be renewed. Cauassú wax is very similar to carnauba and can be used for much the same purposes.

Jojoba Wax. The jojoba bush (*Simmondsia chinensis*), an evergreen shrub of the semiarid regions of the southwestern United States and adjacent Mexico, is unique in having seeds with a 50 per cent liquid wax content. This is suitable for making waxing compounds, polishes, and candles. It can also be used as an illuminant and, after processing, for several types of food products. The oil cake is an excellent stock feed, and the acornlike fruits are edible.

FIG. 108. A bayberry bush (*Myrica pensylvanica*) on Cape Cod, Massachusetts. (*Photo by Walter H. Hodge.*)

Other Waxes. Commercial wax is also obtained from the berries of the Japanese wax tree (*Rhus succedanea*) and allied species and from the leaves of the raphia and licuri palms, sugar cane, and esparto.

SOAP SUBSTITUTES

A considerable number of plants contain natural products that can be utilized as soap substitutes. These are the saponins, a group of water-soluble glucosides. Plants that contain saponins yield a soap froth in water, form emulsions with fats and oils, and are capable of absorbing large amounts of gases, such as carbon dioxide. The use of these plants and their products in industry is correlated with the above properties. In addition to the few that are commercially important, there are numerous wild species that are used locally. The most important saponin-containing plants are:

Soapbark. The soapbark tree (*Quillaja Saponaria*) grows on the western slopes of the Andes in Peru and Chile. The commercial material is the dried inner bark, which is removed after the outer bark has been shaved off. The saponin content of the bark is 9 per cent. Soapbark forms a copious lather in water and is used in washing delicate fabrics. It was one of the best wartime emergency materials for cleaning lenses and precision instruments. In medicine it is used to some extent as an expectorant and emulsifying agent. However, it is a dangerous drug to take internally, as it is very toxic and tends to dissolve the blood corpuscles. For this reason its use to increase the foaming power of beer and other beverages, owing to its ability to dissolve gases, is being discouraged. Soapbark is also a good cutaneous stimulant and is much used in hair tonics.

Soapwort. The familiar old-time garden plant known as Bouncing Bet or soapwort (*Saponaria officinalis*) contains a considerable amount of saponin. This plant, a native of Eurasia, is now naturalized in the United States. When placed in water the leaves produce a lather which is utilized for washing silks and woolens. It not only cleanses, but imparts a luster as well.

Soapberries. Soapberries are the fruits of a tropical American tree, *Sapindus Saponaria*. They are used as soap substitutes and in the preparation of hair tonics.

Soaproot. The bulbs of the California soaproot (*Chlorogalum pomeridianum*) yield a good lather and are much used locally in washing fabrics.

CHAPTER X

SUGARS, STARCHES, AND CELLULOSE PRODUCTS

SUGARS

Sugars are manufactured by green plants and so are to be found in small amounts, at least, in all individuals. So much of the manufactured product, however, is used directly in the metabolism of the plant that comparatively little is accumulated. Storage sugars are to be found in roots, as in the case of beets, carrots, parsnips, etc.; in stems, as in sugar cane, maize, sorghum, and the sugar maple; in flowers, such as the palm; in bulbs like the onion; and in many fruits. Several types of sugar are to be found, chief among which are sucrose or cane sugar, glucose or grape sugar, and fructose or fruit sugar. These sugars obviously serve as a reserve food supply for the plant.

Sugar likewise is one of the most necessary foods for man. The day has passed when it was considered merely as a luxury to be used for flavoring purposes. It constitutes a perfect food as it is in a form that can be readily assimilated by the human body. Its chief value is as an energy producer, and it is particularly well adapted for use after any type of muscular exertion. Although primarily a food, such a vast industry has been built up in connection with the extraction of the sugar from plant tissues, its purification, and refining, that it seems best to consider sugar as an industrial plant product. Moreover, sugar has become an extremely important industrial chemical with some 10,000 different derivatives.

Sugar is one of the most valuable products of the plant world. It is surpassed in importance only by wheat, corn, rice, and potatoes, and over 35,000,000 tons are produced annually with a value of $2,000,000,-000. Considering its importance, it is surprising that the sources from which it is obtained are so few in number. Only the sugar cane, sugar beet, sugar maple, maize, sorghum, and a few palms are of commercial interest. In all of these sucrose is the type of sugar stored.

Sugar Cane

The chief source of sugar at the present time is the sugar cane (*Saccharum officinarum*). This plant is a vigorous and rapid-growing perennial grass, which reaches a height in cultivation of 8 to 12 ft. or more and a diameter of about 2 in. It grows in clumps (Fig. 109), with bamboolike stems arising from large rootstalks and with very ornamental feathery plumes of flowers. The stem is solid with a tough rind and numerous fibrous strands, and contains about 80 per cent of juice, the sugar content of which varies greatly.

The commercial sugar cane is a cultigen and is not known in a wild state. The plant was probably first domesticated in Southeastern Asia or the East Indies from some wild ancestor native to that region. By

FIG. 109. A field of sugar cane (*Saccharum officinarum*).

327 B.C. it had become an important crop in India. It reached Egypt in A.D. 641, and Spain in A.D. 755. Since that time sugar cane has gradually been introduced into all moist tropical and semitropical regions. The Spaniards and Portuguese were the great disseminators of the plant into the New World. They carried it to Madeira in 1420, and to America by the beginning of the following century. Within another hundred years it had spread all over the West Indies and Central and South America. Sugar cane was first introduced into the United States in Louisiana in 1741. The name "sugar" comes from the Sanskrit "sarkara," meaning

gravel, and refers to the crude sugar, which was the only kind known for centuries.

Sugar cane is the greatest export crop of the tropics and is unaffected by many of the conditions that influence other crops. It will grow well in any moist hot region where the average rainfall is 50 in. or more a year, where there is ample sunshine, and where the temperature does not fall below 70°F. The methods of cultivation vary considerably in the different countries where the plant is grown. In general, however, the practice is fundamentally the same. Extensive, flat, low-lying fields are utilized and these are plowed deeply. The sugar cane is propagated by cuttings of varying length made from the upper joints of old canes. These cuttings, known as *seed*, are placed in trenches and nearly covered with soil. They begin to sprout in about two weeks. When the cane is being grown for home consumption, the cuttings are often placed in holes. The crop has to be cultivated, weeded, and fertilized extensively during the first few months. It is harvested from 10 to 20 months after sprouting. Close watch is kept of the sugar content and the canes are cut at just the right stage. This is usually when the flowers are beginning to fade. The stems are cut close to the ground for the lower end of the cane is richest in sugar. Cane knives are ordinarily used, for machines have rarely proved practicable. The rhizomes will normally give rise to two or three more crops, known as *ratoons*, before another planting is necessary. Under exceptional conditions as many as 20 ratoon crops have been obtained.

Many of the cultivated varieties in use today are hybrids of *Saccharum officinarum*, the "noble cane," with other hardier species.

Formerly each small owner of a stand of sugar cane extracted his own sugar in a primitive mill, and even today these simple methods are followed in many places. In general, however, large "centrals" have been established which draw their supply from a wide area. The canes are brought to the centrals by railroad or any available means of transportation (Fig. 110). In the milling process the canes are first carried to crushers, where they are torn into small pieces. They are then passed through three sets of rollers. In the first set two-thirds of the juice is expressed. They are then sprayed with water to dilute what sugar remains, and are passed through the second set. These rollers exert a tremendous pressure and remove nearly all the moisture. After passing through the last set the residue is almost bone dry. This *bagasse*, as it is called, can be used as a fuel for the mills, or as a source of paper or wallboard because of its fibrous nature. It also contains a wax with commercial possibilities.

The juice as it flows from the mill is a dark-grayish sweet liquid full of impurities. It contains sucrose, and other sugars as well, together with proteins, gums, acids, coloring materials, dirt, and pieces of cane. The purification of the sugar involves the separation of the insoluble materials (*defecation*) and the precipitation of the soluble nonsugars (*clarification*). The juice is first strained or filtered to remove the solid particles. It is then heated to coagulate the proteins, a process which the addition of sulphur assists. Next lime is added to neutralize the acids present, to prevent the conversion of sucrose into less desirable sugars, and to precipitate some of the substances in solution. These are removed by a series of filter bags or a filter press. Carbon dioxide is often added to aid in the process. The chemical processes involved in the purification of sugar are of great importance and are under constant supervision. The juice is

FIG. 110. Hauling sugar cane from the field to the mill with a tractor. (*Courtesy of the United Fruit Company.*)

now clear and dark colored, and ready for concentration. It is boiled down to a syrup of such density that the sugar crystallizes out. This operation is carried on in open kettles or vacuum pans. The resulting sticky mass is known as *massecuite*. This is placed in hogsheads with perforated bottoms. The juice slowly percolates through the holes, leaving the crystals of sugar behind. This juice constitutes the familiar *molasses* of commerce. In many modern plants the massecuite is centrifuged, the molasses passing out through fine perforations. The raw or crude sugar thus obtained is brown in color and 96 per cent pure.

It can be seen that there are several by-products of value. The bagasse has already been mentioned. The molasses is a good foodstuff and is much used for cooking and candymaking. It is also used in the manufacture of rum and industrial alcohol. The better grades of molasses are obtained when the cruder methods of sugar milling are employed, for in such cases the sugar content of the molasses is higher. A mixture of bagasse and molasses, known as molascuit, is a valuable cattle food.

The final stage in the preparation of sugar for the market is the refining. This is usually carried on in factories located in the seaboard cities of the United States and Europe. The process involves washing to remove the film of dirt from around the crystals of crude sugar, dissolving the sugar in hot water, the removal of any mechanical impurities by filtering through cloth, decoloring by passing through bone black, recrystallization by boiling, and the removal of the liquids from the granulated sugar by centrifuging or other means. One hundred pounds of raw sugar yields 93 lb. of refined sugar and ¾ gal. of refined molasses. The granulated sugar is washed, dried, screened, and packed. Loaf, cube, and domino sugars are made by treating granulated sugar with a warm concentrated sugar solution and pressing it into molds. Loaf sugar is often sawed into blocks, strips, or other forms. Powdered sugar is made from loaf sugar or imperfect pieces of the other types by grinding, bolting, and mixing with starch to prevent lumping. The refining of sugar is a very old process and was probably derived from the Arabs. The first type of refined sugar was the sugar loaf, which was known in England as early as 1310 and was familiar in America until well into the last century.

In 1947 India led in the production of cane sugar, raising about 25 per cent of the world's crop of 25,653,000 short tons. Cuba, Brazil, Puerto Rico, the other West Indies, and Australia followed in the order named. The United States mainland (Louisiana, Florida, and Texas) produced 376,000 short tons. The United States leads the world in sugar consumption.

Sugar Beet

A second important source of sugar is the sugar beet, a variety of the common garden beet (*Beta vulgaris*) which was derived from the wild *B. maritima*, a species still found on the seacoasts of Europe. The production of beet sugar has equalled, or even exceeded, that of cane sugar at various times. During recent years, however, only about half as much beet sugar as cane sugar has been produced.

The sugar beet was known before the beginning of the Christian era, but was not used as a source of sugar until modern times. The presence of sugar in the roots was first noted in 1590, but Marcgraf in 1747 was the first to realize its possibilities. The first real impetus to the industry came about 1800 in both France and Germany. Napoleon backed it intensively as part of his embargo against British goods. He was subjected to much ridicule because of this, and a famous cartoon was drawn picturing him dipping a sugar beet into his coffee; another showed him offering one to his little son, the King of Rome, with the caption, "Suck, dear, suck, your father says it's sugar." With the decline of Napoleon's power, interest in the crop waned. It was revived in France about 1829, and in Germany in 1835, and since that time has been a crop of steadily increasing importance in many European countries. Attempts were made to cultivate the sugar beet and manufacture beet sugar in the United States as early as 1836, but the industry has been successful only since 1879.

The sugar beet is a white-rooted biennial (Fig. 111) which grows best in regions where the summer temperature is around 70°F. The plant will grow in almost any good soil, and also in semiarid regions that can be irrigated (Fig. 112). The plants are grown from seed and must be thinned out until they are from 8 to 10 in. apart. Thorough weeding and deep cultivation are necessary. The crop lends itself readily to machine cultivation and harvesting, and so is much less expensive than sugar cane. The seeds are sown in April and the roots are kept in the ground until October for the sugar content increases as long as they are intact. Before the ground becomes too hard they are pulled, the tops are removed (Fig. 113) to prevent any sugar utilization, and they are stored. The finest plants are saved for seed. The sugar beet has been greatly improved by selection.

Fig. 111. A typical sugar beet (*Beta vulgaris*). (Reproduced from U.S.D.A. Farmers' Bulletin 1637, *Sugar Beet Culture in the Humid Area of the United States.*)

Extraction of the juice is simpler than in the case of sugar cane, for the roots are soft and pulpy. Formerly they were rasped to a pulp and the juice squeezed out in bags, but at present a diffusion process is almost universally used. The roots are cleaned, cut into thin strips, and heated in running water in a series of tanks. Ninety seven per cent of the sugar can be extracted in this way. The waste beet pulp is removed, and the insoluble impurities in the raw juice are precipitated out by a process known as carbonation. In this process the raw juice is treated with lime,

FIG. 112. Sugar-beet cultivation under irrigation. (*Courtesy of the U.S. Beet Sugar Association.*)

FIG. 113 Harvesting sugar beets. (*Courtesy of the U.S. Beet Sugar Association.*)

which coagulates some of the nonsugars, and carbon dioxide, which precipitates calcium carbonate. This settles out along with the impurities, and the purified juice is separated out by filtration. The process is repeated several times, during which sulphur dioxide is added to adjust the alkalinity. Filtration results in a clear liquid which is concentrated, crystallized, and centrifuged as in cane sugar. The massecuite is reboiled several times. It is impossible to differentiate between raw beet sugar and raw cane sugar for they are identical in composition and appearance. By-products of the industry include the tops, which are used for cattle food and fertilizers; the wet or dried pulp, which is a valuable cattle and sheep food; the filter cake, which is used as a manure; and the molasses, which is used for stock feeding or for industrial alcohol.

In 1947, the total production of beet sugar was 8,806,000 short tons. The U.S.S.R. led with nearly 2,000,000 tons. Germany, France, Poland, and Czechoslovakia also grew large amounts. The output in the United States was 1,835,000 tons, chiefly from California, Colorado, and Idaho.

Sugar Maple

The making of syrup and sugar from the sap of maple trees is confined to northeastern North America, and was discovered and developed in a crude way by the Indians. This was referred to by all the early explorers as far back as 1673. Many interesting legends exist regarding the first discovery of the sap by the Indians.

Several species of maple have a sweet sap. The most important of these are the sugar maple (*Acer saccharum*) and the black maple (*A. nigrum*). The sugar maple is a prominent tree in the northern part of the eastern deciduous forest region. It reproduces naturally and lives to an age of 300 or 400 years. The red and silver maples have such a small yield that they are of little use. The sap begins to flow about the middle of March and continues for a month or more. This is a period of warm, sunny days and cold nights. The best flow comes when the temperature reaches 25°F. at night and 55°F. during the day. The best location for tapping is the first 3 in. of sapwood, about 4½ ft. above the ground.

The Indians made incisions in the bark or large roots and conveyed the sap by reeds or curved pieces of bark into clay or bark receptacles. They boiled down the sap by dropping hot stones into it, and converted the sap into sugar by letting it freeze and skimming off the ice. The early settlers were quick to adopt the custom and they made many improvements, eventually tapping with a 1-in. auger and using spiles to convey the sap into containers. They evaporated the sap in the open in large kettles so the sugar had many impurities. The dark-brown sap was stored and later converted into sugar. This involved the famous sugaring-off, a process in which the syrup was boiled until it became waxy and was then dropped into snow. It was then poured into molds, where it immediately crystallized. These simple methods are still in use for domestic purposes.

In commercial production further advances have been made and today modern evaporators have replaced the furnaces and boiling pans of earlier times. These are able to convert from 25 to 400 gal. of sap into syrup in 1 hour. Great improvements have also been made in cleanliness and in the methods of collecting and transporting the sap. These large operations involve from 100 to 1000 trees. A well managed sugarbush (Fig.

FIG. 114. A maple-sugar orchard in Vermont. The sugar maple (*Acer saccharum*) trees have been tapped and pails put in place for collecting the sap. (*Photo by U.S. Forest Service.*)

114) has about 70 trees to the acre, thus allowing plenty of room for the development of the individual trees.

The maple-sugar industry reached its peak in 1869 when 45,000,000 lb. were produced. With the advent of cane sugar it ceased to be an important commodity. Wholesale adulteration also tended to make it less popular. Today the product is much purer, and the demand is increasing. Vermont leads in both sugar and syrup production, followed by New York and Ohio.

Palm Sugar

The juice of several species of palms constitutes a fourth source of commercial sugar. This, however, is available only in the tropics. The chief species utilized are the wild date (*Phoenix sylvestris*), the palmyra palm (*Borassus flabellifer*), the coconut (*Cocos nucifera*), the toddy palm (*Caryota urens*), and the gomuti palm (*Arenga pinnata*). Several of the oil palms also yield sugar. The date palm is tapped like a maple and the sap is obtained from the tender upper portion of the stem. In the other palms the sap is obtained from the unopened inflorescences. Usually the tip of these is cut off and the sap oozes out and is collected in various sorts of containers (Fig. 115). The yield of this sweet juice, which is known as toddy, amounts to 3 or 4 qt. a day for a period of several months. The sap has a sugar content of about 14 per cent. It is boiled down to a syrupy consistency and poured into leaves to cool and harden into the crude sugar, known as jaggary. Some of this reaches European markets.

FIG. 115. Inflorescences of the coconut palm (*Cocos nucifera*) tapped and bamboo tubes in place for collecting the juice. (*Photo by Philippine Bureau of Forestry.*)

Three quarts of juice yield 1 lb. of sugar. The toddy is often fermented to make the beverage known as arrack. The palm sugar industry is a very old one in India and over 100,000 tons are still produced each year. Palm sugar is also made in many other tropical countries.

Sorghum Syrup

The juice contained in the stem of the sweet sorghum (*Sorghum vulgare* var. *saccharatum*) is much used in making syrup. The difference between a true syrup and a molasses should be borne in mind. A syrup is the product obtained by merely evaporating the original juice of a plant, so that all the sugar is present. Molasses, on the other hand, is the residue left after a juice has been concentrated to a point where much of the sugar has crystallized out and been removed. The sweet sorghum or sorgo is a wild plant of the tropics and subtropics which has long been cultivated in many countries. The juice is a poor source of sugar, but yields a nutritious and wholesome syrup. The stems are easily crushed, and the juice is evaporated in shallow pans. On the average about 11,000,000 gal. are made in the United States for cooking purposes. A similar syrup is now being made from sugar cane by clarifying the juice and evaporating it to a consistency where the water content is 25 to 30 per cent.

Other Sugars

Glucose. Glucose, also known as dextrose or grape sugar, is the first sugar to be manufactured by the plant. It is present in small amounts in many of the organs of higher plants, and is particularly characteristic of fruits. For commercial purposes, however, glucose is prepared from starch, and it will be discussed in connection with starch products.

Fructose. Fructose, sometimes called levulose or fruit sugar, is present in many fruits along with glucose. It is slightly sweeter than cane sugar, and is of value because it can be eaten by diabetics. Fructose is prepared commercially from inulin, a polysaccharide that occurs in the tubers of the dahlia (*Dahlia pinnata*), the Jerusalem artichoke (*Helianthus tuberosus*), and other plants. The latter species is now being cultivated quite extensively as a source of inulin.

Mannose. Mannose does not occur free in nature, but is obtained by hydrolysis from several complex compounds. It is also readily oxidized from the juice of the manna ash (*Fraxinus Ornus*), a tree of Sicily and Southern Europe. This juice oozes from slits made in the bark and dries into a very sweet flakelike substance known as manna. Manna is used chiefly in medicine.

Maltose. Maltose rarely occurs in a free condition in plants, but is readily produced from starch through the activity of the enzyme diastase. The chief use of maltose is in connection with the brewing industry, to be discussed later. Maltose syrup is sometimes used as a substitute for glucose and in medicine. The maltose obtained from rice starch has been used in Japan as a flavoring material for over two thousand years.

Honey

Most showy flowers secrete a sweet substance, known as nectar, which serves to attract the various insects essential for pollination. Nectar is composed chiefly of sucrose, with some fructose and glucose. It is used as food by bees, and some of it, after partial digestion, is converted into honey and stored up for future use. During this process the sucrose is changed into invert sugar, a mixture of fructose and glucose. Honey contains 70 to 75 per cent invert sugar, together with proteins, mineral salts, and water. Often the sugar tends to crystallize out. Honey was probably man's first sweetening material, and it has been used by him as a

SUGARS, STARCHES, AND CELLULOSE PRODUCTS 221

food for countless ages. Beekeeping is one of the oldest of industries. The flavor and quality of the honey vary, depending on the source of the nectar. Flowers that contain essential oils impart a characteristic taste. Although many plants are visited by bees, some are especially favored, and these are often cultivated near the apiaries. Alfalfa, clover, buckwheat, lindens, and several of the mints and citrus fruits are among the favorites. Honey is an excellent food for man, for it is almost pure sugar. It is also used in medicine, in the tobacco industry, and in the preparation of a fermented beverage called mead.

STARCHES AND STARCH PRODUCTS

Starch is one of the most important and widely occurring products of the vegetable kingdom, and constitutes the chief type of reserve food for

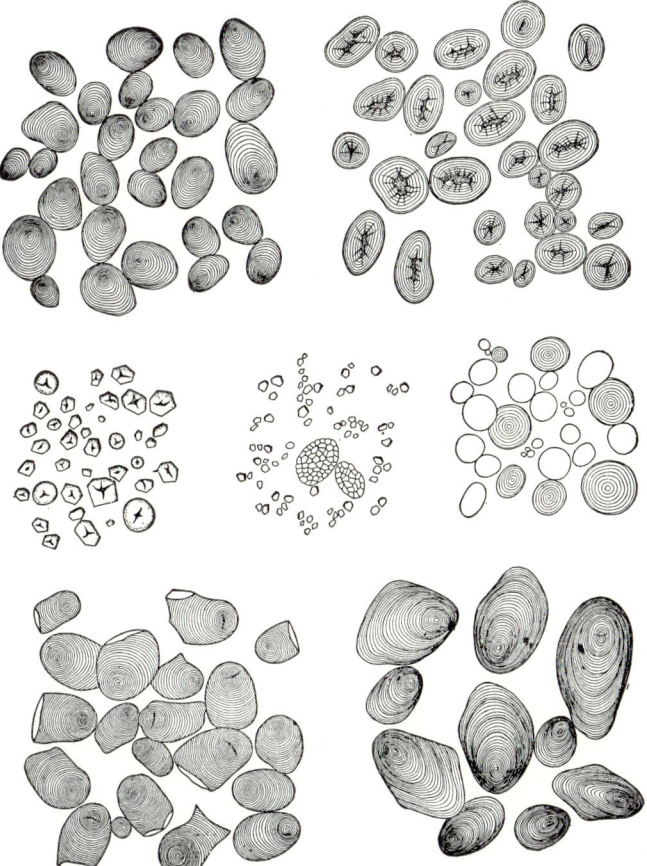

FIG. 116. Starch grains. Upper row, potato and sago; second row, wheat, rice, and corn; third row, bean and West Indian arrowroot. (*Reproduced by permission from Brown, Textbook of General Botany, Ginn and Company.*)

green plants. It is a complex carbohydrate. It is stored up in thin-walled cells in the form of grains. There are several types of starch differing in the size and shape of the grains and other physical and microscopic characteristics (Fig. 116). The most important sources of starch are the cereal grains and underground tubers, although legumes, nuts, and other plant organs may contain appreciable amounts. Although starch serves as the staple food for animals, and the greater portion of mankind as well, it has so many industrial applications that it may well be considered at the present time.

Starch, in the form known as soluble starch, is extensively used in the textile industry to strengthen the fibers and cement the loose ends together, thus making a thread that is smoother and easier to weave. It also gives a finish to the goods. In calico printing it serves as a mordant, a thickener or vehicle for the colors. Starch is likewise used as a sizing agent in the paper industry, in laundry work, in medicine, in the preparation of toilet powders, as a binding material for china clay, and as a source of many derivatives or starch products.

Sources of Commercial Starch

Comparatively few starch-containing plants are used for the commercial production of starch, although the number is constantly increasing. The chief sources are maize, potato, wheat, rice, cassava, arrowroot, and sago.

Cornstarch. The grains of maize or Indian corn constitute the source of five-sixths of the starch made in the United States. The grains are soaked in warm water with a little sulphurous acid to loosen the intercellular tissue and prevent fermentation. Then the corn is ground in such a way as not to injure the embryos. This ground material is placed in germ separators, where the embryos are removed. The starch material is then ground very fine and is either passed through sieves of bolting cloth or is washed in perforated cylinders in order to remove the bran. The resulting milky liquid is run onto slightly inclined tables, where the starch grains settle out and the remaining material flows off. The starch is later collected and dried in kilns, and is then ready for the market. The better grades of cornstarch are used for food; the inferior grades are used for laundry starch and sizing and as a basis for glucose. The United States produces over 400,000 tons of cornstarch.

Potato Starch. Europe leads in the production of starch from potatoes with an output of 300,000 to 400,000 tons annually. In the United States about 30,000 tons are prepared, and some is imported. In this country the industry is carried on in many small factories located in the potato-raising states, chiefly in Maine and Wisconsin. The culls are utilized for starch. These are washed and reduced to a pulp in graters or rasping machines, and the resulting paste is passed through sieves to remove all the fibrous matter. After washing, the solid starch is separated out by sedimentation, the use of inclined tables, or centrifuging, and is then dried. Potato starch is used in the textile industry and as a source of glucose, dextrin, and industrial alcohol.

Wheat Starch. Wheat grains constitute the oldest of the commercial sources of starch. It was known to the Greeks, and was extensively used in Europe in the sixteenth century in connection with the linen industry.

FIG. 117. *Maranta arundinacea*, one of the sources of arrowroot starch, growing at Tela, Honduras. (*Photo by Walter H. Hodge.*)

The presence of the gluten in wheat makes the removal of the starch a more difficult process. It is carried out by extraction with water or by the partial fermentation of the grain. Wheat starch is used almost exclusively in the textile industry.

Rice Starch. Rice starch is obtained from broken or imperfect grains of rice. These are softened by treating with caustic soda, and are then washed, ground, and passed through fine sieves. More alkali is added and after a time the starch settles out as a sediment. This is removed, washed, and dried. Sometimes the starch grains are freed by treating with dilute hydrochloric acid. Rice starch is used for the most part in laundry work, and to some extent for sizing.

Cassava Starch. The cassavas of tropical America are important sources of food and will be discussed later. Cassava flour and the more familiar tapioca, however, are also useful in the industries, particularly as sizing materials and as the source of many starch products.

Arrowroot Starch. This starch is obtained from the tubers of several tropical plants. West Indian arrowroot comes from *Maranta arundinacea* (Fig. 117), Florida arrowroot from *Zamia floridana*, Queensland arrow-

FIG. 118. A plantation of sago palms (*Metroxylon Sagu*) in Singapore. (*Photo by E. H. Wilson; courtesy of the Arnold Arboretum.*)

root from *Canna edulis*, and East Indian arrowroot from *Curcuma angustifolia*. The tubers are peeled, washed, and crushed and the pulp is passed through perforated cylinders. A stream of water carries the starch into tanks where it settles out. Arrowroot starch is very easily digested and so is valuable as a food for children and invalids. It is but little used in the industries.

Sago Starch. This starch is obtained from the stems of the sago palm (*Metroxylon Sagu*), a tall tree of the tropics (Fig. 118). It is cultivated in Malaya and Indonesia. The flowers appear when the trees are about 15 years of age, and just prior to this time the stems store up a large amount of starch. The trees are cut down and the starch pith is removed. This is ground up, mixed with water, and strained through a coarse sieve. The starch is freed from the water by sedimentation and, when washed and dried, is known as sago flour. The sago of commerce is prepared from this by making a paste and rubbing it through a sieve in order to bring about granulation. The product is dried in the sun or in ovens and appears as hard shiny grains, known as pearl sago. Both sago starch and pearl sago are used almost entirely for food purposes.

Starch Products

Soluble Starch. Although starch grains are insoluble in cold water, they readily swell in hot water until they burst, and form a thin, almost clear solution or paste. This soluble starch is much used for finishing textiles and in the paper industry.

Dextrin. When starch is heated directly, or is treated with dilute acids or enzymes, it becomes converted into a tasteless, white, amorphous solid known as dextrin, or British gum. Because of their adhesive properties dextrins are frequently used as substitutes for mucilage, glue, and the natural gums. The United States now uses cornstarch dextrin on its postage stamps. Loaves of bread are painted with dextrin so a crust will form. In the steel industry the sand for the cores used in casting is held together with dextrin. Other uses are in cloth printing, glazing cards and paper, and making pasteboard.

Glucose. If the process of treating starch with dilute acids is carried far enough, the starch is more completely hydrolyzed and is converted into glucose, a valuable sugar. In the United States glucose is made chiefly from cornstarch, and the product is often referred to as corn syrup. Often the same factory extracts the starch and then converts it into glucose. The operation is carried on in large copper boilers under pressure. About 6 lb. of dilute hydrochloric or sulphuric acid are used for each 10,000 lb. of starch. After all the starch has been converted, the free acids are neutralized with caustic soda. The liquid is then decolorized with boneblack and concentrated into a thick syrup. One of the common brands of corn syrup is "Karo." Glucose is often thought of as an inferior substitute for cane sugar. As a matter of fact, it is a real sugar in its own right and is a good food material. It is used as a table syrup, for sweetening, and in candies, jellies, and all kinds of cooking. It is often mixed with maple syrup, brown sugar, honey, or molasses. It is used for making vinegar and in brewing. The United States produces over 1,000,000 lb. annually. In Europe glucose is made chiefly from potato starch. Crystallized corn sugar or grape sugar can be prepared by continuing the hydrolysis of the starch to its completion, particularly if dilute sulphuric acid is used.

Industrial Alcohol. An enormous quantity of industrial alcohol is manufactured from starch. Corn and potatoes constitute the chief sources, although the other starches, and even cellulose, the various products of the sugar industry, and fruit juices are utilized. The process involves the conversion of the starch into sugar by means of diastase and the fermentation of the sugar by yeasts to yield the alcohol. The operations are carried out under different conditions from those followed in the manufacture of alcoholic beverages. After fermentation has stopped, the alcohol is extracted from the mash by fractional distillation. The alcohol thus formed as a result of the fermentation of sugar is known as ethyl alcohol, as distinguished from methyl or wood alcohol, a product of the destructive distillation of wood. In order to render it undrinkable, ethyl alcohol is often "denatured" by the addition of methyl alcohol or other substances. Industrial alcohol is the most important and most widely used solvent and is the basic material in the manufacture of several hundred products. It is also used in medicine, pharmacy, and various industries.

Nitrostarch. Starch is so similar to cellulose in composition that it reacts in similar ways. Just as cellulose reacts with nitric acid to form nitrocellulose, so starch unites with this acid to yield nitrostarch. This is one of the safest of the high explosives, particularly if the ingredients are absolutely pure. Tapioca starch was formerly imported for this purpose, but during the First World War the United States developed cornstarch as a source. By the end of the war 1,720,000 lb. of nitrostarch a month were being made for use in hand grenades.

CELLULOSE PRODUCTS

Cellulose, the most complex of the carbohydrates, is universally present in the cell walls of plants. Because of their strength, cells with thick walls have long been utilized in the industries. We have already discussed the usefulness of the various plant fibers in the textile industry. Here, as in other cases, the natural products were the first to be used. Later the plants were cultivated, and during this period of cultivation it was possible to improve very considerably on nature. As a result of repeated experimentation, longer, stronger, and cheaper fibers were produced. Now we have reached a third stage in which we are no longer entirely dependent on natural products for our fibers, but are able to make them directly from cellulose or from various chemical elements. These artificial fibers are only one example of the countless derivatives of cellulose, all of which are interesting and extremely valuable in our daily lives.

SUGARS, STARCHES, AND CELLULOSE PRODUCTS 227

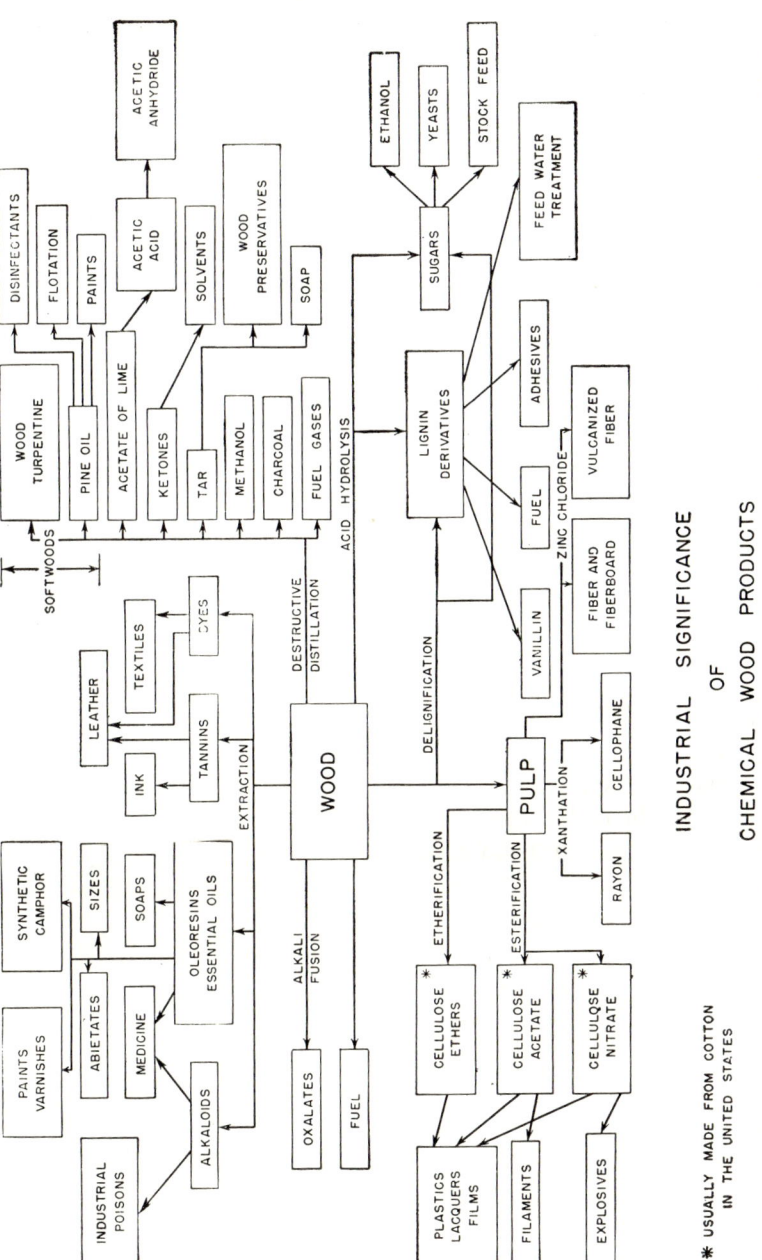

* USUALLY MADE FROM COTTON IN THE UNITED STATES

FIG. 119. Industrial significance of chemical wood products. (*Reproduced from Panshin, Forest Products, McGraw-Hill Book Company, Inc.*)

Cellulose chemistry has become one of the most important phases of organic chemistry.

Cotton is a very pure form of cellulose, and for a long time it was the only source of the cellulose used in the production of the artificial fibers and other cellulose products. The science of wood technology has now made possible the production of high-grade cellulose from wood and its conversion into many different products and raw materials of industry (Fig. 119). When certain woods are treated with concentrated acids or alkalies, the bond between the wood fibers and the lignin which cements them together is broken, and the fibers which are pure cellulose can be removed. These fibers may then be reorganized as *paper*, or they may be treated further with chemicals. If this chemical treatment merely causes the dissolution of the fiber into its component molecules, these molecules may be resynthesized into *artificial fibers* or converted into *cellulose plastics*. However, if the molecules themselves are broken down, their component elements—carbon, hydrogen, and oxygen—may be recombined to form *wood sugar*. In its turn, this wood sugar may be transformed into yeast or alcohol and thus become available for food or as the raw material for numerous industrial products.

Paper and Papermaking

One of the important uses of cellulose is in the manufacture of paper. This is a very old industry. The word "paper" comes from the Latin "papyrus," the name of a sedge, the pith of which was used for paper in Egypt as early as 2400 B.C. The Chinese, however, were the first people actually to make paper. The industry spread from China to India, Persia, and Arabia and thence through Spain to the European countries. The first paper mill in the United States was established near Philadelphia in 1690.

Raw Materials

Paper can be made from any natural fibrous material. The papermaking value of the various fibers depends on the amount, nature, softness, and pliability of the cellulose present in the cell walls. This cellulose may occur alone or in combination with lignin or pectin. The compounds of cellulose that occur in the artificial fibers, however, are not well suited for papermaking. In the past a considerable number of raw materials have been prominent, and many others have been experimented with. At the present time, however, comparatively few are of commercial importance. Chief among these are wood fibers, cotton, and linen.

Wood Fibers

The use of wood as a raw material in the paper industry (Fig. 120) began in 1840 and dates commercially from 1854. Today wood has largely supplanted the other fibers and furnishes 90 per cent of all the paper manufactured in the United States. The industry is still undergoing development in regard to the species utilized and the processes involved. Recent experiments with specially processed and purified wood pulp point to an even greater and more efficient utilization in the future.

Spruce is the most important source of wood pulp and furnishes about 30 per cent of the total supply. It is an ideal material for it has all the requirements of a good pulpwood. The fibers are long and strong with a maximum content of cellulose. The wood is almost free from resins,

FIG. 120. A pile of pulpwood in Somerset County, Maine. (*Photo by U.S. Forest Service.*)

gums, and tannins; and it is light colored, sound, and usually free from defects. Red spruce (*Picea rubens*), white spruce (*P. glauca*), and Sitka spruce (*P. sitchensis*) are all used for pulp.

The second most important pulpwood at the present time is southern yellow pine (*Pinus australis*). This species has had a remarkable rise in prominence in recent years and has outstripped the hemlock, which now occupies third place. The eastern hemlock (*Tsuga canadensis*) continues to be used, chiefly in the Lake States, while the western hemlock (*T. heterophylla*) is increasingly important on the Pacific Coast. The aspens (*Populus grandidentata* and *P. tremuloides*) and the balsam fir (*Abies balsamea*) also furnish a considerable part of the supply.

Other less important species include jack pine (*Pinus Banksiana*), tamarack (*Larix laricina*), white fir (*Abies concolor*), and several hardwoods, among them the beech (*Fagus grandifolia*), sugar maple (*Acer saccharum*), and birch (*Betula lutea*). Sawmill waste is today an increasingly valuable source of wood pulp.

Cotton and Linen

Until the middle of the nineteenth century cotton and linen rags were the only source of paper, and they are still used for the finest grades. Cotton fibers have a high felting power and a high content of cellulose (91 per cent). Rags and raw cotton in the form of fuzz or linters are utilized. Flax fibers, which comprise linen, contain 82 per cent of pecto-cellulose and yield a paper of great strength, closeness of texture, and durability. Textile waste can also be used. In the preparation of rag pulp the material is sorted, cut into small pieces, and freed from dust. It is then boiled in caustic soda to remove the grease, dirt, and dyes, and is washed until perfectly clean. The resulting pulp is ready for the actual operations involved in papermaking.

Raw Materials of Minor Importance

Esparto. The esparto grass (*Stipa tenacissima*) is an important raw material in Great Britain. The plant is a native of Northern Africa, where it flourishes in dry, sandy, and rocky coastal regions. The tufted wiry stems are plucked and pressed into bales for shipment. Esparto is converted into pulp by heating in a caustic-soda solution under pressure. Although the cellulose content is only 48 per cent, the fiber has great flexibility and felting power. It yields an opaque, soft, light paper of uniform grade. The finer printing papers in England are made from esparto, either alone or in mixture. Another grass, *Lygeum Spartum*, is sometimes used as a source of esparto.

Textile Fibers. Many textile fibers in addition to cotton and flax have been used as sources of paper. The waste material of the jute and hemp industry, in the form of old ropes, sacking, sailcloth, and the like, was formerly used extensively for making strong and tough papers. Jute and jute butts were used chiefly for wrapping paper, envelopes, cable insulation, etc., while hemp, after bleaching, yielded ledger and bank-note paper. Manila hemp was an important source of envelopes and wrapping paper. Ramie, sisal, sunn hemp, New Zealand hemp, coir, and many other fibers have been utilized.

Paper Mulberry. The soft, lustrous, and very strong bast fibers of this species, already discussed as the source of tapa cloth, have long been used in Japan for paper lanterns and umbrellas, as well as paper for writing purposes. The fiber is prepared by scraping, soaking, and beating, after which it is mixed with mucilage and spread on frames to dry. When treated with oil this paper becomes strong enough to serve as a substitute for cloth or leather.

Other Sources. The raw materials mentioned above by no means exhaust the list of papermaking plants. The stems of wheat, rye, barley, rice, oats, and other grasses are used for low-grade paper, strawboard, cardboard, and pasteboard. The fibers from these plants have a low cellulose content, and are too short and small to have much tensile strength. They are consequently used in mixture with other fibers. In the United States cornstalks, sugar-cane bagasse, and waste paper have been extensively developed as sources of paper. Even such unusual materials as banana fiber, tree bark, rushes, weeds, licorice, broomroot, tobacco and cotton stalks, beet-pulp waste, and peat have been experimented with.

Still other species are used in oriental countries. Bamboo fiber constitutes an important source of paper in India and China. The papyrus (*Cyperus Papyrus*), baobab (*Adansonia digitata*), and *Daphne cannabina* are much used in India and Africa. The so-called rice paper of China and Japan is made chiefly from *Tetrapanax papyriferum*, *Edgeworthia tomentosa*, or *Wickstroemia canescens*.

The Manufacture of Wood Pulp

Wood pulp plays such an important part in the paper industry that it will be worth while to discuss briefly the methods by which it is manufactured. Either a mechanical process or one of three chemical processes is involved.

Mechanical Process

In this method the wood is barked, washed, and cut into small lengths. These bolts, as they are called, are forced against rapidly revolving stones or grinders, and the fibers are torn apart by abrasion. In the *cold method* a large amount of pure water is used to prevent heating, and a fine, even grade of fiber results. In the *hot method* very little water is used, and coarser, longer fibers are obtained. After grinding, the material is screened to remove any impurities and to grade it. It is then pressed between rollers to remove most of the water. Finally it is passed through a lapping machine which turns out flat layers of pulp. These can be made at once into paper, or they can be dried and shipped to consumers elsewhere.

In the mechanical process resins, lignin, and other undesirable materials are not removed. These substances resist bleaching agents and cause the paper to turn yellow. Furthermore the fibers produced do not felt readily. Mechanical pulp, often called groundwood pulp, is poorest in grade, printing quality, strength, and durability. It is, however, the cheapest kind to manufacture. About 30 per cent of the pulpwood used is reduced by this method. Spruce is the wood ordinarily used. About 85 per cent of the total consumption of this species is converted into mechanical pulp. It is used for newsprint, low-grade wrapping paper, and other types of paper where strength is not required and where the paper needs to be used for only a short time.

Chemical Processes

In these processes the lignin and most of the other materials are removed and the fibers which are almost pure cellulose remain. The wood is split and placed in a chopper, which produces coarse chips. These chips are screened to remove knots and large pieces, and are then subjected to one of three treatments.

In the **sulphite process** carefully selected chips are cooked with steam in a solution of acid calcium sulphite. This is prepared by burning sulphur to sulphur dioxide and passing this gas across broken limestone over which water is trickling. After digestion in the liquid the pulp is washed, screened, lapped, and dried. About 50 per cent of the pulpwood used is reduced by this method. Spruce furnishes two-thirds of the supply and hemlock and balsam fir the remainder. Bleached sulphite pulp is used for the higher grades of paper and, after further purification, for artificial fibers.

The **soda process** lends itself to the reduction of hardwoods and pine. It is comparatively simple, involving the digestion of the wood under pressure in a solution of sodium hydroxide at a temperature of about 240°F., followed by washing, bleaching, and lapping. Aspen is the principal species used, although several other hardwoods and southern pine are utilized.

The **sulphate process** is the most recent method for the reduction of pulpwood, and offers promise of greater development in the future. It is adapted to coniferous woods with a high resin content. First used in 1883, it has always been prominent in Europe, and today it is of steadily increasing importance in the United States. In this country southern yellow pine and the waste from lumbering operations and sawmills supply much of the raw material. Hemlock, spruce, tamarack, jack pine, and balsam fir are also used. In this process the wood is digested in a solution of caustic soda, sulphide of soda, and a little sulphate of soda. After cooking in this mixture, the pulp is washed, screened, and pressed. The greater part of the sulphate pulp manufactured is used in making kraft paper, a strong brown wrapping paper. It may be bleached, however, and yields a white paper that is softer and more pliable than that made from sulphite pulp.

The waste liquor, particularly from the sulphite process, is now being reclaimed and can be used in making many valuable products.

The Manufacture of Paper from Pulp

The actual process of papermaking involves several steps, the first of which are concerned with the further treatment of the pulp. These include bleaching, washing, and the complete separation of the fibers. The latter operation is carried on in a machine known as a beater, or "Hollander." During this process sizing, usually rosin, is added to make the paper water- and ink-resistant. In some cases filling or loading materials are added which result in a more printable surface. Clay is commonly used for this purpose. If colored paper is required, the dye is added during the beating process. Mordants are usually necessary to fix the color, for cellulose is very inactive.

From the beater the pulp passes to the papermaking machine itself. This consists essentially of an endless wire screen revolving around a series of rollers. It is a modification of the original Fourdrinier machine, the first to make paper, which was invented at the beginning of the nineteenth century. The final stages of papermaking are carried out in the presence of enormous quantities of water. They are mechanical and physical rather than chemical in nature and result in the reorganization of the fibers. The pulp is poured on the screen and the water passes through, leaving a continuous sheet of interlacing fibers. This sheet is passed on to a felt and is carried to a series of heavy rollers where the water is squeezed out, and then through heated cylinders where the sheets are dried. Finally the paper is run through the calenders where a smooth finish is put on. The paper is then ready to be cut to any desired size.

Kinds of Paper and Paper Products

Once used only as a means of communication, paper now serves a multitude of purposes, and countless paper products are in daily use.

Different kinds of paper require different raw materials and modifications in the process of manufacturing. *Fine papers*, as they are called, include bond, ledger, book, writing, and similar papers. The best grades are made from rags or sulphite pulp; medium grades are made from mixtures of rag and chemical pulp; the poorer grades contain only wood pulp. *Newspaper stock* usually consists of 85 per cent mechanical wood pulp and 15 per cent unbleached sulphite. *Wrapping paper* comes from kraft pulp for the most part, although various textile fibers are still sometimes used as raw materials. *Blotting paper* is composed of short fibers loosely interwoven and free from loading or sizing. Formerly it was prepared entirely from cotton rags, but today soda pulp is used either alone or in mixture. *Paperboard* is made from any low-grade material—old newspapers, boxes, or other waste material—often in combination with mechanical or chemical pulp. *Papier-mâché* consists of old paper stock boiled to a pulp, mixed with glue and starch paste, and pressed into molds. *Vegetable parchment* is prepared from the purest alpha cellulose stock. Sheets of paper are dipped in sulphuric acid and then thoroughly washed with water. This treatment gives a hard translucent coating to the paper; it is waterproof and greaseproof and is used chiefly in packaging. *Bags* are usually made from unbleached kraft; some bleached kraft and sulphite pulps are utilized. *Carbon paper* is a very strong linen or cotton rag paper coated with a colored wax. Sulphite pulp may also be used. *Facial tissue* is obtained from a soft pulp, usually 100 per cent sulphite. *Cigarette paper* is now made from flax; formerly linen rags were utilized. *Roofing and building papers* and many other special types are manufactured by treating paper with wax, tar, asphalt, oils, resins, or other substances. *Papreg* is paper impregnated with plastics and then laminated or molded into a very strong but lightweight material.

Artificial Fibers

From the Middle Ages until the present time there have been many schemes to make artificial silk and other fabrics. In 1880 Count de Chardonnet made the first artificial fiber, and a few years later the first artificial silk. Shortly after, factories were established for making the product. At the outset this new material was handicapped by its name for the public considered it only as an imitation or a substitute. This condition existed as late as 1908. Now we realize that all the artificial fibers constitute entirely new products with valuable characteristics and properties of their own. Today their production is one of the great industries. In the United States alone over 792,000,000 lb. were manufactured in 1945, a notable increase from the 800,000 lb. that constituted the first commercial production in 1911. At least four different processes have been in use. In view of the fact, however, that about 85 per cent of the output is made by the viscose process and is marketed under the name "rayon," the Federal Trade Commission has decreed that all synthetic fibers should bear this name. The objective of all these processes is to overcome the natural limitations of the fibers and give them more desirable qualities.

The raw material of the rayon industry is high-polymer alpha cellulose, prepared in a pure form from wood pulp or cotton linters. Purification is accomplished by the elimination of mechanical impurities through air separation and then cooking in a $3\frac{1}{2}$ per cent sodium hydroxide solution. This removes all the other organic substances. The pure cellulose fibers that remain are bleached, washed, and dried. The first step in the manufacturing process is to dissolve the cellulose by means of various solvents, thus rendering it sufficiently liquid so that it can be squirted in a fine jet. The solution is then forced by pressure through minute perforations in glass or platinum, and emerges from these "spinnerets" in thin streams. These streams are coagulated into fine, almost invisible filaments in various ways. The solvents are removed, and the filaments are caught up by revolving reels and twisted into threads suitable for spinning. The threads are washed, bleached, and dried in skeins. In the viscose, nitrocellulose, and cuprammonium processes the final product is an almost pure cellulose fiber, known as *regenerated cellulose.* Chemically this is identical with the cellulose in cotton, but it differs in its mechanical properties. The product of the acetate process is a cellulose ester, *cellulose acetate.* This differs from regenerated cellulose in both its chemical and physical properties. The various kinds of rayon will be discussed briefly.

Viscose Rayon. This is the original "rayon" process, and furnishes the greater part of the rayon manufactured (Fig. 121). Bleached sulphite wood pulp and pulp from cotton linters are used, often in equal amounts. The purified cellulose is treated with caustic soda and then with carbon bisulphide. The resulting cellulose xanthate, with the addition of a little water, becomes a foamy orange-yellow mass, known as viscose. This is allowed to age for a while and is then forced through the spinnerets into a regenerating solution. Here the xanthate groups are removed and the filaments of regenerated cellulose coagulate.

Cellulose Acetate Rayon. This product is often called "celanese." Cotton or wood pulp is treated with acetic anhydride, acetic acid, and a little sulphuric acid as a catalyst. When dissolved, the material is poured into water and cellulose triacetate precipitates out. This is dissolved in acetone or other solvents and, when about the consistency of honey, it is forced through the spinnerets into a chamber containing warm air. Here the solvents evaporate and the cellulose acetate filaments coagulate. Acetate rayon is second to viscose rayon in importance.

Cuprammonium Rayon. In this process the cellulose is treated with ammoniacal copper hydroxide. The viscous solution that results is forced through the spinnerets into an acid bath consisting of caustic soda or sulphuric acid, where the threads of regenerated cellulose coagulate.

Nitrocellulose Rayon. This name is a misnomer, and the product should more properly be called cellulose nitrate rayon. It is often called Chardonnet silk, while the process is referred to as the Tubize process. Although it is the oldest type of rayon, the process is virtually obsolete at the present time. Cotton linters are utilized. The cellulose is dissolved in nitric and sulphuric acids and the resulting pyroxylin is further dissolved in an ether-alcohol mixture or some other solvent. It is then forced through the perforations and coagulates in the air. The solvents

236 ECONOMIC BOTANY

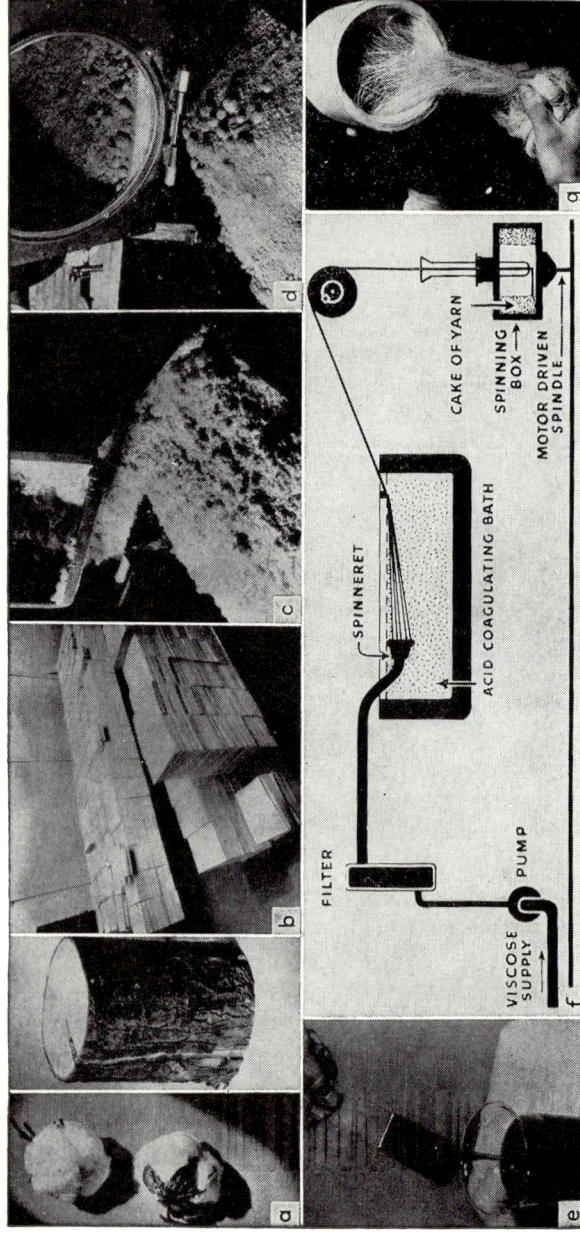

FIG. 121. Steps in the manufacture of viscose rayon: A, cotton and spruce wood; B, sheets of pulp; C, alkali cellulose; D, cellulose xanthate; E, viscose spinning solution; F, spinning operation; G, yarn from spinning machine. (Reproduced from Panshin, Forest Products, McGraw-Hill Book Company, Inc.)

are removed from the cellulose nitrate filaments, and also the nitro groups are removed in order to reduce the inflammability of the product, which is regenerated cellulose.

Rayon fibers vary in tensile strength and elasticity. They may be either lustrous or dull. In general they take dyes more readily than do the natural fibers. One handicap is their tendency to absorb a considerable amount of moisture when wet and to weaken proportionately. They resume their strength, however, on drying out. Improvements are constantly being made which increase the efficiency of the product. Rayon has over 500 uses. It may be utilized alone or in combination with natural silk or fibers of plant origin. Among the products made from rayon may be mentioned hosiery, underwear, shirts, dress and suit fabrics, neckties, ribbons, draperies, and many other household and industrial textiles. An exceedingly important development which greatly increased the usefulness of rayon was the invention in 1926 of the *staple fiber*. This is made by cutting rayon filaments into short, uniform lengths which are then spun into textile yarns. Cellulose acetate rayon does not absorb water as readily as does viscose rayon and so is stronger in the presence of moisture. However, it does not take dyes so readily as the other types. One important use of acetate rayon is in electrical insulation.

Other Cellulose Products

The fact that cellulose will dissolve in various solvents has made possible the development of plastics and many other products of great value and usefulness. It will be possible to mention only a few of these cellulose derivatives.

Cellulose Nitrate Products

When cellulose is treated with concentrated nitric acid in the presence of sulphuric acid, several types of cellulose nitrate are formed. These differ depending on the concentration of the nitric acid, and the consequent degree of nitration; the temperature; and the duration of the action. The higher cellulose nitrates are called guncotton, or, erroneously, nitrocellulose. The lower nitrates constitute pyroxylin, or collodion cotton as it is sometimes called.

Guncotton is prepared from cotton linters, and during the process the cellulose is completely nitrated. It is used as an ingredient of many high explosives. Cordite, for example, is a combination of guncotton and nitroglycerin, while smokeless powder is made from a mixture of guncotton and the lower nitrates. When properly made, guncotton is one of the safest of the explosives to handle, particularly when it is wet.

Pyroxylin is produced by the partial nitration of cellulose. This process is carried on under different conditions from those which result in the formation of guncotton. The product is very valuable in industry. It made modern photography possible, for films usually consist of pyroxylin coated with gelatin. Its use in the rayon industry has already been mentioned. Perhaps the chief value of pyroxylin, however, lies in the fact that it is soluble in a variety of solvents and yields many useful products, such as collodion, celluloid and other plastics, artificial fabrics, and varnishes.

Collodion is a solution of pyroxylin in a mixture of ether and alcohol. If a layer of collodion is spread out and exposed to the air, the solvents gradually evaporate and leave a thin, tough, impenetrable film. This characteristic makes collodion of value as a protective covering for wounds, and "New Skin" is familiar to everyone.

Celluloid, the first plastic, is another well-known cellulose derivative. It consists of pyroxylin dissolved in camphor. Celluloid was first made in 1870 by John Hyatt. He mixed guncotton with camphor and placed the material in a hot press. The result was a clear homogeneous solid. At the present time celluloid can be easily made by nitrating thin tissue paper, or some other form of nearly pure cellulose, up to the point where the cellulose becomes soluble, but not far enough to convert it into guncotton. This pyroxylin is then mixed with camphor, submitted to pressure, and dried. Celluloid can be molded at 100°F. It can be dyed and made to imitate anything from ivory to coral and mosaics. Its chief drawback is its inflammability. Countless other products of a similar nature, such as pyralin, are on the market.

Artificial fabrics are now being made from cellulose nitrate. For many years oilcloth and linoleum were the only materials of this nature, and drying oils were necessary for their manufacture. Now modified cellulose nitrate is combined with various solvents, such as amyl acetate, butyl alcohol, etc., and many new and durable products are obtained. Among them may be mentioned the fabrics used for automobile curtains and cushions, and the leather substitutes, which can be used for shoes, bookbinding, and many other purposes.

The *varnish* industry has been revolutionized by the use of cellulose nitrate. The speed of modern automobile production required a varnish that would dry more quickly than those derived from the natural plant products. In answer to this demand the "lacquer paints," such as "Duco," were developed. Cottons linters are bleached and purified and converted into cellulose nitrate. This is mixed with small amounts of gums, resins, pigments, and various solvents. The resulting product yields in two days a finish more resistant and more durable than the old varnishes formed in 24 days. These new varnishes are adapted to either wood or metal.

Cellulose Acetate Products

Cellulose acetate also has many important industrial uses. It is now extensively employed as a substitute for cellulose nitrate in the film industry for it is much less inflammable. However, the films are more brittle and more expensive. It is also used for automobile goggles, gas masks, automobile windows, coverings for index cards, artificial fabrics, varnish for airplane wings, and many other purposes.

Viscose Products

Viscose products are also of great importance. Perhaps the best known is *cellophane*, so extensively used for wrappings. This is made by forcing crude viscose through tiny slits rather than perforations. It coagulates into a thin transparent film $\frac{1}{1000}$ in. in thickness. These viscose films are now used for countless purposes, even for sausage casings. Viscose fibers have replaced cotton in Welsbach mantles.

Products of Cellulose Hydrolysis

The complete hydrolysis of cellulose by means of acids is a process of saccharification and ultimately results in the conversion of the cellulose into wood sugar, which in its turn yields alcohol and yeast. It is a complement of the manufacture of paper pulp from wood.

During the normal process of pulpmaking 1 ton of wood is converted into ½ ton of pulp and enough wood sugar to yield 10 to 12 gal. of alcohol. If the cooking time is doubled, an edible starchy material, which is an excellent cattle feed, and sufficient sugar to furnish 20 gal. of alcohol are obtained. Using a still longer cooking time all the cellulose can be converted into sugar with an eventual potentiality of 60 to 80 gal. of alcohol.

Several types of wood sugar are produced. Conifers yield glucose, pentose, and mannose; hardwoods glucose and pentose. These sugars are useful because they can readily be fermented.

Glucose and mannose, which comprise about two-thirds of the wood sugar, are converted into ethyl alcohol by the action of ordinary yeast. This industrial alcohol has a wide range of uses. Pentose sugars are not fermentable into alcohol, but they can be converted, through the agency of a bacterium, *Torulopsis utilis*, into the so-called Torula yeast, an edible substance with a 50 per cent protein content. One ton of pulp will yield from 40 to 100 lb. of yeast.

Cellulose hydrolysis should play an increasing role in a well-organized scheme of forest utilization, as it eliminates much of the waste. Small pieces of wood, chips, sawdust, wood flour, and sawmill waste can all be utilized effectively. It is also possible to recover wood sugar from sulphite waste liquor.

HEMICELLULOSE

The seeds of many tropical plants have exceedingly thick, hard, and heavy walls consisting of hemicellulose. This substance, which is a modification of normal cellulose, constitutes a supply of reserve food for the plant. In young seeds the endosperm consists of a milky juice, but as

FIG. 122. The ivory-nut palm (*Phytelephas macrocarpa*) in fruit. The seeds of this palm are the source of vegetable ivory.

the seeds mature this fluid is gradually replaced by the horny material. Hemicellulose cannot be used by animals as a food. It does, however, play a part in the industrial world for it is the source of the vegetable ivory of commerce.

Vegetable Ivory

The chief source of vegetable ivory is the ivory-nut or tagua palm (*Phytelephas macrocarpa*) of tropical America. This palm is a low-growing tree (Fig. 122) characteristic of river banks from Panama to Peru. The drupelike fruits contain from six to nine bony seeds with a thin brown layer on the outside and a very hard and durable endosperm. Great numbers of these seeds are collected by the natives and shipped to Europe and the United States. Ecuador is the chief exporting country. This vegetable ivory can be carved and turned and finds an extensive use as a

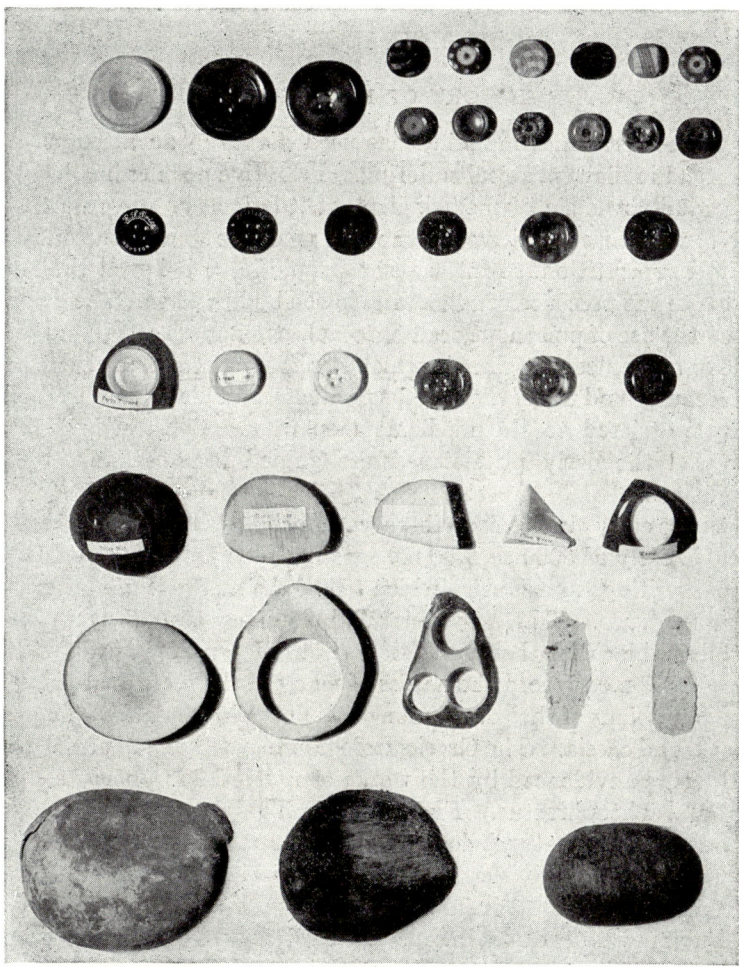

FIG. 123. Stages in the manufacture of buttons from the seeds of the ivory-nut palm. (*Courtesy of the Botanical Museum of Harvard University.*)

substitute for true ivory in the manufacture of buttons (Fig. 123), chessmen, poker chips, dice, knobs, inlays, and similar articles.

Several African and Polynesian palms, as well as other tropical American species, have seeds of a somewhat similar nature. With the exception of *Metroxylon amicarum* of the Caroline Islands, however, these are not important articles of commerce.

CHAPTER XI

MEDICINAL PLANTS

THE HISTORY OF MEDICINAL PLANTS

From earliest times mankind has used plants in an attempt to cure diseases and relieve physical suffering. Primitive peoples in all ages have had some knowledge of medicinal plants, derived as the result of trial and error. These primitive attempts at medicine were based on speculation and superstition. Most savage people have believed that disease was due to the presence of evil spirits in the body and could be driven out only by the use of poisonous or disagreeable substances calculated to make the body an unpleasant place in which to remain. The knowledge regarding the source and use of the various products suitable for this purpose was usually restricted to the medicine men of the tribe. As civilization progressed the early physicians were guided in great part by these observations.

In all the early civilizations there was much interest in drug plants. In China, as early as 5000 to 4000 B.C., many drugs were in use. There are Sanskrit writings in existence which tell of the methods of gathering and preparing drugs. The Assyrians, Babylonians, and ancient Hebrews were all familiar with their use. Some of the Egyptian papyri, written as early as 1600 B.C., record the names of many of the medicinal plants used by the physicians of that day, among them myrrh, cannabis, opium, aloes, hemlock, and cassia. The Greeks were familiar with many of the present-day drugs, as evidenced by the works of Aristotle, Hippocrates, Pythagoras, and Theophrastus. Even in their highly developed civilization, however, the supernatural element was still uppermost. Only a few men were considered able, because of some special power, to distinguish between valuable and harmful plants. These rhizotomoi, or root diggers, were an important caste in ancient Greece. The Romans were less interested in healing plants. However, in 77 B.C. Dioscorides wrote his great treatise, "De Materia Medica," which dealt with the nature and properties of all the medicinal substances known at that time. For fifteen centuries this work was held in high esteem, and even today it is valued by the Moors and Turks. Pliny and Galen also wrote about drug plants.

After the Dark Ages were over, there came the period of the herbalists and encyclopedists, and the monasteries of Northern Europe produced vast compendiums of true and false information regarding plants, stressing in particular the medicinal value and folklore. It was about this time that the curious "Doctrine of Signatures" came into being. According to this superstitious doctrine all plants possessed some sign, given by the Creator, which indicated the use for which they were intended. Thus a plant with heart-shaped leaves should be used for heart ailments, the liverleaf with its three-lobed leaves for liver troubles, and so on. Many of the common names of our plants of today owe their origin to this curious belief. Such names as heartsease, Solomon's-seal, dogtooth violet, and liverwort carry on the old superstition.

From this crude beginning the study of drugs and drug plants has progressed until now pharmacognosy and pharmacology are essential branches of medicine. As an indication of the way botany and medicine have gone hand in hand, even in comparatively recent times, may be mentioned the fact that the great majority of the early botanists in the United States were also medical men.

DRUG PLANTS

For the purposes of economic botany we are most interested in that branch of medical science which deals with the drug plants themselves. This is *pharmacognosy*, and it is concerned with the history, commerce, collection, selection, identification, and preservation of crude drugs and raw materials. *Pharmacology* is the study of the action of drugs. Throughout the world several thousand plants have been and are still used for medicinal purposes. Many of these are known and utilized only by savage peoples, or by herb doctors and dwellers in primitive places who are forced to depend on the native plants of the vicinity.

The most valuable of the drugs and drug plants have been standardized as a result of the Pure Food and Drug Act of 1906. These are referred to as official drugs. Descriptions and other information regarding them are available in many places. The most important of these sources are the "United States Pharmacopoeia," the "Homeopathic Pharmacopoeia," and the "National Formulary." These works are constantly revised and kept up to date. The "Standard Dispensatory" and the "National Dispensatory" are other good reference works for all branches of materia medica. Pharmacopoeias are also issued in most of the larger European countries.

Comparatively few drug plants are cultivated. Most of the supply of drugs is obtained from wild plants growing in all parts of the world, and especially in the tropics. These drug plants are collected and prepared in a crude way for shipment, and eventually reach the centers of the drug trade in this country and abroad. In some instances one country or another has built up a monopoly of some particular drug. Japan, for example, formerly controlled the output of camphor, agar, and pyrethrum, while the Dutch in Java supplied nearly all the quinine that entered the world trade. The United States is an important market for drug plants. From 1920 to 1930 the importation of crude drugs increased 140 per cent. Most of the processing of the crude material is carried on in the United States. In addition, several drugs are produced in this country, either from wild or cultivated sources. These include ginseng, goldenseal, cascara, digitalis, hemp, and wormseed. Still others, such as belladonna, henbane, and stramonium, are grown during periods of shortages.

The medicinal value of drug plants is due to the presence in the plant tissues of some chemical substance or substances that produce a definite physiological action on the human body. The most important of these substances are the alkaloids, compounds of carbon, hydrogen, oxygen, and nitrogen. Glucosides, essential oils, fatty oils, resins, mucilages, tannins, and gums are all utilized. Some of these materials are powerful poisons, so that the preparation and administering of the drugs should be left entirely in the hands of skilled pharmacists and physicians.

CLASSIFICATION OF DRUGS

The classification of drugs and drug plants is difficult for there are many methods of approach. The classification might be based on the chemical nature or the therapeutic value of the plant product, the natural affinities of the various species, or the morphology of the plant organ from which the drug is obtained. For our purposes it seems best to consider the more important drug plants on a morphological basis. In addition to the ones discussed below, many others are referred to elsewhere in the book. In general, we find that the active principles are present in the storage organs of the plants, particularly in roots and seeds, and to a lesser extent in leaves, bark, wood, or other parts of the plant. The total amount of the chemical substances present in any particular organ is so small that it is hard to ascribe any biological significance to it. There may be some slight protective function, but probably these principles, which are so valuable to man in the treatment of disease, are merely waste products of the metabolism of the plant.

DRUGS OBTAINED FROM ROOTS AND OTHER UNDERGROUND PARTS

Aconite. Aconite is obtained from the tuberous roots of the monkshood (*Aconitum Napellus*). Although this familiar garden plant has long been known as a poison, its use in medicine is comparatively recent. The plant is a native of the Alps, Pyrenees, and other mountainous regions of Europe and Asia. It is widely cultivated in temperate countries both as an ornamental and as a drug plant. The commercial supply comes chiefly from Europe. Formerly the leaves and flowering shoots were utilized, but at the present time only the roots are official. These are collected in the autumn and dried. Aconitine is the most important of the several alkaloids that are present. Aconite is used externally for neuralgia and rheumatism, and internally to relieve fever and pain.

Colchicum. Colchicum root is the dried corm of the meadow saffron (*Colchicum autumnale*), a perennial tuliplike herb of Europe and Northern Africa. The active principle is an alkaloid, colchicine, which is used in the treatment of rheumatism and gout. The fresh roots are also used to some extent, and the seeds as well. Colchicine is used in modern genetics to produce doubling of chromosomes.

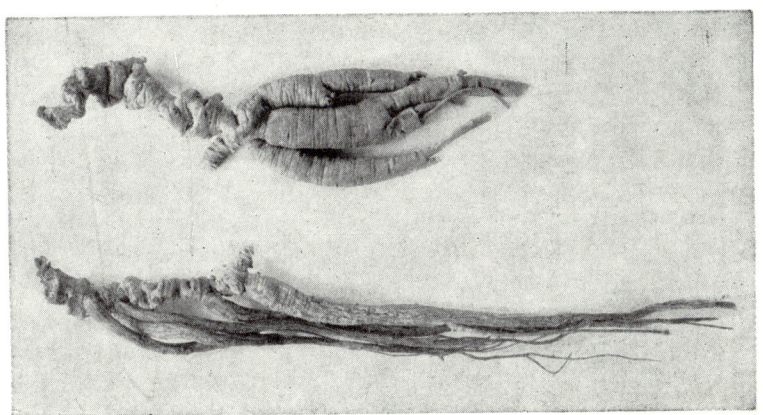

FIG. 124. Ginseng roots (*Panax Schinseng*) from Manchuria.

Gentian. *Gentiana lutea*, a tall perennial herb with conspicuous orange-yellow flowers, is the source of gentian root or bitterroot. The plant is very common in the Alps and other mountains of Europe. The rhizomes and roots are dug in the fall, sliced, and dried. They contain several glucosides, which are valuable as a tonic for they can be used with iron salts.

Goldenseal. Goldenseal (*Hydrastis canadensis*), formerly common in the rich woods of Eastern North America, was a favorite remedy of the Indians and the early settlers. The plant is now cultivated in the Pacific Northwest and North Carolina for it has almost been exterminated as a wild plant by drug collectors. The roots and rhizomes contain several alkaloids. Goldenseal is used as a tonic and for the treatment of catarrh and other inflamed mucous membranes.

Ginseng. Ginseng (Fig. 124) is one of the most important drugs in China, where it is considered to be a cure for a great variety of diseases. The true ginseng (*Panax Schinseng*), a plant of Eastern Asia, was at first the only source of the drug. The demand has been so great, however, that quantities of the American ginseng (*P. quinquefolium*) have been used in recent years. This plant of the eastern woodlands has been almost exterminated by collectors and it is now being cultivated. Some ginseng is used in the United States as a stimulant and stomachic.

Ipecac. Small shrublike plants of the moist rich forests of Latin America are the source of this well-known drug. Several species are utilized, but the official material consists of the dried rhizome and roots of *Cephaëlis Ipecacuanha*. The principal ingredient is emetin, a white, bitter, colorless alkaloid. Ipecac is used chiefly as a diaphoretic, emetic, and expectorant. It is almost indispensable in the treatment of amoebic dysentery and pyorrhea.

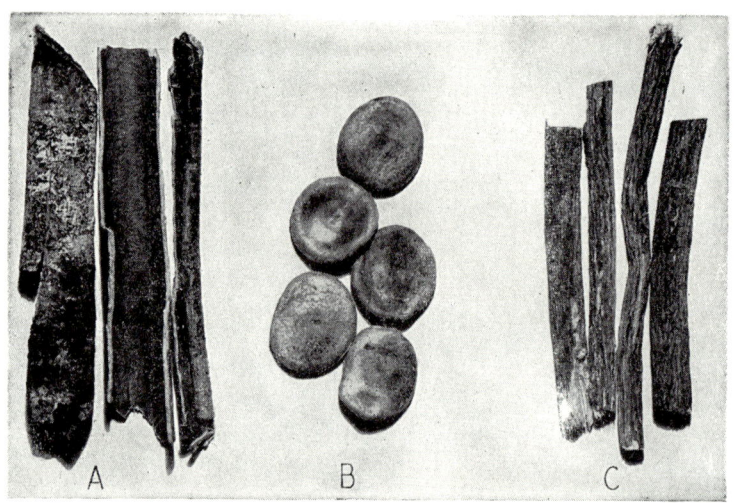

Fig. 125. Three important crude drugs. *A*, cascara sagrada bark (*Rhamnus Purshiana*); *B*, nux vomica seeds (*Strychnos Nux-vomica*); *C*, licorice roots (*Glycyrrhiza glabra*).

Jalap. This resinous drug is obtained from the tuberous roots of *Exogonium Purga*, a twining, morning-glory-like vine of the rich woodlands of eastern Mexico. The plant is cultivated in Mexico, Jamaica, and India. The roots are collected and dried over fires. Jalap is used as a purgative.

Licorice. Licorice has been known from remote times. The licorice plant (*Glycyrrhiza glabra*) is a perennial herb that grows wild in Southern Europe and Western and Central Asia. It is also cultivated in many places within this area. Spain is the largest producer of cultivated licorice root. The United States imports a large amount, chiefly from Asia Minor, the U.S.S.R., and Turkey. The roots (Fig. 125, *C*) are dried in sheds for several months and are shipped in cylindrical pieces. Licorice is used in medicine as a demulcent and expectorant and to disguise the taste of nauseous preparations. Most of the supply, however, is used as a flavoring material in the tobacco and candy industries and in the manufacture of shoe polish. In recent years many other industrial uses have been found for this plant. It furnishes a compound, glycyrrhizin, which is fifty times sweeter than sugar; a solution that is used for etching steel sections in photomicrographic work; and a substance from the waste root which foams readily and is used by brewers to give a head to beer. The fibers are utilized for wallboard and boxboard, under the name "Maftex"; for insulating materials; and for the Jacquard cards used for controlling the designs in the weaving of tapestries and other figured materials.

Podophyllum. The roots and rhizomes of the mandrake or May apple (*Podophyllum peltatum*) yield the drug podophyllum, which has long been used by the country people of the eastern United States as an emetic and cathartic. The plant occurs throughout the Eastern states and Canada, but the commercial supply comes from the southern Appalachians. Mandrake is cultivated to some extent. The roots are collected in the fall or spring and are cut into cylindrical segments and carefully dried. They contain a resin, which is the source of the cathartic principles. Indian podophyllum, an allied drug, is obtained from *Podophyllum Emodi*, a Himalayan species.

Rhubarb. *Rheum officinale* and *R. palmatum*, native shrubs of China and Tibet, are the sources of the drug rhubarb. These plants look much like the garden rhubarb but grow to a greater size. They are extensively cultivated in China. The rhizomes and roots are dug and cut into short pieces or slices. These are threaded on a string and dried in the sun or in kilns. Rhubarb is used as a tonic and laxative and for indigestion. Indian rhubarb, a possible substitute, comes from *R. Emodi* and allied species.

Squills. A white variety of the sea onion or squills (*Urginea maritima*) is the source of this drug. The plant is a native of the sea coasts of the Mediterranean and is also cultivated to some extent. The bulbs are dug up, and the outer scales removed. The fleshy inner scales are then sliced and dried. Several glucosides are present. The drug is used as an expectorant and stimulant. A red variety contains toxic substances as well and is used as a raticide.

Senega. The senega snakeroot or milkwort (*Polygala Senega*), a small herbaceous perennial of Eastern North America, is the source of a glucosidal drug, which is obtained from the dried roots. Its common name is derived from the fact that the Senega or Seneca Indians used the plant as a cure for snake bites. Senega is used as an expectorant, emetic, and stimulant.

Valerian. The dried rhizome and roots of the garden heliotrope (*Valeriana officinalis*) are the source of this drug. This plant, a native of

Eurasia, has long been cultivated in the United States as an ornamental species. The active principle is an essential oil which is used to relieve nervous afflictions, such as pain, coughing, and hysteria.

DRUGS OBTAINED FROM BARKS

Cascara. This familiar drug, which ranks first among the drugs of North American origin, is obtained from the reddish-brown bark of the western buckthorn (*Rhamnus Purshiana*), a tree of the northwestern United States and adjacent Canada. It is occasionally cultivated. It was early used by the western Indians and by the pioneer Spanish settlers, who called it *cascara sagrada*, or sacred bark. For commercial purposes the bark (Fig. 125, *A*) is peeled off in long strips during the summer and dried on racks. It should be kept for a year before use. Cascara is a tonic and laxative.

Curare. The Indians of Northern South America have long used a variety of poisonous extracts prepared from various woody lianas as arrow poisons. The identification of the constituent plant materials in curare has been difficult, as the sources vary in different places. *Strychnos toxifera*, *Chondodendron tomentosum*, species of *Abuta* and *Cocculus*, and other species of *Strychnos* are all utilized. New sources are continually being brought to light. In making curare, portions of the bark, roots, stems, and tendrils are boiled down, the impurities skimmed off, and the residue filtered. Catalyctic agents are added, and the whole mass is boiled to a syrup. This is exposed to the sun and dried to a paste, which is kept in tightly covered gourds or bamboo tubes.

Curare causes progressive paralysis with eventual cardiac failure. The lethal effects are due to several alkaloids. One of these, curarine, has now been made available to medicine for use in shock therapy, as it is an ideal muscle relaxant. Curarine is also used for chronic spastic conditions, in surgical operations and tetanus, and as a powerful sedative.

Quinine. Quinine is one of the most important drugs known, and it has been a great boon to mankind, as it is the only adequate cure for malaria. Although atabrine and similar synthetic products are valuable, they only complement quinine and are not substitutes for it. Quinine is obtained from the hard thick bark of several species of the genus *Cinchona*, handsome evergreen trees native to the Andes of South America. *C. Calisaya* (Fig. 126), *C. officinalis*, *C. Ledgeriana*, and *C. succirubra* have all been utilized.

Cinchona bark was known to the Indians. The first reputed use of the drug by white men was in 1638, when, according to tradition, the Countess of Cinchon, wife of the Viceroy of Peru, was cured of malaria after all other remedies had failed. This often-quoted incident has recently been shown to be a myth. Be that as it may, the Jesuits were familiar with the use of the bark and carried it with them all over the world. Soon Peruvian bark or Jesuit's bark was in great demand. The supply at first seemed inexhaustible but rapidly diminished under the wasteful methods of collection. The trees were cut down, and the bark stripped off and dried in the open or over fires in huts. It soon became evident that attempts at cultivation must be made if this indispensable drug was to be available for future generations. Both the Dutch and the English

Fig. 126. Fruiting branches of *Cinchona Calisaya*. The bark of this species is an important source of quinine. (*Photo by Walter H. Hodge.*)

sent collectors to South America, but the Andean natives guarded the remnants of the cinchona forests zealously. In spite of dangers and hardships, however, a few seedlings and seeds were finally taken out of the country and became the basis of the great plantations of Java and India. Few tropical crops have been studied more intensively. All phases of cinchona production—breeding, culture, harvesting, and processing—were investigated, particularly in Java. The result was that eventually the Dutch developed a virtual monopoly, producing 95 per cent of the world's supply. The Indian output was used chiefly for home consumption.

In recent years the amount of bark produced was regulated in order to keep prices up. Consequently, attempts were made as early as 1934 to establish a cinchona industry in the Western Hemisphere. An experimental plantation was started in Guatemala, and by the outbreak of the Second World War, a considerable mass of data and a nursery of superior

FIG. 127. Typical aspect of the mature bark of *Cinchona Calisaya*, showing the cross fissuring. (*Photo by Walter H. Hodge.*)

clones from all parts of the world were available. When the supply of quinine was cut off by the loss of Java, the United States instituted an extensive program of cinchona procurement in Latin America, utilizing all available wild stands and developing new plantations as well. Several promising new sources were uncovered, among them *Cinchona pitayensis*, a particularly high-yielding species. Between 1942 and 1945 exports of cinchona and quinine from Latin America increased from 207,000 to 7,317,999 lb., with Ecuador and Peru the leading producers. Whether or not Latin America will ever be able to compete successfully with the

Far East is open to question; at least from now on the Western Hemisphere should never again be wholly dependent on the Eastern.

Cinchona bark (Fig. 127) is removed from cultivated trees by uprooting them when about 12 years of age and stripping off the bark from both roots and stems or by cutting the trunks above ground and stripping the felled portion. In the latter case, adventitious roots develop, and later the bark is removed from these in long quills. By far the most important constituent of cinchona bark is quinine, a very bitter, white, granular substance. In addition to its use in the treatment of malaria, it is valuable as a tonic and antiseptic and in the treatment of fevers. Some 29 other alkaloids have been isolated from the bark, including cinchonidine, cinchonine, and quinidine, all of which are useful in medicine, as in totaquina, a standard mixture of all the alkaloids.

Slippery Elm. The inner bark of the slippery elm (*Ulmus rubra*), a large tree of Eastern North America, is the source of this nonpoisonous drug. The bark is removed in the spring and the outer layers are discarded, while the inner portion is dried. Slippery elm bark has a very characteristic odor. It contains mucilage and is used for its soothing effect on inflamed tissues, either in the crude state or in the form of lozenges.

Drugs Obtained from Stems and Woods

Ephedrine. Ephedrine is an alkaloid that is obtained from *Ephedra sinica, E. equisetina*, and other Asiatic species of the genus. These plants are low, dioecious, leafless shrubs with slender green stems (Fig. 128). The entire woody plant is used for the extraction of the drug. Under the name "ma huang," ephedra has been used in China for over five thousand years. In the United States ephedrine has been used extensively in recent years in the treatment of colds, asthma, and hay fever and for other medicinal purposes.

Guaiacum. Guaiacum or gum guaiac is a hard resin that exudes naturally from the stems of the lignum vitae trees (*Guaiacum officinale* and *G. sanctum*), previously discussed. It hardens as round, glassy, greenish-brown tears. It is also obtained from incisions, from the cut ends of logs, or from pieces of the wood. Gum guaiac is used as a stimulant and laxative. It is also a good chemical indicator as it is very sensitive to oxygen. The lignum vitae trees are evergreens native to the West Indies and South America.

Quassia. Quassia is obtained from two different sources. Jamaican quassia comes from *Picrasma excelsa*, a tall tree of Jamaica and other West Indian islands, while Surinam quassia is the product of *Quassia amara*, which grows in tropical America as well as in the West Indies. The latter species is a valuable timber tree with a lustrous, yellowish-white, fine-grained wood. Quassia is shipped in the form of billets, and the drug is prepared by making an effusion of chips or shavings. It has a very bitter taste and is used as a tonic and in the treatment of dyspepsia and malaria. It also serves as an insecticide.

FIG. 128. Left and center, *Ephedra equisetina*, and right, *E. sinica*. These low leafless shrubs are the source of the drug ephedrine. (*Reproduced by permission from Youngken, Textbook of Pharmacognosy, The Blakiston Company.*)

The medicinal use of balsams and other oleoresinous exudations from trees has already been discussed in Chap. VII.

DRUGS OBTAINED FROM LEAVES

Aloe. Several kinds of aloes are on the market. Barbados or Curaçao aloes come from *Aloe barbadensis* of the West Indies, Socotrine aloes from *A. Perryi* of East Africa, and Cape aloes from *A. ferox* of South Africa. Aloes are tropical plants with succulent leaves and showy flowers. They are frequently cultivated in greenhouses. The leaves contain a resinous juice in which there are several glucosides. If the leaves are cut and placed in troughs, the juice slowly exudes and can be collected. It is evaporated in pans to a thick, viscous black mass, which may be solidified. Aloes are used chiefly as purgatives.

Belladonna. This old and important drug is obtained from the dried leaves and tops, and to some extent the roots, of *Atropa Belladonna* (Fig.

129). This plant is a coarse perennial herb, native to Central and Southern Europe and Asia Minor. It is extensively cultivated as a drug plant in the United States, Europe, and India. The leaves are collected during the flowering season and dried. They contain several alkaloids, chief among which are hyoscyamine and atropine. Belladonna is used externally to relieve pain and internally to check excessive perspiration,

FIG. 129. The belladonna plant (*Atropa Belladonna*). The dried leaves and tops, and to some extent the roots, are the source of the drugs belladonna and atropine.

coughs, etc. Atropine is used to dilate the pupil of the eye and for many other medicinal purposes.

Cocaine. The leaves of the coca shrub (*Erythroxylon Coca*), a native of Peru and Bolivia, and related species furnish this drug. The plant is extensively cultivated in South America, where the leaves are used as a masticatory, and also in Java, Ceylon, and Formosa. The leaves mature in about four years and can then be picked three or four times a year. They are carefully dried and shipped in bales. They have a bitter aromatic taste due to the presence of the alkaloid cocaine. It takes 100 lb. of leaves to yield 1 lb. of the drug. Cocaine has been much used as a local anesthetic. It is also employed as a tonic for the digestive and nervous systems, but as it is a habit-forming drug its use should be supervised by physicians.

Buchu. This drug is obtained from the dried leaves of *Barosma betulina*, *B. serratifolia*, and *B. crenulata*, shrubs of the hot dry mountainous regions of South Africa. The active principle is an essential oil

Fig. 130. Foxglove plants (*Digitalis purpurea*). The important drug digitalis is obtained from the dried leaves of this plant. (*Courtesy of Breck and Company.*)

Buchu is used as a disinfectant and to stimulate excretion, and also in the treatment of indigestion and urinary disorders.

Digitalis. This drug, almost indispensable in the treatment of heart disorders, is obtained from the dried leaves of the foxglove (*Digitalis purpurea*), a native of Southern and Central Europe. The plant is a beautiful herbaceous perennial (Fig. 130) with tufted basal leaves and a single, erect, leafy stem which bears a spike of purplish flowers. The foxglove is frequently grown for ornamental purposes, as well as for the drug. Fresh and full-grown leaves are carefully and quickly dried for use as the source of the drug. The medicinal properties of digitalis have been known for a long time. The most active principle is a glucoside, digitoxin. Digitalis is a heart stimulant of the greatest importance because of its powerful action, and it is specific for some types of heart disease. It improves the tone and rhythm of the heart beats, making the contractions more powerful and complete. Consequently more blood

FIG. 131. The blue gum (*Eucalyptus globulus*), an Australian tree, is extensively planted in California. "The Joseph Aram Blue Gum," pictured here, was probably set out in the 1860's. It is about 105 ft. tall and has a diameter of 95 in. at breast height. Oil of eucalyptus is obtained from the leaves of the blue gum. (*Photo by U.S. Forest Service.*)

is sent out from the heart, thus aiding the circulation, improving the nutrition of the body, and hastening the elimination of waste.

Eucalyptus. The scythe-shaped leaves of the older growth of the blue gum (*Eucalyptus globulus*) contain an essential oil that is widely used in medicine. The tree, which is one of the tallest known, reaching a height of from 200 to 300 ft., is a native of Australia. It is extensively cultivated in California (Fig. 131), Florida, and the Mediterranean region. It is commonly supposed that eucalyptus trees aid in ridding a country of malaria. Because of their great height and extensive root system it is assumed that they must remove great quantities of water from the soil, thus tending to dry out mosquito-infested areas. Eucalyptus oil, obtained from dried leaves, is used chiefly in the treatment of nose and throat disorders, malaria, and other fevers. The oil is colorless or pale yellow, spicy, and pungent.

Hamamelis. The common witch hazel (*Hamamelis virginiana*), a shrub of Eastern North America, is the source of this product. The official source is the dried leaves and the commercial supply of these comes chiefly from the southern Appalachians. In New England, however, and elsewhere, the bark, twigs, and sometimes the entire plant are utilized. The active principle, a tannin, is extracted with water and steam and distilled. After 150 cc. of alcohol have been added to each 850 cc. of the distillate, the familiar witch hazel or hamamelis extract is obtained. This is used as an astringent and to stop bleeding.

Henbane. *Hyoscyamus niger*, the source of henbane, is a coarse evil-smelling herb, native to Europe and Asia, but occurring as a weed in many other parts of the world. Normally, the supply of the drug comes from Europe. During the shortages incident to the First and Second World Wars, the plant was cultivated in the United States. The leaves and flowering tops contain several poisonous alkaloids, among them hyoscyamine and scopolamine. Henbane is used as a sedative and hypnotic. Its action is similar to that of belladonna and stramonium, but less powerful.

Hoarhound. The hoarhound (*Marrubium vulgare*) is a native of Europe and Central Asia, but is thoroughly naturalized in America, where it is also cultivated. It is a small herbaceous perennial with white flowers in dense axillary whorls. The dried leaves and flowering tops are used medicinally. Hoarhound is administered as an infusion or in the form of candy or lozenges. It is a favorite domestic remedy for breaking up colds, and is also used for rheumatism, dyspepsia, and other ailments.

Lobelia. This drug is obtained from the dried leaves and tops of wild or cultivated plants of the Indian tobacco (*Lobelia inflata*), a small annual with numerous blue flowers in leafy terminal racemes. The plant is a native of North America and is one of our comparatively few poisonous species. The active principles are alkaloidal in nature. Lobelia is used as an expectorant, antispasmodic, and emetic. It was also used by the American Indians.

Pennyroyal. Pennyroyal (*Hedeoma pulegioides*) is a small aromatic annual common in poor soil throughout the eastern United States. It contains an essential oil that has some use in internal medicine. Pennyroyal is often used as an ingredient of liniments because of its counter-irritant action. Its chief use, however, is as an insect repellent. The oil is obtained commercially from the dried leaves and tops.

Senna. This ancient drug is obtained from the dried leaflets, and also the pods, of several species of *Cassia*, which are indigenous to the arid regions of Egypt and Arabia. Alexandrian senna comes from *C. acutifolia* and Indian or Tinnavelly senna from *C. angustifolia*. Wild plants of the first of these species are still used as a source of the drug in Egypt. Both of them are cultivated in India. The leaves are picked by the natives, dried in the sun, and baled. Senna is used as a purgative.

Stramonium. The Jimson weed or thorn apple (*Datura Stramonium*), one of the most poisonous of plants, is the source of this drug. The plant is a native of Asia, but occurs as a weed in fields and waste places all over the world. It is a coarse, rank annual growing to a height of 4 ft. It is cultivated in the United States and Europe for the drug stramonium, which is extracted from the dried leaves and flowering tops. The active principles are alkaloids, including hyoscyamine, atropine, and scopolamine. This drug is used as a substitute for belladonna for relaxing the bronchial muscles in the treatment of asthma. In many parts of the world stramonium is used for its narcotic effects.

Wormwood. The wormwood (*Artemisia Absinthium*), a perennial species of Europe, Northern Africa, and Northern Asia, is the source of an essential oil obtained by steam distillation from the dried leaves and tops of the plant. The resulting greenish liquid is used to some extent in liniments, but it is no longer official. Wormwood is extremely deleterious when taken in quantity. The chief use of the essential oil is to flavor the liqueur known as absinthe, the use of which is forbidden in many countries. Absinthe contains other aromatics as well as wormwood. Some wormwood is grown in Michigan and Oregon.

Drugs Obtained from Flowers

Chamomile. Chamomile is an old-time remedy obtained from *Matricaria Chamomilla*. This daisylike plant, a native of Eurasia, is cultivated in the United States and elsewhere. The dried flower heads contain an essential oil. Infusions of chamomile are used as tonics and gastric stimulants. The flower heads of the Russian or garden chamomile (*Anthemis nobilis*) are used for similar purposes, and also in poultices for sprains, bruises, and rheumatism.

Hops. The hop (*Humulus Lupulus*) is a native of the north temperate regions of both hemispheres. The plant was known to the Romans and has been grown in some parts of Europe since the ninth century. At the present time hops are extensively cultivated in the United States, Europe, Australia, and South America. The plant is a climbing herb with perennial roots. These send up annually several rough, weak, angular stems with deeply lobed leaves and dioecious flowers. The female flowers are produced in scaly, conelike catkins, which are covered with glandular hairs. These inflorescences contain resin and various bitter, aromatic, and narcotic principles, chief of which is lupulin. In cultivation the hop plants are trained on poles (Fig. 132) or trellises. Hops are harvested in the early fall. The catkins are carefully picked and dried in kilns at a temperature of 70°F. or less. They are treated with sulphur and baled for shipment. Hops are used in medicine for their sedative and soporific properties, and also as a tonic. Sometimes they are used in poultices. Their principal use, however, is in the brewing industry. They are added to beer to prevent bacterial action and consequent decomposition,

Fig. 132. Growing hops (*Humulus Lupulus*) on poles in Canterbury, England.

and also to improve the flavor and impart the characteristic bitter taste to the beverage.

Santonin. The dried unopened flower heads of the Levant wormseed (*Artemisia Cina*) contain a valuable drug known as santonin. The plant is a small semishrubby perennial of Western Asia. The present supply of the drug comes chiefly from Turkestan, although this species is now being grown in the Pacific Northwest. Formerly the crude drug was shipped, but now the santonin is extracted and exported. This drug is one of the best remedies for intestinal worms, and has been used for this purpose for centuries. It was introduced into Europe by the Crusaders.

Drugs Obtained from Fruits and Seeds

Chaulmoogra Oil. For centuries leprosy has been one of the most dreaded diseases of mankind and it was long thought to be incurable. It was known, however, that the natives of Burma and other parts of Southeastern Asia used the seeds and oil obtained from the chaulmoogra tree

(*Hydnocarpus Kurzii*) and related species in the treatment of skin diseases. These tall trees of the dense jungles have velvety fruits (Fig. 133) with several large seeds, which contain a fatty oil with a characteristic odor and acrid taste. The expressed oil is a brownish-yellow liquid or soft solid. Experiments carried on at the University of Hawaii utilized this oil in the development of a successful treatment for leprosy. The crude oil proved unsatisfactory and, moreover, the treatment was very painful. The use of certain acids present in the oil, or of the ethyl esters

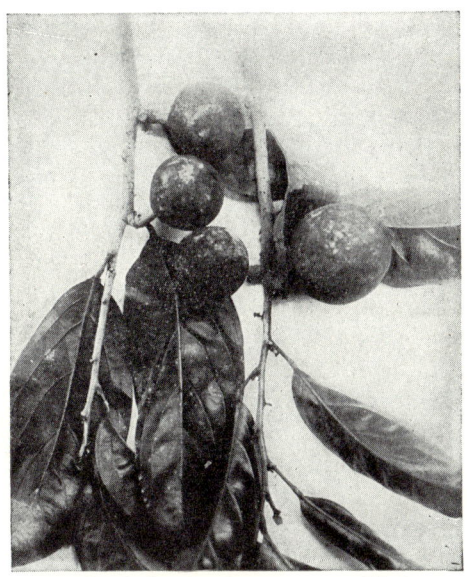

Fig. 133. A fruiting branch of the chaulmoogra tree (*Hydnocarpus Kurzii*). Chaulmoogra oil, which is obtained from the seeds, is used in the treatment of leprosy.

of these acids, was productive of the desired results, and leprosy can now be cured.

Colocynth. The dried spongy pulp of the bitter apple (*Citrullus Colocynthis*) is the source of the glucosidal drug colocynth. The plant is a perennial vine native to the warmer parts of Asia and Africa. It is now widely distributed, and is cultivated in the Mediterranean region. The fruits resemble oranges in appearance. The rind is removed and the white bitter pulp is dried and shipped in balls. Colocynth serves as a powerful purgative.

Cubebs. Cubebs are the dried unripe fruits of *Piper Cubeba*, a climbing vine of eastern India and Malaya. It is cultivated in Java, Thailand, Ceylon, and the West Indies. The berries resemble black pepper, but are stalked. They have a warm, bitter aromatic taste and a strong odor, due to the presence of an oleoresin. Cubebs are used in the treatment of catarrh, often in the form of cigarettes, and as a kidney stimulant. They were formerly used as a spice or condiment.

Croton Oil. Croton oil is a fatty oil obtained from the dried ripe seeds of *Croton Tiglium*, a shrub or small tree of Southeastern Asia. It is cultivated in India and Ceylon. Croton oil is a yellowish-brown liquid with a burning taste and nauseous odor. It is one of the most powerful purgatives known. The flowers and crushed leaves are used in India to poison fishes.

Nux Vomica. This valuable drug is obtained from the seeds of *Strychnos Nux-vomica*, a tree native to Ceylon, India, Cochin China, and Australia. The large fruits contain from three to five grayish seeds (Fig. 125, *B*) which are very hard and bitter. Ripe seeds contain two important alkaloids, strychnine and bucine. Nux vomica is used as a tonic and stimulant; strychnine is utilized in the treatment of nervous disorders and paralysis, always in small doses, for it is a virulent poison. The use of this drug dates back to the sixteenth century.

Opium. Opium, one of the most helpful, and at the same time vicious, drugs, is the dried juice or latex obtained from the unripe capsules of the opium poppy (*Papaver somniferum*). This poppy is an annual herb with large showy white flowers. A native of Western Asia, it now occurs in most countries as a weed. It is cultivated extensively in India, China (Fig. 141) Asia Minor, the Balkans, and elsewhere. Shortly after the petals fall the capsules are incised with a knife, and the white latex exudes and soon hardens in the air. It is scraped off and shaped into balls or cakes, which are often wrapped in the poppy petals. Crude opium is a brownish material containing as many as 25 alkaloids, the most important and most powerful of which are morphine and codeine. Because of their narcotic and sedative action opium and its derivatives are used to relieve pain, induce sleep, and relax spasms. They should never be administered except under a physician's direction. Because of the flagrant misuse of the drug and its deleterious effects physically, mentally, and morally, the opium trade is very carefully supervised and restricted.

Psyllium. The drug known as psyllium has changed from a position of obscurity to one of the foremost importance comparatively recently. Although used locally in Europe for some time, it was slow in entering world trade. Commercial psyllium is the seed of several of the fleaworts, European and Asiatic species of plantain, which are cultivated in France, Spain, and India. French psyllium comes from *Plantago indica*, Spanish psyllium from *P. Psyllium*, and blonde psyllium, the Indian product, from *P. ovata*. Two crops are raised each year with a yield of 7000 to 8000 lb. per acre. Psyllium seed contains a tasteless mucilaginous substance which acts as a mild laxative and is comparable to agar and mineral oil for use in chronic constipation. When extracted, the mucilage can be used as a cosmetic and in stiffening fabrics.

Strophanthus. The dried, greenish, ripe seeds of *Strophanthus Kombe* and *S. hispidus*, woody climbers of the African forests, are the source of the drug strophanthus, which is used as a heart stimulant. It has a direct and powerful action. The active principles include the glucoside strophanthin and several alkaloids. Recently another glucoside, sarmentogenin, has been isolated from another species of *Strophanthus*, probably *S. sarmentosus*. This substance can be transformed into cortisone, the rare and expensive drug which holds much promise for the successful treatment of arthritis, rheumatic fever, and various heart disorders.

Wormseed. The American wormseed (*Chenopodium ambrosioides* var. *anthelminticum*) is a native of the West Indies and Central and South America. It is naturalized in the United States, occurring as a weed in waste places. Wormseed has also been cultivated in many areas, and it is now extensively grown in Carroll County, Maryland, for its essential oil. This oil is obtained by distillation from the fruits and is used in the treatment of hookworm infections.

Drugs Obtained from the Lower Plants

Antibiotics

Antibiotics are substances produced primarily by certain harmless microorganisms that are injurious to the growth and activity of various pathogenic bacteria. Although known to occur previously, antibiotics were not considered of importance until 1939. Since then, extensive investigations have been carried on, and a considerable number have been isolated and their therapeutic action studied. Molds, actinomycetes, and bacteria are the chief sources, although antibiotics are also present in higher plants.

Penicillin. Penicillin is the best known antibiotic. It was discovered by chance in 1929. Reexamined in 1937, it shortly proved to be extremely valuable in combating staphylococcus, streptococcus, and gas gangrene infections. It is obtained chiefly from *Penicillium notatum*, a blue-green mold occurring in floccose masses with a white margin. When grown on gelatin, the mycelium excretes penicillin into the substratum, which becomes liquid. The crude penicillin is recovered, purified, and dehydrated. It is an organic acid and readily forms salts and esters. Superior strains which yield greater quantities of the drug have been developed. Other species of *Penicillium*, especially *P. chrysogonum*, also furnish the antibiotic. Penicillin is highly selective in its action and is effective against gram-positive bacteria. It is valuable because of its lack of toxicity; it is particularly useful in the treatment of bacterial endocarditis, mastoiditis, gonorrhea, local infections, and certain types of pneumonia.

Streptomycin. This antibiotic comes from *Streptomyces griseus*, isolated in 1944 after testing soils from all over the world. The organism is one of the actinomycetes and is grown in deep submerged cultures. Streptomycin is particularly effective against gram-negative bacteria and is used in the treatment of tularemia, empyema, urinary and local infections, and some forms of tuberculosis, meningitis, peritonitis, and pneumonia.

Aureomycin. Aureomycin is produced by *Streptomyces aureofaciens*, which was isolated in 1948 from ordinary soils. It is more versatile than penicillin or streptomycin, attacking not only gram-positive and gram-negative bacteria, but also the *Rickettsiae*, which had previously been immune to chemical attack. It is used to combat forms of virus pneumonia, undulant fever, osteomyelitis, whooping cough, and eye infections and in cases where a patient has developed a resistance to the other antibiotics or to sulphur drugs. Recently, aureomycin has been found to be one of the greatest growth-producing substances yet to be discovered.

Chloromycetin. This antibiotic is a pure crystalline substance produced by *Streptomyces venezuelae*, which was isolated in 1948 after a search involving the study of thousands of soil samples from all over the world. It is the only antibiotic which has also been produced synthetically. Chloromycetin, like aureomycin, is effective against the *Rickettsiae*. It is useful in the treatment of undulant fever, bacillary urinary infections, primary atypical pneumonia, typhoid fever, typhus fever, scrub typhus, parrot fever, and Rocky Mountain spotted fever.

Terramycin. This recently discovered antibiotic is secreted by *Streptomyces rimosus*, isolated from a piece of Indiana dirt after an exhaustive search involving scores of thousands of soil samples. It is valuable in treating common forms of pneumonia, typhoid fever, streptococcic and many intestinal- and urinary-tract infections. It is effective against gram-positive and gram-negative bacteria, *Rickettsiae*, and large viruses. Although little different in its therapeutic action from the other antibiotics, it will be valuable as an extra weapon of defense.

Neomycin. An organism resembling *Streptomyces Fradiae*, isolated from soil in 1949, is the source of neomycin. This antibiotic has a complex composition and a wide range of experimental uses. It may prove to be most effective in the treatment of tuberculosis.

Other Antibiotics. Many antibiotics are known to be produced by bacteria. Of these, gramicidin and tyrothricin from *Bacillus brevis*, bacitracin and subtilin from *Bacillus subtilis*, and polymixin from *Bacillus polymixa* seem to have the best therapeutic possibilities.

Agar

Agar is an almost pure mucilage obtained from various species of red algae. Prior to the Second World War, Japan had a virtual world monopoly of this product, utilizing *Gelideum corneum*, *Eucheuma spinosum*, *Gracilaria lichenoides*, and other species found off the eastern coast of Asia. Some agar has been produced in the United States since 1919. During the Second World War, however, production was greatly expanded until a peak of 165,954 lb. was reached in one year. The principal species used were *Gelidium cartilagineum* on the Pacific Coast and *Gracilaria confervoides* on the Atlantic Coast. Agar industries have also been developed in the U.S.S.R., Australia, South Africa, and other countries. The algae are collected, bleached, and dried, and the mucilaginous material is extracted with water. Agar reaches the market in flakes, granules, or strips which are brittle when dry but become tough and resistant when moist.

The medicinal value of agar lies in its absorptive and lubricating action. It is often used in a granular condition to prevent constipation. Perhaps its greatest use is as a culture medium for bacteria and other fungi. In dentistry it is valuable for making impressions for plates and molds. The cosmetic, paper, silk, and other industries find it of value, and it is also extensively used as a food.

Ergot

Ergot is the dried fruiting body of a fungus, *Claviceps purpurea*, which is parasitic on rye and other grasses. The disease attacks the young fruit, and when mature, a purplish structure, the sclerotium, replaces the grain. Commercial ergot comes chiefly from Europe, where it is picked from the rye plants by hand or after the rye has been threshed by special machinery. Considerable ergot is now being produced in Minnesota. Wheat ergot is equally good as a drug, but it is not official. Ergot is used chiefly to increase the blood pressure, particularly in cases of hemorrhages following childbirth and other uterine disturbances.

Kelp

In Japan and several northern European countries various of the larger brown algae have been used as a source of iodine, potash, and other salts. More recently a considerable industry has been built up on the west coast of the United States for the extraction of iodine and potash from the giant kelps of the Pacific, particularly *Macrocystis pyrifera*. During the Second World War these kelps were also used as a source of acetone and kelp char, a bleaching carbon. Some attention has been given to the medicinal value of these seaweeds. Several products are on the market which make available the surprising wealth of mineral salts and vitamins which these plants contain.

Other kelps, chiefly *Laminaria digitata* and *L. saccharina* of the Atlantic and *Nereocystis Luetkeana* of the Pacific, have been exploited as a source of algin, a valuable colloid extensively used in the drug, food, and other industries. From 2,000,000 to 3,000,000 lb. a year are produced. Algin or its salt, sodium alginate, is used as a suspending agent in compounding drugs; in lotions, emulsions, and hand pomades; as a sizing for paper and textiles; and in ice cream.

Lycopodium

The infinitely small and almost impalpable spores of *Lycopodium clavatum* and other club mosses contain about 50 per cent fixed oils and so are but little affected by water. They are much used as a covering for pills, as a diluent for insufflations, and as a dusting powder for abraded surfaces. In industry they are utilized in making pattern molds and, because of their inflammability, in flares, fireworks, and tracer bullets. Formerly obtained chiefly from Europe, a considerable amount of *Lycopodium* is now being produced in Maine.

Male Fern

The rhizomes and stalks of the male fern (*Dryopteris Filix-mas*) of Eurasia and North America and the marginal shield fern (*Dryopteris marginalis*) of Eastern North America yield the drug known as male fern or aspidium. This is an oleoresinous substance that has been used for centuries for expelling tapeworms. The commercial supply usually comes from Europe.

INSECTICIDES AND RATICIDES

Plants have been used to combat insects, rats, and other vermin for many years in various parts of the world. Over 1200 species have been recorded as insecticides. Some 250 have been utilized for this purpose in the United States, but only 5 per cent of these are of any commercial value. Nicotine, obtained from tobacco, has long been familiar. In more recent years other botanical insecticides have increased greatly in importance, particularly in the United States, and this country is today the leading market of the world for these products. Pyrethrum or insect flowers, which yield pyrethrum, and derris and cubé, which contain rotenone, are in greatest demand. Promising new sources, however, are continually being discovered. Red squill is the principal raticide.

Pyrethrum

Three principal sources of pyrethrum are recognized in the United States: Dalmatian insect flowers from *Chrysanthemum cinerariaefolium*,

Persian insect flowers from *C. coccineum,* and Caucasian insect flowers from *C. Marshallii.* The first of these is by far the most important. *C. cinerariaefolium* is a slender, glaucous, pubescent perennial 18 to 30 in. in height with pinnate leaves and small daisylike flowers. It is a native of Dalmatia in Jugoslavia, where it has been cultivated for many years. Prior to the Second World War, Japan was the leading producer of pyrethrum flowers, and they constituted one of that country's most valuable exports. Great care was exercised in gathering, drying, and packing the crop. At the present time this species is being cultivated in California and other parts of the United States, British East Africa, Italy, Australia, Brazil, Peru, and Ecuador.

Pyrethrum is noninflammable, nonpoisonous and leaves no oily residue. During the last war the army found it highly effective against flies, fleas, body lice, and yellow fever and malarial mosquitoes. Pyrethrum bombs were standard equipment in malaria-infested areas. They contained the insecticide in a solvent under a pressure of 90 lb. per sq. in. A mechanical release allowed the vapor to escape through a valve like a fog. A 3-sec. application permanently paralyzed all insects. Pyrethrum ointment is used in the treatment of scabies.

Rotenone

The poisonous properties of rotenone-containing plants have been appreciated by native peoples for centuries. The use of these unimpressive climbers and creepers of the *Leguminosae* as fish poisons was noted by De Rochefort in 1665 and Aublet in 1775. Derris was in commercial use in the United States as early as 1911, but with variable and uncertain results. Many years of research resulted in standardizing the product and made possible an astonishing increase in its use. Even in 1930 there was almost no trade, but by 1940 the United States alone imported 6,500,000 lb.

Rotenone, a colorless crystalline compound, and related substances occur as solids in the dried roots. The content may be as high as 12 per cent. Rotenone is fifteen times more toxic than nicotine and twenty-five more than potassium ferrocyanide. It has little or no effect on humans or other warm-blooded animals. There are two principal sources: species of *Derris* in the Far East and of *Lonchocarpus* in Latin America.

Derris. Derris or tuba roots (Fig. 134) have long been used by the natives of Malaya and Borneo for fish and arrow poisons. The various species of *Derris* are climbing vines characteristic of the jungle undergrowth from India to Indonesia and the Philippines. The plants have a short trunk, 3 or 4 ft. in height and 4 in. in diameter, with numerous long branches that climb over the vegetation. The two most important species are *D. elliptica* and *D. trifoliata.* Derris may be propagated by cuttings, and it is cultivated in some regions. It is now a commercial crop in Guatemala and Ecuador, where it grows well at low altitudes in deep, well-drained, fertile soils. The dust made from ground roots has marked insecticidal properties but is nonpoisonous to man, at least when taken through the mouth. The active ingredients, rotenone and a resin, may be extracted and used directly or in the form of a soap.

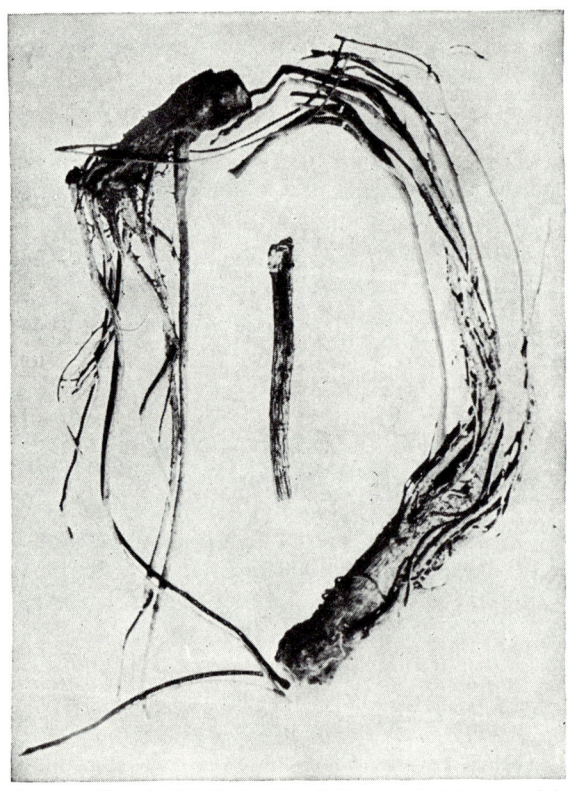

Fig. 134. Derris roots (*Derris elliptica*). Derris has marked insecticidal properties, due to the presence of an active ingredient known as rotenone. (*Courtesy of H. W. Youngken and the American Journal of Pharmacy.*)

Lonchocarpus. The roots of several species of *Lonchocarpus*, chiefly *L. Urucu* in Brazil, *L. utilis* in Peru, and *L. Nicou* in Guiana, constitute an important source of rotenone. These plants, known variously as cubé, timbo, and barbasco, are used by the natives as fish poisons. At first they are bushlike but later resemble vines and climb tall trees. They thrive in the tropical forests at low altitudes where there is an 80-in. rainfall and well-drained soil. When the plants are two or three years old, the tops are cut away and the roots are dug up, dried, bundled, and exported for processing. They are ground into a powder and mixed with talc or clay for dusting or with a liquid for spraying. Cubé contains more rotenone than does derris and, like the latter, is an ideal insecticide for crop plants, as there is no residue. Lonchocarpus is easily propagated by stem cuttings, and there are now many small commercial plantations, although most of the supply still comes from wild plants. Cubé first entered world trade in 1934. In 1946, the United States imported 11,000,000 lb., chiefly from Peru and Brazil.

Red Squill

Red squill is obtained from the bulbs of the red variety of *Urginea maritima*, a native Mediterranean species which is cultivated in Algeria. Used as a raticide from ancient times, red squill has come into prominence within the last 20 years as a result of more active rat-control measures. The toxic substance, a glucoside, has little effect on other animals. In 1947, 628,000 lb. were imported.

CHAPTER XII

FUMITORIES AND MASTICATORIES

In all ages and in all countries human beings have smoked or chewed various substances for pleasure, for some physiological effect, in connection with their religious ceremonies, or in an attempt to seek "some flight from reality," as Norman Taylor puts it. Some of these substances, such as spruce gum and chewing gum, are perfectly harmless, and are chewed merely because of their pleasant taste. The only benefit to be derived is possibly a psychological one brought about by the mechanical act involved.

The majority of the materials that are chewed or smoked, however, have a distinct stimulating or even narcotic effect, due to the presence of various alkaloids. Tobacco, betel, and cola are the least harmful of these drugs. At best they act only as mild stimulants and produce no effects on the consciousness of the user. Possibly the combustion products of tobacco, formed during smoking, are more harmful and may cause deleterious effects. It is a different matter, however, when we consider the great rivals of tobacco—coca, opium, and cannabis. These true narcotics contain alkaloids that are detrimental even in small amounts. When used in quantity, they may lead the addict to the lowest depths of depravity and degradation and cause stupor, coma, convulsions, and even death. The drug habit, brought on by the continued use of these narcotic plant products and their derivatives, is such a serious problem socially, as well as physiologically, that it is indeed fortunate that the sources of such drugs are relatively few in number.

TOBACCO

Tobacco is an important commodity the world over. It is used in one form or another by both civilized and uncivilized peoples. The tobacco plant is a native of tropical America. The original wild ancestor is not known for tobacco is very old and has been under cultivation for centuries. The use of tobacco originated with the Indians in connection with their religious rites. The plant had spread over North America before the coming of the white man. Columbus was the first to note the use of tobacco by the Indians. The plant was introduced into Europe in 1556 and at first was grown only for its ornamental and medicinal value. Jean Nicot, for whom the plant was named *Nicotiana*, was chiefly responsible for its exploitation in France. From there it rapidly spread over the rest of Europe, and to Africa, Asia, and Australia as well. The practice of smoking was not very general in Europe until 1586, but from that time on its popularity increased in spite of the opposition of the clergy and the governments, who almost taxed tobacco out of existence.

The narcotic and soothing properties of tobacco are due to the presence of the alkaloid nicotine. This active principle is extracted by the mucous membranes of the nose and throat. The aroma and flavor are caused by essential oils and other aromatic substances developed during the curing and fermentation process.

Kinds of Tobacco

Some 50 species of tobacco are known, but only two of these are of commercial importance. *Nicotiana Tabacum* is the source of most of the

Fig. 135. A tobacco plant (*Nicotiana Tabacum*) in flower.

tobacco in use at the present time. This was the native tobacco of the West Indies and most of Central and South America. Although originally a tropical species, the plant has become adapted to cultivation in both subtropical and temperate regions, and it is now grown wherever the climate permits. Over 100 horticultural varieties are known. Tobacco is a handsome unbranched annual, growing to a height of 3 to 6 ft., with large oval leaves that are usually sessile with clasping bases. The branching inflorescence (Fig. 135) bears pink flowers. The fruit is a capsule with numerous very small seeds. The leaves and stems are covered with glandular hairs, which secrete a resinous fluid and are very sticky to the touch.

Nicotiana rustica is a smaller hardier plant with yellow flowers. It probably originated in Mexico and still grows wild in parts of North America. This species was cultivated and smoked by all the eastern Indians before the coming of the white man. It was the first tobacco to be grown in Virginia, but was soon replaced by *N. Tabacum.* It is, however, still grown to some extent in parts of Central Europe, Northern Asia, and the East Indies. This tobacco is used for insecticidal purposes as well as for human consumption.

Cultivation of Tobacco

Tobacco is a highly specialized, costly, and temperamental crop. The several types differ in their cultural requirements as to temperature,

Fig. 136. Tobacco growing in the Connecticut valley, Massachusetts. (*Courtesy of the Agronomy Department, University of Massachusetts.*)

moisture, sunlight, soil, fertilizers, etc., and also in the methods of harvesting, curing, and handling. Tobacco seeds are so small that it takes 400,000 to make an ounce. They are planted in seed beds, and the seedlings are transplanted when they are 4 to 6 in. high. The best soil is a light sandy loam, rich in humus, and well fertilized with potash, lime, and other essential elements. Considerable cultivation of the fields (Fig. 136), both before and after transplanting, is necessary. Cigar-wrapper tobacco is often grown under shade, thus insuring a more uniform environment and product. After the plants get a good start, the terminal bud is removed, an operation known as topping, in order to keep the strength in the leaves. Often inferior leaves and suckers are also taken off. When fully ripe, as indicated by a change in the color of the leaves, either the entire plant is cut off or the leaves alone are harvested, one by one, as they mature. This latter method is called pruning and is virtually restricted to flue-cured and shade-grown cigar-wrapper tobacco.

Curing of Tobacco

Freshly harvested plants or leaves are allowed to wilt and then are suspended in an inverted position from a framework in specially con-

Fig. 137. Interior of a tobacco shed on a farm near Mitchell, Ind. (*Reproduced by permission from MacDonald, Then and Now in Dixie, Ginn and Company.*)

structed curing barns. There are two principal methods of curing: air-curing and flue-curing, and two others, fire-curing and sun-curing, of minor importance.

Air-curing is a slow process carried on under virtually natural conditions in well-ventilated barns (Fig. 137) in which the temperature and humidity can be carefully controlled. Artificial heat is resorted to only in unfavorable weather. *Flue-curing*, a much quicker process, is carried on in small barns, and artificial heat is supplied through flues from a small furnace. Flue-cured tobacco develops a characteristic bright-yellow color. It is a comparatively new method. In *fire-curing* the leaves are dried over fires of charcoal or hardwood. They are virtually smoked without much increase in temperature, and during the process they develop a creosotelike odor. This is the oldest method of curing and was practiced by the Indians. *Sun-curing*, carried on in the open, is almost never utilized in the United States at the present time. It is an old method and is still used for Turkish and oriental tobaccos.

Curing is essentially an oxidation process or dry fermentation, during which the leaves lose most of their water and green color and become tougher. Certain changes in chemical composition also take place, which are essential for the development of the desired quality. Curing usually takes from three to six months. Freshly cured leaves must be sorted, fermented or "sweated," and aged before manufacturing. This is usually carried on in the warehouses after the leaves have been graded. They are either piled up in large heaps or pressed into special containers. The process may take from six months to three years, during which time the aroma and odor are developed, harshness, bitterness, and other undesirable qualities are eliminated, and the color and burning qualities are improved.

Grades of Tobacco

The proper grading of tobacco is a very specialized task and can be done only by experts. A single crop may yield as many as 50 different grades. The Department of Agriculture recognizes *classes*, based on the method of curing and, in the case of cigar tobacco, on the use; *types*, usually designated by the regions in which they are produced; and *grades*, based on use, texture, color, quality, and other characteristics. There are seven classes as follows:

Class 1. Flue-cured. This so-called bright tobacco is used for cigarettes, pipe and chewing tobacco and constitutes 75 per cent of the tobacco exported.

Class 2. Fire-cured. This is used for export, snuff, and plug wrappers.

Class 3. Air-cured.

a. Light. Two types of light air-cured tobacco are grown, the *Burley*, used for cigarettes, pipe and chewing tobacco, and the *Maryland* used for cigarettes and export.

b. Dark. This is used for chewing and plug tobacco, export, and snuff.

Class 4. Cigar Filler.

Class 5. Cigar Binder.

Class 6. Cigar Wrapper.

Class 7. Miscellaneous. Among the miscellaneous tobaccos is perique, grown only in St. James Parish, Louisiana, and an ingredient of many mixtures of smoking tobaccos.

Utilization of Tobacco

The cured and aged tobacco leaf is manufactured into various forms for use by the consumer, either as snuff, chewing tobacco, or smoking tobacco, all three of which were used by the Indians, or as cigars or cigarettes, which are later developments. Throughout all the phases of manufacturing, particularly in the case of cigarettes, it is necessary to maintain the proper moisture content. This is done by adding various hygroscopic agents, such as glycerin. These are known as "humectants."

A great variety of flavoring and conditioning materials are also used in making chewing and smoking tobacco. These affect the taste and smoking qualities. They are utilized as a "sauce," in which the leaves are immersed, or as a "spray." Among them may be mentioned licorice paste, sugar, honey, molasses, rum, and tonka beans. Deer's tongue, an old-time flavoring, consisting of the powdered leaves of *Trilisa odoratissima*, the wild vanilla of the southeastern United States, is still in use. As in the case of tonka beans, the flavor is due to the presence of coumarin.

Blending, the use of different grades of leaf, is also an important feature of tobacco manufacture. Perique and Latakia, imported from Syria, are common ingredients.

Snuff is made by grinding up dark air- and fire-cured leaves to a powder. The poorer grades and waste are often utilized. Although snuff was in use before the time of Columbus, it has never been of great importance.

Chewing tobacco is prepared from Burley, dark air-cured, and flue-cured tobacco. It requires leaves that are rich in flavor, tough, gummy, and highly absorptive to the various flavoring materials that are added. It was an early development of the industry, reaching its maximum production in the early 1900's. Navy plug is very sweet and thick and consists of a filler with a wrapper.

Smoking tobacco is made from heavily sauced blends of Burley, flue-cured, and miscellaneous tobaccos or from mildly flavored straight Burley. Granulated tobacco, the oldest type, is blended, whereas the plug cuts, which lack wrappers, consist of Burley.

Cigars are older than cigarettes and had reached their maximum consumption by 1930. Formerly made by hand, they are now machine-made, save for the very finest. Three grades of tobacco are utilized: fillers, binders, and wrappers, all of which are air-cured. Tobacco for fillers must have a sweet pleasant flavor and burn evenly with a firm white ash. For wrappers leaves are required that are free from flavor, thin and elastic, with small veins, and uniform in color. The individual leaves are picked.

Cigarettes require light-colored leaves that lack gummy substances and have been either air- or flue-cured. The most spectacular phase of the tobacco industry has been the amazing development of the cigarette. This has been due primarily to the perfection of the modern efficient cigarette machine and to the use of blended tobaccos. The first cigarettes were made from straight Virginia, flue-cured, or Burley tobacco.

As recently as 1880 only a few were made, chiefly Richmond Straight Cuts and Sweet Caporals. These were consumed almost entirely in the East, for manufactured cigarettes made slow progress in the Western states. In 1894, the Egyptian cigarette made from oriental tobaccos appeared and slowly made headway, even though it was expensive. In 1913, manufacturers began to add Turkish tobacco to improve the burning qualities of their own product, and cigarettes became more popular. With the outbreak of the First World War, they came into great demand and their use became nation-wide. The figures tell the story. In 1913, 3,000,000,000 cigarettes were produced; in 1914, 15,000,000,000; in 1915, 46,000,000,000; in 1921, 52,000,000,000; in 1930, 123,000,000,000; in 1935, 139,000,000,000; in 1940, 189,000,000,000; in 1942, 257,000,000,000; in 1945, 332,000,000,000; and the estimated production for 1948 was 387,000,000,000.

In present-day cigarette manufacture, properly aged leaves are utilized and the stems are removed by hand or by machine. The moisture content is then increased to from 18 to 20 per cent, and the various grades are blended. Although the actual formulas are trade secrets, in general cigarettes contain about 53 per cent flue-cured tobacco, 33 per cent Burley, 10 per cent oriental, and 4 per cent Maryland. The leaves are then run through a cutting machine, where they are shredded and dried. During the process, the "casing," consisting of licorice, sugar, glycerin, and various flavorings, is added. Cigarette paper is made chiefly from flax fiber.

The 1947 tobacco crop was utilized in the following manner: 98,440,000 lb. for chewing tobacco, 104,680,000 lb. for pipe tobacco, 39,163,000 lb. for snuff, and 507,826,000 lb. for export; 5,567,346 cigars and 369,683,306,000 cigarettes were manufactured.

Production of Tobacco

The United States has always led the world in the production of tobacco. This crop was first grown in 1612 and was first exported in 1618 from Jamestown. From the very outset tobacco was the backbone of the Virginia colony, even serving as currency. Around its cultivation in tidewater Virginia a culture grew up which has never been equaled in America and which flourished for two centuries. After the Revolution the industry declined, owing partly to the competition of other countries and partly to the exhaustion of the soil. Gradually the industry moved westward into the Piedmont region of Virginia and North Carolina, and the great estates gave way to small farms. Tobacco has been grown in New England to some extent from the earliest days, but the crop has been important only since 1795. Today the tobacco industry is specialized, since certain areas are better suited to one kind of tobacco than to others.

The crop is grown commercially in 21 different states, with Kentucky and North Carolina producing about 60 per cent of the total amount. In 1946, 1,963,400 acres were devoted to tobacco, and the crop amounted to 2,321,596,000 lb. divided among the several grades as follows: *flue-cured*, 1,317,466,000 lb., grown in Virginia, North Carolina, South Carolina and Georgia; *Burley*, 484,704,000 lb., grown in Kentucky, Indiana, Ohio, West Virginia, Tennessee, North Carolina, and Virginia; *Maryland*, 37,762,000 lb., grown in southern Maryland; *dark air-cured*, 37,195,000 lb., grown in eastern Kentucky and Tennessee, southern Indiana, and north-central Virginia; *fire-cured*, 85,850,000 lb., grown in western Kentucky, northwestern Tennessee, and central Virginia; *cigar filler*, 63,160,000 lb., grown in Pennsylvania, Ohio, and Kentucky; *cigar binder*, 70,255,000 lb., grown in the Connecticut valley, Wisconsin, Minnesota, New York, Georgia, and Florida; *cigar wrapper*, 13,480,000 lb , grown in the Connecticut valley, Georgia, and Florida; *Perique*, 249,000 lb., grown in Louisiana.

In spite of the large domestic production, a considerable quantity of tobacco is imported, principally oriental types for use in cigarettes.

Other large tobacco-producing countries are China and India, each of which grow approximately a billion pounds, the U.S.S.R., Indonesia, Brazil, Turkey, Italy, and Japan. Many other areas produce high-grade tobaccos in smaller quantities. This is particularly true of Cuba.

Other Tobacco Products

A tobacco extract, made from low grades of fire- and dark air-cured leaves, is exported for use in making chewing tobacco. *Nicotiana rustica* is grown as the source of nicotine, which is extracted for use as an insecticide. Rutin, ordinarily obtained from buckwheat, is also present in tobacco leaves, but the extraction process has so far been too costly for it to be of commercial value.

BETEL

Probably more people chew betel nuts than any other masticatory. The number has been estimated as over 400,000,000. The desire for betel is very great, and it is chewed by all classes of people at all times. The habit is indulged in from Reunion and Zanzibar to India, Burma, Malaya, Indo-China, and southeastern China and in the East Indies, the Philippines, and some of Oceania. The widespread occurrence of the habit is indicative of its antiquity. It was first described by Herodotus in 340 B.C. Over 100,000 tons of the nuts are used in India alone, and the consumption of *pan*, as it is called, plays an integral part in the daily life of the inhabitants.

Betel nuts or areca nuts are the seeds of the betel-nut palm (*Areca Catechu*). This palm is a native of Malaya but is extensively cultivated wherever the nuts are used. The process of betel chewing may be quite complex. The simplest and most usual method, however, involves the use of only three ingredients (Fig. 138): betel nuts, betel leaves, and lime. Slices of cured (semiripe) or ripe nuts are placed in the mouth. Then

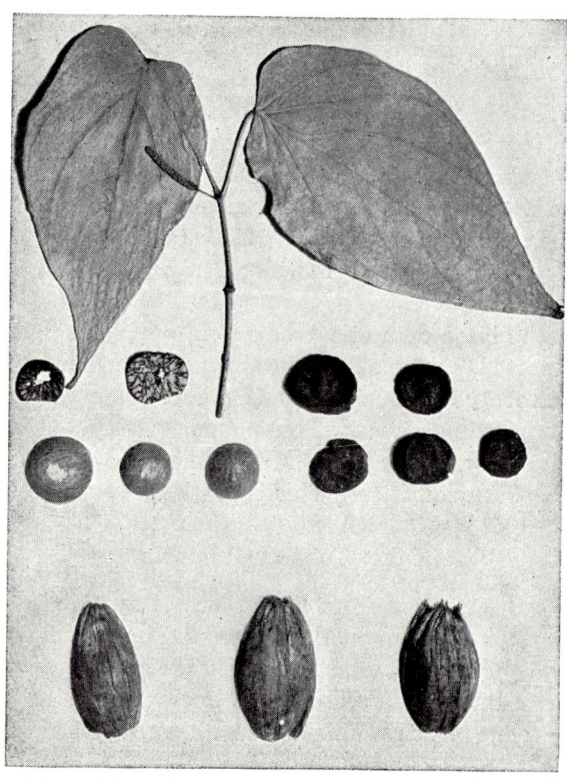

FIG. 138. Materials used in betel-nut chewing. Above, leaves of the betel pepper (*Piper Betle*); center, betel nuts (*Areca Catechu*) with husk removed and some of the nuts cut; below, betel nuts with the husk surrounding the fruit. (*Courtesy of the Botanical Museum of Harvard University.*)

fresh leaves of the betel pepper (*Piper Betle*) are smeared with lime and chewed with the nuts. This practice is always indulged in after dinner as a breath sweetener. It is not in the least harmful and may even aid in digestion. Often mixtures of the nuts with cloves, cinnamon, cardamom, nutmeg, or other spices are utilized. Another type of betel chewing involves the additional use of tobacco. This may be habit forming, but as in the case of tobacco, the narcotic principles are not especially harmful in that they do not affect consciousness in any way but merely produce a mild stimulation and feeling of well-being.

COLA

The seeds of the cola tree (*Cola nitida*), known as cola or kola nuts, are extensively used in many parts of tropical Africa as a masticatory. This tree, a tall species with a straight trunk reaching a height of 50 to 65 ft., grows wild in the forests of tropical West Africa. It is also cultivated in this region and the adjacent Sudan, and has been introduced into Jamaica, Brazil, India, and other parts of tropical Asia. The fruit consists of star-shaped follicles which contain eight hard, plano-convex, fleshy seeds with a reddish color and the odor of roses. These "nuts" (Fig. 139) are marketed fresh and are usually chewed directly, although powdered nuts may be used.

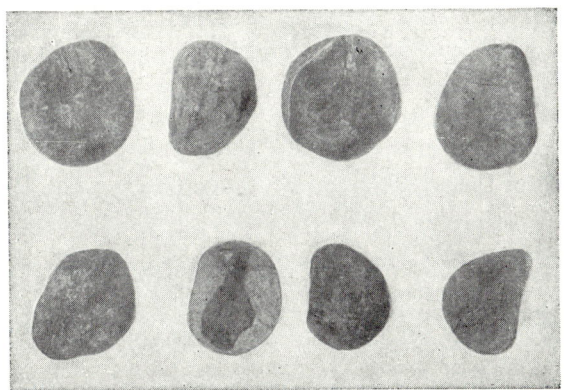

Fig. 139. Cola nuts, the seeds of *Cola nitida*.

Cola has long played an important part in the social, religious, and commercial life of the African natives. It was first reported in the twelfth century. The nuts are in great demand and the natives will go to considerable trouble to obtain them. Although bitter at first, they leave a sweet taste in the mouth. The use of this masticatory results in a slight stimulation and temporary increase in physical capacity. It is said to lessen hunger and fatigue. Cola is the most complex of the caffeine-containing products. It not only contains 2 per cent caffeine, but an essential oil and a glucoside, kolanin, as well. The stimulating effect of cola is due in part to the caffeine and in part to the kolanin, which acts as a heart stimulant. Old nuts tend to lose their kolanin, and so are less invigorating. The chewing of cola nuts has no effect on the consciousness and produces no other deleterious results.

THE TRUE NARCOTICS

We have already seen that the narcotic plants contain alkaloids that are valuable in medicine, when used in exceedingly small amounts. They are used to relieve pain, produce sleep, and quiet anxiety and fears. It is so easy to produce serious physiological effects, however, that they must be used with the utmost discretion and only under a physician's direction. There is no possible excuse for their use under any other conditions.

The narcotic drugs vary considerably in their effects on the human system. Cocaine and opium act as sedatives on mental activity and bring about a state of physical and mental comfort. This is accompanied by a diminution, and even suspension, of emotion and perception, together with a lowering and often complete suppression of consciousness.

Cannabis, peyote, fly agaric, caapi, and the solanaceous narcotics, on the other hand, cause cerebral excitation and bring about hallucinations, visions, and illusions. Their use causes intoxication and may be accompanied or followed by unconsciousness or other symptoms that indicate that the brain is no longer functioning normally.

Kavakava is a sleep-producing drug and tends to induce a hypnotic state.

Coca

The chewing of the whole or finely powdered leaves of the coca plant (*Erythroxylon Coca*), already mentioned as the source of the drug cocaine,

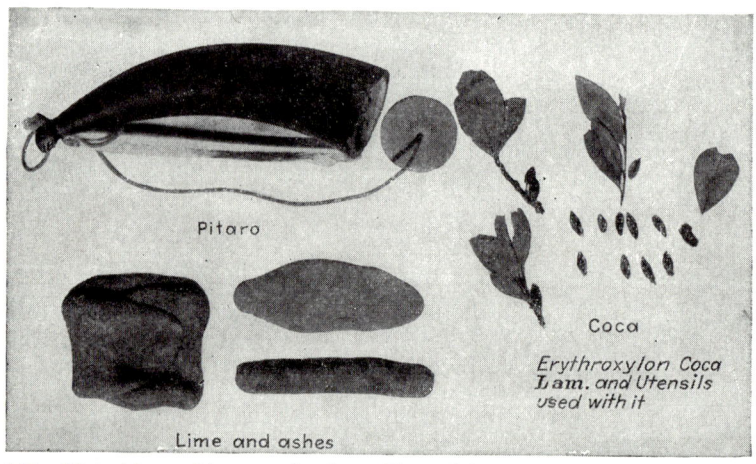

FIG. 140. Materials used in coca chewing. Upper left, the pitaro, a container for lime; lower left, lime and ashes; right, leaves and fruit of the coca shrub (*Erythroxylon Coca*). (*Courtesy of the Botanical Museum of Harvard University.*)

is a centuries-old custom among the Indians of the Andes and the western half of the Amazon basin. The discovery of the plant and its properties is shrouded in mystery. It was highly esteemed by the Incas, who used it as an emblem of royalty. The use of coca gradually spread from the higher classes among the common people, and Pizarro found it in general use in 1553.

People who use coca can resist physical and mental fatigue and work for long periods without food or drink. The average consumption is 25 to 50 grams daily. The chewing of coca is followed, after a short period of rest, by greatly stimulated activity. The narcotic acts directly on the central nervous system, causing immediate psychic exaltation to such an extent that the consumer is enabled to forget his hunger or other pains. The custom may become a habit and lead to physical deterioration, sickness, and even death, as its favors malnutrition. The leaves are chewed with lime and the highly alkaline ashes of some plant, such as quinoa or cecropia leaves. Coca chewing (Fig. 140) is so indispensable and widespread in South America that very little coca is exported, although the plant is widely cultivated on the eastern slopes of the Andes from Colombia to Argentina. It is also cultivated to some extent in Java and India. The use of coca, and its derivative cocaine, is subject to the strictest regulations in most countries.

Opium

Opium is a very old narcotic. As we have seen, it is the dried juice that exudes from injured capsules of the opium poppy. Originating probably in Asia Minor, the use of opium soon spread westward. The drug was known to the Greeks, Romans, and Egyptians and perhaps to the earlier Lake Dwellers of Switzerland. In its eastward dispersal it had reached Persia, India, and China by the eighth century, and since then has spread all over the world. When properly utilized, opium and the alkaloids derived from it are valuable medicinally and have proved a great blessing to mankind in the relief of pain. Excessive use of the drug, however, and the resulting opium habit have been and still are the cause of unbelievable suffering and evil. No other drug has caused so much corruption and tragedy. In spite of every effort to stamp out the habit, it seems to be increasing. This is particularly true of the Orient, where opium has taken a toll of millions of lives. It has been estimated that over 900,000,000 people are still apt to use it.

It is an all too easy step from the use of a small quantity of opium as a soporific, or for the pleasure of the moment, to the point where it becomes necessary for the very existence of the unfortunate addict. The immediate effects of taking opium are pleasurable, and alluring dreams and visions are induced. Continued use, however, leads to delirium and death. The opium addict soon loses the will power necessary to resist the craving. Even if he develops sufficient moral resistance, the withdrawal pains are so severe that it becomes virtually impossible for him to continue his abstinence.

In India opium is usually eaten and the habit is common to all classes of society. So great is the demand that the cultivation of the opium poppy is one of the most profitable industries in the country. In earlier days opium was openly exploited by the Portuguese, the Dutch, and the English in turn. Today attempts are being made by the various governments to regulate the production and trade in opium. Unfortunately, because of the large revenues that can be derived, such attempts are likely to be only half-hearted and consequently only partially successful. In India immense quantities of opium are available in the open market, although the government has a monopoly on the drug and attempts to restrict its sale to accredited buyers only.

Fig. 141. A field of opium poppies (*Papaver somniferum*) in southern Manchuria. (*Photo by F. N. Meyer; courtesy of the Arnold Arboretum.*)

Conditions in China are much worse. Here the usual method of consumption is opium smoking—placing a small pellet in the bowl of a special pipe and inhaling the fumes. In this way more morphine is said to be absorbed and the effects on the system may be greater. Until recently few attempts have been made to stamp out the practice. In the past the nation as a whole has shown in its mental and physical characteristics the effects of the opium habit. What the future holds for China is problematical for it is claimed that more opium is grown and used today than ever before. It has been estimated that the consumption is eight times that of the rest of the world together. Every province grows the poppy (Fig. 141) and the crop may be so profitable that little food is raised, with the result that famine is always hovering over the land.

In Europe and the United States, although opium, morphine, heroin, codeine, and the other derivatives are used to a much less extent than in the Orient, the smuggling and use of these narcotics are matters of grave concern. Laws relating to narcotics cannot be too rigidly enforced. In countries where the habit has a strong foothold, opium is not only a social problem but often the cause of international complications as well. China and India have had serious differences over the opium trade. The suppression of the production and use of opium and similar narcotics is one of the greatest problems facing the world today. The League of Nations rightly devoted considerable attention toward setting up prohibitive regulations and other methods of dealing with the situation.

Cannabis

The hemp plant (*Cannabis sativa*) (Fig. 10), already discussed as the source of a textile fiber and a drying oil, also yields a narcotic drug. The dried flowering tops of the female plants, pressed together into solid masses, constitute the official "cannabis indica." This drug is used in medicine to relieve pain and in the treatment of hysteria and various nervous disorders. The active principle is resinous in nature and contains three or four very powerful alkaloids. Hemp is often cultivated solely as a drug plant. This is especially true in India, where hemp growing is almost a science. The use of hemp as a narcotic stimulant is very old, extending back to 3000 B.C., first in China and later in India. The plant was used by the Assyrians and was known to Herodotus. At the present time the habit is indulged in chiefly in Mongolia, India, and other parts of Southern Asia, Asia Minor, and Northern and tropical Africa.

Indian hemp is consumed in various ways. The pure, undiluted, sticky yellow resin which is naturally exuded from the flowering tops of cultivated female plants is known as *charas* or *hashish*. Formerly the resin was obtained by rolling or treading on the leaves or by having natives run violently through a mass of the plants. The resin stuck to the body or clothes of the runner and was easily removed. Now it is carefully pressed out of the flowering tops between layers of cheesecloth and then scraped off. Charas is smoked. It is the most powerful form of the drug. *Bhang* consists of the tops of wild plants, which have a lower resin content, in a water or milk decoction. It is also smoked. In America this type of hemp is known as *marijuana*. *Ganja* is a specially cultivated and harvested grade of hemp used for smoking and in beverages and sweetmeats. It has a high resin content.

Cannabis in its various forms often produces serious results for the consumer. It causes a stupefying and hypnotic effect, accompanied by hallucinations, agreeable and often erotic dreams, and a general state of ecstasy. The addicts, while under the influence of the drug, are apt to be happy and noisy and may even become fanatical and commit murder. When used in moderation, however, it is apparently not too harmful. Cannabis has been of little importance in the United States, but recently the use of marijuana cigarettes, illicitily made from hemp, seems to be increasing, in spite of all repressive measures.

Peyote

A cactus, *Lophophora Williamsii*, is the source of peyote or mescal buttons. This species is indigenous to Mexico and the southwestern

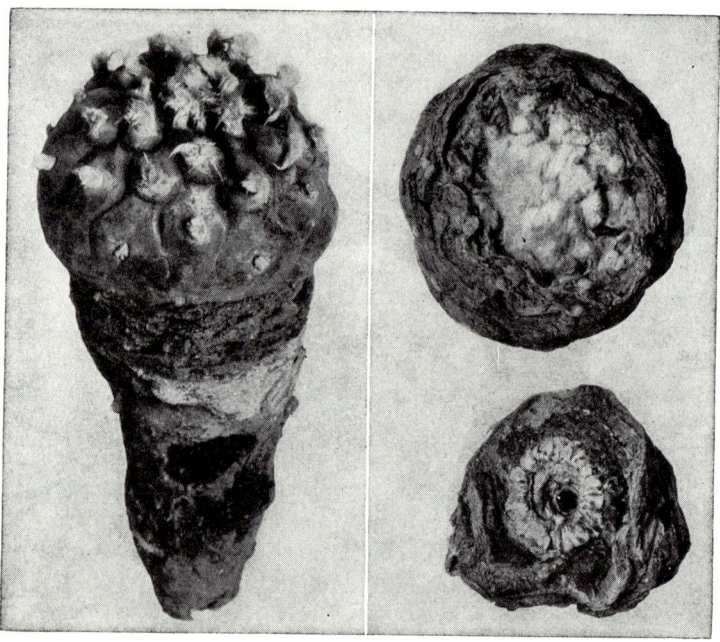

FIG. 142. Peyote (*Lophophora Williamsii*). *Left*, an entire plant of this diminutive cactus; *right*, the dried crowns, known as mescal buttons. These have powerful narcotic properties. (*Photo by Richard E. Schultes.*)

United States, occurring on dry, arid plateaus in a limited area bordering on the Rio Grande. The cactus resembles a huge carrot (Fig. 142) with all parts of the plant, except a buttonlike top, below ground. These tops are round and flattened and have a wrinkled surface. They do not bear spines, but only little tufts of silky hairs. Some of the buttons are very tiny, while others may measure 3 or $3\frac{1}{2}$ in. in diameter.

Mescal buttons contain several very powerful alkaloids with narcotic properties. The native Indians of the region have used peyote for hundreds of years in connection with baptisms and other religious ceremonies. In spite of serious opposition the habit is still actively spreading among the southwestern Indians, owing primarily to the existence of a sacred cult devoted to its use. The buttons in either a fresh or dried state are chewed until they are soft. They are then rolled up in the hand into little pellets which are swallowed. Sometimes a beverage is prepared by boiling the buttons. Peyote produces a state of ease and well-being, accompanied by visions and hypnotic trances. The addict finds himself in a world full of new sensations and pleasures. Peyote is also used extensively by the Indians for its supposed medicinal value. It is considered by them to be a panacea for most bodily ills.

Fly Agaric

The fly agaric (*Amanita muscaria*) is known in Europe, Asia, and America as one of the most poisonous of fungi. In spite of this fact, quantities of this mushroom are consumed in Siberia and other parts of Northeastern Asia for its intoxicating effect. Dried agarics are chewed until soft, rolled into sausage-shaped pieces, and swallowed. At other times they are used in milk, water, or the juice of berries as a beverage. The use of this fungus produces hallucinations and illusions, accompanied by giddiness, involuntary words and actions, and finally unconsciousness. Two small fungi are sufficient to produce an intoxication that will last for a whole day. The habitual use of the narcotic will completely shatter the nervous system, yet the craving for the drug among its addicts is too great to be denied.

Ololiuqui

The Indians of southwestern Mexico were using a narcotic plant in their religious ceremonies and as an intoxicant prior to the Spanish Conquest. Ololiuqui, a decoction made from the seeds of *Rivea corymbosa*, a woody vine resembling a morning glory, is still used, especially in Oaxaca, in medicine, as a means of divination, and as a narcotic. Ololiuqui induces a hypnotic sleep or coma with hallucinations and a feeling of ecstasy, but with no unpleasant aftereffects.

Caapi

It has been estimated that most of the Indian tribes of the northwestern part of the Amazon basin utilize *Banisteriopsis Caapi, B. inebrians, B. quitensis*, and species of the related genus *Tetrapterys* (Fig. 143) as the source of narcotic beverages. These plants are known as *caapi* in Brazil, *ayahusca* in Peru and Ecuador, and *yaje* in Colombia. They are lianas which grow naturally in the virgin forests and are sometimes cultivated in Indian villages. In preparing the beverage, the lower part of the stem is cut off, cleaned, macerated, and boiled or utilized as a cold decoction. Caapi is used in the religious ceremonies of the natives. It produces visions, dreams, and other mental disorders. It is also an excitant and induces an unnatural courage.

Fig. 143. The interior of a forest in the Rio Uaupés basin, Brazilian Amazon. Caapi, a narcotic beverage, is prepared from the bark of lianas such as this species of *Tetrapterys*. (*Photo by Richard E. Schultes.*)

Solanaceous Narcotics

Several members of the *Solanaceae* contain alkaloids that produce disorders of the brain and marked excitation when smoked or eaten. These narcotics are often responsible for some of the incomprehensible acts of fanatics in the East and elsewhere.

The genus *Datura* has been and still is extensively used in all the continents except Australia for its narcotic and hypnotic properties. Long before there were written records various species were used in folklore, religion, and medicine. The Jimson weed (*D. Stramonium*), the source of the drug stramonium, was known as a narcotic as early as A.D. 37. It still is one of the favorite sources of "knockout drops" in the tropics. The maikoa (*D. arborea*) and *D. sanguinea*, subtropical shrubs of South America, are much used by various tribes of the westernmost Amazon region for their narcotic properties. *D. innoxia*, used by the Aztecs as a medicine and in their religious ceremonies, is still utilized to some extent by the Mexican Indians. Other species are similarly used elsewhere, all of them producing comparable effects, such as sense illusions and motor disturbances, together with senseless activities and loss of memory.

Other important solanaceous narcotics include the henbanes and pituri. The black henbane (*Hyoscyamus niger*) has long been used as a poison and a sorcerer's potion, and to bring about visions and prophecies. *H. muticus* is smoked in Northern Africa and India for its intoxicating effect. Large quantities of the pituri (*Duboisia Hopwoodii*) are chewed and smoked by the natives of Australia for its stimulating and narcotic effects. It contains the poisonous alkaloid scopolamine and is also used as a fish poison.

KAVAKAVA

The use of kavakava, which is almost universal throughout Oceania, produces different results from those hitherto discussed. The beverage acts as a sedative, a soporific, and a hypnotic, bringing about pleasant dreams and sensations. Excessive use is apt to produce skin diseases and weaken the eyesight. The active principle is a resinous substance that is stimulating in small amounts. The source of kava, as it is sometimes called, is a bushy shrub 6 to 8 ft. tall, with rounded or cordate leaves. This species, *Piper methysticum*, is indigenous to Fiji and other Pacific islands, but is now grown everywhere in the South Seas. The thick, knotty, grayish-green roots are the important part. These are dug up and the bark is removed. After the roots are thoroughly cleaned, they are cut up into small pieces. These pieces are chewed until they are fine and fibrous and are then placed in a bowl with water and allowed to ferment. Formerly while the roots were being chewed the saliva was ejected into bowls and this constituted the beverage. After straining, kava is a grayish-brown liquid and is very refreshing. It is closely connected with the entire social, political, and religious life of the people. It is used in connection with festivals and religious observations, as a soporific, as a token of good will, and as a daily beverage.

CHAPTER XIII

THE HISTORY AND NATURE OF FOOD PLANTS

THE HISTORY OF FOOD PLANTS

The most remarkable fact concerning the food plants in use in the world today, and for that matter the industrial plants as well, is their great antiquity. Most of them were domesticated from wild ancestors long before the beginning of the historical period, and all available records indicate that they were as familiar to the peoples of the ancient world as they are to us. Comparatively few new plants have been developed during the last 2000 years, although the older ones have been greatly altered and improved in response to the increasing complexity of man's existence.

The history of our useful plants and their influence on civilization has always been of interest to botanists and ethnologists. Many investigations have been carried on in an attempt to determine their age and place of origin, as well as their cultural history.

The Work of De Candolle. The classic work dealing with this phase of botany is De Candolle's "L'origine des plantes cultiveés," which appeared in 1883. So careful and painstaking was his work that few of his conclusions regarding geographic distribution have had to be altered in the light of more recent studies. His dates, however, are of little or no value. De Candolle (Fig. 144) based his conclusions on a great variety of evidence: the works of Theophrastus, Dioscorides, and other old historians; Chinese writings; archeological and ethnological data, such as the monuments of Egypt, the ruins of Pompeii, the remains of the Lake Dwellers of Europe, and the Inca ruins of South America; philological indications, involving the names of plants in Hebrew, Sanskrit, and other ancient languages; and botanical conclusions based on distribution, number of varieties, presence or absence of wild types, length of cultivation, and similar matters. He arranged the useful

Fig. 144. Alphonse De Candolle (1806–1893), from a photograph taken in 1866. (*Courtesy of the Gray Herbarium.*)

plants in six classes, and it will be interesting to give a few examples of each of these groups:

A. Old World Species Cultivated for Over 4000 Years

almond	date	millet	rice
apple	eggplant	mulberry	sorghum
apricot	fig	olive	soybean
banana	flax	onion	tea
barley	grape	peach	turnip
broad bean	hemp	pear	watermelon
cabbage	mango	quince	wheat
cucumber			

B. Old World Species Cultivated for Over 2000 Years, and Perhaps Longer

alfalfa	chestnut	mustard	poppy
asparagus	cotton	nutmeg	radish
beet	garden pea	oats	rye
breadfruit	grapefruit	orange	sugar cane
carrot	lemon	pepper	walnut
celery	lettuce	plum	yam
cherry	lime		

C. Old World Species Cultivated Probably for Less Than 2000 Years

artichoke	endive	okra	raspberry
buckwheat	gooseberry	parsley	rhubarb
coffee	horseradish	parsnip	strawberry
currant	muskmelon		

D. New World Species of Ancient Cultivation, More Than 2000 Years

cacao	maize	sweet potato	tobacco
kidney bean	maté		

E. New World Species Cultivated before the Time of Columbus, Antiquity Not Known

avocado	peanut	pumpkin	squash
cotton	pineapple	quinoa	tomato
guava	potato	red pepper	vanilla
Jerusalem artichoke			

F. New World Species Cultivated Since the Time of Columbus

allspice	cinchona	gooseberry	plum
blackberry	cranberry	pecan	rubber
black walnut	dewberry	persimmon	strawberry
blueberry			

From these examples it can readily be seen that our most valuable economic plants, including the cereals, most of the vegetables and fruits, tea, coffee, cocoa, and the fiber plants, were discovered, utilized, and cultivated by man thousands and thousands of years ago.

It is even more difficult to determine the native homes of our cultivated plants. Obviously they must have been derived at some time in the remote past from wild ancestors, which originally had a restricted distribution. In most cases these wild forms no longer exist, or they have been carried by man far from their original home. For these and many other reasons it is a well-nigh impossible task to come to a definite conclusion as to the place of their origin.

The Work of Vavilov. An important work which throws some light on this point of cultivated-plant origins is that of Vavilov, which appeared in 1926. His conclusions are based on a variety of facts obtained from sources different from those of his predecessors. He considers such features as the anatomy, genetics, cytology, distribution, and diseases of the plants concerned.

A valuable conclusion in Vavilov's work is that many of our cultivated species of first rank, the primary crops as he calls them, had a diversified rather than a single origin. In the case of wheat, for example, the author points out that there were at least two distinct centers of distribution. The soft wheats came from Southwestern Asia, while the hard wheats originated in the Mediterranean region. Similarly barley was derived from Southwestern Asia, North Africa, and Southeastern Asia.

Another point is concerned with the so-called secondary crop plants. It is the contention of the author that these were originally weed companions of the primary crops. These weeds could not be eliminated and were either ignored or tolerated by the farmer. In regions that were favorable for the primary crops the weeds were of little importance. In unfavorable areas, however, the weeds tended to become more and more prominent, gradually replacing the primary crop, and eventually becoming established as a cultivated crop. Rye and oats are conspicuous examples of such plants.

A final contention is that the great centers of distribution of our cultivated crops were always in mountainous regions, and that the greatest amount of diversity occurred in such areas. Vavilov in general recognizes the four centers of distribution to be discussed below, with the addition of a fifth area in Abyssinia and adjacent parts of Northern Africa. He also suggests the possibility of a sixth center in the Philippine Islands where rice and coix may have originated.

ORIGIN AND DISTRIBUTION OF FOOD PLANTS

At the present time the available data seem to establish the fact that there were at least four chief centers in which our economic plants originated, and from which they were later dispersed all over the world: (1) Southwestern or Central Asia, the mountainous region from India to Asia Minor and Transcaucasia; (2) the Mediterranean region; (3) Southeastern Asia; and (4) the highlands of tropical America.

The parallelism between the history of mankind and the history of his domesticated plants (and animals as well) is obvious. It was in this same Central Asian plateau that scientists tell us man had his origin, and from which the human race was dispersed. Thus from the earliest beginnings man had at his disposal various food plants, and he must have been dependent on them to a great extent for his existence. For countless ages he was a nomad, wandering from place to place, content merely to gather the edible fruits, grains, seeds, and tubers as he needed them, possibly for temporary storage in small amounts. At some later period in his history he began to make primitive attempts at cultivating these useful plants by sowing seeds in some favorable location. Whether these first attempts at agriculture were accidental or purposeful, they were of profound importance for they changed the whole nature of his existence. Of necessity he had to forsake his nomadic life and remain in one place at least long enough to harvest his crop. In so doing he took the first step toward becoming civilized, for agriculture is the only mode of existence that has enabled men to live together in communities and accumulate the necessities of life. The establishment of agriculture was of the utmost importance to man and probably represents the most significant single advance in his development.

Gradually these first simple types of plant culture were replaced by an agriculture of a much higher grade, which eventually led to the building up of the great nations of antiquity. These ancient civilizations were restricted in area, for they developed only in those regions where the useful plants that were the foundation stones of their existence were native. And so we find that Asia Minor and adjacent areas in Southwestern Asia, the Mediterranean region, Southeastern Asia, and the tropical American highlands were the locations of the older civilizations. The presence of valuable plants in all these regions was the most important factor in the successful development of agriculture, although in all these areas climate and soil conditions were very favorable. The climate was equable, with no extremes of heat and cold; the soil was fertile; and there was either ample rainfall or irrigation could be practiced.

In *Central Asia* the native plants included alfalfa, apple, barley, broad bean, buckwheat, cherry, flax, garden peas, garlic, hemp, lentil, mulberry, olive, onion, pomegranate, plum, quince, radish, rye, and spinach.

The *Mediterranean region* was the home of the artichoke, asparagus, cabbage, cauliflower, cotton, fig, horseradish, millet, parsnip, parsley, and rhubarb. Common to both these areas were the almond, carrot, carob, celery, chestnut, grape, lettuce, mustard, turnip, and walnut. Wheat is also a native of some part of this combined area. Whether it was indigenous to Syria and Palestine, to Turkestan or Mesopotamia, or perhaps had a multiple origin, it was early available for all the nations of the region.

In *Southeastern Asia* the banana, breadfruit, millet, peach, persimmon, orange, rice, soybean, sugar cane, and yam were native; in the *American area* cacao, American cotton, kidney and lima beans, maize, potato, squash, tobacco, and tomato were indigenous.

It is interesting to note that a cereal was available in all these cultural areas. Ancient agriculture was based chiefly on these cereals, just as modern agriculture is. Their highly nutritious seeds were the staff of life 5000 and 10,000 years ago, and have remained so up to the present time. It was the cultivation of wheat in the fertile valleys of the Tigris and Euphrates rivers that made possible the great nations of Biblical time, Chaldea, Assyria, and Babylonia. Egypt, Greece, and Rome had both wheat and barley available. Rice was the basis of the restricted civilization that developed in the valleys of the Hwang Ho and Yangtze Kiang rivers and led to the development of the great Chinese empire. The primitive peoples of the highlands of tropical America cultivated the native maize, the foundation of the remarkable civilizations that persisted until overthrown by the Spanish invaders. In all these cultural areas the history of agriculture has been the same: first, the gathering of the edible portions of wild plants; then the primitive cultivation of certain species best adapted to man's needs; and finally the evolution of a highgrade agriculture, which involved the breeding of new varieties, improvements in cultivation, irrigation, and similar features. Because of this similarity in the development of agriculture, particularly between the Old and the New World, many authorities have believed that the American civilizations must have had some contact with those of the Old World and been influenced by them. The evidence, however, seems to show that agriculture in this continent has had an entirely separate development and that the resemblances which occur are only chance ones.

THE NATURE OF FOOD PLANTS

Food is necessary for the existence of all living things. The various substances that constitute the food of plants and animals are used by them either in the formation of the living protoplasm, the building up of their bodily structure, or as a source of energy. We have already pointed out, in Chap. I, that green plants alone are actually able to manufacture food from raw materials. Man and the other animals must take their food readymade, and so are dependent, either directly or indirectly, on plants. Fortunately for the animal world, plants manufacture much more food than they can utilize immediately, and they store up this surplus as a reserve supply for future use. It is this supply of reserve food that man appropriates for his own use. The essential foods, carbohydrates, fats, and proteins, each valuable in its own way in man's metabolism, are all available in plants. So, too, are mineral salts, organic acids, vitamins, and enzymes, which are also necessary for his well-being. Thus it is possible for man, if he so desires, to live entirely on a vegetarian diet.

Plants utilize roots, stems, leaves, fruits, and seeds, to a greater or less extent, for the storage of reserve food. The most important of these from the standpoint of man are the dry fruits and seeds. In this category are found the cereals and small grains, the legumes, and the nuts. All these contain a very large amount of nutritive material and have a proportionately low water content. This latter fact enhances their value to man, for they can be stored and transported easily. Roots, tubers, bulbs, and other earth vegetables are next in importance as sources of food for human beings, and the lower animals as well. Their value is lessened by their high water content. The leafy parts of plants, the greens, salad plants, and other herbage vegetables, contain comparatively little stored food. However, they are necessary because of the vitamins and mineral salts they contain and the mechanical effect of the indigestible cellulose material. The same is true of the fleshy fruits, which may also contain various organic acids. In the present discussion the food plants will be considered under the following headings: cereals, small grains, legumes, nuts, earth vegetables, herbage vegetables, fruit vegetables, and fleshy fruits.

It will obviously be impossible to discuss, or even list, all the plants used for food throughout the world. Hundreds of species, both wild and cultivated, are used only by primitive races or in restricted areas. An attempt will be made to consider the outstanding food plants of the United States and Europe, together with a few of the more conspicuous ones of other countries.

Before proceeding to this discussion, which will be concerned primarily with the higher plants, some reference will be made to the lower plants as sources of food.

THE LOWER PLANTS AS SOURCES OF FOOD

Fungi

The use of mushrooms, truffles, and other fungi as sources of food is very ancient, possibly as old as man himself. The first records go back as far as the fifth century B.C. Mushrooms were well known to the Greeks and were highly prized by the Romans. During the Middle Ages the consumption of these edible fungi was enormous. Today they are eaten by both primitive and civilized peoples. Not only are wild forms utilized, but the cultivation of mushrooms is extensively carried on in Europe, the United States, and many parts of the Orient.

Mushrooms

Mushrooms occur naturally in fields, pastures, and woods. They represent the reproductive stage of certain of the higher fungi. The vegetative stage of these fungi consists of masses of fine threads, or hyphae, which constitute the mycelium. This mycelium extends in all directions through the soil, deriving its nourishment saprophytically from decaying organic matter. Sooner or later, depending in part on favorable environmental conditions, the visible spore-bearing stage is produced. It may take years for this to develop.

Space will not permit a discussion of the many edible wild mushrooms. These are more delicate in flavor and more palatable than the cultivated forms. However, great caution is necessary in distinguishing them from the poisonous species, familiarly known as toadstools, for the resemblance is often very close. No hard and fast rules can be laid down which

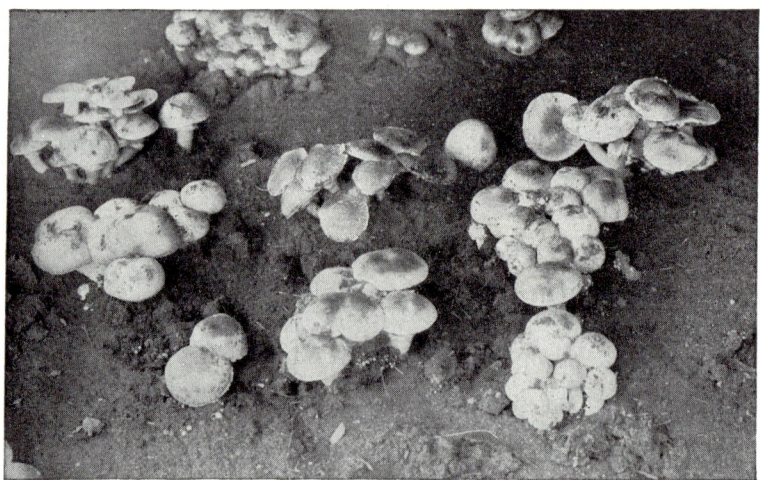

FIG. 145. Mushrooms (*Agaricus campestris*) grown under cultivation. (*Courtesy of the University of Massachusetts.*)

absolutely separate the two groups. Definite and accurate knowledge as to the identity of any particular species is necessary. It is wisest to discard any mushroom that resembles a poisonous form, even though it is known to be edible.

Cultivation of Mushrooms. The cultivation of mushrooms dates back to the beginning of the seventeenth century. Today it is carried on with a high degree of efficiency, particularly in France, England, and the eastern United States. The most important species utilized is the common meadow mushroom (*Agaricus campestris*). Propagation is by means of spores, or more usually by using spawn. Spawn consists of masses of mycelium compressed into little bricks. Suitable environmental conditions are essential for development. There should be little or no light, plenty of moisture, and a constant temperature, with 55 to 58°F. the optimum. The soil should be rich in organic matter, stable manure serving as an excellent medium. Cellars, tunnels, and caves afford ideal situations for growing mushrooms. The important French industry makes use of abandoned quarries and mines.

Under these conditions spores germinate, or the mycelium resumes its growth, ramifying through the soil and in about six weeks forming little buttons on the surface of the soil. Eventually the buttons enlarge to form a chamber in which the gills develop. This chamber is raised up on a short fleshy stalk and, when mature, opens out into the characteristic umbrellalike pileus with the gills on the underside. The innumerable spores are produced on the gills. Mushrooms (Fig. 145) are gathered in the button stage, or before they are fully mature. The mycelium continues to bear from six to eight months.

The food value of mushrooms is low, as nearly 91 per cent of the flesh is water. Proteins make up 3.75 per cent, carbohydrates 3.50, and fats only 0.20 per cent.

Truffles

Truffles have been famous almost as long as mushrooms and today are considered as greater delicacies and so are higher priced. Truffles differ from most other fungi in producing their fruiting bodies underground. They are solid, with a firm black or grayish-brown flesh and an agreeable odor and taste. Truffles are common in England and on the Continent. The chief commercial area is southern France. These fungi prefer a light, porous limestone soil in oak, beech, or birch forests. They are usually collected in the wild state, although crude attempts at cultivation have been made. These consist for the most part of stimulating natural production in favorable areas. Truffles are harvested with the aid of specially trained dogs or pigs, whose keen sense of smell enables them to locate the fungi. Truffles are collected when comparatively mature. Several species of the genus *Tuber* are utilized, chiefly *T. melanosporum*, *T. aestivum*, and *T. brumale*.

Other Fungi

The *morel* (*Morchella esculenta*) is a familiar wild edible fungus in the United States (Fig. 146). This species and several allied ones are grown to some extent in France, and other fungi elsewhere in Europe. In Japan, where enormous quantities of wild fungi are eaten, several native species are cultivated, chiefly the *shiitake* (*Cortinellus edodes*). This aromatic species is grown on logs. It can be kept for a long time in a dried condition and is a favorite ingredient of soups. Over 1000 tons are grown annually.

Algae

Only a few species of algae are utilized in Europe and America for food purposes and seaweeds are of little economic importance. In Japan,

China, and the Pacific islands, however, algae constitute one of the chief articles of diet. So great is the demand in Japan that the natural supply is insufficient and many species are cultivated. It is not unusual to find six or seven different kinds served at a single meal. Over 70 varieties are eaten in Hawaii, and a few of these are cultivated. The nutritive value of algae is high. They have a high carbohydrate content, around 50 per cent, with small amounts of proteins and fats. Moreover, they

FIG. 146. The morel (*Morchella esculenta*), an edible wild mushroom. (*Photo by D. H. Linder.*)

are rich in vitamins and also contain a greater variety of mineral salts than any other food.

Only three species of algae are of any importance as sources of food in the United States. These are Irish moss, dulse, and agar.

Irish Moss (*Chondrus crispus*) is a perennial species found from Maine to North Carolina. The fresh plant is greenish purple in color with densely tufted fronds from 2 to 10 in. in length. These are narrow and cylindrical at the base but soon become flattened and repeatedly forked. This species is especially common in New England and the collection of the "moss" commercially is carried on in eastern Massachusetts. The plants are gathered with iron rakes at ebb tide and are then spread out on the beach to dry and bleach (Fig. 147). After a while they are soaked in salt water and again bleached. This process is repeated four or five times. The final commercial product is yellowish white and has a hard, horny consistency. Irish moss has a high mucilage content and is used chiefly in making blancmanges, farinas, and similar desserts. The colloidal material, carrageenin, is extracted and purified. It is an excellent emulsifying and suspending agent. It is used in the baking and dairy industries and in a variety of other products ranging from hand lotions and tooth paste to beer. This species also grows on the Atlantic Coast of Europe and is a favorite food in the west of Ireland.

FIG. 147. Drying and bleaching Irish moss (*Chondrus crispus*) on the beach at Scituate, Mass. (*Courtesy of the E. L. Patch Company.*)

Dulse (*Rhodymenia palmata*), a red alga found on rocky shores on both sides of the Atlantic, is often dried and used for food. It is sometimes marketed as "sea kale."

Agar, which has already been discussed under drug plants, is a favorite food in China and Japan, where it is used in jellies, soups, sauces, etc. In the United States and Europe its use by bakers and in making ice cream, candy, cheese, mayonnaise, jellies, and desserts is increasing. It is also used in canning fish, clarifying liquors, and various other industrial operations.

Other species of algae, which are used in Scotland, Ireland, and Iceland for food, include the green laver (*Ulva lactuca*), pink laver (*Porphyra laciniata*), and murlins (*Alaria esculenta*).

CHAPTER XIV

THE MAJOR CEREALS

THE NATURE OF CEREALS

The cereals are without question the most important sources of plant food for man and the lower animals. This is not only true today, but has been so since earliest time. During their long period of cultivation their original wild ancestors have been lost sight of, and countless new species and varieties have been evolved. Much of this evolution occurred prior to the historical period for the older civilizations were already familiar with several kinds of wheat, barley, and other grains. Moreover, the actual origin of these useful plants had been so long forgotten that they were given supernatural powers and played a part in the religious ceremonies of the various nations of antiquity.

Long before the Christian era the ancient Romans held festivals at seed time and harvest in honor of the goddess Ceres, whom they worshipped as the giver of grain. At these festivals they brought offerings of wheat and barley, the *cerealia munera*, or gifts of Ceres, a fact responsible for the modern name, "cereals." The Greeks had similar religious observations. In the New World the Mexican natives worshipped an agricultural deity to whom they brought the first fruits of their harvest. In fact, nearly every primitive race has worshipped some deity who presided over its crops.

The cereals are all members of the great grass family, *Gramineae*, and are alike in possessing the characteristic fruit of that family, the karyopsis. In this fruit the wall of the seed becomes fused with the ripening ovary wall to form the husk. The term "grain" is applied either to this type of fruit or to the plant that produces it. The true cereals are six in number: barley, maize, oats, rice, rye, and wheat. Of these wheat, maize, and rice are the most important, and each has played an important part in the development of civilization. Sometimes the millets, sorghums, and even buckwheat are erroneously referred to as cereals.

The reasons for the importance of cereals as food plants are many. One or more of these grasses are available for each kind of climate. The northern regions have barley and rye, the temperate regions wheat, and the tropics and warmer temperate areas maize and rice. Cereals also have a wide range of soil and moisture requirements. They can be cultivated with a small amount of labor, and have a large yield. The grains are easy to handle and store because of their low water content, and they are very high in food value. Cereals contain a higher percentage of carbohydrates than any other food plants, together with a considerable amount of proteins and some fats. Even vitamins are present.

WHEAT

Wheat is the chief cereal of temperate regions and the one most widely grown and so is by far of the most importance to the white race. It is very old and its native home is in doubt. Some of the more recent investigations point to the highlands of Palestine and Syria as the place of origin, although the Central Asian plateau and the Tigris and Euphrates valleys have been suggested. It has already been pointed out that Vavilov considers wheat to have had a multiple origin, the soft wheats coming from the mountains of Afghanistan and the southwestern Himalayas; the durum wheats from Abyssinia, Algeria, and Greece; and einkorn from Asia Minor. The archeological evidence seems to indicate that this grain has been cultivated for at least 6000 years. Wheat was the basis of the Babylonian civilization and it was cultivated by all the other nations of antiquity. Numerous varieties were known to Aristotle, Theophrastus, Pliny, and the other Greek and Roman writers. It was grown in China as early as 2700 B.C., and was used by the Lake Dwellers of Switzerland and Hungary who go back to the Stone Age. Wheat was first introduced into the New World in 1529 when the Spaniards took it to Mexico. Gosnold, the English explorer, sowed wheat in New England in 1602. It reached Virginia in 1611, California in 1769, and Minnesota in 1845.

Characteristics of Wheat

Wheat is an annual grass, belonging to the genus *Triticum*, which comprises a considerable number of wild as well as cultivated species. The wild forms are often troublesome as weeds, but are valueless as food plants. Cultivated wheat (*T. aestivum*) grows to a height of from 2 to 4 ft. The inflorescence is a terminal spike or head consisting of from 15 to 20 spikelets borne on a zigzag axis. The individual spikelets are sessile and solitary and consist of from one to five flowers each. The mature grain (Fig. 4) consists of the embryo (6 per cent), a starchy endosperm (82 to 86 per cent), the nitrogenous aleurone layer (3 to 4 per cent), and the husk or bran (8 to 9 per cent). This last structure is composed of the remains of the nucellus, the integuments of the seed coat, and the ovary walls or pericarp.

Kinds of Wheat

The long period of cultivation of wheat has resulted in the production of innumerable species and varieties, as a result of the conscious or uncon-

Fig. 148. Heads of the eight kinds of wheat. The heads from left to right are: common, club, Polish, durum, emmer, spelt, poulard, einkorn. (*Reproduced by permission from Etheridge, Field Crops, Ginn and Company.*)

scious selection on the part of man of forms that possessed some particularly desirable qualities. In general eight kinds of wheat are recognized (Fig. 148). These were differentiated as species by Hackel.

			monococcum..	einkorn
			polonicum...	Polish wheat
		dicoccum.............................	emmer
Triticum		*spelta*................................	spelt
	(*sativum*)		*vulgare*...........................	common wheat
		(*tenax*)	*compactum*.......................	club wheat
			durum............................	durum wheat
			turgidum.........................	poulard wheat

More recently, the various groups of wheat have been classified on the basis of their chromosome numbers. The most important species fall into the following groups.

Diploid (7 pairs)	Tetraploid (14 pairs)	Hexaploid (21 pairs)
T. monococcum	*T. dicoccum*	*T. compactum*
	T. durum	*T. aestivum*
	T. polonicum	*T. Spelta*
	T. Timopheevi	
	T. turgidum	

The most primitive of these species are einkorn, emmer, and spelt. Like the wild grasses of the genus, these have a fragile jointed head, which breaks during threshing, and the grain does not separate readily from its enclosing envelopes.

Einkorn. This is also called one-grained wheat as it has only one fruit in each spikelet. It is one of the oldest of wheats, going as far back as the Stone Age. It is a plant which can utilize very poor soil and will grow where other types cannot. Einkorn is a small plant, rarely 2 ft. in height, with a very low yield. It is still cultivated to some extent in the mountainous regions of Southern Europe, especially in Spain. It is rarely used for bread, but chiefly for fodder. In the United States the plant is grown for experimental purposes only.

Emmer. This species, also known as starch wheat, rice wheat, or two-grained spelt, has a flattened head with bristles or awns. It is another very old type and was grown in Babylonia and by all the early Mediterranean nations and the Lake Dwellers of Europe. It is still cultivated in the mountainous parts of Spain, Italy, Switzerland, Germany, and the U.S.S.R. The plant thrives in dry soil. Emmer was introduced into the United States from Russia. It has some use for livestock and breakfast foods and in breeding programs.

Spelt. Spelt, another primitive wheat of antiquity, is very hardy and can be grown on the poorest soils. It has been cultivated in the Mediterranean region for centuries, and is still grown in Spain. In the United States it is grown somewhat as a food for livestock.

The remaining wheats have a stout unjointed head, which does not break, and the ripe grain separates easily and cleanly.

Polish Wheat. This plant, also known as giant rye, has a very characteristic appearance, due to the long papery bracts surrounding each spikelet. The stems are solid, and the bluish-green ears are flattened. The species is of comparatively recent origin. In spite of its name, it is not a native of Poland. It is grown chiefly in Spain, and also in Italy, Turkestan, and Abyssinia. The plants are large, but have a small yield of little value. Polish wheat has been exploited in the United States, but it is not well adapted to our conditions and gives a very unsatisfactory return.

Poulard Wheat. This species, known also as English wheat or river wheat, is an old form, which probably originated in the dry areas of the eastern and southern Mediterranean region. The heads are large, but the yield is small and the plant is of but little importance save in England. It has been much exploited in the United States, but has no real commercial value.

Club Wheat. The club wheats, often called dwarf or hedgehog wheats, differ from all the other types in having short compact heads and small kernels. The plants are small and have a very stiff and strong straw. They are well adapted to poor soil and are grown chiefly in the mountainous districts of Central Europe, Turkestan, and Abyssinia. These wheats have been introduced into this hemisphere and are grown in

Chile and in the Pacific and Rocky Mountain states. The grains are soft and have a low protein content, and so are not well adapted for bread making. Their chief use is for pastry flour and for export purposes.

Durum Wheat. These wheats have thick heads with long stiff beards and large, very hard, amber or red grains, which are rich in gluten. This type has been cultivated for a long time in the Old World in arid regions. It is the principal wheat in Spain, and is also grown in Algeria, India, and the U.S.S.R. Durum wheats have been introduced into the United States from Russia and have proved extremely valuable. The low rainfall and high temperature, characteristic of much of the Great Plains, renders the region unsuitable for most crops. These wheats are very hardy and drought-resistant and are grown with great success in this area. Because of the high gluten content, the flour is used chiefly for macaroni, semolina, and similar pastes. When mixed with other flour it can be used for bread. The red durum wheats are used chiefly for livestock feeding.

Common Wheat. These common wheats are the chief source of bread flour. They occur in innumerable varieties differing in both external morphological and physiological characters. There are bearded and beardless wheats, red and white wheats, and hard and soft wheats. The hard wheats are richer in proteins and usually have small grains; the soft wheats produce large grains, which are richer in starch. The physiological characters include such features as yield per acre; lateness or earliness of maturing; resistance to drought, cold, or disease; behavior in milling and baking; and the seasonal aspect, *e.g.*, spring or winter wheat. Spring wheat is sown in the spring and harvested in late summer. Winter wheat, on the other hand, is planted in the fall and develops a partial root system before the coming of cold weather. In the spring it has a vigorous early growth, and it is harvested in early summer. Winter wheat has a higher yield, is more resistant to disease, and matures earlier.

Timopheevi. *Triticum Timopheevi*, a recently discovered Russian species, is very disease resistant and is being used in breeding experiments with standard varieties.

Grades of Wheat

The United States Department of Agriculture recognizes seven grades or classes of wheat in its official grain standards. The five most important comprise the following:

Hard Red Spring Wheat. This class constitutes 20 per cent of the United States crop. These wheats are grown chiefly in Minnesota, the Dakotas, and Canada, where the winters are too severe for winter wheat, and are used for bread flour. Marquis wheat has been the chief variety.

Durum Wheat. The amber durum wheats are all spring wheats and include 10 varieties, of which Kubanka is the best known. They comprise about 6 per cent of the wheat crop, and are grown chiefly in the Dakotas and Minnesota. They are used almost entirely for macaroni and similar products.

Hard Red Winter Wheat. This class is grown chiefly in the central and southern Great Plains where hot summers and severe dry winters prevail. Kansas, Nebraska, Texas, and Oklahoma lead in production. This type of wheat constitutes 47 per cent of the total crop. Turkey is the most widely grown variety. The flour is of high breadmaking quality.

Soft Red Winter Wheat. This class yields about 30 per cent of the wheat grown. It is the chief wheat east of the Mississippi River, centering in Ohio, Indiana, and Illinois, and it is also cultivated in the Pacific Northwest. It is adapted to a more humid climate than the other wheats. The grains are more starchy, and the flour is used for pastry, cake, breakfast foods, and home baking. This class includes the red club wheats.

White Wheat. This class, which makes up about 5 per cent of the wheat crop, comprises all the white-grained forms, whether common wheat or club wheat, and includes both hard and soft and spring and winter wheats. They are grown in the Pacific Northwest and in the northeastern United States. The flour is well suited for pastry and breakfast foods, and is blended with hard-wheat flour for breadmaking. There is also a large export trade in this wheat.

Cultivation of Wheat

Wheat is adapted to all moderately dry temperate climates, but is not grown in warm humid regions. Areas with a growing season of at least 90 days and an annual rainfall of not less than 9 in. are required. Over 30 in. of rain is detrimental. In general, regions with a cool moist spring merging into warm, bright, dry harvest periods are best, but the various kinds of wheat differ somewhat in their requirements. The proper climatic conditions for wheat are found in nine different areas in the world, and these are the chief wheat-producing regions. They are the plains of southern Russia and the Danube, the Mediterranean countries, Northwestern Europe, the central plains of the United States (Fig. 149) and Canada, the Columbia River basin in the Pacific Northwest, northwest India, north central China, Argentina, and Southeast Australia.

The best soils for wheat are clays and loams, although a light sandy soil can be utilized. If the ground is too wet, the plants lack vigor and produce a small yield. On the other hand, a porous soil does not hold enough moisture. Lime is an essential element and must be added if the calcium content of the soil is low. Nitrogen, phosphorus, and potassium are also necessary. The best fertilizer is barnyard manure.

The land must be thoroughly cleared, for wheat is easily choked out by weeds. Crop rotation is often practiced, and wheat is planted after a crop like beets, turnips, or tobacco, which kill out weeds. The methods of cultivation naturally vary depending on the kind of wheat and the character of the soil and climate. The time of sowing depends on whether the plant is a winter or summer annual. For a good crop the seed must be heavy, well developed, and fully ripe. Only the finest ears are used for seed. The grains are winnowed to remove dust and light grains, are

Fig. 149. A field of Thatcher wheat growing in the Red River valley near Crookston, Minn. (*Courtesy of the Agronomy Department, University of Minnesota.*)

then sifted and bolted, and are treated with chemicals to kill any fungus spores. Wheat may be sown broadcast, either by hand or by sowing machines, the former method being used only on small farms. On large farms two kinds of machines are used: one which sows broadcast, and the other which drills furrows and buries the seed at once. Germination begins immediately and the first leaves appear within a fortnight. In the case of spring wheat growth continues unchecked until maturity, but in winter wheat it is halted with the advent of frost. If the cold is too severe, or if the roots are exposed, winter wheat may be killed. Weeding is constantly necessary. On the largest farms machines are used which plow 24 furrows at one time. The various stages of the ripening grain are known as milk-ripe, yellow-ripe or dough, full-ripe, and dead-ripe. Wheat is not always allowed to mature fully for it is then more valuable for fodder. Wheat is attacked by several insect and fungus pests. The latter include bunt, smut, and rust. Wheat rust causes enormous losses, often wiping out a whole crop. Many attempts have been made to import rust-resistant varieties, as well as drought-resistant varieties, and also to breed them. The efforts to produce rust-resistant strains of wheat comprise one of the most thrilling stories of modern plant breeding.

Harvesting of Wheat

The harvesting methods vary with the size of the farm. Simple reaping hooks, scythes, or reaping machines are used to cut the culms; and

FIG. 150. Harvesting wheat with a combine in Washington. (*Reproduced by permission of the Philadelphia Commercial Museum.*)

binding machines bind them into sheaves. The wheat is then housed and must be kept dry. In many parts of the world, a privilege, as old as man himself, allows the poor of the neighborhood to come in and glean the ears left in the field. Threshing is the next process, and this involves the separation of the grain from the spike. This is usually done by hand, using a flail. This is a long tiresome process, but is less damaging to the grain than a threshing machine. Rows of wheat all pointing the same way are laid on the threshing floor to the depth of 1 in. These are struck at regular intervals with the flail, and then the wheat is turned and the process repeated. A cart, which traces a spiral course over the stalks, is much used in Europe. After threshing, the wheat is winnowed and sifted. Threshing machines are often used. These are either horizontal or vertical and consist of rapidly revolving drums of hard wood, provided with barbed beaters which strike the ears with great force and with a frequency sometimes as high as 800 r.p.m. The most complicated harvesting machines have been developed on the great wheat ranches of the United States (Fig. 150). These are the *combines*, which reap, clean, thresh, winnow, and sift the grains, separate the wheat from the chaff, eliminate foreign seeds, sort into grades, and bag the grain, leaving the bags behind, and finally binding the straw. These huge portable factories are drawn by horses or tractors, and can cut a swath 40 ft. wide. It is possible with the aid of only eight men to harvest 120 acres daily.

Wheat must be stored in firmly built structures to keep out grubs and small pests, and it must be well ventilated. Buildings with a concrete wall and floor are best, although iron is much used. In the tropics subterranean silos are built. The great grain elevators at the world ports are a familiar sight. There are over 40,000 of these in the United States alone.

The Milling of Wheat

In the earliest times the grains were "brayed" between two stones; then a mortar and pestle were used, and later millstones operated by wind or water power. In most of the old mills there was a fixed lower stone upon which a movable upper one revolved. The grains were dropped into openings in the upper stone and gradually worked out between the stones, which had grinding surfaces cut in radiating lines. The whole grain was used.

Within the last 65 years the roller process of milling has been perfected. The first step in this process comprises cleaning and scouring. This consists of screening, which removes all foreign seeds, dust, sticks, straw, and pieces of bran, which might drop off later and get into the flour. The grains are then thoroughly washed and scoured. The next step is tempering. This gets the grain into the best condition for milling. A little water is added, which toughens the bran and prevents its breaking up, so that it will flake out all in one piece. Finally the conditioned and tempered wheat is submitted to breaking, grinding, and rolling. The grains are first ground between corrugated iron rollers, the so-called "first break." This cracks the grain and partially flattens it. A small amount of flour, the "break flour," is separated out by sieves, while the main portion goes to the "second break" for more complete flattening and the partial separation of the bran and embryo. This process is repeated until five sets of rollers, each moving at a different speed, have been utilized. In each case bolting separates the ground material from the coarse bran. Eventually all the bran is removed and the purified material is passed to smooth rollers for final granulation. It is finally bolted with silk cloth, containing 12,000 meshes per square inch, and is then ready for packing (Fig. 151). This final product is the best grade of flour, the First Patent. Material that has been separated out is known as middlings. This may be processed and made into inferior grades of flour, or used for other purposes. Granular particles, midway in size between the grain and flour, are known as semolinas. Durum-wheat semolina is used for macaroni, and ordinary wheat semolina for farinas.

FIG. 151. Interior of a flour mill in Minneapolis, Minn. Packing the flour for shipment. (*Reproduced by permission of the Philadelphia Commercial Museum.*)

The process described above results in white flour. In the milling of graham flour the entire grain is used, while in whole-wheat flour only a part of the bran is removed.

Production and Consumption of Wheat

The world production of wheat in 1947 amounted to 5,775,000,000 bu., with the United States raising over one-quarter of this amount, and Kansas and North Dakota the leading states. The domestic output was 1,367,186,000 bu., and 74,389,000 acres were cultivated. Other large wheat-producing nations are U.S.S.R., China, Canada, India, France, Italy, Germany, Argentina, Turkey, and Australia. Of these Canada, Argentina, and Australia share with the United States in the export trade. Formerly Russia was an important exporter, but, while her production is increasing under the soviet government, her export trade has fallen off. The economic aspects of the wheat industry, both as regards domestic and international trade, are of the utmost importance, and attempts have been made by the largest wheat-growing nations to regulate the production and exportation of this, the world's most important crop.

France leads in the per capita consumption of wheat, followed by New Zealand, Australia, the United States, Great Britain, Germany, and Canada. In the United States the annual consumption of wheat has been estimated at 4½ bu. per person yearly.

Wheat Products

Wheat products are probably the most widely used articles of human diet. In this country they furnish about one-fifth of the total food materials of the average family. The flour is used chiefly for making bread, and "bread" always means wheat bread. Where other cereals are used, the product is called corn bread or rye bread, etc. The hard wheats furnish bread flour, while the flour from soft wheats is used for cakes, crackers, biscuits, pastry, and similar articles. Other edible by-products are breakfast foods, like "Shredded Wheat," "Puffed Wheat," "Bran Flakes," and the various farinas; and the pastes, such as macaroni, spaghetti, and noodles. In the manufacture of macaroni, semolinas are used. These are separated from the flour and bran and mixed with 30 per cent water. The resulting dough is kneaded and put in a hydraulic press. The dough is squeezed out through holes in the bottom. Each hole has a little pin in the center, with the result that a hollow tube of dough is formed. Strings of dough are cut into 3-ft. lengths and are dried and cured at a temperature of 70°F. Spaghetti and vermicelli are merely small types of macaroni. Noodles are made by rolling out the dough into thin strips. Durum wheat is used for macaroni, and it is grown chiefly in the U.S.S.R. and the United States.

Wheat is also extensively used in the manufacture of beer and other alcoholic beverages and industrial alcohol. It is an excellent feed for livestock. A special kind is grown for the preparation of starch for use in the sizing of textile fibers. Wheat straw excels all other kinds because of its very great strength. It is used for seating chairs, stuffing mattresses, and the manufacture of such diverse articles as straw carpets, string, beehives, baskets, and wickerwork. Leghorn hats are straw hats made from the bearded wheat of Tuscany. Wheat straw is also used for packing and thatching and as a fodder and manure. The wheat plant is also a valuable source of fodder.

MAIZE

The Indian corn or maize plant (*Zea Mays*) is America's only contribution to the important group of the cereals. This species probably originated in a wild state in the lowlands of tropical South America, whence it spread to the Andes where its cultivation goes back to prehistoric time. Grains of maize found in the tombs of the Incas in Peru represent several different varieties, so that the plant must have been grown for many centuries prior even to the period of the Inca civilization. Thence maize proceeded northward and played a prominent part in the civilization of the Mayas and Aztecs. It was grown by the Indians in New Mexico as early as 2000 B.C. By the time that America was first visited by European voyagers maize was growing all the way from the Great Lakes and the lower St. Lawrence valley to Chile and Argentina. Introduced into Europe by Columbus, and into Asia by the earlier Portuguese explorers, maize has taken hold wherever the climate would permit and has now spread all over the world. Even under primitive conditions of agriculture large yields are possible and the plant has always been a popular one. Because of a confusion of terms, it is more desirable to use the word "maize" than "corn" in referring to this plant. In the United States "corn" always means maize, but in other countries "corn" is used for all the cereals, and may mean any hard edible seed, grain, or kernel. In England, for example, an ear of corn means a head or spike of wheat. It was only natural that the early colonists in America should have called maize "Indian corn." Maize is known as Turkish wheat in Holland, as Spanish corn in France, Egyptian corn in Turkey, Syrian durra in Egypt, and mealies in Africa.

Characteristics of Maize

Maize is the largest of the cereals (Fig. 152), a tall annual grass attaining a height of from 3 to 15 ft. The jointed stem is solid and contains a considerable amount of sugar when young. The leaves are large and rather narrow, with wavy margins. In addition to the extensive fibrous root system, aerial prop roots are usually formed at the base of the stem. Two kinds of flowers are produced. The tassel, at the top of the stem, bears the staminate flowers, while the cob or ear with the pistillate flowers is produced lower down on the stalk and so is protected by the leaves. Each ovary has a long silky style, the corn silk. The ovaries, and consequently the mature grains, are produced in rows on the cob. The cob is surrounded by a husk composed of leafy bracts. The grains consist of the hull (6 per cent), protein or aleurone layer (8 to 14 per cent), endosperm (70 per cent), and embryo (11 per cent). Two kinds of endosperm are usually present: a hard, horny yellow endosperm and a soft white starchy endosperm.

Kinds of Maize

No wild species of the genus *Zea* are known today. The original ancestor was probably a pod corn and gave rise through a process of evolution, which probably involved hybridization with some species of *Tripsacum*, to the present-day maize plant, and its nearest relative, the teosinte (*Euchlaena mexicana*) of Mexico, as well. Maize is well adapted to breeding experiments, and even the Indians had learned how to select, produce, and preserve the best varieties, which have given rise to the easily cultivated and rapidly maturing types of today. No other cereal has so many forms. These fall into seven quite distinct classes, (Fig. 153) which breed true to type. Although they hybridize readily, there are surprisingly few intermediate types in nature. These classes, differing chiefly in the nature of the endosperm and the shape of the grain, have

FIG. 152. A field of maize (*Zea Mays*) in Connecticut. (*Courtesy of the Connecticut Agricultural Experiment Station in New Haven.*)

been considered by some authorities to be species and by others to be varieties (see appendix). They should more properly be considered as merely agronomic groups. They include:

Pod Maize. In this interesting type each grain is covered with a husk, in addition to that which covers the whole ear. The plant is exceedingly leafy and the tassels are very heavy. The grains may resemble those of any of the succeeding types, suggesting that pod maize must be very close to the primitive form from which the others have been derived. Obviously this type of maize is of no commercial value owing to the presence of the individual husks.

Pop Maize. The grains in this type are usually elongate and oval, and, although small in size, they are very hard and flinty with a tough hull. The endosperm is mostly of the hard glossy variety. When the dry grains are exposed to a high temperature, they explode, forming a snow-white fluffy, palatable mass. This phenomenon is called "popping" and is caused by the sudden expansion of the soft endosperm, which results in the grain's turning inside out. Several theories have been advanced to explain this. It is probably due to the expansion of the moisture content of each individual starch grain after partial hydrolysis during the heating of the grain. For a time the swelling endosperm is confined by the flinty protein layer, but eventually this breaks and the sudden release of pressure causes the endosperm to become everted about the embryo and hull. The presence of too much white endosperm prevents popping.

FIG. 153. Representative ears of the six kinds of maize. From left to right the ears are: pop, sweet, soft, flint, dent, pod.

Two kinds of popcorn occur: rice popcorn, in which the grains are pointed and tend to be imbricated; and pearl popcorn, in which the grains are rounded and very compact. The plants produce a large number of small ears. This type of maize was probably grown in prehistoric time. Today some 25 different varieties are grown for human consumption. Most of the crop is local, and there is but little commercial production.

Flint Maize. In flint maize the embryo and white endosperm are entirely surrounded by the hard endosperm so the grain is not dented. The plants reach a height of from 5 to 9 ft., and have a tendency to be two eared. The ears are long and cylindrical with hard smooth grains in from 8 to 16 rows. The grains are likely to be of different colors. Flint maize matures early, and so is grown in New England and adjacent areas,

Wisconsin, and other Northern states. It is a very old type. Some 70 varieties are grown.

Dent Maize. In dent maize the white endosperm extends to the top of the grain, and the hard endosperm is present only on the sides. This results in an indention of the mature grain at the top, due to the shrinking of the soft material. This is the largest type, the stems attaining a height of from 8 to 15 ft. The plants produce a single ear. The ears are very large, up to 10 in. in length and weighing $\frac{3}{4}$ lb., and with sometimes as many as 48 rows. The deep wedge-shaped grains are usually yellow or white. Dent maize is the principal type grown in the Corn Belt as it has an enormous yield. It is the source, not only of most of the commercial grain, but of most of the fodder and ensilage as well. About 325 varieties have been developed.

Soft or Flour Maize. In soft maize the hard endosperm is entirely lacking. This type is very old and was extensively cultivated by the Indians because of the ease with which it could be crushed. The grains resemble flint maize in shape and appearance, but the size varies from small forms to the large Cuzco variety of Peru, which are $\frac{3}{4}$ in. or more in diameter. About 27 kinds are known. Soft maize matures very late in the season. It is not grown on a commercial scale in the United States.

Sweet Maize. Sweet maize has the entire endosperm translucent or horny, and the starch has been more or less changed to sugar. The grains are broad and wedge-shaped with a characteristically wrinkled surface. The plant is adapted to the cooler areas, and is the chief type grown in the North Atlantic and Central states for canning purposes. The grain is used in the unripe state. More than sixty varieties have been developed.

Waxy Maize. In this type the endosperm is of a waxy nature and the carbohydrate material occurs in a different form from that in the other varieties and furnishes a satisfactory substitute for tapioca. The starch is entirely amylopectin, whereas ordinary cornstarch is a mixture of amylopectin and amylose.

The Cultivation and Harvesting of Maize

Maize is distinctly a summer annual and requires very definite environmental conditions for its proper development. The best soil is a fertile, friable, well-drained alluvium, such as the deep, warm, black loams along river bottoms and in drained swamps. These soils must have a high organic and nitrogen content, and must not bake out. In addition to soil, temperature, sunshine, and moisture are limiting factors. The temperature of both the air and soil is important, especially during the growing season from May to September. A mean average summer temperature of 75°F. is the best, and it should not fall below 66°F. Sunshine is essential, and too many cloudy days are bad. Moisture is also very necessary. The optimum amount is a 20-in. annual rainfall coming mostly in summer. A great difference in the habit of growth is correlated with different climatic conditions, and there are varieties adapted to each type. A continental climate is most favorable. The growing season of maize varies from 90 to 160 days, depending on the locality. Maize does not mature north of 50°, although it can be grown as a fodder crop beyond that latitude.

Comparatively few regions have the right combination of the necessary environmental conditions so that maize can be raised as a commercial

FIG. 154. "Corn shocks" in a Connecticut field. (*Courtesy of the Connecticut Agricultural Experiment Station in New Haven.*)

crop on a large scale. The principal maize-growing regions of the world include the east central and middle western United States; the Mexican plateau; the Argentine pampas; the highlands of Brazil; the basins of the Danube, Dnieper, and Po rivers in Europe; northern India; China and Manchuria; French Indo-China; Java; the Nile valley; and South Africa. The one outstanding area is the great Corn Belt of the United States, located in the Mississippi valley in the states of Illinois, Indiana, Missouri, Iowa, Kansas, and Nebraska. Here the optimum conditions for development are found: a mean summer temperature of from 70 to 80°F., with night temperatures above 58°F.; no frost for 140 days; and an annual precipitation of from 25 to 50 in., at least one-fourth of which comes in July and August.

In growing maize the fields must be well plowed and harrowed. The seed is planted to a depth of 1 to 3 in. in regularly spaced rows. Constant weeding and hoeing are necessary, care being taken not to injure the roots. The use of fertilizers and the rotation of crops are advisable. Maize has comparatively few enemies. The corn borer is the worst insect, and corn smut is the most serious of the fungus pests. Drought, however, may cause very serious damage.

Considering the long period of cultivation, until recently there have been very few changes in the method of harvesting (Fig. 154). On small farms the ears are still husked by hand directly in the field, and then cattle are turned in to graze. On the larger farms the maize is cut with a corn knife or a machine. The stalks are stacked to permit the grain to ripen further. After a month of this curing process, the ears are husked by machine. Modern mechanization has made it possible for one man to plant, cultivate, and harvest 120 acres. Maize must be stored in well-ventilated bins, so that excess moisture will evaporate, and proper protection must be taken against rodents and other small pests.

Uses of Maize

Maize is utilized in more diversified ways than any other cereal. The chief use is as a food for livestock, about one-half the crop being used for this purpose. The grain is very nutritious, with a high percentage of easily digested carbohydrates, fats, and proteins and very few deleterious substances. The pork industry in the United States is dependent almost entirely on maize and uses about 40 per cent of the total amount raised. Cattle, horses, and other domestic animals are also fed on maize. It has been estimated that 10 to 12 lb. of corn is converted into 1 lb. of beef, while 5 or 6 lb. yields 1 lb. of pork. About 85 per cent of the maize crop is used on the farms where it is raised. Not only is the grain valuable as a stock feed, but the plant as a whole is an important fodder crop. It is used green, dried, or as *silage*. For the latter purpose the leaves and stems are cut into small pieces and placed in silos, large receptacles with airtight sides and bottoms. Here a slight fermentation takes place, and the resulting product is more palatable for cattle. *Stover*, the residue after the ears have been removed, is also fed to cattle or used for silage.

Maize is not very important as a food for man. Cornmeal is a poor breadstuff, owing to the absence of gluten, and corn bread is very crumbly and cannot be baked in loaves. The meal was first prepared by merely pounding the grain; later millstones were used, and now a milling process, involving the use of rollers, has been substituted. The whole grain was formerly used in milling, but the fatty oil, present in the embryo, gave an unpleasant odor and taste to the meal. In modern processes the embryo and hull are removed. Both white and yellow meal are milled. Cornmeal has many uses in other countries, and in the southern United States. When boiled with water, it becomes mush, or hasty pudding, the Italian polenta. It is often baked in cakes, such as johnny cakes, ash cakes, hoe cakes, corn pone, and the Mexican tortillas. For corn bread the meal is usually mixed with wheat or rye flour. Scrapple is cornmeal that has been boiled with scraps of pork, liver, and kidney, and then seasoned and fried. Hominy or samp and hulled corn are prepared by soaking the grains in the lye of wood ashes to remove the hull, and then cooking until soft. Small portions of the hard endosperm obtained during the milling process constitute hominy grits. The grain is also used in the preparation of many breakfast foods. In the United States much corn on the cob is eaten, and sweet corn is extensively canned, the 1948 pack amounting to over 750,000,000 cans.

The **industrial uses of corn and corn products** are of increasing importance, 100,000,000 bu. being used annually for this purpose. The manufacture of cornstarch and its derivatives, glucose or corn syrup, corn sugar, dextrins, and industrial alcohol; and the production and uses of corn oil, obtained from the embryo, have already been discussed. The grain is used for making various alcoholic beverages. The fibers in the stalks have been used for making paper and yarn; the pith for explosives, as a light packing material, and formerly for upholstery; the inner husks for cigarette papers, after being boiled in sugar, pressed smooth, and dried; and the cobs for fuel, smoking pork products, pipes, and as a source of charcoal and furfural, the latter a raw material used in making solvents, explosives, plastics, synthetic rubber, and nylon. Zein, the protein occurring in maize, is now being made into artificial fibers with good tensile strength and woollike properties.

Production of Maize

The United States normally raises nearly one-half of the world's supply of maize. Iowa and Illinois are the leading states. In spite of the fact that its commercial production is restricted, maize has the widest range of any crop and is grown in every state and on 75 per cent of the farms of the nation.

One of the most important innovations in the history of American agriculture has been the utilization of hybrid seed which has greatly increased the production of maize in recent years. In 1947, Iowa's entire acreage and 72 per cent of that of the rest of the country were planted with this type of seed.

Hybrid seed (Fig. 155) is produced by crossing two carefully selected superior inbred strains. The first generation single crosses are uniform in size like the parents, but the ears are small. They are especially suitable for the production of sweet-corn seed. Double-cross hybrids, however, resulting from combining two single crosses, are exceptionally vigorous and produce larger, though still uniform ears, with from 15 to 20 per cent more kernels. The yield from hybrid seed is increased from 5 to 25 bu. per acre. The raising of hybrid corn and the production of the seeds have become an important new enterprise.

The maize production of the world in 1946 was 5,265,000,000 bu., of which the United States was responsible for 3,249,950,000 bu., grown on 88,489,000 acres. China was the second largest producer, followed by Argentine, Brazil, India, Mexico, South Africa, Italy, and the U.S.S.R.

RICE

In tropical countries rice replaces all other cereals as the staff of life and dominates the economic and social existence. As a matter of fact rice is an indispensable food of over half the population of the world. This statement may seem to be extreme until we remember that the most important rice-eating countries are all densely populated. China with 400,000,000 people, India with 300,000,000, Japan with 50,000,000, and Java with 40,000,000 are only four of the countries in which the growing of rice is the chief agricultural industry. Ninety-five per cent of the world crop is produced in the Orient.

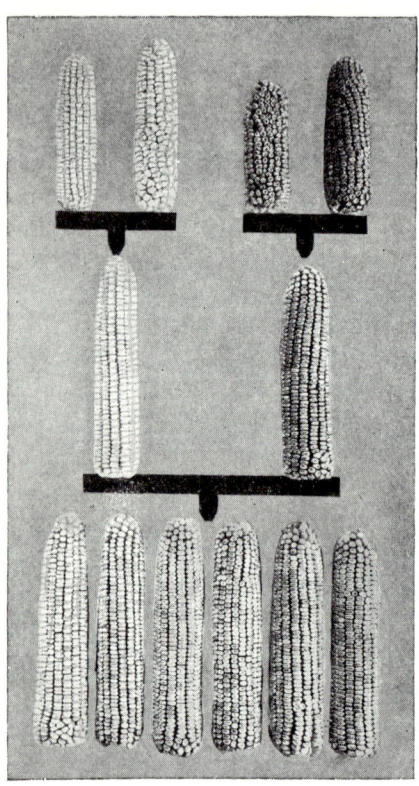

FIG. 155. Hybrid corn. *Upper row*, ears of inbred strains; *middle row*, single-cross hybrids, resulting from crossing pairs of inbred strains; *lower row*, double-cross hybrids, obtained by crossing two single crosses. These ears are large, very uniform, and exceedingly productive. A plentiful seed source is available in the single-cross parents. (*Courtesy of the Massachusetts Horticultural Society.*)

The cultivation of rice extends back into the dim past with no authentic records as to when it began. The plant originated somewhere in Southeastern Asia, but during its long period of cultivation it has spread to every warm region of the world. The history of rice and the history of China are closely interwoven. The Chinese were the first to cultivate rice, and their records go back for 4000 years. They have always considered it of great importance, and many ceremonies have arisen in connection with planting and harvesting. In classical Chinese the words for agriculture and rice culture are synonymous, indicating that rice was the staple crop at the time when the language was taking form. In other languages the words for rice and food are identical. Rice was introduced into India before the time of the Greeks, and very early reached Syria and Northern Africa. The first rice was grown in Europe in 1468 in Italy. The first rice in America was grown in South Carolina in 1694 from seed brought from Madagascar.

Characteristics of Rice

The rice plant is a large annual grass growing to a height of from 2 to 4 ft. Instead of bearing an ear, rice produces a panicle, an inflorescence

FIG. 156. Herbarium specimen of rice (*Oryza sativa*). The inflorescence of rice is a panicle.

(Fig. 156) composed of a number of fine branches, each terminating in a single grain surrounded by a husk. The grains are readily detached together with this brown husk. In this condition it is known as paddy. Common rice is *Oryza sativa*. This species has developed countless varieties under cultivation. There are said to be 1107 in India alone, and 1300 more in other countries. These differ in the color, shape, size, flavor, and other characteristics of the grain. One of these types contains a sugary substance instead of starch, which forms a soft, sticky, palatable mass on boiling. This plant is grown to some extent. Other species of *Oryza* occur as wild plants in the tropics of both hemispheres.

Cultivation of Rice

Rice is primarily a plant of the hot moist tropics. It prefers a climate where the average summer temperature does not go below 77°F. It

FIG. 157. Terraced rice paddies at Leptano, Bontoc, Philippine Islands.

grows best on damp soils underlaid with a semi-impervious subsoil in places where it can be flooded. The delta and flood plains of the monsoon region are particularly favorable. One type of rice, the upland or hill rice, can be grown without irrigation like any other crop, but this ordinarily is of little importance. In Central and South America, however, it is the only kind grown. Elsewhere the lowland rice, which has to be flooded during part of its development, is grown almost exclusively. Rice culture in wet areas is much the same the world over. In many countries where land is cheap, primitive methods of agriculture are used. In other countries rice cultivation has had a remarkable development, and even steep hillsides are utilized by the construction of terraces (Fig. 157). The fields are plowed or hoed and the rice is sown broadcast or transplanted from seed beds when 9 or 10 in. high. The young plants are covered with water, at first only at night, but later all the time, and the water is kept in circulation. As soon as ripening starts, the water is drawn off and the field is allowed to dry out. Rice is harvested much as wheat is and the stalks are stacked up to dry. There may be two or three crops a year. In the United States it is possible to use machines in harvesting the crop, which greatly reduces the expense.

Milling of Rice

The grains of rice are removed by threshing or by drawing the stalks through narrow slits. When rice is to be used directly for daily consumption, it is left in the "paddy" condition, since it keeps much better. The grains are husked just before they are to be used, and are then pounded in a mortar with a wooden mallet and winnowed. The resulting grain is very nutritious for it contains considerable protein and fat as well as starch.

In the commercial preparation of rice the impurities are removed and the paddy is passed between millstones to break up the husk. This chaff is removed by blowers. The grain is then pounded in huge mortars and a portion of the bran layer and embryo is removed. The waste is known as rice bran. The white rice is then scoured by friction and polished and a coating of glucose, talc, or chalk is added. During these last processes the outer, more nutritive parts of the grain are removed. The rice polish which is left as a residue is twice as nutritious as the finished product.

Uses of Rice

The chief use of rice is as a food, and more people use it than any other grain. Because of an insufficient protein content, rice should be supplemented by legumes or some other food rich in proteins. A diet of rice and soybeans constitutes the food of millions of Asiatic natives. The polished rice, which reaches the world markets, is much less nutritious, but its use is increasing. Rice hulls and rice polish are valuable as stock feed and are of increasing importance in industry. The straw may be plaited and made into hats, shoes, and other articles. Rice starch is much used in Europe. In several tropical countries intoxicating beverages are made from rice.

Production of Rice

In 1946 the world production of rice was 6,972,000,000 bu. China was responsible for 2,253,921,000 bu., and India and Pakistan 2,063,673,000 bu. Other important rice-producing countries were Japan, Java, Indo-China, Thailand, Burma, Brazil, Korea, the Philippine Islands, and the United States. Rice is also grown in Egypt and other African countries, while Italy leads in European production.

In the United States rice is grown chiefly in Louisiana, Texas, California, and Arkansas. Formerly the Carolinas produced the best rice, but the crop has not been important since the Civil War. Louisiana, however, has developed a great industry and today produces almost one-

FIG. 158. A field of rice in California.

half the total domestic crop. California has grown rice only since 1903. The heavy clay soils of the Sacramento and San Joaquin valleys have proved to be well adapted to the crop and it is of increasing importance (Fig. 158). The 1946 output of rice in the United States amounted to 72,216,000 bu., sufficient to supply all domestic demands and leave a considerable amount available for export.

CHAPTER XV

THE MINOR CEREALS AND SMALL GRAINS

BARLEY

Barley is of great antiquity as a cultivated cereal and was used for bread even before wheat. Pliny claimed that it was man's most ancient foodstuff, and some modern scholars consider it the oldest of all cultivated plants. It was known to all the ancient civilizations of the Old World, and the Lake Dwellers of Europe were familiar with no less than three kinds. Barley, according to Vavilov, probably originated in the arid

FIG. 159. Harvesting winter barley with a cradle.

lands of Southwestern Asia, Northern Africa, and also in Southeastern Asia. It reached the Western Hemisphere in the sixteenth or seventeenth century.

Barley is an annual plant, tending to become perennial. It is related to wheat and resembles the latter in many respects. It seldom reaches more than 3 ft. in height (Fig. 159). The inflorescence is a dense head with three sessile spikelets alternating at each joint of the straight axis. Most barleys are conspicuously bearded, though some are naked. The grains, which are often colored, remain enclosed in a husk formed by the subtending scales. The structure of the grain is similar to that of the other cereals.

The genus *Hordeum*, to which barley belongs, contains about 20 species, mostly weeds of temperate regions. As in the case of other plants that have been cultivated for centuries, there are a great number of present-day forms of our cultigen, *H. vulgare*. The classification of these different species, or varieties, of barley is particularly difficult, and no two authorities agree on all points. However, there seem to be two well-defined groups, the six-rowed forms and the two-rowed forms. In the former (*H. vulgare* and *H. intermedium*), all the spikelets are fertile. In the latter (*H. distichon* and *H. deficiens*), the lateral spikelets are sterile, so that only two rows develop. The ancestor of barley may have belonged to either of these two types. The wild barley (*H. spontaneum*) of Western Asia is considered by many to be the progenitor of our cultivated forms; it is also possible that there may have been two ancestral species.

Barley is very hardy and has a short growing season so it can be grown at high altitudes and latitudes. It is adapted to a wide variety of soils and climates. It is grown in the Rocky Mountains at an altitude of 7500 ft. and in the Andes at 11,000 ft. In Alaska it reaches a latitude of 65°N.L., while in the U.S.S.R. it is grown up to the shores of the Arctic Ocean. Barley, however, is not confined to cold regions, for it is an important crop in France, Turkey, and even semitropical California. Both winter and spring barley are grown. The principal barley centers are China and Japan; the U.S.S.R., Turkey, and Rumania; Western Europe; northern India; California and the northern prairie states.

Until the sixteenth century barley was the principal source of bread flour, and it has remained a staple food in Northern countries until comparatively recently. In the most prosperous countries it has now been supplanted by the more palatable wheat. The nutritive value of barley is considerable, though a deficiency in gluten makes it a poor breadstuff. Unleavened barley cakes, however, are a favorite food of the rural classes in Scotland and other Northern countries. The husk is ground off, yielding pot barley, and, if more of the grain is ground, the familiar pearl barley results. This is much used for soups. Barley is also used for breakfast foods and children's food. The six-rowed kinds, which have a higher protein content, are used for food purposes, both for man and for animals.

The chief utilization of barley today is as a feed for livestock. From 20 to 25 per cent of the crop is used as a source of malt to be used for alcohol, whisky, beer and similar beverages and various malt extracts and breakfast foods. For this purpose the two-rowed types, with a low protein content, a softer, mealy endosperm, and thin hull, are preferable.

The preparation and use of malt will be discussed later. Barley is also used for hay and pasturage and as a smother crop to kill out weeds. The straw is used for stock feed and bedding.

The Soviet Union usually leads the world in the production of barley, with China, the United States, or Germany as its closest competitor. The 1946 world production was 2,087,000,000 bu. The United States furnished 262,258,000 bu. The northeastern section grows the six-rowed kind, the Northwest the two-rowed, and the rest of the country both kinds. North Dakota, California, and South Dakota are the leading states. India, Spain, Canada, and Japan are also important producers of barley.

RYE

Rye (*Secale cereale*) is of more recent origin than the other cereals. *S. montanum*, a wild species of Afghanistan and Turkestan, is thought by some authorities to be the wild ancestor. Others consider *S. anatolicum* of Asia Minor to be the progenitor. In any event rye is probably a native of the Black and Caspian Seas region of Central Eurasia. It has been cultivated for a much shorter time than the other cereals. There are no traces of it among the ruins of Egypt or the Lake Dwellings, although the plant was known to the Greeks and Romans.

Rye (Fig. 160) is related to both barley and wheat, and resembles the former in habit; its grain looks like that of wheat. The stalks are slender and tough, reaching a height up to 6 ft. The leaves are somewhat bluish. The heads consist of a large number of spikelets, which are produced singly at the joints of the axis, and each one contains two fertile flowers. The grains have a normal structure. Rye is a very constant plant, and there are only a few varieties.

Rye is the most adaptable of the cereals. It will produce satisfactory crops in regions of severe winter temperatures and at high altitudes. It does well on poor soil and in arid areas, and has been called the "grain of poverty." In spite of this, rye thrives best on more fertile soil and in a mild climate.

Rye is primarily a plant of Europe, where 90 per cent of the world's crop is produced and consumed. It is used there chiefly for bread, as the grain contains gluten. Rye bread, known as *schwartzbrot* or black bread, is dark colored and soggy and has a bitter flavor. Until the middle of the last century it was the chief food of a third of Europe's population. Rye was an important bread crop in America until after the Revolution.

Rye is also used for hay and pasturage, as a winter cover to prevent erosion and leaching, as a sand binder, and in crop rotation. The straw is valuable, for it is very tough, and is in demand for hats, bedding, packing purposes, and the manufacture of paper and various other straw products. The grain is used for stock feed and as a source of whisky and alcohol. The 1946 world output was 1,380,000,000 bu. The U.S.S.R. has long been the chief producing country, even when only primitive methods of cultivation and harvesting were in vogue. Germany, Poland, and Czechoslovakia also produce large amounts. In the United States rye is ordinarily of little importance, the production in 1946 amounting to only 18,879,000 bu. The annual output, however, has been as high as 56,938,000 bu. North Dakota, South Dakota, Nebraska, and Minnesota, are the leading states. Two-thirds of the crop is fed to animals and

FIG. 160. A field of rye (*Secale cereale*) in South Carolina.

the remainder is used for flour, whisky, or alcohol. With us rye flour is mixed with wheat for breadmaking. In some years considerable amounts are exported.

OATS

The oat plant (*Avena sativa*) has never been found in a wild state, although it frequently escapes from cultivation and becomes established with such ease that it may appear wild. Consequently it is difficult to determine the native home of oats or the wild ancestor. Oats probably had a multiple origin, some coming from Abyssinia, others from the Mediterranean area, and still another from China. They were grown by the Lake Dwellers, though not by the early Mediterranean nations. There are several wild species of *Avena*, two of which, *A. fatua* or *A. brevis*, may be the ancestors of the cultivated forms.

Oats vary in height from 2 to 5 ft. The leaves are abundant and bluish green in color. The inflorescence is a one-sided or spreading panicle

(Fig. 161) which may be either erect or drooping. The panicles contain some 75 spikelets, which are two- to many-flowered and which are protected by long pendant outer scales. The grain, which is surrounded by a hull formed by the inner scales, except in the so-called naked varieties, contains two aleurone layers.

Several species of oats are cultivated. The most important is *Avena sativa*. This species is quite variable and has been much improved by

FIG. 161. Heads of oats (*Avena sativa*) showing types with a spreading and with a one-sided panicle. (*Reproduced from U.S.D.A. Farmers' Bulletin 424, Oats: Growing the Crop.*)

breeding and selection. Other cultivated forms are the side oat (*A. orientalis*), the red oat (*A. byzantina*), the naked oat (*A. nuda*), and the short oat (*A. brevis*).

Both spring and winter oats are grown, the latter in regions of mild winters, such as California and the Mediterranean area. Spring oats are best adapted to cool moist climates, such as occur in Northern Europe and the northeastern United States. Oats can be grown with profit farther north than any other cereal except rye. They reach a latitude of 69 degrees in Alaska and 65 degrees in Norway. An island climate is particularly favorable. In Scotland a third of all the cultivated land is devoted to this crop, and in Ireland half the land. Oats prefer heavy soils, but they can be grown on any tillable soil, even poor and exhausted ones. They have a high water requirement and so are not profitable in regions of high temperature unless plenty of water is available. Drought often causes great damage.

Oats are sown broadcast and are cultivated and harrowed. They are often planted with other crops. They may be harvested when the leaves are still green and when the grain is not fully ripe. The quality is improved by stacking and capping the stalks. Oats are cut, like hay, with a scythe or a machine.

Oats are the most nutritious of all cereals for human use as they have a high fat, protein, and mineral content. Oatmeal has long been popular. It is a good food for muscle building, and because of its high energy content is especially well adapted for use in cold weather, and by people who lead an active outdoor life. The protein material does not occur in the form of gluten, so oatmeal is not a good bread flour. It is used chiefly in cakes, biscuits, and breakfast food. Oatmeal is prepared by grinding the grains rather coarsely between stones. The more popular rolled oats are made more carefully. Grains are thoroughly cleaned and kiln dried, and are then graded by size and run through millstones which grind off the husk. After the pieces of husk have been removed by suction, the groats, as they are called, are softened and crushed by rollers in steam chests.

Oats constitute the chief grain food for horses, and may be used for any other domestic animal except pigs. Over 70 per cent of the domestic crop is used directly on the farms where it is produced. Oats are also grown for hay or green forage, and are frequently used in crop rotation or as nurse plants.

The United States usually leads the world in the production of oats. Few plants are more widely grown and the area of cultivation is increasing. Of the 3,920,000,000 bu. produced in 1946, the United States accounted for 1,497,904,000 bu. Iowa, Minnesota, Illinois, and Wisconsin lead, although oats are grown in every state in the union. The Soviet Union, which sometimes outstrips the United States, Canada, Germany, France, the United Kingdom, and Poland are important oat-producing countries.

THE SMALL GRAINS

SORGHUM

The sorghums include a large number of widely cultivated grasses, known under a confusing variety of common and scientific names. These plants were among the first of the wild species to be domesticated by man. Sorghum was grown in Egypt prior to 2200 B.C., and has continued to be an important crop in that country ever since. It was cultivated in China and India at an early date. From their native home in Africa and Asia the sorghums have been dispersed to all warm countries, in temperate regions as well as in the tropics. Although somewhat less nutritious than maize, they constitute a staple food for millions of native peoples in Africa and Asia. Sorghums are also used for stock feed and forage; in the manufacture of brushes, syrup, and paper; and in the Orient for many other purposes.

The plants are tall coarse annuals, growing to a height of from 3 to 15 ft., and resembling maize in habit. The inflorescence is a dense head or panicle, and the grains are smaller and rounder than those of the true cereals. The root system is shallow and twice as extensive as that of maize, and the leaf area is only half as great. These facts, together with the highly absorptive nature of the roots and the ability of the leaves to roll up in dry weather, enable the plants to withstand a great amount of heat and consequent evaporation. Their low water requirement renders them exceedingly drought-resistant, so that they are well adapted to semiarid and arid regions where maize will not grow.

In the United States the cultivated sorghums are usually referred to a single species, *Sorghum vulgare*. This has been derived from the perennial Johnson grass (*S. halepensis*), an Old World species grown as a forage grass in the warmer areas of both hemispheres. Elsewhere many of the varieties are likely to be considered as distinct species. Four types of sorghum are grown in this country. These include the grass sorghums, such as the Sudan grass (var. *sudanensis*) and the Tunis grass (*S. virgatum*), used entirely for hay and pasturage; the broomcorns, used in the manufacture of brushes; the sweet or saccharine sorghums, used for forage and making syrup; and the grain or nonsaccharine sorghums, which are cultivated for the grain and to some extent for forage. There is some evidence that both the sweet and grain sorghums were known to the early colonists, but the plants failed to become established. They have, however, become increasingly important since the middle of the last century and today rank in seventh place among American crops. About 20,000,000 acres are devoted to sorghums.

Broomcorn

In the broomcorn (var. *technicum*) the stems are dry and the inflorescence is a long, loose, much-branched panicle (Fig. 18) with a short axis. The spikelets are small and produce reddish-brown seeds. The use of the elongated branches of the panicle in the manufacture of brushes and brooms has already been discussed. The broomcorn was probably derived by selection from a sweet sorghum. It has been cultivated in Italy, Hungary, and other European countries for centuries, and has been grown in the United States since 1797. At the present time the greatest acreage is in Illinois, Oklahoma, Kansas, and Colorado.

Sweet Sorghums

The sweet sorghums (var. *saccharatum*), also known as sorgos or forage sorghums, are tall leafy plants with an abundant sweet juice. Their utilization as a source of syrup has been referred to. They are also extensively used for forage and silage. The black amber sorgos are of Chinese origin and were introduced into the United States in 1853 from France. The other types, prominent among which are sumac, gooseneck and orange, originated in South Africa and were brought into the Carolinas and Georgia from Natal in 1857. Although fairly widely cultivated, sorgos are most important in the Great Plains and Gulf States areas. Sumac is the leading variety.

Grain Sorghums

The grain sorghums are more stocky than the sweet sorghums and have a dry or only slightly juicy pith. Their commercial production in the United States dates from 1874 when durra was introduced. This was followed by kafir in 1876, milo in 1880–1885, shallu in 1890, the kaoliangs from 1898–1910, feterita in 1906 and 1908, and hegari in 1908. Today some fifty different kinds are grown, including many hybrids. These sorghums are particularly well adapted to the conditions of soil and climate which prevail in the southern Great Plains and parts of the Southwest. In 1946, 106,941,000 bu. were produced, principally in Texas. The grain is fed to all kinds of livestock, and the plants are also used for forage. In other countries these sorghums furnish a staple food for man, as well as for other animals, and have many industrial uses as well. During the Second World War a large amount of sorghum was used in making industrial alcohol, whisky, beer, an edible oil, and starch. The utilization of sorghum for such purposes is also a part of our present-day economy.

Durra. The durras are the chief type of grain sorghum in Northern Africa, Southwestern Asia, and parts of India, and millions of acres are cultivated. They are of comparatively little importance in this country, however. The plants have dry stems; compact, goosenecked, bearded heads; and flattened seeds; they mature early. The classification of this group has been especially chaotic for milo, feterita, and other forms have often been included. The three forms grown in the United States at present probably represent two distinct varieties. The white durra (var. *cernuum*) was at one time very popular and was grown under the name of Egyptian corn or Jerusalem corn. The plant is still grown in California where the seeds are a favorite poultry feed. A dwarf form is somewhat more in demand. The brown durra (var. *durra*) is also grown to some extent in California.

Kafir (var. *caffrorum*). Kafir corn, as it is often called, is a native of tropical Africa but has spread all over the world. It is an important food plant and many forms are grown. Its peculiar and characteristic flavor is not appreciated in this country, but it is highly nutritious and is similar in maize in composition and digestibility. The kafirs are stout, stocky plants from 4 to 7 ft. in height. The leafy stems have a slightly acid juicy pith, and so are valuable for forage. The inflorescences are

FIG. 162. A field of milo (*Sorghum vulgare* var. *subglabrescens*) in Texas. The plants are from 4 to 5 ft. in height. (*Reproduced from U.S.D.A. Farmers' Bulletin 322, Milo as a Dry-Land Grain Crop.*)

long, slender, cylindrical, beardless heads, which produce small, oval, white or colored seeds that are late in maturing. Some 12 varieties of kafir are grown in the United States, with Kansas, Texas, and Oklahoma the leading states. Standard Blackhull kafir is the most important of all the grain sorghums.

Milo (var. *subglabrescens*). The milos (Fig. 162), also of African origin, have slightly juicy stems; compact, usually bearded heads, which are usually recurved or goosenecked; and large soft yellow or white seeds. They sucker very freely. The plants are very adaptable to moisture conditions and respond readily to irrigation. Although very drought-resistant and able to produce some crop even under severe conditions, the yield is unusually large when conditions are favorable. Dwarf yellow milo ranks second in importance among the grain sorghums. About 12 kinds of milo are grown. The plants mature rather late but earlier than kafir.

Shallu (var. *Roxburghii*). This distinctive late-maturing sorghum was introduced from India, where it is extensively grown as a winter crop. It has tall, dry, slender stems and long open panicles. The small, hard, white seeds are exposed at maturity. Shallu is grown to some extent in the Gulf States.

Kaoliang (var. *nervosum*). The kaoliangs are Chinese sorghums, and constitute one of the oldest and most important crops in that country. They have furnished grain, sugar, and forage for thousands of years, and all parts of the plant have some economic value. They have dry slender stalks with but few leaves, loose or compact erect heads, and small brown or white seeds. Although they mature early and so can be grown farther north than the other grain sorghums, and although they are very drought-resistant, the yield is low and they have never become popular in the United States.

Feterita (var. *caudatum*). Feterita, an importation from the Sudan, has dry stalks; erect, compact, oval heads; and very large, soft, white seeds. It matures early and produces a crop in seasons with a limited amount of moisture. Three kinds are grown, chiefly in Kansas and Texas.

Hegari. This sorghum is probably a form of kafir. It produces leafy juicy stems which sucker freely, and in other respects it seems to be intermediate between kafir and feterita. Hegari is very variable as to time of maturity and yield. It is grown in Texas, Arizona, and New Mexico.

MILLETS

The term "millet" is loosely applied to a large number of cultivated grasses with very small seeds. The millets are used for forage and as a food for both man and domestic animals. The real importance of millet is not appreciated in the United States or Europe, but fully one-third of the world's population uses these grains as a regular article of food. In India over 40,000,000 acres are cultivated and the crop is comparable in importance to wheat in the United States. Japan normally produces 35,000,000 bu. annually. The plants have abundant foliage and are much used for forage. Millets are very drought-resistant and are extensively grown in the Great Plains area. They are sensitive to cold and cannot be planted until all danger of frost has passed. In the United States millets are used chiefly as hay crops, pasturage, and for birdseed, although some varieties are used for grain. Millets are among the most ancient of food grains, and have been grown in China since 2700 B.C. They probably originated in Eastern Asia. The most important cultivated varieties include:

Foxtail Millet (*Setaria italica*). Some dozen varieties of foxtail millets are grown, and these are very likely to occur spontaneously as weeds. They are known by many different names, such as Italian, German, Hungarian, and Siberian millet. The plants are smaller than in the other cultivated grasses and have a dense spike for an inflorescence, with innumerable long or short bristles. In one group the heads are short, thick, and erect; in another they are long and drooping. The origin of these millets is in doubt, although some authorities are inclined to the belief that they have been derived from *S. viridis*, a common wild grass of the Old World. The native home was probably Eastern Asia, and not Europe as the common names of the plant would seem to indicate. Millet must have been domesticated in the Orient ages ago, for it was one of the five sacred Chinese plants as early as 2700 B.C. Millet seeds abound in the Lake Dwellings, but the plant was apparently unknown in Syria and Greece. At the present time the foxtail millets are extensively grown in Japan, China, India, the East Indies, and other parts of Asia; in temperate Europe; in North Africa; and in Canada and the United States. They are used for human food everywhere, except in North America, and are also an important forage crop. When used for food, the grains are boiled or parched. Millet is an important hay crop, and is good for silage. It is much used in crop rotation, and as a supplementary or catch crop after some other crop has failed. This is possible because of the rapidity with which it matures, only 6 to 10 weeks being necessary.

Several millets belong to the genus *Echinochloa*. The **Japanese or sanwa millet** (*E. frumentacea*), an erect awnless species with turgid purplish seeds, is much cultivated. It is used in the United States entirely as a forage crop, and is very desirable for it produces as many as eight crops a year and has a large leaf area. In the Orient it serves as a food plant and is eaten as a porridge or with rice. In Japan it is cultivated in areas where rice will not grow. A smaller species, the **shama millet** (*E. colona*), is another valuable forage and food crop, particularly in the East Indies and India. It is a favorite food of laborers, and is eaten by the Hindus on fast days. The common **barnyard millet** (*E. crusgalli*), which occurs with us as a weed, is cultivated in India and the Far East as a forage and food crop under the name "bharti."

Proso Millet (*Panicum miliaceum*). This is the true millet, the *milium* of the Romans. It is also known as broomcorn millet, hog millet, Russian millet, and Indian millet. Its native home was probably India or the eastern Mediterranean region. It has been cultivated in Europe for a long time, and was largely used by the Lake Dwellers. The plant grows to a height of 2 or 3 ft., with an open, branching, compact or one-sided panicle (Fig. 163). The grains are variously colored and are closely surrounded by the scales of the spikelet. This millet is extensively grown in the U.S.S.R., where 80,000,000 bu. are produced each year. China, Japan, India, and Southern Europe also produce large amounts. Proso millet was introduced into the United States many years ago, but was grown only in the Northwest. Recently there has been a great revival of interest in the plant, and it is now extensively cultivated in the northern prairie and Great Plains regions. Although of some value as a hay crop, its greatest use is as a forage grain as it is very nutritious. In addition to carbohydrates, the grains contain 10 per cent protein and 4 per cent fat. It is an excellent hog feed, and is much used as a substitute for maize or sorghum. A palatable bread can be made from fresh grains.

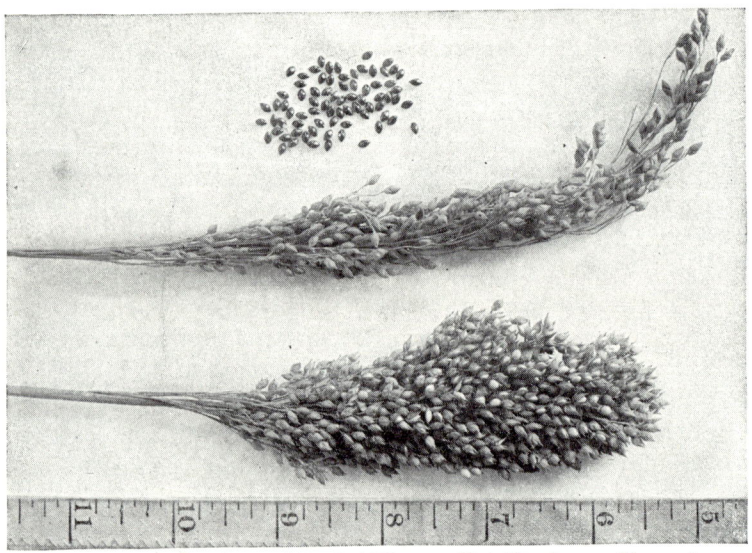

FIG. 163. The grains and heads of Proso millet (*Panicum miliaceum*).

Pearl Millet (*Pennisetum glaucum*). This is a tall plant, growing from 6 to 15 ft. in height, with three to eight compact cylindrical spikes that bear white grains. Pearl millet is grown in India, Egypt, and Africa as a rainy-season crop. It furnishes an important food for the lower classes, and is particularly valuable in cold weather because of its heating qualities. The flour made from this millet is very nutritious and is used for bread or cake. It has an enormous yield of forage, which is succulent when young, and it can be cut several times in a season. Pearl millet has been introduced into the United States as a forage crop. The wild form of the plant is unknown.

Ragi (*Eleusine coracana*). This tall grass, which is also known as finger millet, African millet, or korakan, has tufted stems, each with four to six spikes. Both upland and irrigated forms are grown from Indonesia to Northern Africa. Ragi has an exceedingly heavy yield, sometimes as much as 1500 lb. to the acre. Even on poor soil the yield is abundant. It is one of India's major crops, particularly during the rainy season, and is an important food. The grain is free from insects and can be stored for a long time. Ragi flour is used for puddings and cakes, and a fermented beverage is made from the grain. New and improved varieties are being developed by the Indian Agricultural Department.

OTHER GRAINS

Wild Rice

Wild rice is the only edible grain used in the United States which is not cultivated. The supply is obtained from wild plants of *Zizania aquatica*

FIG. 164. Wild rice (*Zizania aquatica*) in flower at the Arlington Experimental Farm, Virginia. (*Reproduced from U.S.D.A. Department Circular 229, Wild Rice.*)

and the variety *angustifolia*. These tall annual grasses (Fig. 164) grow partially submerged along the margins of lakes and sluggish streams inland, and even in brackish areas along the coast, oftentimes covering hundreds of acres. Formerly an important food of the eastern Indians, wild rice is still used by the tribes of the Great Lakes and upper Mississippi valley region. The grains are borne in slender panicles and drop off very readily when mature. The Indians continue to use primitive methods of gathering the crop. Just before the grains are fully ripe, they push their canoes in among the plants and beat off the grains into the bottom of the canoe with a stick. Later the grains are dried in the sun or over fires, and the husk is pounded or charred off. Wild rice is very nutritious and palatable and is often served with fowl and game. It can be obtained in the markets of the larger cities, although the Indians only sell what they cannot use themselves. Wild rice also serves as one of the most important foods for ducks and other waterfowl. Attempts have been made to plant the grass in the great coastal swamps. The seeds, however, are easily killed and germinate only if they have been constantly kept under water. Wild rice occurs both in Eastern North America and in Western Asia. In the latter area the young stems and leaves are used as a vegetable and the straw is used for paper.

Job's-tears

This grass, *Coix Lachryma-Jobi*, is a native of Southeastern Asia, but is cultivated in nearly all tropical countries. It has large, shining, pear-

FIG. 165. The inflorescence of Job's-tears (*Coix Lachryma-Jobi*). (*Photo by W. H. Weston, Jr.*)

shaped fruits (Fig. 165), which bear a fanciful resemblance to tears. These grains are used as food by the poorer classes, and are also said to have medicinal properties. Some varieties, especially the Philippine adlay, are good for forage. The chief use of the fruits is for ornamental purposes, and they are made up into necklaces, mats, rosaries, etc. A wholesome beer can also be made from them.

PSEUDO CEREALS

In various parts of the world many other plants are used in a manner similar to the cereals and smaller grains as sources of human food. For this reason, even though they are not grasses, they are often erroneously referred to as cereals. Two of the most important of these are buckwheat and quinoa.

Buckwheat

Buckwheat (*Fagopyrum sagittatum*) is a native of Central Asia and still grows wild in Manchuria and Siberia. As compared with most of the

FIG. 166. A field of buckwheat (*Fagopyrum sagittatum*) in blossom.

cereals, it is of recent use, the earliest records occurring in Chinese writings of the tenth and eleventh centuries. Buckwheat was introduced into Europe during the Middle Ages, and was first cultivated in 1436. It is widely used on the Continent, especially in the U.S.S.R., where it constitutes one of the chief foods of the peasants, and over 5,000,000 acres are cultivated. France and Germany also raise a large crop. It was brought into the United States by the early settlers, and is now widely grown (Fig. 166). In 1946, 7,124,000 bu. were produced, chiefly in Pennsylvania and New York. Buckwheat is normally a plant of cool, moist, temperate regions and thrives best in a sandy well-drained soil. However, it will grow in dry and arid regions and areas with very poor soil and drainage. The plant is a small branching annual. The stems are smooth and succulent with alternate hastate leaves. The inflorescence is a raceme bearing small white or pinkish flowers. The fruit is a three-cornered achene that resembles a beechnut, hence the name *buchweizen*. The seeds, or groats, are hulled and ground and the starchy flour is used for porridge, soups, and in this country for pancakes. The whole grains, middlings, or flour are also fed to livestock and poultry, and the straw is used for feed and bedding. Buckwheat is also grown as a fertilizer crop, cover crop, and catch crop; the flowers are an important source of honey. The crop is planted late in the spring, so as to avoid frosts, and harvested in August and September. Buckwheat has recently come into prominence as a source of *rutin*, a glucoside which is used in the treatment of capillary fragility associated with hypertension or high blood pressure.

Quinoa

Quinoa (*Chenopodium Quinoa*) is the staple food of millions of South American natives. The plant is an annual herb, which grows to a height of 4 to 6 ft. and resembles the common pigweed. It is a native of Peru, and was used in great quantities by the ancient Incas. The Spanish explorers found nearly all the nations using it. At present it is grown chiefly in Ecuador, Peru, and Bolivia, where it is cultivated at altitudes up to 13,000 ft. The plants produce a large crop of white, red, or black seeds (Fig. 167), which mature in 5 or 6 months. They are very nutritious, containing 38 per cent starch, 5 per cent sugar, 19 per cent protein, and 5 per cent fat. Whole seeds are used in soups, or are ground into flour which is made into bread or cakes. The seeds are also used in making beer, in medicine, and as a poultry feed. The ash is often mixed with coca leaves to give more flavor to the latter. Quinoa has been introduced into the United States, where the thin leaves are used as a substitute for spinach.

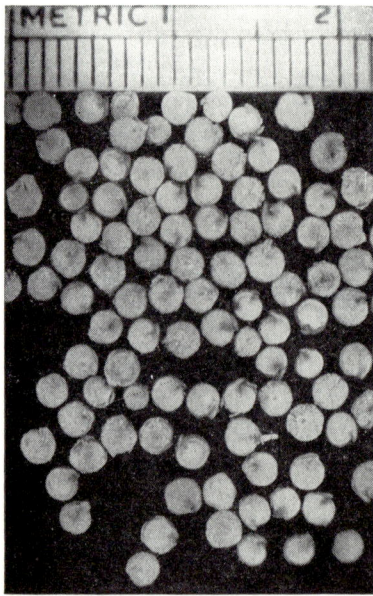

FIG. 167. Seeds of quinoa (*Chenopodium Quinoa*) from the market at Huancayo, Peru. (*Photo by Walter H. Hodge.*)

CHAPTER XVI
LEGUMES AND NUTS

LEGUMES

Legumes are next in importance to cereals as sources of human food. They contain more protein material than any other vegetable product, and so are nearer to animal flesh in food value. Carbohydrates and fats are also present. The proteins occur as small granules in the same cells with the starch grains. The high protein content is correlated with the presence, on the roots of many legumes, of tubercles that contain nitrogen-fixing bacteria. These bacteria are able to utilize free atmospheric nitrogen and convert it into nitrates, thus augmenting the supply of nitrogenous material available for the plants.

The legumes or pulses all belong to the great family *Leguminosae*, which is characterized by having a special kind of fruit, a legume, which is a pod that opens along two sutures when the seeds are ripe. Nearly 11,000 species of legumes are known, and many are of importance as industrial, medicinal, or food plants. They have been cultivated and used for food for centuries all over the world. The seeds (Fig. 168) are of greatest importance; as in the case of the other dry seeds, the low water content and impervious seed coats enhance their value for storage purposes and increase their longevity. Legumes are easily grown, mature rapidly, and are highly nutritious. They are not only rich in proteins but also in minerals and vitamin B. They are an absolute necessity in countries where little meat is eaten. Before the advent of the potato, they constituted a great part of the food of the poorer classes in Europe. Legumes have a high energy content and are particularly well adapted for use in cold weather or where physical exertion is involved. The immature fruits also serve as a food.

Owing to the fact that not only the seeds but all other parts of the plant are rich in protein, legumes are very valuable as field and forage crops. When plowed under they are an excellent fertilizer and greatly increase the nitrogenous content of the soil.

Peas

The common pea (*Pisum sativum*) is a native of Southern Europe and has been cultivated since before the beginning of the Christian era. Peas were well known to the Greeks and Romans. Although an old crop, they were not grown in Europe to any great extent until the middle of the seventeenth century. They were brought to America by the earliest colonists. Peas are annual, glaucous, tendril-bearing, climbing or trailing plants, with white or colored flowers and pendulous pods (Fig. 169). Although natives of warm regions, they thrive where there is a cool summer temperature and abundant moisture. Canada and the northern United States are particularly well suited to pea growing.

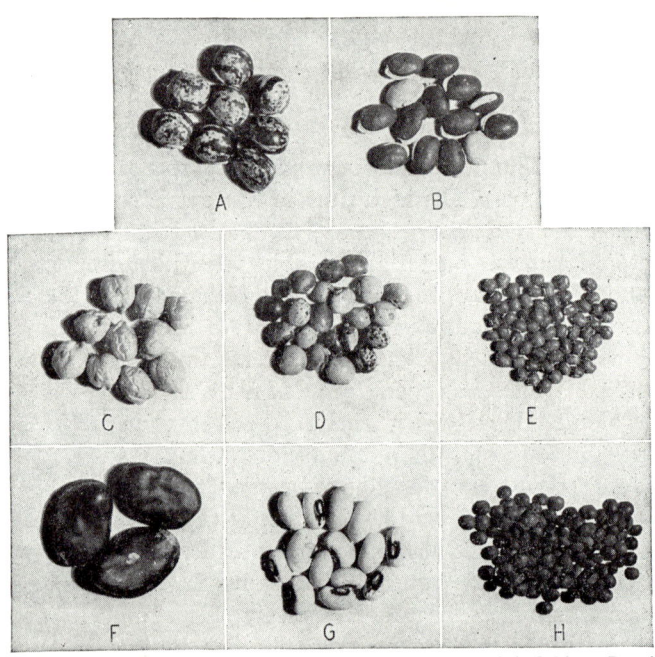

FIG. 168. Edible seeds of various legumes. *A*, velvet bean (*Stizolobium Deeringianum*); *B*, lablab (*Dolichos Lablab*); *C*, chick pea (*Cicer arietinum*); *D*, cajan pea (*Cajnus Cajan*); *E*, mung bean (*Phaseolus aureus*); *F*, broad bean (*Vicia Faba*); *G*, cowpea (*Vigna sinensis*); *H*, lentil (*Lens culinaris*).

Ontario alone has produced over 14,000,000 bu. annually. There are two groups of peas:

Field Peas. These may have originated from the gray pea, which still grows wild in Greece and the Levant. They have colored flowers and angular colored seeds, and are very hardy, withstanding frost and altitudes up to 8000 ft. Field peas are grown for the seed, which is used for human consumption in the form of pea meal or split peas. They are also unexcelled as part of the grain ration for livestock. The plants are used for forage, silage, and green manuring.

Garden Peas. These have white flowers and round smooth or wrinkled seeds, which are yellow or white in color. They contain more sugar than do field peas, and the seeds are eaten green or are used for canning, a great industry in the United States. In 1947, 33,119,000 cases were packed. Wisconsin leads in the production of peas for canning, and California in fresh peas. For canning, peas are usually harvested with a mowing machine. Pea-cannery refuse is a valuable stock feed. In some types the pods are fleshy and crisp and can be eaten as well as the seeds.

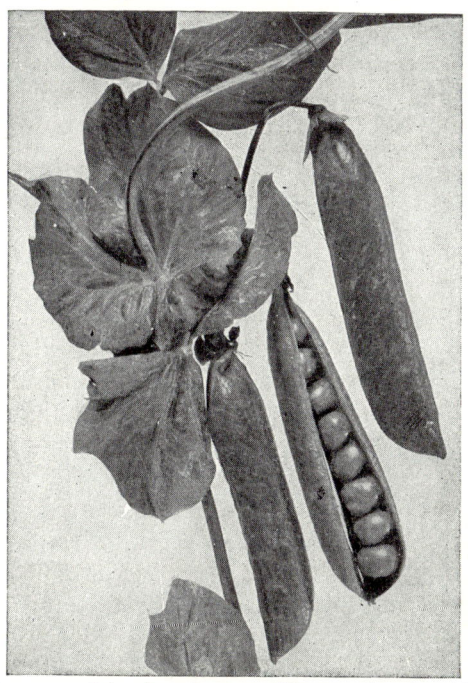

FIG. 169. Early Morn, a variety of garden pea (*Pisum sativum*). (*Courtesy of Breck and Company.*)

Garden peas are famous as the plants with which Gregor Mendel carried on his epoch-making experiments in plant breeding.

Chick Peas

The chick pea (*Cicer arietinum*) is a native of Southern Europe, where it is still extensively grown. It is also an important food plant in many parts of Asia, Africa, and Central America although but little used in the United States. India grows an amount equal to the sugar cane acreage of the whole world, and the crop is valued at $100,000,000 The plant is a branching, bushy annual, which matures in 90 days. It is well adapted to arid and semiarid regions. The chick pea is one of the best legumes for human consumption as the seeds (Fig. 168, *C*) are very nutritious. It has been cultivated for a long time and was known to the Egyptians,

Hebrews, and Greeks. The sparse foliage is poisonous so the plant cannot be used for forage. The green pods are sometimes eaten, and the seeds are used as a substitute for and as an adulterant of coffee.

Garden Beans

The common field, garden, or kidney beans (*Phaseolus vulgaris*) are natives of the New World. They were probably domesticated by the Incas, and were early used by the Indians of both South and North America. Today the young pods (string or snap beans), the unripe seeds (shell beans), and the dried ripe seeds are all used for human consumption, while the whole plant is used for forage. Beans are low, erect, or twining annuals with small white or colored flowers, trifoliate leaves, and slender pods (Fig. 170). They are grown as either bush or pole beans, and over a thousand varieties are cultivated. Both groups have green-podded and wax-podded varieties. The commercial dried bean is of recent origin. It was first grown in 1836 in New York, when 1 pt. was raised. Today over 170,000 tons are produced, chiefly in New York and Wisconsin. Much of the crop is canned. Although beans will grow on a great variety of soils, a fertile soil, rich in lime, is necessary for a good yield. A warmer climate than for peas is desirable, and crop rotation should be practiced. On large farms the crop is harvested by machines and the beans are dried, stored, and threshed before marketing. The culls are fed to livestock and the straw is used for forage. These beans were unknown in Europe until after the discovery of America. They are now grown all over the world.

FIG. 170. Supergreen snapbeans, a variety of the common garden bean (*Phaseolus vulgaris*). (*Courtesy of the Massachusetts Horticultural Society.*)

Lima Beans

The lima bean (*Phaseolus limensis*) has been called the aristocrat of the bean family. It is a native of Peru and Brazil and has been grown in South America for centuries. Originally a perennial, it is usually treated as an annual. Lima beans require warmer weather and higher humidity than ordinary beans. The original types were pole beans; the bush limas (Fig. 171) arose later as mutations. Either green or dried beans are eaten, and a large quantity is processed. In addition to the familiar large lima bean, a smaller form, the sieva bean (*Phaseolus lunatus*), is also grown. It, too, is a native of tropical America.

Other species of *Phaseolus* are often cultivated. These include the **scarlet runner beans** (*P. coccineus*), which have a thickened root and

(FIG. 171. Bush lima beans (*Phaseolus limensis*). (*Courtesy of the Massachusetts Horticultural Society.*)

ornamental flowers, and the **mung bean** (*P. aureus*). The latter plant is one of the very ancient legumes of India and is still an important crop. The small oval seeds (Fig. 168, *E*) are highly nutritious, and the green pods are also eaten. Over a hundred kinds are grown in China and other Asiatic countries. The mung bean is grown in the United States as a forage plant. The **adsuki bean** (*P. angularis*) is next to the mung bean in importance in China and Manchuria, while the **rice bean** (*P. calcaratus*) is extensively cultivated in Southern Asia.

Cowpeas

The cowpea (*Vigna sinensis*) in spite of its name is more closely related to the beans than to the peas. It is a vigorous bushy or trailing summer annual, with curious, cylindrical, pendant pods. The plant grows indefinitely as long as the environmental conditions are favorable. The cowpea is a very old crop, probably native to Central Africa, although it has been grown in Southeastern Asia for over 2000 years. It has been introduced into the tropics and subtropics of both hemispheres, reaching the West Indies in the seventeenth century and the United States in the eighteenth. The seeds (Fig. 168, *G*) are fed to cattle and poultry, and are sometimes used as a coffee substitute. However, its chief value is as a forage crop, as a cover crop to prevent erosion, and as a green manure. The cowpea is an important crop in China, India, and the southern United States, where it is increasing in prominence. It is susceptible to frost and so is grown only in warm humid areas in a sandy or loamy soil.

Soybeans

The soybean (*Glycine Max*) is a small, bushy, erect or prostrate annual (Fig. 100) resembling the cowpea. It is a much better crop, however, for it does not become tangled, matures earlier, has a larger yield, produces a better seed, and can be threshed. The seeds all mature at the same time. The soybean is one of the oldest crops grown by man. It was cultivated in China centuries before the first written records in 2838 B.C. It is a native of Southeastern Asia, where over 1000 varieties are grown, Manchuria leading in commercial production, followed by Korea, Japan, China, and Indonesia. It is the most important legume in the Far East, where soybeans, fresh, fermented, or dried, are used everywhere in the daily diet of the people to supplement rice. The seed is the richest natural vegetable food known. In tropical countries, particularly Indonesia, soybeans are boiled and then fermented through the agency of a mold to yield the food known as *tempe*. Soybean sauce, made from cooked beans, roasted wheat flour, salt, and a ferment, is widely used. The flour, with a low carbohydrate and high protein content, is an excellent food for diabetics. Soybean milk, extracted from the seed, is used in cooking and is recommended for infants and invalids. Soybean sprouts are a favorite food.

The soybean has manifold other uses, both in the Orient and in temperate regions, and has become a highly essential and vital crop. It is an important aid to agriculture, a valuable commercial crop, a good feed for livestock, and the source of numerous raw materials for use in industry. Soybean oil, an important drying oil, has already been discussed. Soybean protein is extensively used to produce the foam liquid used for extinguishing oil fires and as the source of a synthetic fiber, similar to the casein fibers.

In the last 20 years the soybean has assumed a position of great importance in the agriculture of the United States. The 1935–1939 average was 56,167,000 bu.; the estimated yield for 1948 is 220,201,000 bu. Although grown in some 30 states, it is of greatest importance in the North Central states, with Illinois producing over 50 per cent of the total output. Soybeans are grown for hay, silage, and green manure, as well as for the seeds, and hundreds of varieties are cultivated. It can be grown under a variety of soil and moisture conditions but requires a fairly warm temperature and is susceptible to frost.

Broad Beans

The broad bean (*Vicia Faba*), also called the Windsor, horse, or Scotch bean, is grown as a forage crop, as well as for the seeds, which furnish food for both man and the domestic animals. The plant is a strong erect annual, 2 to 4 ft. in height, with flat pods and large seeds (Fig. 168, *F*). It has been cultivated since prehistoric times and probably originated in Algeria or Southwestern Asia. Over 100 varieties are grown, chiefly in the Old World. The broad bean was the only edible bean known in Europe before the time of Columbus. It is still an important crop in England. Its growth is restricted by dry hot summers, so it is not grown to any great extent in North America, except in Canada. The broad bean is sometimes used as a cover crop, in crop rotation, and for fodder and silage, as well as for the seeds.

Peanuts

The peanut or groundnut (*Arachis hypogaea*) is a true legume rather than a nut, for the shuck is merely a shell-like pod. The plant is a bushy or creeping annual with the peculiar habit of ripening its fruit underground. The peanut is a native of South America but was early carried to the Old World tropics by the Portuguese explorers and is now grown extensively in India, East and West Africa, China, and Indonesia. It was brought to Virginia from Africa by the slaves and is now one of the most important crops of the South. In 1946, 2,187,985,000 lb. were grown, with Georgia the leading state. The commercial development has come about since the Civil War. As many as 20 kinds of peanuts are grown, differing in habit and the size of the pod. The cultivation of peanuts presents many difficulties. They require ample warm sunshine and a moderate rainfall and can be grown successfully only south of 36°N.L. A sandy soil is best, although any but a low soil can be utilized. The soil must be friable so that the ripening fruit can be buried, and it must be well fertilized. In harvesting the crop (Fig. 172), the rows are plowed and the plants are lifted out with forks, shocked, and capped to cure. Later the fruits are removed, cleaned, and polished. The plants may be used for forage, stock feeding, or as soil renovators. The nuts or seeds are used for roasting or salting, in candy, and for the preparation of peanut butter. For the latter purpose the seed coats and embryo are removed and the nuts are roasted either dry or in oil, and are then ground to a paste. Peanuts are a very nutritious food. One lb. yields 2700 cal., whereas 1 lb. of beef furnishes only 900 cal. Peanut oil, an important food oil, has already been discussed. The oil cake is fed to

FIG. 172. A field of Virginia bunch peanuts (*Arachis hypogaea*). (*Reproduced from U.S.D.A. Farmers' Bulletin* 1656, *Peanut Growing*.)

livestock. The protein in peanuts is used in the manufacture of ardil, a synthetic fiber.

Lentils

The lentil (*Lens culinaris*) is one of the most ancient of food plants and also one of the most nutritious. It is a native of Southwestern Asia, and was early introduced into Greece and Egypt. Lentils are frequently referred to in the Bible. The plant is a slender, tufted, much-branched annual with tendrils. The pods are short and broad, with small lens-shaped seeds (Fig. 168, *H*). The seeds are widely used, chiefly in soups. They are more digestible than meat and are used instead of meat in many Catholic countries during Lent. The plants are somewhat used for fodder.

Cajan Peas

The cajan pea or pigeon pea (*Cajanus Cajan*) is one of the most promising legumes at the present time. It was first domesticated in Asia or Africa and is now widely cultivated in the tropics and subtropics, particularly in the East Indies, India, and the West Indies, where over 30 kinds are grown. The plant is an erect shrub. Both the immature and ripe seeds (Fig. 168, *D*) have been used for human and animal food since earliest time. In recent years the plant has been developed as a forage crop and rivals alfalfa in importance. It is drought resistant, grows well in any soil, matures rapidly, and in many other ways is highly desirable. Livestock and poultry are particularly fond of it.

Lablab

The Lablab or bonavist bean (*Dolichos Lablab*) is an important legume in many tropical countries. The plant is normally a woody climber with a large yield of pods, continuing over several years, but may be grown as an annual. Both the pods and the seeds (Fig. 168, *B*) are eaten, and the plant is also used for hay and forage, chiefly for horses and cattle.

Horse Beans

The horse beans or Jack beans (*Canavalia ensiformis*), natives of the West Indies, are grown in nearly all tropical countries for their seeds. The plants are bushy annuals with long sword-shaped pods, which may contain as many as 12 large beans. The unripe seeds and pods are used for human consumption, and the whole plant can be used for green forage. The plants are hardy, drought-resistant, and immune to most pests. They are grown to a considerable extent in the southern United States.

Velvet Beans

The velvet bean (*Stizolobium Deeringianum*) is widely cultivated in the tropics for its edible seeds (Fig. 168, *A*) and for fodder. The plant, an annual herbaceous climber, exceeds all the other legumes in the rapidity and extent of growth. It is becoming of increasing importance in many of the Southern states, particularly in Georgia and Alabama.

FORAGE CROPS

As soon as man began to domesticate animals, the need for forage crops arose. At first, wild grasses probably sufficed, but eventually other sources were utilized. As time went on, these became more numerous and more varied. The present-day extensive cultivation of grasses and legumes as forage crops is chiefly the product of the European and American civilizations.

In the United States, as has already been noted, various cereal grasses and food legumes are also utilized for forage purposes. In addition, many other species are grown entirely as forage crops and they have no value as human food.

These include such grasses as timothy (*Phleum pratense*), orchard grass (*Dactylis glomerata*), redtop (*Agrostis alba*), brome grass (*Bromus inermis*), Johnson grass (*Sorghum halepense*), Tunis grass (*S. virgatum*), and the very valuable Sudan grass (*S. vulgare* var. *sudanense*).

Forage Legumes

In many parts of the United States annual legumes rather than grasses are cultivated for forage. These have the additional value of furnishing green manuring. The most important include alfalfa, clover, sweet clover, kudzu, the lespedezas, and the vetches. New species are constantly being introduced.

Alfalfa. Alfalfa (*Medicago sativa*) was probably the first cultivated forage plant. A native of Southwestern Asia, it was grown by the Persians, Greeks, and Romans. It was early introduced into China and Europe and reached the United States during the colonial period. Many new varieties have since been developed. Alfalfa is the most important forage crop now grown in the United States, as it is unsurpassed for general feeding. Some 15,000,000 acres are devoted to it. Although it is grown to some extent in every state, it is primarily a Middle Western and western crop. It is useful for pasture, hay, and silage and for improving the soil. Dehydrated alfalfa or alfalfa meal is now in use.

Other species of *Medicago*, such as bur clover (*M. hispida*) and medic (*M. lupulina*), are grown to some extent.

Clover. Clover is widely grown, particularly in the Northeastern and North Central states. It is valuable in crop rotation and is often grown in mixtures with forage grasses. The most important species is red clover (*Trifolium pratense*). Others include alsike clover (*T. hybridum*), ladino or white clover (*T. repens*), and crimson clover (*T. incarnatum*).

Sweet Clover. Sweet clover has become an important forage crop only within the last 40 years. It is particularly valuable for pastures and for soil improvement. Production is centered in the Corn Belt. Both the white sweet clover (*Melilotus alba*) and the yellow sweet clover (*M. officinalis*) are utilized.

Kudzu. The kudzu bean (*Pueraria lobata*), a drought-resistant perennial legume native to Japan and Eastern Asia, has been introduced comparatively recently and is now an exceedingly valuable crop in the Southeastern states. It has a long tap root and produces runners up to 50 or 100 ft. in length. Kudzu yields good hay and forage; it is particularly important, however, in erosion control and the rebuilding of agriculture, as it will grow well on depleted soils. The pods can be used, and a valuable starch is obtained from the large roots.

Lespedezas. These legumes have also become prominent in recent years, especially in the Southeastern states. They are useful in soil conservation for renewing worn-out soils, and they also furnish excellent hay and pasturage. Three species are important: the perennial Chinese lespedeza (*Lespedeza cuneata*), the annual common lespedeza (*L. striata*), and Korean lespedeza (*L. stipulacea*).

Vetches. Numerous species of *Vicia*, both native and introduced, are utilized as forage crops, particularly in the Southern, Atlantic Coastal, and Pacific states. Vetches are for the most part weak-stemmed viny annuals and are useful for cover, green manure, and soil improvement, as well as for hay, pasture, and silage. They are very palatable to livestock. Vetches are often grown in mixture with small cereal grains. Perhaps the most important species are the common vetch (*V. sativa*) and the hairy vetch (*V. villosa*).

TREE LEGUMES

Although most of the leguminaceous food plants are herbaceous, there are a few trees that bear edible pods. These tree legumes and their importance to the future of agriculture have been brought to the attention of the public in recent years largely through the efforts of Professor J. Russell Smith. Erosion, one of the greatest enemies of agriculture, is greatly increased by excessive cultivation of the soil. This cultivation, however, is necessary in growing cereals and many other herbaceous plants. To remedy this condition, which, if unchecked, might lead to the destruction of all our arable lands, Professor Smith advocates the use of tree crops in place of cereal grasses. Many trees are already grown for their edible fruits and nuts, and many more could be domesticated and improved. There are, moreover, a few tree legumes that are excellent substitutes for wheat, corn, and other cereals in stock feeding. The most important of these are the algaroba, carob, and honey locust.

Algaroba

The algaroba, mesquite, or keawe (*Prosopis juliflora*) is a native of the arid regions of the West Indies, Mexico, and Central America. A single tree was introduced into Hawaii and from this small beginning the keawe spread until it now covers 50,000 acres and has become one of the most valuable trees on the island. The flowers are a source of honey, and the pods, and ground beans as well, are an important stock feed. The tree (Fig. 173) has an enormous yield, from 2 to 10 tons per acre, and over 500,000 bags are produced annually. It has been estimated that 1 acre of algaroba produces 1600 lb. of beef, while 1 acre of corn or alfalfa pro-

Fig. 173. A large specimen of algaroba (*Prosopis juliflora*) near Honolulu, Hawaii. (*Photo by U.S. Forest Service.*)

duces only 450 lb. The tree grows very rapidly, is drought-resistant, and can utilize arid, barren ground, where no other crop will grow. Algarobas are now being cultivated in many other parts of the world with a similar climate.

Carob

The carob bean (*Ceratonia Siliqua*) is a native of Syria and has been cultivated from antiquity in the Mediterranean countries. It constituted the "locusts" that were the food of John the Baptist and the Prodigal Son. The pods have been fed to farm animals for ages. More recently carob beans were almost the only food available for the cavalry horses in Wellington's peninsula campaign and Allenby's campaign in Palestine. The tree is a small evergreen with glossy green foliage. It blooms in the autumn and carries the young fruit until late the next summer. It prefers rocky dry soil. The carob is very productive with an exceedingly high yield. The dried pods (Fig. 174) contain 50 per cent sugar, and are often sold in the cities to be eaten like candy. The United States imported 1,234,316 lb. in 1935 for use in flavoring dog biscuit and chewing tobacco and for other purposes. The trees are now being grown with success in California. The ground-up seeds yield a highly nutritious meal. Several bakeries are now making a bread which contains 25 per cent carob flour. The pods also contain a valuable gum known as tragasol.

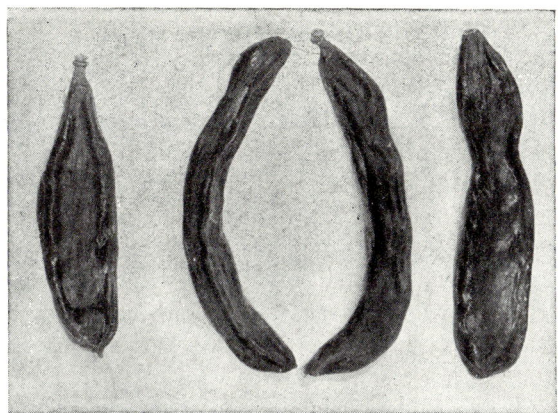

Fig. 174. Dried carob beans, the pods of *Ceratonia Siliqua*, purchased in a Boston market.

Honey Locust

The honey locust (*Gleditsia triacanthos*) is a native tree of the humid eastern United States, a region with hot summers and cold winters. The pods contain 29 per cent sugar and are readily eaten by animals. As it is also ornamental and a valuable timber tree, it should be given serious attention as a cultivated crop.

Other Tree Legumes

Other woody plants with edible pods occur in many parts of the world. Another mesquite (*Prosopis glandulosa*), a spiny shrub of the arid southwestern desert regions of North America, has pulpy pods that have long been used as food by man and beast. The rain tree (*Samanea Saman*), a huge tree of tropical America, bears curved black pods full of a sweet pulp, which is an excellent stock feed. The nittas (*Parkia biglobosa* and *P. filicoidea*) of West Africa have large pods with a 31 per cent sugar content. The seeds yield a very nutritious flour with 36 per cent protein, 23 per cent fat, 15 per cent starch, and $12\frac{1}{2}$ per cent sugar. They are much used by the natives when traveling, as they constitute a concentrated natural ration. In many parts of tropical America *Inga edulis* and other species of the genus are cultivated for their pods, which have a sweet and agreeable pulp, and as shade trees in coffee plantations.

NUTS

Very few botanical terms are used more loosely than "nut." Technically a nut is a one-celled, one-seeded dry fruit with a hard pericarp (shell). A few of the so-called nuts of commerce answer this description. Among them are the acorn, chestnut, filbert, and hazelnut. The others may be seeds, as the Brazil nut; legumes, like the peanut; or dry drupes from which the outer parts of the fruit have been removed, such as the almond, coconut, pecan, and walnut. For convenience all these "nuts" will be considered together, regardless of their morphological nature.

Nuts are a valuable food material, and have been used as such for a long time in many parts of the world, although in the United States they have been considered more as a confection until comparatively recently. Nuts are inexpensive, for they are produced in abundance and are easy to obtain. Because of their low water content they are a concentrated food, and also keep and handle well. They can withstand transportation, rough handling, and low temperature. If kept cool they rarely spoil; otherwise they may deteriorate by becoming wormy, rancid, or musty.

The food value of nuts is due chiefly to a high protein and fat content. However, they also contain starch, and sometimes sugar, and so furnish a well-balanced diet. They are also rich in mineral elements. For some reason, chiefly lack of knowledge, nuts have been considered indigestible. As a matter of fact, the reverse is true, and they are used as food by thousands of people, especially in the tropics where meat is scarce. Unless nuts, like any other food, are eaten in too large quantities, they are harmless. They may be eaten raw or cooked, or in the form of nut butters and pastes. They are often ground up to serve as coffee substitutes. Nuts are also valuable for feeding livestock. Nuts are marketed in the shell or shelled. The former are often bleached, polished, or stained for the sake of appearance. The latter are apt to collect dirt in the crevices, and should be thoroughly cleaned before using. Improved methods of processing and packing are stimulating the use of nuts.

Owing to the increasing value of nuts and the readiness with which nut-bearing trees can be grown on nonagricultural land, considerable attention is being directed to them with a view to improvement. Experiments are being carried on in selection and hybridization, which promise important results.

In general, three groups of nuts can be recognized: those with a high fat content, those with a high protein content, and those with a high carbohydrate content.

Nuts with a High Fat Content

Brazil Nut

The Brazil-nut tree (*Bertholletia excelsa*) is a rough-barked giant of the Amazon forests in South America. It bears from 18 to 24 hard, brown, spherical, woody fruits from 4 to 6 in. in diameter and weighing 2 to 4 lb. Each fruit contains from 12 to 24 seeds (Fig. 175) with a hard bony covering. These are the Brazil nuts of commerce, also called cream nuts or niggertoes. They have long been used by the natives for food. Their food value is high, as the fat content is 65 to 70 per cent and there is also 17 per cent protein content. The collecting and shipping of these nuts is an important industry in South America. The trees are never cultivated and the entire output, over 50,000 tons, is obtained from the wild trees. Most of the supply is shipped to Europe or the United States. Attempts have been made to cultivate the tree in the Southern states, but it is too sensitive to cold. Similar nuts of finer quality and more delicate flavor

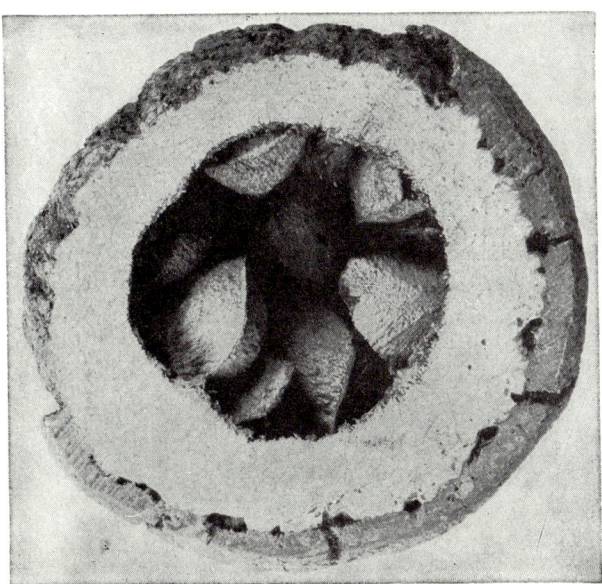

FIG. 175. Section of the fruit of the Brazil-nut tree (*Bertholletia excelsa*), showing the seeds. These seeds constitute the "Brazil nuts" of commerce. (*Photo by S. J. Record.*)

are obtained from the sapucaia or paradise-nut trees (*Lecythis Zabucajo, L. usitata*, and other species).

Cashew Nut

The cashew nut is obtained from *Anacardium occidentale*, a handsome native tree of Brazil, now extensively cultivated in tropical countries from Mexico to Peru and Brazil, in the West Indies and southern Florida, the Mediterranean area, Mozambique, India, and the East Indies. The tree bears a thin-skinned, pear-shaped, yellow or reddish, juicy "fruit" known as the cashew apple (Fig. 176). This is really the swollen peduncle and disk. The true fruit, a small curved or kidney-shaped structure, is borne on the outside of the "apple" at the distal end. This is the cashew "nut." The very rich kernel is delicately flavored and these nuts have become justly popular in recent years. A nutritious oil can be extracted from the seed. The grayish-brown coat, or shell, contains an oil that blisters the skin. The ripe fruit, which has a characteristic aroma, is eaten in many countries or used for preserves; the fermented juice makes a wine, kajú, which is sometimes bottled. The leaves, the light close-grained wood, the sap, and the bark are also useful. Between 2,500,000 and 3,000,000 lb. of cashew nuts are exported from

FIG. 176. A glass model of the leaves, flowers and fruit of the cashew (*Anacardium occidentale*). (*Courtesy of the Botanical Museum of Harvard University.*)

South America, and the vast stands in Brazil have scarcely been touched. India also exports a considerable amount.

Coconut

The coconut palm (*Cocos nucifera*) is one of the world's most important economic plants and is indispensable in the daily life of millions of natives of the South Seas and other tropical countries. The tree is a native of the Malay archipelago, but it has been carried to tropical and subtropical regions in all parts of the world. It usually grows near the seashore, but may occur at altitudes up to 5000 ft. It is one of the most graceful and beautiful of the palms, often with a characteristic leaning habit. The bases of the slender trunks are swollen. The large pinnate leaves, from

6 to 12 ft. long and 18 in. wide, are borne in a cluster at the tip of the stem. The flowers are produced in a large compound spadix, enclosed by a spathe (Fig. 115). The fruit is a three-sided dry drupe (Fig. 177). It consists of a smooth rind, or exocarp; a reddish-brown fibrous mesocarp; and a hard stony endocarp, or shell, which encloses the seed. The so-called coconut meat and milk represent the endosperm of the seed; the embryo is embedded in the hard endosperm. Coconuts, as they reach the world's markets, consist of the endocarp and its contents.

Few plants have more varied uses. The fibrous husk yields coir, which has already been discussed. The hard shell, or endocarp, is used for fuel, vessels and other containers, and a fine grade of charcoal. The water

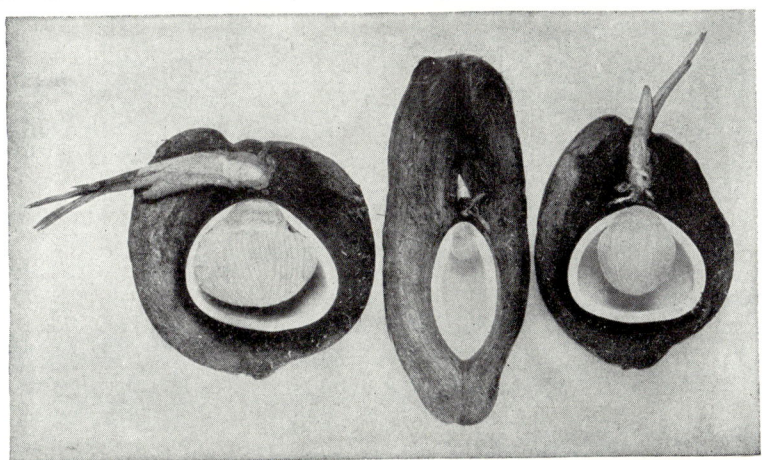

Fig. 177. Sections of coconut fruits, showing stages in germination.

makes an agreeable and refreshing drink. The meat may be eaten raw or may be shredded and dried to form desiccated coconut. It is often ground up and pressed through a cloth after water has been added. The resulting coconut milk is very palatable and a good substitute for cow's milk, as it contains several vitamins. The chief use of the meat, however, is for copra, the source of coconut oil and oil cake. The unopened inflorescences yield a sweet liquid, which is converted into palm sugar or fermented to make palm wine, arrack, or vinegar. The leaves are used for thatching, baskets, hats, mats, and curtains; the petioles and midribs are used for fence posts, canes, brooms, needles, and pins. The trunk furnishes a strong, durable wood for houses and bridges. Some of the porcupine wood of commerce, much used for cabinetwork, is obtained from the coconut. The bud or heart at the apex of the stem, is eaten in salads or is cooked. The bark contains a resin and the roots a drug.

The coconut is essentially a tropical plant and thrives best within the actual tropics. It will grow in any kind of soil, but naturally prefers a fertile area. Wild trees are still an important source of coconuts, but for commercial purposes plantations are usually established. Mature nuts are planted in a nursery and barely covered. They germinate in a few months and the seedlings are transplanted when about a year old. Coconut growth is improved by proper spacing, clean cultivation, and intercropping. Cover crops, fertilization, and irrigation also help to

Fig. 178. The crown of a coconut palm showing the clusters of fruit and the large pinnate leaves.

maintain the yield. Flowering and fruiting go on continually, and ripe nuts can be obtained during every month in the year (Fig. 178). They are usually picked every two months. The yield and size of the nuts vary with the spacing and the variety planted. It takes from 3500 to 7000 nuts to produce 1 ton of copra, which yields 1200 lb. of coconut oil and 800 lb. of oil cake. One thousand nuts will yield 165 lb. of coir fiber.

Coconuts are husked by driving them against a sharp spike fastened in a piece of wood and wrenching them apart. An experienced man can husk from 1200 to 2500 nuts a day. The nuts are broken into two halves with a blow of a heavy dull knife. The dried meat or copra, the most important commercial product, is prepared in several ways. About half the supply is dried by simple native methods, using the sun or drying on racks over fires made from coconut shells. After a few days the meat curls away from the shell and can readily be detached. Copra prepared in this manner is dark colored and has an oil content of around 50 per cent. Plantation copra is dried within 24 hours by utilizing the sun during the day and heat from fires in drying houses during the night. This copra is white and has a higher oil content, from 60 to 65 per cent. The best grade of copra comes from Ceylon, where the coconut industry is highly developed and efficient.

Desiccated coconut, which is used by confectionery and candy makers and in cooking, is prepared from the best grade of nuts. These are cured for several weeks and then are carefully cracked and the meat is removed while fresh. This is washed and cut into threads and dried in a vacuum for an hour at 160°F. Ceylon produces most of the desiccated coconut. The Philippines and India are also important.

The preparation and use of coconut oil, oil cake, and palm sugar have already been discussed.

At the present time Indonesia leads in the production of coconuts, followed by the Philippine Islands, India, Ceylon, and the islands of the southwest Pacific. The Western Hemisphere has been slow in developing the coconut as a crop, but the production center may be moved to the Atlantic in order to lessen the distance required for transportation. There are about a million acres of coconut palms in the American tropics, with some plantation culture in Brazil, Mexico, and the West Indies. The United States is the chief consumer, with Europe second.

Hazelnut

Hazelnuts are common in the cool temperate regions of both hemispheres. The native American shrubs, *Corylus americana* and *C. cornuta*, produce small nutritious and palatable nuts of no commercial importance. The larger European species, *C. Avellana* and *C. maxima*, are the source of filberts, cob nuts, and Barcelona nuts. Filberts are extensively cultivated in Southern Europe and are now being grown in Oregon.

Hickory Nuts

The hickories are native American trees common throughout the eastern deciduous forest region. They belong to the genus *Carya*. One group, the bitternut hickories, contain a large amount of tannin and are not good for food, although they are eaten by animals. A second group has sweetish edible nuts. The best of these is the shagbark hickory (*Carya ovata*). Good hickory nuts are among the finest of the wild nuts of the United States and have excellent keeping qualities. The trees show great promise under selection and experimentation. They can be grafted and crossed, and many new varieties, adapted to a wide range of soil and climatic conditions, have been produced. The nuts yield a good salad oil, and the wood is a valuable timber. The shellbark hickory (*Carya laciniosa*) also has nuts of excellent quality.

Macadamia Nuts

Macadamia or Queensland nuts are obtained from *Macadamia ternifolia*, a native of Australia. The tree has been introduced into other subtropical areas in both hemispheres. It has become of considerable commercial importance in Hawaii, where over 60,000 trees have been planted. Both thin-shelled and thick-shelled varieties are grown. The kernels have a sweet flavor and are rich in oil.

Pecan

The pecan (*Carya illinoensis*) is a native of the southeastern United States and Mexico. Originally obtained from wild trees, pecans have so increased in popularity that the trees are now being extensively cultivated in the Southern states, particularly in Texas and Oklahoma. With the introduction of new varieties, the area of production has been extended northward into Virginia, Indiana, and the upper Mississippi valley. The pecan industry is a profitable one for the trees are easily grown and begin to bear in three or four years after setting out, and the nuts command a high price. Factories are often located near the supply, and here the nuts are cracked and the meats picked out and prepared for shipment. Paper-shelled varieties, with a thin shell that can be broken with the fingers, have been developed recently. One-half the crop is now marketed in the shell. Pecans have a higher fat content than any other vegetable product, over 70 per cent. Pecans are used as dessert nuts and in candy, ice cream, cakes, etc. By-products include pecan oil and a tannin obtained from the shells.

Pili Nut

Pili nuts, often found in mixed nuts, are the seeds of *Canarium ovatum*, a native tree of the Philippine Islands, which has recently been introduced into the United States. These nuts and the very similar Java almonds from *C. commune* of Eastern Asia and the East Indies are highly appreciated in the Orient, where they are eaten raw or roasted. They are spindle shaped and triangular in cross section with a very thick hard shell. In the tropics a fatty oil is expressed and used for eating and in lamps. The plumlike fruit of the pili is also edible. The tree yields a resin.

Pine Nuts

The pine nuts or piñons (Fig. 179, *A, B*) are the edible seeds of several species of *Pinus* that are native to the Rocky Mountain and Pacific Coast region. Among these are the nut pines, *P. cembroides* var. *edulis* and var. *monophylla;* the Digger pine, *P. Sabiniana;* and the Torrey pine, *P. Torreyana.* These nuts, which are about the size of a bean, with a thin brownish-red shell, are very popular. Over 1,000,000 lb. are consumed annually. The American Indians have always eaten piñons. They roast the cones so that the scales will fall apart and allow the seeds to separate.

FIG. 179. Pine nuts, the seeds of various species of *Pinus*. *A*, Torrey pine (*P. Torreyana*); *B*, piñon (*P. cembroides* var. *edulis*); *C*, stone pine (*P. Pinea*), a European species. These last seeds constitute the pignolia nuts of commerce.

The pignolia nut (Fig. 179, *C*) is the seed of *Pinus Pinea*, a species of Southern Europe. It is longer and yellower than our native pine nuts and has a rich delicious flavor. Large quantities are imported. Edible pine nuts also occur in India and Eastern Asia.

Walnuts

Walnuts are important for both their nuts and their timber. They are found in the eastern United States and in Europe as native trees.

The **black walnut** (*Juglans nigra*) is one of the chief trees of the eastern deciduous forest region of the United States. It is a tall handsome tree (Fig. 51) often cultivated for ornamental purposes. The large spherical fruits are green when ripe, and the outer covering has to disintegrate or be beaten off to free the nuts. The walnut kernels are rich in oil and were a favorite food of the Indians. They retain their flavor when cooked and have a food value four times greater than meat. Nevertheless they are not much used as table nuts, probably because the stout thick shell is so hard to crack. The chief market is in the candy and ice-cream industry. The tree is very productive and can be grown anywhere, attaining its best growth in rich alluvial soil. Walnuts furnish valuable timber, and also a brownish-black dye.

The **butternut** (*Juglans cinerea*) occurs as a native tree in the limestone areas of the eastern United States and adjacent Canada. It is a smaller tree with elliptical nuts, which have a deeply corrugated shell. Butternuts are rich in fats and are thought by many to have a finer and richer flavor than walnuts, although they have an oily taste. The kernels are obtained more readily. Butternuts are much used in the candy industry.

The **English walnut** (*Juglans regia*) in spite of its name is a native of Iran. It is extensively cultivated in Southern Europe, particularly in France, in China, and other parts of Asia, and is now being grown in California and Oregon with great success. The English walnut has been cultivated for a long time, and many varieties are known. The beautiful trees are usually planted in rows. Only the outer limbs produce perfect nuts. The kernels are easily freed from the pericarps and are often bleached and polished. The characteristic furrowed kernels are the cotyledons of the seed, no endosperm being present. Walnuts yield an excellent oil for table use, and the oil cake is a good stock feed.

NUTS WITH A HIGH PROTEIN CONTENT

Almonds

Almonds are the most popular nuts and enter the world trade in the largest quantities. They are obtained from a small tree (*Prunus Amygdalus*), which is related to the peach and closely resembles it in blossom and young fruit. Almond trees are often cultivated as ornamentals. The almond fruit is an inedible drupe, with a tough fibrous rind surrounding the stone (shell) and seed (nut). Two types of almonds occur.

The **sweet almonds** (var. *dulcis*) have an edible seed and are the source of the commercial product. The almond is a native of the eastern Mediterranean region, where it has been cultivated for many centuries, and is grown throughout Southern Europe and also in California, Australia, and South Africa. The seeds are particularly delicious when eaten green. Usually, however, they are roasted or salted or made into a paste to be used in making bread or cake. Almond extract is also prepared. There are innumerable varieties, some with paper shells and some with hard shells. Jordan almonds are a hard-shelled type with a thinner integument on the seed and a finer flavor. They are imported from Spain, and are much used for confections. Imported almonds are marketed shelled, while American-grown ones are always in the shells.

The **bitter almonds** (var. *amara*) contain a bitter glucoside, amygdalin, which readily breaks down into prussic acid and so prevents their use as food. However, they are much grown in Southern Europe as a source of the oil of bitter almond. During the extraction process the prussic acid is eliminated, and the oil can be used for flavoring. Bitter almonds are also used as a stock upon which to graft sweet almonds.

Beechnuts

Beechnuts are the seeds of the beech tree (*Fagus grandifolia*), one of the most characteristic trees of the eastern deciduous forest region of North America. The nuts are small, triangular, and very sweet. They are of little importance for human food, but are eagerly eaten by cattle, pigs, squirrels, poultry, and birds. They give a fine flavor to pork, and razorback hogs are fed on mast, which is a mixture of beechnuts, chestnuts, and acorns. The European beech (*F. sylvatica*) yields slightly larger nuts, which are eaten and used for the edible oil of beechnuts.

Pistachio Nuts

The pistachio or green almond (*Pistacia vera*), a small tree native to Western Asia, has been cultivated in the Mediterranean countries for nearly 4000 years. It is also grown in Iran, Afghanistan, the southern United States, and California. The fruit is a drupe. The seeds contain two large green cotyledons with a reddish covering. These "nuts" are salted in brine while still in the shell. They are highly prized for their color and resinous flavor, and are much used in mixed nuts and as a flavoring material in ice cream and candy.

Nuts with a High Carbohydrate Content

Acorns

Acorns are the characteristic fruits of oaks (*Quercus* sp.). They are true nuts. In this country acorns have been used for ages as a fattening food for animals, particularly hogs. They are equally good for man and should be used more. The white oak (*Q. alba*) and the live oak (*Q. virginiana*) are the best of a dozen species with edible acorns. The North American Indians have always used acorn flour. They ground up the nuts, leached them to remove the tannin and other bitter principles, pounded them into a meal, and used them in porridge, mush, and other ways. In other countries acorns have been eaten from prehistoric time. Acorns furnish 25 per cent of the food of the poorer classes in Italy and Spain in the form of acorn bread or cake. This is highly nutritious and cheap and keeps indefinitely. Any species of acorn is edible after the tannin has been removed, but the holm oak (*Q. Ilex*) is the chief source. Oaks are very productive, are precocious and rapid growers, and are well adapted to poor soil. They should be cultivated as commonly in this country as they are in the Mediterranean region.

Chestnuts

Chestnuts occur in the eastern United States, Japan, and Europe. The American species (*Castanea dentata*) was formerly very abundant in the deciduous forest region. It was a handsome tree and furnished valuable timber, as well as the familiar nuts (Fig. 180), which had served as food, either raw or roasted, for over 200 years. The ravages of the

FIG. 180. A fruiting branch of the chestnut (*Castanea dentata*), showing the nuts surrounded by the burlike involucre. (*Courtesy of the Arnold Arboretum.*)

chestnut blight disease have almost wiped out this tree in the last 35 years. The European chestnut (*C. sativa*) with larger fruits has been extensively cultivated in Southern Europe for centuries, and several hundred varieties have been established. The nuts, or marrons as they are called, are a standard article of food, and are as important as wheat or corn is with us. They are grown everywhere, often on dry hillsides that are unfit for other purposes. The nuts are eaten raw or are roasted, boiled, or used for stuffing or flour. The Japanese chestnut (*C. crenata*) is immune to the chestnut blight and is being introduced into this country. Its nuts are edible, and are often cooked like potatoes.

CHAPTER XVII
VEGETABLES

In a technical sense all plants are vegetables. The term, however, is usually applied to edible plants which store up reserve food in roots, stems, leaves, and fruits and which are eaten cooked, or raw as salad plants. Vegetables constitute a large and varied group of considerable importance in the world's commerce. Most of them are very old, and their origin as food plants is lost in antiquity. The food value of vegetables is comparatively low, owing to the large amount of water present (70 to 95 per cent). Even so, they rank next to cereals as sources of carbohydrate food. This is usually present in the form of starch, although occasionally sugar, pectins, or other substances may occur. Proteins, save in legumes, are rarely available, and fats are stored only in very slight amounts. The nutritive value of vegetables is increased greatly, however, by the presence of the indispensable mineral salts and vitamins, while the roughage value of the various tissues aids digestion. For convenience the vegetables may be classified as earth vegetables, herbage vegetables, and fruit vegetables.

EARTH VEGETABLES

The earth vegetables include all forms in which food is stored in underground parts. The storage organs may be quite different morphologically. Some are true roots, while others represent modified stems, such as rootstalks, tubers, corms, and bulbs. All these structures are especially well adapted to storage because of their protected position. Many wild, as well as cultivated, species have fleshy underground parts, and these have played a role in the development of civilization and agriculture second only to the cereals and legumes. From earliest time roots and tubers have furnished food for man and beast. Even though the amount of stored material is less than that in dry fruits and seeds, these are extremely valuable since they are readily digested and have a high energy content. One objection to them is the high water content, which not only reduces the amount of available food material but impairs their keeping qualities as well. Their bulk, too, makes it impossible to transport and store them as efficiently as cereals, legumes, or nuts. Root crops, as these earth vegetables are often called, are an important phase of agriculture all over the world. In most countries they are grown fully as much for stock feed as for human consumption. In the United States root crops have been neglected more than they should have been, because it has been easier and cheaper to grow cereals which do not require so much hand labor. The various earth vegetables will be grouped according to their morphological origin. Of the many hundreds that are used for food throughout the world, only the most important can be considered.

ROOTS

Beets

The various beets now in cultivation—chard, mangels, sugar beets, and common beets—are all referred to a single species, *Beta vulgaris*. This has without doubt been derived from the wild beet (*B. maritima*) of the seacoasts of the Mediterranean regions and adjacent Europe. The beets are biennials, producing the first year a large cluster of leaves from a crown at the tip of a fleshy taproot.

Chard (var. *Cicla*) is the oldest type of beet. It was known as early as 300 B.C. For several centuries the roots were used, both as a vegetable and in medicine. Later the tender leaves were utilized. Under cultivation the leaves have been developed at the expense of the roots, and now chard has large leaves with thick stalks and roots that are very little enlarged. Chard is used almost entirely as a pot herb.

Mangels or mangel-wurzels have been developed from chard. They have the roots and the lower part of the stem thickened, with a crimson, golden, or white sap. Mangels have been an important cattle feed since the sixteenth century. They are extensively grown in Europe and in Canada. Although early introduced into the United States, they have never been of much importance. Mangels contain 3 to 8 per cent sugar. They are fed to cattle dry or are used for silage.

Sugar beets have been developed from the mangels. They are smaller and have a higher sugar content. Their cultivation and utilization as a source of sugar have already been discussed.

Common beets have been known since the beginning of the Christian era. Many varieties are grown, differing in size, shape, color, sugar content, and time of maturing. The early red beets are most desirable. Beets are boiled, pickled, or canned, and are often used for salads. The leaves and young beets, the familiar beet greens, are a favorite pot herb.

Carrots

The Carrot (*Daucus Carota*) has been cultivated for over 2000 years. It was known to the Greeks and Romans, and reached Europe early in the Christian era. It was a favorite vegetable in England in the time of Queen Elizabeth, and was brought to Virginia in 1609 and New England in 1629. The Indians carried it over the rest of America. Carrots now are found all over the world, often as weeds with long dry taproots that have lost their fleshiness. They are usually biennials, but may mature in one year. The pinnately compound leaves are very characteristic. The numerous varieties of carrots differ in size, shape, color, and quality, and are correlated with differences in the soil. A deep sandy loam gives the best results. The roots are harvested just before the ground freezes and are stored in cellars. Most of the food is stored in the outer cortical portions of the taproot. The central portion remains somewhat woody and unpalatable. Carrots are eaten raw or cooked and are often used for flavoring soups and stews. They are a valuable food for animals, particularly horses. The yellow coloring matter, carotin, is sometimes extracted and used for coloring butter.

Oyster Plant

The oyster plant or salsify (*Tragopogon porrifolius*) is a hardy biennial with a large fleshy taproot, sometimes as much as 1 ft. in length. It is a composite, and when mature has large purple heads, with fruits resembling those of a dandelion. The plant is a native of Southern Europe, but is cultivated in many parts of the world, and often becomes established as a weed. The roots are cooked or are used as a relish. They have a flavor suggestive of oysters, which gives the name to the plant.

Parsnips

The parsnip (*Pastinaca sativa*) was used by the Greeks and Romans and has been cultivated in Europe ever since. It is figured in nearly all the old herbals. A native of Europe, it reached the West Indies in 1564, Virginia in 1609, and by the beginning of the eighteenth century was grown by all the American Indians. The plant tends to escape from cultivation and revert to its ancestral habit with tough dry roots. Seedlings from the wild forms, when brought into a more favorable environment, gradually resume the cultivated form. Parsnips are difficult to harvest for the roots are entirely below ground. They have a considerable sugar content and even some fat. They are used as food by man and livestock, and also for making wine.

Radishes

Radishes (*Raphanus sativus*) are annual or biennial plants with a fleshy taproot and rosette of small leaves, which later are replaced by the erect flowering and fruiting part of the plant. They have been cultivated for over 2000 years, but are still close to the ancestral type and often revert to a form with a dry woody root. Radishes are grown all over the world, and are highly esteemed because of their pungent flavor. Many varieties are in cultivation differing greatly in size, shape, and color of the roots. Early, summer, and winter types occur, and they are often forced in hotbeds. Radishes are usually eaten raw, but may be cooked like any other vegetable.

Turnips and Rutabagas

The turnip and the rutabaga are very closely related, and are sometimes considered as varieties of a single species, *Brassica campestris*.

In the **turnips** (*Brassica Rapa*) both the root and the lower part of the stem are fleshy and rough. The flesh varies in texture. Those with a fine texture are used by man, while the coarser forms are fed to stock. Turnips have been grown for nearly 4000 years, and have spread from the original home in temperate Europe all over the world. They reached Mexico in 1586, Virginia in 1610, and New England in 1628. Many types are grown differing chiefly in the shape and color of the root. Turnips thrive best in cool climates. They are used more in Europe than in the United States. Turnip tops are used for greens and for forage and manuring.

The **rutabagas,** or swedes (*Brassica Napobrassica*), have a larger smooth root, with a short neck composed of stem tissue. Their flesh is more solid, and so they keep better. They grow best in northern regions where the cool climate favors the development of the characteristic sweet flavor. Rutabagas are fed to stock and are also eaten by man in large quantities.

Sweet Potatoes

The sweet potato (*Ipomoea Batatas*) is a native of tropical America, where it was probably first grown by the Indians centuries before the coming of the white men. By the beginning of recorded time it was cultivated in the tropics of both hemispheres. Although today it is widespread in all tropics and some parts of the temperate zone, and is particularly abundant in the South Seas, China, Japan, and Indonesia, it is definitely of American origin. The sweet potato, together with the cassava, yam, and taro, are practically indispensable to the tropical natives. These humble root crops are cheap, are available throughout the year, and grow in every soil. Although of little commercial importance and rarely marketed, they are found in every native garden patch.

The sweet potato is a twining, trailing perennial vine with adventitious roots that end in swollen tubers. These contain both starch and sugar, and even a little fat. The sweet potato requires a sandy soil and a warm moist climate. In the United States the Atlantic coastal plain, from the Gulf States as far north as New Jersey, is the chief producing area, with Louisiana, North Carolina, and Georgia the leading states. The plants are grown as annuals and propagated vegetatively by using the roots, or pieces of roots, or vine cuttings. The yield is low, and attempts should be made to improve it. This is particularly true in the poorer tropics where the sweet potato is such an essential plant. Two types are grown (Fig. 181). The first type has a dry, mealy yellow flesh and is much preferred in the North. The second type, often erroneously called yams, has a more watery, soft, gelatinous flesh, which is richer in sugar. This is

FIG. 181. Varieties of sweet potatoes (*Ipomoea Batatas*). Above, sweet moist-fleshed varieties; below dry, mealy-fleshed varieties. (*Reproduced by permission from Etheridge, Field Crops, Ginn and Company.*)

the favorite type in the South, where sweet potatoes are a staple crop and next to white potatoes in importance. Sweet potatoes are not only a common table vegetable, but they are used for canning, dehydrating, flour manufacture, and as a source of starch, glucose syrup, and alcohol. They are fed to horses, cattle, and hogs. The green tops are used for fodder. Because of their high water content, sweet potatoes spoil readily and about a third of the crop is usually lost.

Yams

The true yams belong to the genus *Dioscorea*. There are a great many species in the tropics and subtropics of all countries, and they are difficult to distinguish. The most commonly cultivated species is *D. alata*. The yams are all climbing vines with large storage roots, often weighing as much as 30 or 40 lb., and sometimes with aerial tubers as well. They require a deep soil, but are quite drought-resistant. Yams are the chief food of millions of people in the West Indies, South America, the South Seas, and the Asiatic tropics. They are baked or boiled or ground into flour. Yams are becoming a crop of increasing importance in the southern United States, where they are used to feed hogs and other livestock.

Cassava

Next to the sweet potato, the cassava (*Manihot esculenta*), is the most important of the tropical root crops and furnishes the basic food for millions of people. It is a native of South America, but is widely grown in all tropical and subtropical regions. Over 150 varieties are known, most of which are used by the native peoples for food. Two main groups are usually recognized—the bitter cassavas and the sweet cassavas. The cassavas (Fig. 182) are shrubby perennials with stems that reach a height of 9 ft., deeply 3- to 7-parted leaves, and roots that end in very large tubers. All the varieties contain a glucoside related to prussic acid, and some are quite poisonous. In most cases, however, a slight amount of heat is sufficient to drive off the volatile acid and render the material harmless. Cassava, also called manioc, mandioc, or yuca, is one of the most wholesome foods. It is eaten by all classes of people in South and Central America at least twice a day throughout the year, and it has been estimated that over 7000 tons are produced in these countries annually. The crop is easily grown with but little labor. It is propagated by stem cuttings, using pieces 6 to 10 in. long. The plant matures in from six months to a year. The yield is large, 1 acre producing upward of $6\frac{1}{2}$ tons. The roots of a single plant may weigh from 25 to 50 lb.

The tubers are eaten raw or are cooked. Sweet cassavas are usually boiled. A flourlike meal, known as *farinha*, is prepared by peeling, washing, and scraping or grating the tubers, and then placing the material in a bag, or press, where the liquids are squeezed out. After drying and sifting, the meal is baked into thin cakes, known as cassava bread. This has a high food value and replaces wheat bread in the diet. The poisonous milky juice is concentrated to a thick consistency by boiling, and constitutes cassareep, or West Indian pepper pot, which is used as the basis of many sauces. Raw cassava starch is used for healing purposes, and is fermented into an intoxicating beverage. It is also fed to livestock and used for laundry work and sizing.

Much of the raw cassava is used in making tapioca. For this purpose the roots are peeled and grated and the milky juices are expressed. The starchy material is then soaked in water for a few days, is kneaded, and is strained to remove any fibers and impurities. After sifting and drying in the sun, it is heated gently on hot iron plates. This partially cooks the starch and causes it to ball up into the familiar little round lumps, which

FIG. 182. Mature cassava plants (*Manihot esculenta*). (*Courtesy of the Minute Tapioca Company, Division of General Foods Corporation.*)

constitute the tapioca of commerce. The United States imports tapioca chiefly from Brazil and the East Indies.

UNDERGROUND STEMS

Jerusalem Artichoke

The Jerusalem artichoke (*Helianthus tuberosus*) is a native plant of North America and has been cultivated by the Indians for centuries. It is a hardy perennial sunflower 6 to 12 ft. in height. The name is derived from the Italian word for sunflower, *girasole*. The plant was introduced into Europe in 1616, and has always been cultivated more extensively there than in this country; many new and improved varieties have been developed. The Jerusalem artichoke is adapted to almost any climate but does best in the Northern states and Canada. The tubers, which somewhat resemble potatoes, but with larger eyes, are cooked, pickled, or eaten raw. The carbohydrate food is in the form of inulin, which is a good food for diabetics, and is also used as a source of levulose and industrial alcohol. The plants are also grown as a forage crop and weed eradicator.

Potato

The white or Irish potato (*Solanum tuberosum*) is one of the most important food plants of the world. It is a native American species and was cultivated from Chile to New Granada at the time the Spanish explorers reached this continent. The first mention of the potato in literature was in 1553 in Pedro de Leon's "Chronica del Peru," while the first published illustration appeared in Gerard's Herbal in 1633. The potato was introduced into Europe soon after 1580 by the Spaniards, and by the end of the seventeenth century it had spread all over Europe and the British Isles. The potato was first introduced into New England by Irish immigrants in 1719. Possibly it had been carried to Virginia and the Carolinas at an earlier date, but it did not spread farther.

The potato is an erect, branching, more or less spreading annual from 2 to 3 ft. in height. It has pinnately compound leaves, fine fibrous roots, and numerous rhizomes which are swollen at the tip to form the familiar tubers. The flowers are white, yellow, or purple, with a tubular corolla, while the fruit is a small brownish-green or purple inedible berry.

Potatoes are adapted to many soils and many climates. They are, in fact, grown the world over, except in low tropical regions. They are hardy and mature rapidly, and so can be grown as far north as 60°N.L. and at altitudes up to 8000 ft. The best environment is a cool moist climate, with a mean annual temperature of 40 to 50°F., and a rich light soil.

Potatoes are usually propagated vegetatively by means of tubers, or parts of tubers, the so-called "seed potatoes." They may, however, be grown from seed. The famous Burbank potato was obtained in 1871 from seed. The 500 or more varieties now in cultivation have been obtained by selection and hybridization and by the utilization of mutations, which are of frequent occurrence. The essential parts of the seed potatoes used for propagation are the eyes. These are really groups of buds situated in the axils of aborted leaves. There is usually a central bud in each eye, surrounded by smaller lateral ones. The eyes are more numerous toward the apex of the tuber. Pieces of the tubers are cut at right angles to the main axis so as to remove the inhibiting effect of the terminal bud. The larger the piece, the more vigorous is the vegetative growth, and there is a correspondingly greater yield. In any case, at least one eye must be present. The tubers have a rest period of several weeks' duration after they have matured, during which they will not sprout. This is a period of afterripening in which various physiological changes take place. The duration of the rest period can be controlled by the use of cold and by various gases and chemicals.

Within the tubers several regions can be noted. These include the skin or periderm varying in color, texture, and thickness; the narrow cortex, a dense area with small starch cells; a ring of fibrovascular bundles; the external medulla, which contains most of the starch; and the internal medulla, which has a much greater percentage of water and less starch. Branches of the internal medulla extend outward toward

FIG. 183. Harvesting potatoes in Aroostook County, Maine. (*Courtesy of the Maine Development Commission and the Massachusetts Horticultural Society.*)

each eye. In all these areas the starch occurs in characteristic oval grains of varying sizes in thin-walled parenchyma cells. The mealiness of the potato is due to the swelling of the grains and the rupturing of the cell walls. If the external layers are poor in starch, the walls do not burst and the tuber is soggy. Potatoes contain about 78 per cent water, 18 per cent carbohydrates, including a little sugar as well as starch, 2 per cent proteins, 0.1 per cent fat, and 1 per cent potash. They are well adapted to storage if a cool dark place is provided. The water loss over winter amounts to about 11 per cent.

Over 90 per cent of the potato crop of the world is grown in Europe, where it exceeds the wheat crop of the world in volume and value. Of the estimated 1948 world production of 8,956,601,000 bu., Germany produced 1,396,750,000 bu., the U.S.S.R. 2,880,000,000 bu., Poland 983,105,-000 bu., France 576,204,000 bu., and Great Britain 451,099,000 bu., while the United States produced 445,850,000 bu. Potatoes are grown more universally than any other crop, and are cultivated in all but seven counties of the United States. They are grown in the Northern states in the summer and in the Southern states in the late winter and early spring. They are a valuable small farm and garden crop for they will grow in any soil and have a high yield. The commercial production of potatoes is often concentrated in areas where both the climate and market conditions are most favorable. Chief among these are the Aroostook region in Maine (Fig. 183), Long Island, New York, southern New Jersey, Pennsylvania, and the eastern shore of Maryland. Other important states are California, Idaho, North Dakota, and Colorado.

In the United States potatoes are used chiefly for human consumption and are a universal table food. Small tubers are utilized for the production of starch and industrial alcohol, and they are also fed to stock. In Europe a considerable portion of the crop is used for industrial purposes.

South American Tubers

The high Andes of South America produced not only the common potato but many other species of *Solanum*, some of which are cultivated by the Andean natives. In addition, three other tubers (Fig. 184) are native in this area, and they have been important food plants for centuries. The most important is **oca** (*Oxalis tuberosa*), an upright succulent herb with trifoliate leaves and orange-yellow flowers. Several varieties are grown. The tubers contain calcium oxylate crystals and have to be mellowed in the sun before they are eaten. The **ullucu** or **melloco** (*Ullucus tuberosus*) is second in importance. Its tubers resemble small potatoes. The plants are very resistant to frost and have a high yield. The **añu** (*Tropaeolum tuberosum*), a twining plant related to the garden nasturtium, is not so valuable but is grown to some extent.

The **arracacha** (*Arracacia xanthorrhiza*), also a native of the Andes, has spread northward to Venezuela. This robust herb with large fleshy roots is extensively cultivated and is an important source of starchy food in many parts of tropical South America. *Canna edulis*, known as **achira**, has an edible tuber, which is also one of the sources of arrowroot starch. The **yam bean** (*Pachyrrhizus erosus*) likewise has tubers which are eaten either raw or cooked.

Taros and Dasheens

The taros and dasheens are next to yams in importance in the Orient, and they constitute the staple food of millions of people. Some 1000 horticultural varieties are grown. They are among the few edible aroids

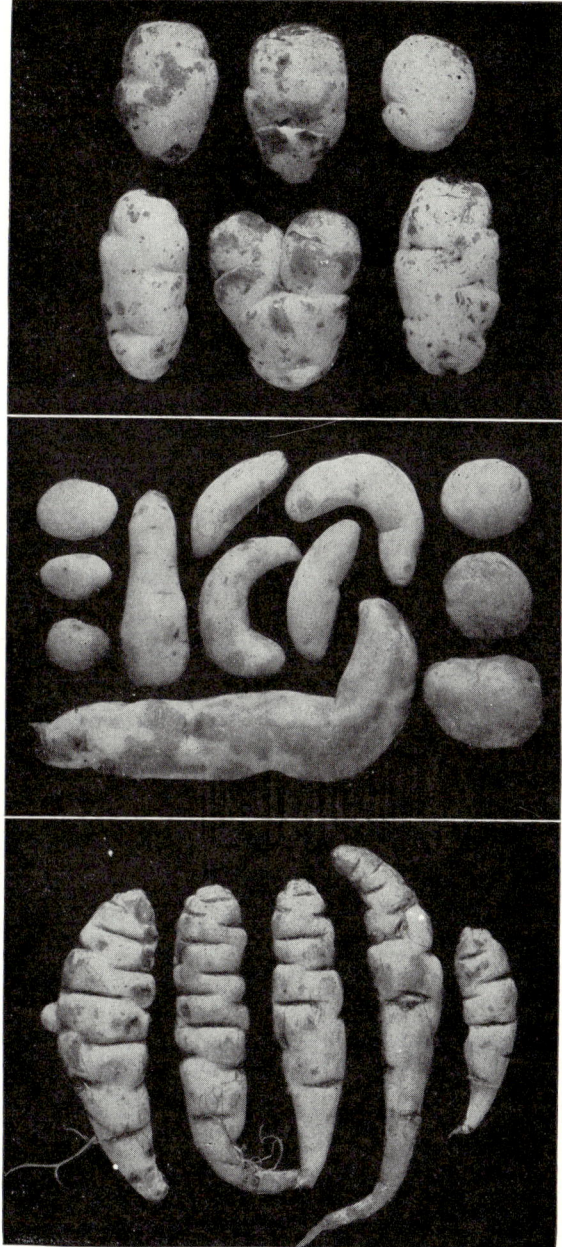

Fig. 184. Three Andean tubers. *Upper*, oca (*Oxalis tuberosa*); *middle*, ullucu (*Ullucus tuberosus*); *lower*, añu (*Tropaeolum tuberosum*). (*Photo by Walter H. Hodge.*)

and belong to the genus *Colocasia*. Taros and dasheens are very similar in appearance and are sometimes considered to be variants of a single species. There is no erect stem, but a cluster of large leaves from 4 to 6 ft. in length.

The **taro** (*C. antiquorum*) is native to Southeastern Asia, whence it has spread throughout Polynesia and the Pacific islands. Today it is of paramount importance in Hawaii, and it has been introduced into tropical Africa and America. The taro has huge peltate leaves. It has been cultivated for so long that it never flowers. Probably no civilized man has ever seen a taro seed. The tops of the corms are used for propagation.

FIG. 185. Tubers of the dasheen (*Colocasia esculenta*). (*Reproduced from U.S.D.A., The Dasheen as a Root Crop for the South*, 1913.)

Taros require a wet rich soil and a long season in which to mature. The yield is high and the starch is exceedingly wholesome and readily digested. It is an interesting side light that there is no word for indigestion in the Kanaka language. The taros are baked or boiled to destroy the acrid calcium oxylate crystals in the raw tubers. One of the principal dishes of the Polynesians is *poi*, a thin pasty mass of taro starch. This has a faintly acid taste and is very palatable. It is often made into cakes and baked or toasted.

The **dasheen** (*C. esculenta*) has large tubers, with smaller ones on the side (Fig. 185). The flesh is mealy with a nutty flavor, and has more carbohydrates and proteins than the potato. Since 1913 dasheens have been grown as a commercial crop in the southern United States, and the tubers are used as a potato substitute.

Yautias

The yautias are among the least known but are probably the oldest of the root crops. These plants are found only in tropical America. The commonest species is *Xanthosoma sagittifolium*. The yautias resemble the taros, to which they are closely related. They are taller, reaching a height of 7 or 8 ft., with arrow-shaped leaves, and produce both corms and tubers. These plants are common in the West Indies, especially in Puerto Rico, and many varieties are grown. The tubers are twice as nutritious as the potato.

Onions

The onion (*Allium Cepa*) is the chief food plant in which the food is stored in a bulb. It is very old, its use going back over 4000 years

FIG. 186. Onions (*Allium Cepa*) drying in windrows. (*Courtesy of the University of Massachusetts.*)

beyond the beginnings of authentic history. It is unknown in the wild state. The onion was probably a native of Southern Asia or the Mediterranean region. It has long been valued in China and India for its flavoring. In Egypt it was worshiped before the Christian era, and it also played a part in the Druid rites. Onions are cultivated over large areas in temperate and even tropical climates. They prefer cool moist regions with a sandy soil. They are grown from seeds or sets, small bulblets that are produced instead of flowers. Onions have to be dried (Fig. 186) and cured before they are stored in order to develop the characteristic flavor and taste, which are due to an acrid volatile oil, allyl sulphide. They are both food plants and flavoring agents. Over 250 species of *Allium* are known, some of them native to boreal America. Many occur as weeds. The commonest forms in cultivation are garlic, leeks, chives, shallots, and the true onions.

The **garlic** (*Allium sativum*) is a perennial plant with narrow flat leaves and several small egg-shaped bulbs, known as cloves, enclosed in a white skin. The inflorescences produce both seeds and bulblets. The latter, together with the cloves and the leaves, have been used from earliest time for flavoring soups, sausages, and salads. Garlic also has a medicinal value as it possesses antiseptic and bactericidal properties.

The **leek** (*Allium Porrum*) is a very old plant. It is a robust perennial of the Mediterranean region with rather thick, flat, broad leaves and small bulbs. The bases of the leaves are mild flavored and edible and they are often blanched like asparagus. Leeks are also used in seasoning.

Chives (*Allium Schoenoprasum*) have hollow and cylindrical leaves with very small clustered bulbs and dense umbels of rose-colored flowers. They are hardy perennials growing in dense clumps. The young leaves and bulbs are used for seasoning.

Shallots (*Allium ascalonicum*) likewise have cylindrical hollow leaves, but the plants are not cespitose. They are perennials with large clustered bulbs, which are much used for pickling. The leaves are short and awl shaped.

The **true onions** (*Allium Cepa*) are biennials with a single large bulb and long, hollow, cylindrical leaves. From each bulb arises a single leafless scape as much as 2 or 3 ft. in height, with numerous small flowers. There are many different forms with either round, flat, white or colored bulbs. The foreign onions, like the Spanish and Bermuda onions, are larger, less hardy, and milder than the domestic types. Onions were brought to America by the early colonists. They are propagated by division, bulblets, and seeds. They are used for flavoring, as a vegetable, for pickles, and in medicine. They are easily digested and are good for constipation. An extract of onion is available for flavoring purposes.

HERBAGE VEGETABLES

Herbage vegetables have the nutrient materials stored in parts of the plant developed above ground. They are the familiar pot herbs and salad plants. Almost any part of the shoot system of the plant may be utilized for storage. The leaves are used in kale, cabbage, spinach, and lettuce; stems in asparagus and kohlrabi; buds in Brussels sprouts; leaf-stalks in celery and rhubarb; and immature flowers and flower stalks in cauliflower and broccoli. The chemical composition and food value of herbage vegetables are close to those of the earth vegetables. There is more water, however, and a correspondingly smaller amount of carbohydrates. They contain more proteins, since the leaves are the workshop of the plant, and also a considerable amount of mineral salts and vitamins, which makes them an essential part of man's diet. There is also some roughage value.

Artichoke

The globe artichoke (*Cynara Scolymus*) is a native of the Mediterranean region and the Canary Islands. The plant resembles a thistle in size and habit. The flower stalks terminate in globular inflorescences with numer-

FIG. 187. A field of artichokes (*Cynara Scolymus*) showing plants in full flower.

ous subtending involucral bracts (Fig. 187). The immature heads together with the fleshy bases of the involucral leaves and the thickened receptacle are eaten either raw or cooked. Artichokes grow best in low ground near the seacoast. They are extensively cultivated in Central and Southern Europe, where they have been more appreciated for food purposes than in the United States. However, the plant is now important in California, where increasing numbers are grown. Artichokes are canned or used as a fresh vegetable.

Asparagus

Asparagus (*Asparagus officinalis*) is a native of temperate Europe and Western Asia, and still grows in a wild state in saline areas. It has been known and prized by epicures since Roman times and is widely grown and used throughout Europe. It was early introduced into the United

States. The plant has perennial roots, which send up each year an erect branching stem several feet in height. Instead of true leaves, modified branches known as cladophylls occur. This is characteristic of the entire genus, which includes the so-called asparagus fern of the florists. The axillary flowers are small and the fruit is a berry. The new shoots are very juicy and succulent, and these constitute the asparagus of commerce. If the shoots are allowed to develop, the plant soon becomes bushy and woody. Asparagus is an important truck crop, both for use as a fresh vegetable and for canning purposes. Although it can be grown under a wide range of soil and climatic conditions, it thrives best in fertile well-drained soil in moist temperate regions with an abundance of sunshine. It can be grown from seed or from year-old crowns. Once started, asparagus will continue to yield for 15 to 20 years. The shoots may be used either green or blanched. The former are more delicate and are usually eaten fresh, while the latter, which are thicker, are used chiefly for canning. For the best flavor asparagus should be cooked within 12 hours of picking. The food value is low, as the water content amounts to 94 per cent, but there is more protein present than in most vegetables. The pulp is sometimes dried, or canned as a paste. Asparagus has a definite medicinal value. Although widely grown for domestic use, it is a commercial product in only a few areas. California leads in the production of both green and canned asparagus. Other areas include: Washington, New Jersey, Illinois, and South Carolina.

Cabbage and Its Allies

One of the most ancient and most important herbage vegetables is the cabbage (*Brassica oleracea*). The wild ancestor, the colewort, a stout, weedy perennial of the seacoasts of Great Britain and Southwestern Europe, is still in existence. From this plant have arisen by selection or mutation the great variety of cultivated forms. Although best adapted to the Mediterranean type of climate, they will grow from the arctic to the subtropics. Cultivation of the cabbage is very old, at least since 2500 B.C., and several varieties, true cabbage, cauliflower, and broccoli, were known to the Greeks and Romans. The ancient Germans, Saxons, and Celts were the first to grow cabbage in Northern Europe, and it became important at a very early date in Scotland and Ireland. Today the plants are grown the world over, except in the low tropics. Cabbage is one of the best protective foods, for it contains the antiscorbutic vitamin and is also rich in sulphur. No other cultivated plant has varied so much. The commonest forms include the kales or collards, Brussels sprouts, head cabbage, kohlrabi, cauliflower, and broccoli. In the kale and Brussels sprouts the stem of the first year is elongated, while in the others it is very short. These types will be discussed very briefly.

Kales and Collards (var. *acephala*). These erect branching forms, which are also known as borecole or marrow cabbage, are closest to the wild form. They have numerous large broad leaves, which are used as a boiled green vegetable or for stock feed. Kale is grown more in other countries than in the United States. The plants are resistant to cold, heat, and drought. The giant cabbage kales of England reach a height of 8 or 9 ft. and the stout stems can be used for rafters or canes.

Brussels Sprouts (var. *gemmifera*). In this variety the axillary buds on the erect stem develop little heads, instead of forming branches. These miniature cabbages are a favorite vegetable. Both tall and dwarf forms occur. Brussels sprouts are cool-season plants, and are more tender and delicate than cabbage.

Cabbage (var. *capitata*). In the familiar cabbage the stem is so short that the great mass of thick overlapping leaves tends to form a head. The older leaves surround the younger, smaller, more tender leaves and the miniature stem, so that in section the cabbage resembles a huge bud. There are many varieties, some with smooth leaves and others with curled leaves. The latter, the Savoy cabbages, have the best flavor. Both green and red cabbages are grown. Cabbage is one of the best market garden crops in cool climates. It can be grown on heavy soil. The plant is very old and was introduced into England by the Romans. It is still so important in that country that it is often jokingly referred to as the national flower of England. Cabbage contains 91 per cent water, with some sugar and starch, considerable protein, and valuable lime salts. It is eaten raw (slaw) or cooked. Steaming is preferable to boiling for the nutrients are retained. Cabbage is not very easily digested. It is used for feeding stock and chickens. Sauerkraut is really human silage. Small pieces of cabbage are fermented in their own juice together with salt. Lactic acid bacteria act on the sugar to produce lactic acid, which is responsible for the sour taste. Sauerkraut originated in Asia and is now a favorite food in the U.S.S.R. and Germany, and in this country as well. In most European countries cabbage, in some form, is an important part of the daily diet of the poorer people.

Kohlrabi (var. *gongylodes*). In kohlrabi no head is formed, but the short stem is transformed into a juicy mass of edible tissue, which stands out of the ground (Fig. 188). It is large, spherical, and turniplike, white or purple in color, with large leaf scars. It is much used for human consumption abroad and by the foreign population of the United States. In this country it is used chiefly for stock feed. Kohlrabi is an early spring or fall crop, as it does not like the heat of summer. The kohlrabi is perhaps better considered to be a distinct species, *Brassica caulorapa*.

Cauliflower and Broccoli (var. *botrytis*). In these forms a short erect stem is produced with an undeveloped inflorescence. The whole inflorescence forms a large head of abortive flowers on thick hypertrophied branches. The leaves are often tied around the mass of flowers to keep them white. In broccoli the heads are smaller and the leaves larger, and the whole plant is green. Cauliflower and broccoli are very old.

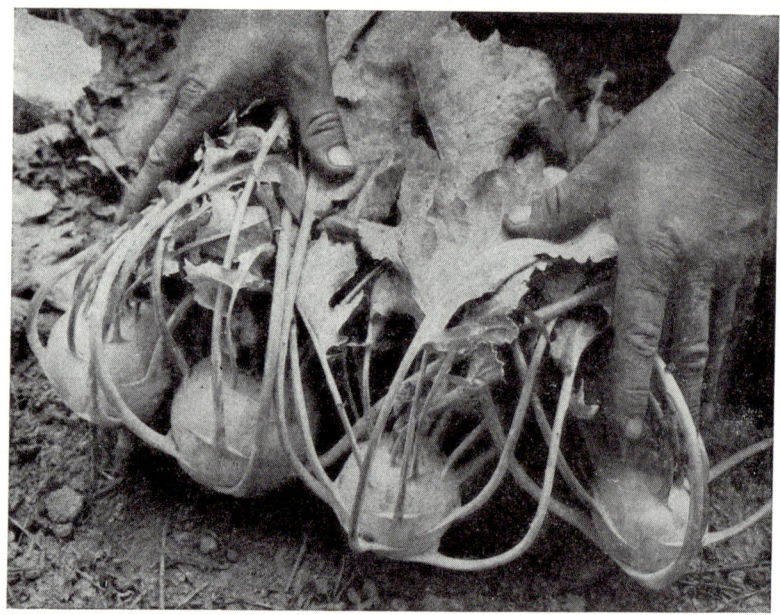

FIG. 188. Kohlrabi plants (*Brassica oleracea* var. *gongylodes*), showing the distinctive habit. (*Courtesy of the Massachusetts Horticultural Society.*)

They are more delicate and easier to digest than cabbage, and are a favorite vegetable in all temperate regions.

Celery

Celery (*Apium graveolens*) is a native of temperate Europe from England to Asia Minor. In a wild state the plant is tough and rank with an acrid and poisonous juice. It grows in marshes, ditches, and other wet places. Under cultivation (var. *dulce*) it is a biennial forming a fleshy root and clump of compound leaves with long leafstalks. These leafstalks constitute the celery of commerce. They are large and succulent, and their quality is much improved by blanching. This is done by placing boards, dirt, or paper around them to shut off the light and so prevent chlorophyll from developing (Fig. 189). Celery requires a rich sandy loam and lots of water. It is grown in the south of the United States in winter, and in the north in summer. Commercial celery growing has become an important industry with an annual production of around 20,000,000 crates. California, Florida, Michigan, and New York lead. Celery was formerly grown in the Old World for its foliage, which was used for flavoring and as a garnish and for medicinal purposes. The roots are often boiled. The outer stalks, which are too tough to eat, are used as a basis for cream of celery soup. Celery seeds are grown to be

FIG. 189. Harvesting late celery (*Apium graveolens* var. *dulce*). Note the way in which the rows are banked with dirt. This blanches the leaf stalks by preventing chlorophyll from developing. (*Courtesy of the University of Massachusetts.*)

used as a savory. The larger turniplike roots of a European variety, **celeriac** (var. *rapaceum*), are often used for soups and flavoring.

Chicory and Endive

The chicory (*Cichorium Intybus*) and the endive (*C. Endivia*) have been cultivated for over 1000 years, but are still very wild looking and show but slight effects of their association with man.

Chicory is a perennial with a long taproot, a coarse branching stem, and numerous basal leaves. The flowers are usually blue. The plant is a native of Europe, but is a common weed in the United States. It is much used abroad as a salad plant or for greens. The roasted root is an important adulterant of coffee.

Endive is a native of India. It was a favorite salad plant of the Egyptians and Greeks, and is grown today in every European country. It is becoming more important in the United States. The young basal leaves, which often have curled margins, are used as a salad. They are frequently blanched. The plant is an annual or biennial.

Lettuce

Lettuce (*Lactuca sativa*), a native of Southern Europe and Western Asia, is descended from the wild lettuce (*L. Scariola*), a common weed of roadsides and waste land in both the Old and New Worlds. It is another herbage vegetable of great antiquity. The Greeks cultivated three varieties, the Persian kings used it as early as 300 B.C., and the Moors developed many types, among them the present-day romaine. Nearly all the forms now grown seem to have been known in Europe prior to the Middle Ages. The plant produces a basal rosette of leaves, and later in the season a stalk with flowers and fruits. Lettuce has a milky juice. It is of little value as a food, except for its vitamins and iron salts. Several hundred varieties are grown. Formerly cultivated only in home gardens, lettuce is now an important commercial crop. It thrives best in a sandy or loamy soil and requires cool weather and not too much sun. Among the principal types are head, cos, romaine, and cut-leaved forms.

Rhubarb

The succulent acid leafstalks of the rhubarb (*Rheum Rhaponticum*) are much used for pies and sauce. The plant is a native of Asia, where it still grows wild. It is a perennial with large rhizomes and produces early in each season a number of very large leaves, and, later, elongated flower stalks which bear dense masses of tiny whitish flowers. It is extensively cultivated as a food plant, and sometimes as an ornamental plant, in temperate portions of Europe and America. Rhubarb has nearly 95 per cent water with a little sugar and fat and salts of oxalic and malic acids. A wine is sometimes made from the juice.

Spinach

Spinach (*Spinacia oleracea*) is perhaps the commonest herbage vegetable used for greens. It is a native of Southwestern Asia and is widely cultivated in cool regions where there is an abundance of water. Early in the season it produces a large number of basal leaves, and later the flowering part. It is an annual, occurring in several forms. In addition to its use as a pot herb, spinach is now canned in large quantities.

Minor Herbage Vegetables

Among cultivated herbage vegetables of lesser importance may be mentioned Chinese cabbage (*Brassica pekinensis* and *B. chinensis*), an annual plant used for greens and salads; the dandelion (*Taraxacum officinale*), long popular for greens in a wild state, and now grown commercially; water cress (*Nasturtium officinale*), an aquatic perennial much used for salads; New Zealand spinach (*Tetragona expansa*), a warm-weather crop in which only the tender young leaves are harvested; and tampala (*Amaranthus gangeticus*) (Fig. 190), a recent introduction with both red and green varieties.

FIG. 190. Tampala (*Amaranthus gangeticus*), a recently introduced herbage vegetable. (*Courtesy of the Massachusetts Horticultural Society.*)

Many wild species are used by country people for pot herbs, particularly in the spring. The leaves of beets, turnips, and mustard are also utilized.

FRUIT VEGETABLES

Fruit vegetables are technically fruits, but they are rarely eaten in the raw state except in salads and usually require cooking. In other words, they are used as vegetables rather than as fruits. In their food value and other properties they resemble the other vegetables.

Avocado

The avocado or alligator pear (*Persea americana*) is a small tree native to Mexico or South America and grown extensively in Central America and the West Indies for many years. The use of avocados is increasing in the United States and the plant is now of commercial importance in California and Florida. They are often marketed as "calavos" (a registered trademark of the Calavos Growers of California). The brownish-green pear-shaped fruit, from 4 to 6 in. in length, is really a one-seeded berry. The pulp surrounding the large seed has a buttery consistency and contains up to 30 per cent fat, considerable carbohydrate material, and more proteins than any other fruit. The vitamin content is also

380   ECONOMIC BOTANY

high.   Over 500 varieties are known.   There are three races: the Guatemalan, West Indian, and Mexican, differing in shape, size, and hardiness, Avocados constitute a valuable human food and may be eaten as a salad fruit or cooked.

Breadfruit and Jack Fruit

The **breadfruit** (*Artocarpus altilis*) is one of the most important food fruits in the world.   It is a native of Malaya, but is now widespread in

FIG. 191.   Leaves and fruit of the breadfruit tree (*Artocarpus altilis*).   The breadfruit is one of the most important fruits of the tropical world.

the tropics, particularly in Polynesia. It has been cultivated since antiquity. The tree is very handsome, reaching from 40 to 60 ft. in height, with deeply incised leaves. The prickly fruits (Fig. 191), which are about the size of a melon, are brownish yellow when ripe, with a fibrous yellow pulp. They are often borne in small clusters. The breadfruit is eaten fresh or cooked. It is either baked, boiled, roasted, fried, or ground up and used for bread. During the few months when the fruit cannot be obtained, a paste that has previously been made is used. Over 100 varieties are known, some with seeds and others seedless. Few plants furnish a more wholesome food for man and beast, or have a greater yield. An eight-year old tree will produce between 700 and 800 fruits. The carbohydrate content is especially high.

The **jack fruit** (*Artocarpus heterophyllus*) has a similar use. It is an Indo-Malayan species, now widely dispersed in the tropics. The handsome tree reaches a height of 60 to 70 ft. It has entire leaves and huge fruits, 1 to 2 ft. long and weighing from 20 to 40 lb., which are borne on the trunk.

Chayote

The chayote (*Sechium edule*) is a trailing vine of tropical America, which produces a gourdlike fruit much prized as a vegetable. The applelike fruits are technically pepos, berries with a spongy pulp and a hard firm rind. The plant is a perennial with large tuberous roots. Both the fruits and tubers were among the principal foods of the Aztecs, Mayas, and other primitive peoples of Central America. It has been grown in the Southern states for many years, but only recently has become of commercial importance. Chayotes grow vigorously and have a prolific yield. The plant has many uses. Not only are the tubers and fruits valuable for food, but the foliage can be used for greens or forage and the young shoots serve as a substitute for asparagus. The straw is valuable for making baskets and hats. It can also be grown as a bee plant or for ornamental purposes.

Cucumber

The cucumber (*Cucumis sativus*) is another gourd fruit, probably indigenous to southern India. It has been cultivated for 4000 years. The earliest writings of the Hebrews, Egyptians, Greeks, and Romans have numerous references to this plant. It had reached Europe by the seventeenth century and is now widespread. The cucumber is a rough-stemmed trailing vine with yellow axillary flowers and round to elongate prickly fruits. The water content is high, 96 per cent, so there is but little food value. The English forcing cucumbers have a smooth, usually seedless fruit, sometimes 2 ft. in length. Cucumbers are eaten raw, pickled, or cooked.

Pickles are made from small cucumbers or special varieties, such as the West Indian **gherkin** (*Cucumis Anguria*), which has tiny fruits with a thin flesh and numerous seeds. The fruits are soaked in brine tanks and treated with boiling vinegar. Sometimes spices, like dill, are added to give a special flavor. About 5,000,000 bu. are used for pickling. The northeastern United States is the center of the industry.

Eggplant

The eggplant or aubergine (*Solanum Melongena*) is a native of India, but is widely grown in the warmer regions of both hemispheres, especially in the West Indies and southern United States. Florida, New Jersey, Louisiana, and Texas are the leading states. Several taxonomic varieties may be recognized. The plant is an erect branching herb, several feet in height. The fruit is a large, ovoid, whitish or purple berry about the size of a coconut. The plant is cultivated as an annual and requires a high temperature. The fruit is usually cut into slices and fried or broiled.

Okra

Okra (*Hibiscus esculentus*) is a native of tropical Africa. It was cultivated by Europeans as early as A.D. 1216, and has now been introduced into most warm tropics and subtropics. The plant is a stout annual, much resembling cotton in its habit. The young pods are very mucilaginous and are much used in soups under the name "gumbo," the Spanish word for okra. Okra is also cooked in various ways. It is often dried or canned. The stems and mature pods yield a fiber used in papermaking and for textiles.

Pumpkins and Squashes

Pumpkins and squashes (Fig. 192) are gourd fruits belonging to the genus *Cucurbita*. There has been considerable speculation as to whether they are natives of America or Africa. The present evidence indicates that they are all definitely of American origin. The five cultivated species have never been found in the wild state and have been important in primitive agriculture for centuries. At least two were grown in Peru as early as 2000 B.C.; *C. Pepo* was cultivated by the North American Indians at about the same time, and *C. moschata* by A.D. 312.

The plants are coarse annual vines with large yellow flowers and fruits that rest on the ground. The numerous varieties are insect pollinated and readily cross. Immature fruits are used as a fresh vegetable, stewed, boiled, or fried, while mature fruits are baked, canned, or fed to livestock. The seeds are rich in fats and proteins and can be utilized as the source of an edible vegetable oil. Pumpkin seeds, fried in deep fat and salted, are now in the markets under the name of *pepitos*.

Cucurbita Pepo includes the field pumpkins used for pies, canning, and cattle feed; the summer or crookneck squashes; acorn squashes; the scallop squashes, pattypans, or cymlings; zucchinis; vegetable marrows; and small inedible gourds grown for ornamental purposes. *C. moschata* includes autumn and winter varieties such as the butternut. *C. mixta*, a species recently separated from *C. moschata*, comprises the cushaw squashes and several gourds; *C. maxima* includes such autumn and winter squashes as the buttercup, mammoth, Hubbard, and turban. *C. ficifolia*

FIG. 192. Four varieties of squashes. From left to right, buttercup; Hubbard (*Cucurbita maxima*); Des Moines (*C. Pepo*); butternut (*C. moschata*). (*Courtesy of the Massachusetts Horticultural Society.*)

the only perennial species, has had a long period of cultivation in Latin America. The Malabar and other ornamental gourds belong here.

Tomato

The tomato (*Lycopersicon esculentum*) was probably originally confined to the Peru-Ecuador area, from which it spread northward in pre-Columbian times to Mexico, where it was first domesticated. The Spanish explorers carried the plant to Southern Europe where it was eaten for a long time before it was utilized by the people of Northern Europe and the United States. For many years it was considered to be poisonous, and was grown only for ornamental purposes under the names "tomatl," "love apple," or "*pomme d'amour.*" Today it ranks next to potatoes and sweet potatoes in importance. Tomatoes are coarse, branching, erect or trailing herbs, with a true berry for a fruit (Fig. 193). They differ greatly in habit depending on the environmental relations and the kind, over 175 of which are grown. Several taxonomic varieties may be involved. The Mediterranean region and the United States are the most favorable regions for tomato growing. In California the plant is a perennial, but in the rest of the country it is grown as an annual. In the Northern and Central states the plants are started in a hotbed and later transplanted. Tomatoes are eaten raw or cooked, and are also preserved. They are especially rich in vitamins. Only the pulp, which retains its characteristic flavor, is used in canning. The waste material, consisting of skins, cores, seeds, and unripe parts, was formerly discarded. This is now utilized and a fixed oil is expressed, which can be used for

FIG. 193. A fruiting spray of greenhouse trellis tomatoes (*Lycopersicon esculentum*). (*Courtesy of the University of Massachusetts.*)

food, soap, or as a drying oil. The oil cake is of value as a stock feed. Ripe tomatoes are also used for chili sauce, ketchup, tomato juice, and tomato pastes, while green tomatoes are used for pickles and preserves.

CHAPTER XVIII

FRUITS OF TEMPERATE REGIONS

Ancient man must have learned early in his career to appease his hunger and eke out his existence by eating the numerous wild fruits, which everywhere attracted his attention. He soon became aware of the inedible nature of some of these fruits, and these he avoided. At first the nomadic tribes were content to gather in a region where the edible wild fruits were most abundant and to linger long enough to harvest them before moving on. Later man began to cultivate these fruits, choosing those with the best taste and the largest yield, as well as those which were easiest to grow. Finally, as civilization progressed and as the nutritive qualities and physiological action of these foods were studied, our present-day fruits were gradually developed. Most of the changes and improvements have been brought about by selection and hybridization. In many cases wild fruits are still used, both by primitive peoples and by civilized nations as well. In America the wild fruits used by the Indians were first cultivated after the advent of the white man.

It is interesting to note that a great many of the fruits grown today had their origin in the same part of Asia that was the earliest home of man. Thus from the very beginning man has had close contact with them, and his history has closely paralleled theirs. This is particularly true of the rose family, which includes a large number of our most familiar fruits, such as the apple, pear, cherry, plum, apricot, raspberry, blackberry, and strawberry. Apples and plums still grow wild in great profusion in the mountains of Central and Western Asia.

In his gradual dispersal over the surface of the earth man carried his food plants with him. He early reached the Mediterranean region, and here he found an area particularly well adapted for the growing of fruit. In this region, with its dry summers, its mild winters, and its fertile soil, a multitude of plants, once widely scattered, were brought together from their native homes and improved and perfected. Many varieties were known to the Greeks, Egyptians, and Romans, and fruit growing was an important part of their civilization. The Dark Ages failed to wipe out the knowledge and experience gained by these ancient people, and, once revived, agriculture and horticulture have been carried on continuously in most of the regions with a high degree of perfection. Fruit growing is now carried on all over the world, often in areas not suitable for the cultivation of cereals. The early settlers in the United States brought fruit seeds and plants with them, and eventually these have spread over the country. Commercial fruit growing has become increasingly important, especially on the Pacific Coast. Today, California, Washington, and Oregon constitute one of the largest fruit-producing areas of the world.

Food Value of Fruits. In temperate regions fruits are considered more as an agreeable addition to the diet than as a staple food. In the tropics, however, the reverse is true, and fruits may often be the chief, and even the only, source of food, as in the case of the banana, plantain, date, fig, coconut, and breadfruit. Our temperate fruits have only a slight nutritive value. The water content is about 80 per cent, and the remainder consists of cellulose, which may have some roughage value, and a solution of sugars, starches, pectin, and organic acids, flavored with essential oils and aromatic ethers. Carbohydrates are most abundant, the exact amount and kind of sugar being determined by the stage of ripeness. Fats and proteins are negligible. Organic acids, however, are present in greater amounts than in any other plant products. These are chiefly malic, citric, and tartaric acids. The various pectin compounds are important for they have the property of forming a jelly under proper conditions. Mineral salts are also present in considerable quantities.

Preservation of Fruits. Because of their perishability, many ways of preserving fruits have been devised. These include drying, salting, and smoking; sweetening with sugar, honey, and spices; preserving in alcohol or chemicals; pickling in vinegar; packing in fats; sterilization; canning; and freezing. The most important processes at present are drying, canning, and freezing. Drying has long been a favorite method. Bacteria fail to develop when the water content is below 25 per cent. The common drying agents are the sun and hot air. Often the evaporation plants are on a huge scale. Some fruits are cooked with sugar before drying. Canning has always been important, for both domestic and commercial purposes. A strong solution of sugar, honey, or glucose will keep out agents of decomposition; consequently great quantities of fruit are preserved as jams, jellies, marmalades, and candied fruits. The quick-freezing or cold-pack method of preservation is very effective. The fruit is placed in small containers, and the heat is extracted rapidly from both the top and the bottom of the container. This results in the formation of smaller ice crystals as the material freezes and the consequent better quality, color, flavor, and vitamin content of the product. Quick-freezing is equally practicable for commercial or home use. Although it has been utilized since about 1905, it has become important only recently. In 1936, 70,000,000 lb. were frozen, but by 1946, the amount had increased to 450,000,000 lb. However, the development of improved cold-storage methods and increased facilities for transportation in recent years have made it possible to use fresh fruit to a greater extent than ever before. Today the production of fresh fruit has become an important industry, and it is possible to draw on all parts of the world for the supply, which is virtually unlimited. Of the thousands and thousands of edible fruits only about 100 are cultivated, and of these not more than 50 per cent reach the Northern markets.

Classification of Fruits. The classification of fruits is very varied. Technically a fruit is the seed-bearing portion of the plant, and consists of the ripened ovary and its contents. Usually the ovary alone is involved in the formation of the fruit, but in the accessory fruits other structures, such as the calyx and receptacle, are involved. Simple fruits are derived from a single ovary, and compound fruits from more than one. In the latter case the aggregate fruits are formed from numerous ovaries of the same flower, while multiple fruits come from the ovaries of different flowers. All these may be either dry or fleshy. In our discussion of food plants we have already considered the edible dry fruits; the grains, legumes, and nuts; and also certain fleshy fruits that are usually classed as vegetables. The present chapter will be concerned only with those fruits which are usually eaten without cooking, to which the term "fruit" is restricted in common usage. For convenience the fruits of temperate regions will be considered first, followed in the next chapter by the more important of the tropical fruits.

POME FRUITS

The pome fruits are simple accessory fruits in which the ovary is surrounded by a fleshy outer portion derived from some other part of the flower. Authorities differ as to the morphological nature of this edible portion. Some consider it to be a fleshy calyx, but the most recent studies indicate that it is an enlarged receptacle. In either case the ripened ovary forms the core only. In most of the pomaceous fruits the flesh entirely surrounds the carpels, but in the medlar the carpels are exposed at the top. Although the word "pome" is restricted to this type of fruit, "pomology" retains its original Latin significance and refers to the whole subject and practice of fruit growing.

Apple

The apple (*Pyrus Malus*) occupies the first place among fruits of temperate regions in importance and extent of cultivation. It is a native of Eastern Europe and Western Asia and has been grown for over 3000 years. Apple seeds have been found in the remains of the Lake Dwellers. Twenty-two kinds were known to the Romans, and today there are over

6500 horticultural forms. This great number is due in part to the readiness with which apples hybridize and to their great variability.

The apple is a low round-crowned tree rarely exceeding 40 ft. in height. It may reach an age of 100 years. The wood is hard and dense and is used for tool handles and firewood. The attractive pink and white flowers and the leaves are borne together, usually at the ends of short twigs, known as spurs. Apples are adapted to many different soils and climates. The best yield is obtained where the soil contains a slight amount of lime.

FIG. 194. A fruiting branch of russet apples. (*Courtesy of the Massachusetts Horticultural Society.*)

Although apples are hardy and can be grown as far north as 65°N.L., they are subject to frost injury. They are not grown in the tropics. North America, Western Europe, Australia, New Zealand, and South Africa are the chief apple-producing regions.

The apple was brought to America by the early colonists as soon as it was found that the native species were of little value. By 1750 there were many well-established orchards. At the present time apples are grown from Nova Scotia to Georgia and west to the Great Plains, and from the Rocky Mountains westward, except in the Southwest. The most important regions are the Northeastern, Central Atlantic, and Pacific Coast states. Apples can be grown from seed, but they are usually propagated by grafting or budding. Summer, fall, and winter varieties (Fig. 194) are cultivated. Washington, New York, Virginia, Pennsylvania, California, Michigan, Ohio, and West Virginia lead in production in the order named. Canada is also a large producer. Apples are picked when fully ripe in order to allow ample time for all the chemical changes that take place during ripening. These involve an increase in the amount of sugar and a corresponding decrease in starch and acidity.

Until the end of the last century apples were practically the only fresh fruit available from November to June. This was due to their exceptional keeping qualities. Apples are dried in large quantities, as well as eaten raw and cooked. A considerable amount is canned, usually as applesauce. The juice is converted into cider and vinegar. The sugar in fresh apple juice is readily changed into alcohol by the action of wild yeasts. When the alcoholic content is at its maximum, hard cider is the result. Later acetic acid bacteria convert the alcohol into acetic acid or vinegar. Applejack is an alcoholic beverage made from cider. Other by-products include apple concentrate, apple powder, apple pumice, and apple syrup, the last used in bread, cigarettes, and smoking tobacco to maintain the proper moisture content.

Crab apples produce a small yellowish or reddish fruit about 1 in. in diameter. Although there are several native American species, their fruits are of little value, and they are rarely grown, except for their attractive flowers. The crab apples of cultivation are usually hybrids between the common apple and the Siberian crab apple (*Pyrus baccata*). Many oriental species have been introduced for ornamental purposes and are grown for their showy flowers.

Pear

The **common pear** (*Pyrus communis*) is likewise a native of Eurasia, and it was known to all the ancient peoples. It resembles the apple in general appearance, although it is taller and more upright. The flowers are usually white and appear with the leaves. The characteristic pyriform fruit (Fig. 195) has a persistent calyx. The pear is sweeter and more juicy than the apple, and the flesh contains numerous grit or stone cells, a specialized type of cell with exceedingly thick walls. Pears are less hardy and have a more restricted range. They prefer heavy soils with considerable humus and good drainage and regions with a fairly equable climate, such as occur near large bodies of water. They are propagated from seed and by grafting. Pears are extensively grown in Europe, where over 5000 kinds are recognized. France is the leading country. The United States produces about 23 per cent of the world output. Argentina, South Africa, Australia, and New Zealand are other

important growers. Pears are picked by hand before they are fully ripe
The European pear is the source of most of our eating pears.

The **Chinese** or **sand pear** (*Pyrus pyrifolia* var. *culta*), the source of many
of the varieties grown in this country for cooking purposes and storage,
is a native of China. The flowers appear just before the leaves. The
large fruit has a deciduous calyx and a very gritty, hard flesh with excel-

FIG. 195. A fruiting branch of the pear (*Pyrus communis*). (*Courtesy of Breck and Company.*)

lent keeping qualities. This pear is frequently used in hybrids and
grafting with the common pear.

In 1947 the United States produced 35,312,000 bu. of pears. California, Washington, Oregon, and New York are the leading states.
There is a large export trade with Europe. Pears are used for table fruits
and great quantities are canned. A beverage, perry, similar to cider is
made from the juice.

Medlar

The medlar (*Mespilus germanica*) is a small tree with spreading
branches borne at right angles. It is a native of Europe and several
varieties are cultivated, chiefly in England. The brown apple-shaped
fruit has a harsh flesh and rather acid taste. It is used for jams and
jellies, and somewhat as a table fruit.

Quince

The quince (*Cydonia oblonga*) has also been cultivated from very ancient times. It was much esteemed by the Romans. It is a native of Western Asia from Iran to Turkestan, and still grows wild. In spite of its long cultivation it has been but little changed from the wild form. It is a small tree from 15 to 20 ft. in height with many crooked branches. Under cultivation it is sometimes a large bush. The fruit is round or pear-shaped and quite large. The leaves are densely tomentose beneath, and even the fruit is wooly when young. The golden-yellow flesh is hard and rather unpalatable. The seeds have a mucilaginous covering and are of value in medicine. The fruit is rarely eaten raw, but is used for jelly and marmalade, often combined with apples and pears. It is also canned. The chief producing states are California, Ohio, Pennsylvania, and New York.

STONE FRUITS

The stone fruits or drupes are fleshy fruits with a single seed enclosed in the hard inner portion of the ripened ovary wall. There are three regions in the fruit: the outer skin or epicarp; the fleshy edible mesocarp; and the stone or endocarp, which contains the seed. Most of the drupaceous fruits of temperate regions belong to the genus *Prunus*. They are trees and shrubs, which often exude a natural gum. The leaves, bark, and seeds contain a glucoside, amygdalin, which is readily converted to prussic acid and may cause poisoning. The stone fruits and pome fruits are often classed for horticultural purposes as orchard fruits.

Apricot

The apricot (*Prunus Armeniaca*) is a native of Asia, where it still grows wild over a wide area. It was cultivated in China as early as 2000 B.C., and soon reached India, Egypt, Persia, and Armenia. It was introduced into Europe in the first century. The apricot is a small tree 20 to 30 ft. in height with pink flowers produced before the leaves. The fruit, which is peachlike in color and shape, is velvety when young and has a yellowish-orange flesh. The stone is smooth and flattened. The apricot is susceptible to frost and is grown only in warm temperate regions, chiefly in China, Japan, and Northern Africa. In the United States over 90 per cent of the crop is grown in California and the balance in Washington and Utah. In 1946, 306,000 short tons were produced. Apricots are used as a table fruit in the regions where they are grown. They are also dried, frozen, canned, candied, and made into a paste. A substitute for almond oil is expressed from the seeds.

Cherry

The cherries are trees with a birchlike bark, white or pinkish flowers produced in clusters, and small, smooth, long-stemmed fruits with a round smooth stone. They are natives of Eurasia and were cultivated long before the Christian era. They have been grown in the United States since colonial times. The over 1100 varieties which have been in cultivation are referred to two distinct species. The fruits of the native American cherries are of no commercial value.

The **sweet cherry** (*Prunus avium*) is a tall long-lived tree with yellow or greenish fruit. It has a rather restricted range in this country and is grown only in regions with an equable climate, such as New York, the Lake States, and the Pacific Coast. There are some 600 varieties in cultivation. It is used chiefly for fresh fruit.

The **sour cherry** (*Prunus Cerasus*) is a smaller tree, with a heavy wood and red fruits. It is grown in the Atlantic States and westward to the Mississippi valley. New York, Wisconsin, and Michigan lead in the production of sour cherries, which are used chiefly for canning purposes and freezing. Three hundred varieties are grown.

Cherries are used as table fruits, in pies, for glacé fruits, and in canning. For the last purpose they are often bleached in fumes of sulphur and treated with brine and sodium sulphite to harden the flesh. Cherry brandy or kirschwasser and maraschino are distilled from cherry juice. Maraschino cherries are grown in Dalmatia. The juice is also used for syrup, cherry cider, and jelly. A fixed oil can be obtained from fresh seeds.

Cherries are widely grown in temperate regions, and are especially important in Germany, Italy, and other Southern European countries. The United States in 1947 produced 172,140 small tons. New York, Michigan, California, Washington, and Oregon are the leading states.

Several species of Japanese flowering cherries, chiefly *Prunus serrulata*, are cultivated for ornamental purposes.

Peach

The peach (*Prunus Persica*) is the second most important fruit in the United States. The tree is a native of China, where it has been grown for thousands of years. Numerous varieties have been developed there, and many legends have sprung up in regard to the fruit. The peach reached the Mediterranean region very early in history and the Romans knew at least six kinds. It was brought to the United States by the early colonists. Today it is cultivated in most temperate countries, particularly in Southern Europe, the United States, South Africa, Japan, and Australia.

Between 2000 and 3000 varieties are grown, few of which reach the markets.

The peach is a low tree, rather short-lived and susceptible to frost injury and low temperatures. The attractive pink flowers are produced before the leaves. The round fruits (Fig. 196) have a velvety skin and a compressed, pitted or furrowed stone. The peach does best in a sandy soil. The commercial orchards are usually near large bodies of water. In 1947, 82,270,000 bu. were grown, with California, South Carolina, Georgia, Michigan, and Washington leading in production.

FIG. 196. Leaves and fruit of the peach (*Prunus Persica*). (*Courtesy of Breck and Company.*)

Peaches are a favorite table fruit. However, they are very perishable and so are hard to transport and store. They are the most popular fruit for canning. A considerable quantity is also dried or frozen. Both fixed and volatile oils, akin to almond oils, are obtained from the seeds. During the First and Second World Wars peach stones were used as a source of charcoal for gas masks.

Nectarines (var. *nectarina*) are closely allied to the peach. They have a smooth skin and are smaller, and are grown chiefly in California and Texas.

Plums and Prunes

Plums are shrubs, or small trees, with white flowers and large, smooth, clustered fruits with a bloom. The smooth stones are flattened. The commercial plums of the United States are derived from three main sources: the European plums, native American species, and Japanese species. These three types have such diverse climatic requirements that plums of some sort can be grown all over the United States and well into Canada, except in the colder regions. The Pacific Coast, however, has 80 per cent of the plum trees.

European Species. The European plum (*Prunus domestica*) is a native of Eurasia and still grows wild in that region. It has been cultivated for

FIG. 197. The beach plum (*Prunus maritima*) in fruit. (*Courtesy of the Massachusetts Horticultural Society.*)

more than 2000 years, and was known to the Lake Dwellers and the Greeks and Romans. Today it is the best known and most widespread of all the plums. It was brought to America by the colonists, and is now grown on the Atlantic and Pacific coasts and in the Lake States. It is a large tree, 30 to 40 ft. in height, with variously colored fruits. Over 900 varieties are cultivated, particularly in Europe. These include the green gages, egg plums, and prunes. *Prunus insititia*, a smaller and more hardy plum, has also been grown for over 2000 years and occurs wild in Europe and Asia. It includes the damson and bullace plums. Another variety, the sloe (*P. spinosa*) is much used in Europe in making liqueurs.

American Species. The various American plums have been derived from native species in recent times. They are hardy and are grown in the Mississippi valley and in the South, where the European types do not thrive. Several species have been domesticated, chiefly *Prunus americana*, with some 260 forms, *P. hortulana*, and *P. nigra*. The fruits of the last two species are small and not very palatable, and so are used chiefly in preserves and marmalades.

FIG. 198. A branch of California prunes (*Prunus domestica*). (*Courtesy of Ginn and Company.*)

The beach plum (*Prunus maritima*) (Fig. 197), which grows in sandy soil from southern Maine to Virginia, is used locally for preserves and may eventually become of commercial importance, as it is attracting the attention of plant breeders.

Japanese Species. *Prunus salicina* and other Japanese species were introduced into California at the end of the last century. Nearly 100 new varieties and hybrids have been developed by Burbank and others. These plums are not much used but they have a wider range of cultivation than even the European types.

Plums are used for fresh fruit, cooking, canning, and jams. They are picked when mature, but not quite ripe, if they are to be used as table fruit. For use in canning and jams they are allowed to ripen longer, and for prunes they must be fully ripe. The 1947 production of plums was 78,000 tons, 74,000 of which were grown in California, the balance in Michigan.

Prunes. Prunes are plums with a high sugar content, and can be cured without removing the stone. Large fruits (Fig. 198) of European varieties are picked carefully and the skin is ruptured. They are then dried, either in the sun or by artificial heat, after which they are allowed to "sweat" for a few weeks, and are finally graded and "glossed." This process, which consists of heating in steam or salted boiling water, glycerin, or fruit juice, gives a glossy appearance to the surface and also sterilizes the skin. The prune industry is a very important one on the Pacific Coast. The 1947 production of prunes was 594,500 tons. California alone has 6,500,000 trees and produces about 500,000 tons annually. Europe also produces prunes in large amounts.

GOURD FRUITS

The gourd fruits are trailing tendril-bearing herbs, often of great size. The fruit is a pepo, a modified berry with a hard and firm rind. They include several edible forms, such as the squashes, pumpkins, and cucumbers, which have already been discussed under fruit vegetables; the melons and watermelons; and the ornamental gourds.

Melons

The melon (*Cucumis Melo*) probably originated as a wild plant in Southern Asia. It is very old and was known to the Egyptians and Romans. The melon reached Europe in the seventeenth century. It is now cultivated in most warm temperate countries. Several kinds of melons are grown.

The netted or nutmeg melon is the type grown to the greatest extent in the United States. It is also called muskmelon and, erroneously, cantaloupe. This melon has a soft rind and netted markings on the surface. There are many varieties, one of which is the Rocky Ford, a melon with delicious taste and odor, which was first grown under irrigation in Colorado. The true cantaloupe is a European melon that is not grown in America. It has a hard warty rind and dark yellow flesh. The winter melons, such as the casaba and honeydew, are larger, smoother, and more spherical types. They require a longer season for ripening, but have good keeping qualities.

Melons as a rule require a fertile soil and long growing season, with a high temperature and plenty of moisture and sunlight. Muskmelons are nearly ripened on the vines, as this increases the sweetness and flavor. The winter melons are ripened in storage. The leading melon-producing states are California, Arizona, and Colorado. In the cooler parts of the country melons are grown under glass.

Watermelon

The watermelon (*Citrullus vulgaris*) is a native of tropical Africa, where it has long been used by the wild tribes. It has been cultivated for centuries and reached India and Egypt very early in history, as evidenced by its having a Sanskrit name and figuring in Egyptian paintings. It is still important in these countries, and also in Southern Europe and Central and Southern United States. Georgia, Texas, and Florida have over 50 per cent of the acreage devoted to this crop. The watermelon is an annual plant with extensive vines, which may cover a whole field, and large fruits that may weight 50 lb. or more. The reddish or pink pulp is very sweet and juicy, with white or black seeds. The varieties differ in the shape of the fruit, its color, and the thickness of the rind. The plant requires a fertile sandy soil with abundant sunshine. The fruits are picked when fully ripe and stand shipment well. In 1947 the United States grew 81,063,000 watermelons.

A variety (var. *citroides*) with a white, more solid flesh is called the "**citron**" or preserving melon. It is used in jams, jellies, and preserves, and, because of its high pectin content, is added to fruit juices that do not jell readily. This plant must not be confused with the true citron, which is one of the citrus fruits.

GRAPE

Although technically a berry, the grape is so different in appearance and of so much more importance than the other berries that it warrants special consideration. Grapes grow wild in many temperate portions of Europe, Asia, Africa, and America. They must have been widely dispersed by birds even before the existence of man. The cultivated grapes of the present day have been derived from European and American species. Grapes are grown in home gardens in every state and commercially in 37 states. California, however, is responsible for 90 per cent of the output.

The **European** or **wine grape** (*Vitis vinifera*), the "vine" of the ancients, is one of the oldest of cultivated plants. It probably originated in the Caspian Sea region of Western Asia. Grapes are frequently mentioned in the Bible. They have been grown in Egypt for 6000 years, and were highly developed by the Greeks and Romans. They spread all over Europe with the Roman civilization, and now are found in all temperate regions. The grape is a woody, climbing, tendril-bearing vine with large palmate leaves; small, insignificant, sweet-smelling flowers; and large clusters of fruits. The European grape has ellipsoidal fruits with a rather solid flesh, a high sugar content, and a relatively thin skin that does not slip off the flesh readily. In nature the vines grow rapidly and reach a considerable length, but in cultivation they are pruned back until they are short stout stumps, 3 or 4 ft. in length. Grapes require a loose, well-drained soil, and hillsides are often utilized. They are usually propagated by cuttings. The European grape is the source of most of the wine grapes. It is common all over Europe, especially in the Mediterranean region. This species is very susceptible to various fungus and insect pests, particularly the root louse (*Phylloxera*), which at one time threatened the very existence of the grape industry. Fortunately the American grapes are not so susceptible and they are now used as stocks on to which the European varieties are grafted. *Vitis vinifera* was introduced into the United States as early as 1616 by Lord Baltimore but it did not thrive. In spite of many attempts it has never been grown successfully in the eastern part of the United States, owing probably to its susceptibility to cold and pests. The story is different west of the Rocky Mountains and the growing of the European grape has become one of the chief industries, particularly in California. This grape is used for wine, for raisins, for cultivation under glass, and as a table grape, for it has good keeping qualities.

There are several **native American grapes** that have been domesticated, and numerous horticultural varieties of these are grown in the Eastern states. Hybrids between these native species and the wine grape also occur. The northern fox grape (*Vitis Labrusca*) of the eastern United States has given rise to the greatest number, including such well-known types as the Concord, Catawba, Delaware, and Niagara, which are grown for the most part in the Great Lakes region; the muscadine grape (*V. rotundifolia*) has given rise to the scuppernong, a long-lived, vigorous variety extensively grown in the Atlantic and Gulf States. *V. vulpina* and *V. aestivalis* are also in cultivation. The American grapes (Fig. 199) are larger and more hardy than the European. The fruit is round with a more watery flesh and a thin skin that slips off very easily. They are used for eating and for making grape juice, jams, and jellies and to some extent for wine. Wild grapes are harvested extensively and by many people are preferred to the cultivated forms as sources of jelly.

Grapes are the source of raisins and the dried currants of commerce. **Raisins** are dried grapes, prepared from wine grapes with a high sugar content and firm flesh. Both seeded and seedless varieties, like the Sultana, are marketed. The best quality is used for table raisins, which are merely dried in the sun. Cooking raisins are prepared from poorer grades and treated with lye and sulphur before drying. In the United States all the raisins are produced in California.

Currants are small dried grapes prepared from a variety that grows in Greece. It is a very old type, dating back as far as A.D. 75. Currant growing has always been an important industry in Greece.

In recent years most of the waste products of the grape industry have been utilized in some way. Fertilizers, stock feed, acetic acid, cream of tartar, a fixed oil, and tannin are all prepared from one part or another of

FIG. 199. Clusters of Sheridan grapes. This variety may be a hybrid between *Vitis Labrusca* and *V. vinifera*. (*Courtesy of the Connecticut Agricultural Experiment Station in New Haven.*)

the fruit. Grapes are sometimes grown for ornamental purposes, and the wood has some value.

The chief grape-growing countries are Europe, the United States, Argentina, Chile, Australia, and South Africa. The United States produced 3,036,400 short tons in 1947, with California responsible for 2,836,000 short tons. New York and Michigan are also important states.

BERRIES

The term "berry," as in the case of "nut," "fruit," and "vegetable," is used in several senses. Technically a berry is a thin-skinned one-celled fleshy fruit with seeds scattered through the flesh. According to this definition the tomato, grape, eggplant, and many other fruits are berries, while such fruits as the strawberry, raspberry, and blackberry are not berries at all, but rather aggregate compound fruits. For the purposes of the present discussion the term will be used to include the common bush fruits, or berries, of cultivation and the mulberry as well. In nearly all these fruits wild plants serve as an important source of the crop today, just as they have from time immemorial, and the cultivated forms have been domesticated in comparatively recent years. It will obviously be impossible to list all the species involved in the several cases.

Blackberries and Raspberries

The blackberries and raspberries belong to the genus *Rubus*, which includes many species and a vast number of hybrids. Comparatively few of these so-called brambles are of economic importance.

The **blackberries** are erect, decumbent, or creeping shrubs, usually armed with thorns and prickles. The erect "canes" die down to the ground every few years and are renewed from the rootstalks. The

FIG. 200. Red raspberries ready for market. (*Courtesy of the University of Massachusetts.*)

velvety black fruits are aggregate fruits, consisting of numerous ovaries of the flower ripened into small drupelets. When picked the fruit does not separate from the somewhat fleshy receptacle. Although blackberries can be grown anywhere except in regions with severe winters or extreme heat or drought, they are nowhere of great importance. The blackberry is almost entirely an American fruit. The cultivated forms have been derived chiefly from *Rubus alleghaniensis*, *R. argutus*, and *R. frondosus*. Trailing species are known as dewberries, and these include *R. flagellaris*, *R. trivialis*, and *R. vitifolius*. Blackberries are used fresh and for jams, cordials, preserves, and canning.

The **loganberry** (*Rubus loganobaccus*), which originated in California, has very large fruits, but is rather poor in flavor. It is grown for canning purposes, both the fruit and the juice being used. The loganberry is usually considered to be a hybrid between a blackberry and a raspberry, but it may be a distinct species, or a variety of *R. ursinus*.

The **raspberries** are smaller shrubs, usually with a vigorous, erect, bushy habit. They have small prickles or bristles. The aggregate fruit separates from the receptacle when ripe, thus leaving a cavity on one side (Fig. 200). The black raspberries are derived from *Rubus occidentalis* of eastern North America, while the red raspberries come from *R. idaeus* of Europe, or its American var. *strigosus*. The European species has been cultivated from great antiquity and was highly esteemed by the Greeks and Romans. The raspberry is one of the hardiest of fruits and can be grown as far north as Alaska and northern Canada. It is used fresh or cooked, and is utilized for jams, jellies, vinegar, and as a flavoring material. Large quantities are canned and frozen.

Blueberries and Huckleberries

Wild plants are still the source of most of our supply of blueberries and huckleberries. These plants are low ericaceous shrubs, common on acid

FIG. 201. Fruiting spray of a high-bush blueberry (*Vaccinium corymbosum*). (*Photo by F. M. Dearborn.*)

soil throughout eastern North America. In the huckleberry the fruit is a berrylike drupe, while in the blueberry it is a true berry. The former occurs only in the wild state, with *Gaylussacia baccata* the chief species. Blueberries are now being cultivated in many areas on sandy or clayey acid soil and show a remarkable increase in size and yield over the wild fruit. In the so-called blueberry barrens of Maine and adjacent New Brunswick the plants are so abundant in the sterile acid soil that they can be subjected to a crude sort of cultivation. The yield is kept up by frequently burning over the area, and, if other shrubs are kept out, an almost pure stand of blueberries results. Often the berries are so numerous that they can be harvested with a cranberry rake. The industry is of increasing importance in Maine and in many places even forested areas are being cut over and converted into blueberry land. Blueberries are eaten fresh or cooked, chiefly in pies, and large quantities are canned and frozen. The principal eastern low-bush species are *Vaccinium myrtilloides*, *V. angustifolium*, and *V. vacillans;* the high-bush species of most importance include *V. atrococcum* and *V. corymbosum* (Fig. 201). Valuable western species include *V. ovatum* and *V. membranaceum*.

Cranberries

Cranberries are low trailing woody plants, characteristic of bogs and wet acid soil throughout Northeastern North America and Northern Europe. The fruit is a true berry. The American cranberry (*Vaccinium macrocarpon*) has been an important cultivated plant since 1840. It is grown in acid sandy or peaty bogs, which can be flooded during the winter and spring. The berries are picked with a scoop, the cranberry rake, and foreign material is removed by winnowing. Cranberry growing is an important and highly specialized industry in Massachusetts, with the Cape Cod area producing 70 per cent of the crop. Much of it is canned as sauce or made into a beverage. Cranberries are also grown in Wisconsin, New Jersey, and to a lesser extent in other states. Wild plants of this species, and the smaller *V. Oxycoccus*, furnish some fruit for home consumption. The smaller highland or mountain cranberry (*V. Vitis-Idaea*) with a firmer, more spicy fruit is abundant in Northern Europe and is highly esteemed in Scandinavian countries; the American form (var. *minus*), a boreal and arctic-alpine species, is equally desirable.

Currants and Gooseberries

Currants and gooseberries are usually classified in the genus *Ribes*, although the gooseberries are sometimes segregated in *Grossularia* by some authorities. They are low bushy plants, very hardy, and well adapted to cold climates. The currants are usually smooth with the flowers and fruits in racemes (Fig. 202). In the gooseberries the stem is usually armed with spines or prickles, and the flowers and fruits are solitary.

Currants are native to both the New and the Old Worlds. The common red and white currants (*Ribes sativum*) are natives of Eurasia. They have been grown in Europe since the Middle Ages, and were early brought to America, where they have become naturalized in many places. Several varieties are grown, chiefly for domestic use, and the plants are likely to suffer from neglect for ordinarily they are allowed to take care of themselves. They attain their best development in cool humid regions. The

European black currant (*R. nigrum*), also a native of Eurasia, is but little grown in the United States. There are several wild species in America with edible fruit, the most important of which is *R. americanum*. Currants are too acid to make a good table fruit, but are much used for jams, jellies, sauces, pies, and wine.

FIG. 202. Leaves and fruit of the red currant (*Ribes sativum*). (*Courtesy of the University of Massachusetts.*)

The **European gooseberry** (*Ribes Grossularia*), another species of Eurasian origin, is grown in the cooler parts of both Europe and America. The tart round fruits may be red, yellow, green, or white, and hairy or smooth, depending on the variety. Gooseberries were formerly much more important than they are today. The most important native American species is *R. hirtellum*.

Mulberry

The mulberry is not a berry in any sense of the word. It is a multiple accessory fruit derived from a whole inflorescence. The actual fruits, little achenes or nutlets, are surrounded by the fleshy sepals and grouped together with the fleshy axis to form the so-called syncarp. The mulberries are trees native to both Asia and America. The fruits are very juicy and are much eaten in some parts of the world. They also afford valuable food for animals, particularly hogs and poultry.

The black mulberry (*Morus nigra*), an ornamental tree 40 to 60 ft. in height, is a native of Asia Minor and Persia. It has been cultivated for a long time. Mulberries are frequently mentioned in the Bible, and the tree was familiar to the Greeks and Romans. It reached Europe in the twelfth century, and is now naturalized in both Europe and America, where it is planted chiefly in the Southern states. The fruit is black or dark red, and is used as a dessert fruit.

The red mulberry (*Morus rubra*), the largest of the genus, is a native of eastern North America. The bright red or blackish fruits are not much eaten, except by animals. The wood, however, is valuable.

The white mulberry (*Morus alba*), with white or pinkish fruits, is a small tree. It is much less hardy than the other species. It is a native of Asia and was introduced into both Europe and America for its leaves, which serve as food for the silkworm. The fruits are of but little value. In Europe the wood is used and a yellow dye is obtained from the roots.

Strawberries

The strawberry is one of the most important of the small fruits and is a favorite in all temperate countries. Its only drawback is its perishable nature. The fruit is not a berry, but an aggregate accessory fruit, consisting of a number of small dry achenes embedded on the surface of a large fleshy receptacle. The strawberry (Fig. 203) is a low perennial herb with a very short thick stem and trifoliate leaves. It produces numerous runners, which root at the tip and are used to propagate the plant. Strawberries have been grown in Europe since the fourteenth century, and in America since colonial days. They need only good soil, a temperate climate, and lots of sunshine, and so are widely grown. In the United States their cultivation has been of commercial importance since 1860. Harvesting begins in the South in the winter and progresses northward with the advancing season until summer finds them fruiting in the Northern states. There are three main sources of cultivated strawberries, of which there are hundreds of varieties. The native plant of Eastern North America, *Fragaria virginiana*, was grown by the early settlers and taken by them to England, where it has been cultivated since the seventeenth century. The European *F. vesca* is the source of the everbearing types, grown both here and abroad, and often escaping from cultivation. The majority of the cultivated forms, however, are derived from *F. chiloensis*, a native of Western America from Chile northward to the

FRUITS OF TEMPERATE REGIONS 405

FIG. 203. Flowers and fruit of a cultivated strawberry. (*Courtesy of the Massachusetts Horticultural Society.*)

mountains of Mexico. This species is less hardy than the others. Strawberries are a dessert fruit primarily, but are also canned, frozen, and used in jams and preserves and as a flavoring material. The United States produced 8,895,000 crates in 1947.

CHAPTER XIX

TROPICAL FRUITS

The number of edible tropical fruits is legion. Thousands of them are used daily by the native peoples, and most of these are unknown to the white man. There are 250 edible fruits in the Philippine Islands alone. Sooner or later the people of the United States as well as of other temperate regions will have to turn to a greater extent to the tropics to supplement their own food resources, and fruits are among the best products that the tropics have to offer. Improved methods of transportation will make it possible to use fruits actually grown in the tropics, while Florida and California have a climate favorable to the cultivation of many tropical species. Several tropical fruits have already been exploited. The banana and pineapple are now as familiar as the apple and pear, and citrus fruits are known the world over. In comparison with the fruits of temperate regions, tropical fruits have been much neglected horticulturally and few improvements have been made over the wild product. This condition is now being remedied and with the use of scientific methods of fruit growing the products of the tropics should be all the more valuable. While edible fruits occur in a vast number of families, they are particularly important in the *Anacardiaceae*, *Annonaceae*, *Myrtaceae*, *Rutaceae*, *Sapotaceae*, and *Sapindaceae*. Of these the *Rutaceae* is the best known and most important, for it is the source of the citrus fruits.

CITRUS FRUITS

The citrus fruits were domesticated from wild ancestors in Eastern and Southern Asia in very early time. Some of them have been cultivated for over 3000 years. They were sometimes grown for other purposes than eating. Citron, for example, was planted in the famous Hanging Gardens of Babylon for use in toilet water and pomades. These fruits were early introduced into the Mediterranean region, where they have always been an important crop. As many as 100 species of *Citrus* have been described, many of which are probably of hybrid origin. Only a few, however, are of commercial importance.

The citrus fruits are thorny aromatic shrubs or small trees. The leathery evergreen leaves are glandular dotted and, although they appear to be simple, are actually unifoliate-compound leaves with a joint between the leaf blade and stalk. The white or purplish flowers are solitary, but produced in great profusion, and often very fragrant. The fruit is a modified berry known as a hesperidium. This type of fruit has a thick leathery rind with numerous oil glands. The flesh is very juicy with many juice sacs. A peculiar feature of these plants is the fact that they do not develop root hairs, and are dependent on mycorrhiza—fungi, which are closely associated with the roots—for the absorption of liquids.

Cultivation of the citrus fruits is on a high scale. They are usually grown at sea level where sufficient moisture is readily available. Any well-drained soil, except an extremely sandy one, can be utilized. The various species differ in their resistance to cold, but in general a temperature ranging from 24° to 120°F. is best. They are propagated chiefly by budding. The species hybridize very readily both in the wild state and in cultivation, and there is a great tendency to form "sports."

The citrus fruits are grown around the world. Although they are tropical plants, most of the commercial groves are in subtropical regions. The fruits ripen throughout the year. Oranges and grapefruit are allowed to ripen on the trees, while lemons and limes are picked green.

The United States leads the world in the production of citrus fruits. Florida and California are the principal states, while Texas, Arizona, Alabama, Louisiana, Mississippi, and Georgia grow a small amount. The Mediterranean countries are second, with Spain, Portugal, Italy, and Palestine the most important. The West Indies, where the trees grow in great luxuriance, and Central America, are of increasing importance. Brazil, Argentina, South Africa, Australia, China, and Japan are also large producers.

Although the citrus-fruit industry has been of commercial importance for only a little over 50 years, it nevertheless has been at times exceedingly profitable. Oranges and grapefruit alone have brought a return of nearly $400,000,000, while the industry as a whole has been a sesquibillion one. The outlook for the future is also very bright, for the European market, particularly in the case of canned products, has scarcely been touched.

Oranges and grapefruit are highly esteemed as table fruits, and, apart from their palatable nature, constitute a valuable addition to our diet. Like all the citrus fruits they contain considerable amounts of the essential vitamin C, the antiscorbutic vitamin, as well as fruit acids. They are also used for marmalade and various confections. Canned products, particularly in the case of the grapefruit, have been developed in recent years and are of increasing importance. Both the flesh and the juice are preserved in this manner. Often the juice of citrus fruits is more important than the fruit itself and it is extensively used in both alcoholic and nonalcoholic beverages. Frozen juice has recently become an outstanding product. The rind of nearly all these fruits yields a valuable essential oil. The dried waste pulp is an excellent cattle feed. Other by-products include citric acid, pectin, and various glucosides.

Oranges

Sweet Orange. The sweet orange (*Citrus sinensis*) is a native of Southeastern Asia, probably China or Cochin China. It was first cultivated between 1500 and 1000 B.C. It reached India very early in history, and was carried to Europe by Genoese traders early in the fifteenth century. The Spaniards were responsible for its introduction into the New World. It reached Florida in 1565 and California in 1769.

The sweet orange is a small evergreen tree with slender blunt spines, growing upward to 20 ft. in height when in cultivation. The leaves have narrow-winged petioles and the flowers are white and very fragrant. The fruit is nearly round, with an abundant, sweet, solid pulp and spindle-shaped juice sacs. Seeds may be present or absent. This orange is the hardiest member of the genus and can be grown in any warm dry climate where the soil is fertile and well irrigated. At first it was grown from seed, but now the plants are budded or grafted. Several types of sweet oranges have been developed: Spanish oranges, with large coarse-grained fruits; Mediterranean varieties, with fine-grained fruits; blood oranges, with a red pulp, or streaked red and white; and the navel oranges, which are seedless, and characterized by the navel at one end, formed by the protrusion of additional carpels produced inside the flesh. Oranges contain from 5 to 10 per cent sugar, 1 to 2 per cent citric acid, and vitamin C.

Orange growing in the United States began in Florida, in a small way, as early as 1809. The commercial industry dates from 1821. In 1894 the output was 4,000,000 oranges. Excessive frosts the following winter almost exterminated the crop and reduced the output in 1895 to 75,000 oranges. In spite of the danger of frost and the more recent ravages of the fruit fly, the industry has persisted and in 1947, 58,400,000 boxes were produced in that state. The Florida orange is medium sized, russet or light yellow in color, thin skinned, and very sweet. It is picked when the skin is green and the yellow color develops later. Gases or chemicals are often used to destroy the chlorophyll and bring out the yellow color more rapidly. California is at present the chief producer of oranges, with a 1947 output of 45,830,000 boxes. The cultivation of two different varieties makes it possible to harvest oranges the year round. The most important of these is the Washington Navel or Bahia orange. This seedless orange originated in Brazil, but in the favorable California climate (Fig. 204) has become the great commercial orange of the world. It is the largest variety, with a thick bright-orange skin, and bears during the winter months. In the summer the California growers utilize the smaller Valencia orange, a Spanish type with seeds, which is also grown in Florida. Texas, Arizona, and Louisiana also grow oranges. The total output in the United States in 1947 was 110,510,000 boxes. The leading foreign countries in sweet-orange production are Brazil, Spain, Italy, Palestine, and Mexico.

The chief use of sweet oranges is for fresh fruit. Once a luxury, they are now in reach of every purse and are widely used. Orange juice and orangeade are favorite beverages. The peel is candied, and oil of orange is extracted from the rind. This essential oil, which is expressed in

FIG. 204. A California orange grove. (*Courtesy of the California Fruit Growers Exchange and Ginn and Company.*)

Southern Europe and the West Indies, is used in the perfume and soap industries, in medicine, and for flavoring. Orange extract is made by dissolving a small amount of the oil in alcohol. Orange trees are sometimes grown for ornamental purposes.

Sour Orange. The sour orange (*Citrus Aurantium*), sometimes known as the bitter, bigarade, or Seville orange, is also a native of Southeastern Asia. Curiously enough, it was brought to Spain by the Arabs and cultivated there several thousand years before the sweet orange. It is a small tree, 20 to 30 ft. in height, with blunt spines. The petioles have broad wings. The flowers are exceedingly fragrant and are the source of the oil of neroli, used in perfumery. The large, globose, orange-red fruits are rough and have a very acid pulp. The flesh, however, is of good quality with small spindle-shaped juice sacs. A hollow core develops at the center when the fruit is ripe. The sour orange is grown in the United States for ornamental purposes and to be used as a stock in grafting, as it is too acid to be used as a fresh fruit. It is grown in Spain extensively, however, and the fruits are used for marmalade, orangeade, and candied orange peel. The essential oil obtained from the rind is used in perfumery, medicine, and in the manufacture of the liqueur curaçao.

The bergamot (*Citrus Aurantium* subsp. *Bergamia*) is a small spiny tree with golden-yellow pear-shaped fruit. The pulp is very acid and inedible. The plant is grown in the Mediterranean region as a source of the oil of bergamot, which has already been discussed under essential oils.

Mandarin Orange. The mandarin orange (*Citrus reticulata*) comprises the so-called glove oranges: the orange-yellow mandarins and the reddish-orange tangerines. It is a native of China and Cochin China. The small round fruits have an easily removable peel and segments that separate readily. They are extensively grown in Japan and also in Southern Europe and the Gulf States, chiefly Florida, Alabama, and Mississippi. The Satsuma orange is a small, very hardy type with a deep-orange pulp and few if any seeds.

The King orange, often designated as *Citrus nobilis*, is probably a hybrid between a sweet orange and a mandarin. It was introduced into the United States in 1882. It bears heavily, is frost resistant, and the sweet or slightly acid flesh with broad blunt juice sacs is very palatable. The fruit, however, has never been very popular.

Grapefruit

Few important fruits have a more doubtful origin or status. It is probable that the grapefruit or pomelo (*Citrus paradisi*) originated in the West Indies as a sport of the shaddock or possibly as a hybrid with the sweet orange. However, it may have been introduced from the Old World. Although many authorities consider it to be merely a variety or horticultural form of the shaddock, it seems better to consider it as a distinct species.

It is a vigorous tree, 20 to 40 ft. in height, with winged petioles. The round or pear-shaped, pale-yellow, smooth fruits are produced in clusters (Fig. 205). They are the largest of the edible citrus fruits, weighing from 2 to 12 lb. and with a diameter of 4 to 6 in. The skin is thin with many inconspicuous oil glands. The flesh is acid or subacid and mildly bitter, with large spindle-shaped juice sacs.

The grapefruit was brought to Florida from the West Indies in 1809 and was used as an ornamental tree until 1880. The production of grapefruit has increased enormously since 1905. In 1935 the United States produced 18,347,000 boxes, in 1947, 61,630,000. Of these 33,000,-000 boxes were grown in Florida, 23,000,000 in Texas, and the remainder in California and Arizona. Although grapefruit was first canned only in 1917, in recent years the pack has been as high as 5,500,000 cases of segments and 26,000,000 cases of juice. The waste rind, pulp, and seeds are fed to cattle. Puerto Rico also cans a considerable amount. Grape-

FIG. 205. A cluster of grapefruit (*Citrus paradisi*). (*Reproduced by permission from Smith and Walker, Geography of Texas, Ginn and Company.*)

fruit is grown in the West Indies, Palestine, and South Africa to some extent.

Lemon

The lemon (*Citrus Limon*) is probably a native of Southeastern Asia and has been grown there for ages. It must have reached India at an early date for there is a Sanskrit word for it. It has been grown in the Mediterranean region since the days of the Greeks and Romans, and has always been particularly well adapted to that area, growing everywhere in moist fertile soil. The lemon is a small tree, 10 to 20 ft. in height, with short spines and large white and purple flowers. The small, light-yellow, oval fruits end in a blunt point. The fruit is picked green, as it deteriorates if allowed to ripen on the tree. Regardless of the state of maturity, it is removed when about 2¼ in. in diameter. It is then cured, and colored and ripened in storage. Lemons contain 0.5 per cent sugar and 5 per cent citric acid. The juice is used for lemonade and other beverages and as a flavoring material, bleaching agent, and stain remover. Although widely grown, commercial production is restricted to warm regions, for it is not very hardy. In 1947, the world output was 29,091,-000 boxes. Of these, California was responsible for 12,870,000 and Italy for 8,137,000. Spain and Argentina also grow a considerable amount. Lemons were formerly grown in Florida, but the trees were too often injured by frost to be of commercial importance. The rind is the source of oil of lemon. The expression of the oil is an important industry in Sicily, where the sponge method is used. It takes from 800 to 1100 lemons to yield 1 lb. of the oil. Lemon oil is used in perfumery and soap and lemon extract, which is next to vanilla in importance as a flavoring substance. The extract is made by dissolving 5 parts of the oil in 95 parts of strong alcohol. The utilization of cull lemons has recently been developed in California with a large production of citric acid, lemon oil, and pectin.

Lime

The lime (*Citrus aurantifolia*) was domesticated in the East Indies. It is very susceptible to cold and is a distinctly tropical plant. It is a low straggling shrub or small tree with numerous very sharp spines and small white flowers. The fruits are small, from 1¼ to 2½ in. in diameter, and are greenish yellow in color. They are thin skinned, with an abundant acid pulp and oval pointed juice sacs. The lime is one of the sourest fruits on the market and is not suitable for eating. It is grown chiefly for the juice, which is often extracted and shipped in a raw or concentrated form. Lime juice is used in beverages, as a source of citric acid, and medicinally to prevent scurvy. Although long famous for the latter purpose, lime juice actually contains only one-quarter as much vitamin C as either oranges or grapefruit. Limes are grown to some extent throughout the tropics and are of commercial importance in Mexico, Egypt, the West Indies, and Florida, where 12,870 boxes were produced in 1947. Oil of lime is expressed from the rind.

Kumquat

The kumquats are the smallest of the citrus fruits. They are small evergreen shrubs with aromatic white flowers and golden-yellow fruits produced in clusters. The fruits are from 1 to 1¼ in. in diameter with a thick spicy rind, acid flesh, and small seeds. They are grown for ornamental purposes and for eating, either whole or preserved. They are usually placed in the genus *Fortunella*, a segregate of *Citrus*. Two oriental species are grown to some extent in the orange belt. These are *F. japonica* with globose fruits and *F. margarita* with oval fruits.

Citron

The true citron (*Citrus Medica*) is the oldest of the citrus fruits and the first to be known by Europeans, going back as far as the fourth century

FIG. 206. A citron tree (*Citrus Medica*) growing in Dominica, B.W.I. (*Photo by Walter H. Hodge.*)

B.C. It probably originated in Northern India and has long been cultivated in Southeastern Asia. It was described by Theophrastus from Babylon. The citron is a small thorny tree with attractive purple and white flowers and a fruit resembling a large lemon (Fig. 206). It is fragrant, greenish yellow in color, oblong in shape, and from 6 to 9 in. in length. The thick skin is tough and warty and the acid pulp is scanty. Commercial citron is the candied rind. It is prepared by treating the fruit with brine to remove the bitter oil. This also brings out the flavor and aroma and prevents decay. Then the rind is candied in a sugar and glucose solution. Citron is one of the best and most expensive of the condiments. About 6000 tons are produced annually, of which the United States uses half. Citron is cultivated chiefly in Corsica, Sicily, Greece, and the West Indies. The essential oil of cedrat used in perfumery is expressed from the rind.

Citrus Fruits of Minor Importance

The **shaddock** or pomelo (*Citrus grandis*) is a native of Malaya and Polynesia. Several kinds are grown in Southern Asia. The fruit is like a grapefruit, but much larger, growing to the size of a watermelon and weighing 10 to 20 lb. It is more pear-shaped, with larger juice sacs and a hollow core, and has a coarse thick rind and thick leathery septa. The reddish flesh is aromatic and spicy, but quite bitter. This species was introduced into the West Indies by a Captain Shaddock, for whom it has been named. The shaddock has given rise to the grapefruit.

The **deciduous orange** (*Poncirus trifoliata*) is a native of China and Japan. It is a low tree with large spines and trifoliate deciduous leaves. For this reason it has been segregated from *Citrus*. The white flowers are produced before the leaves. The rough hairy orange fruits have a bitter, gummy, inedible pulp. This species is very hardy and is used in hybrids and as a stock for grafting the other citrus fruits. It is cultivated as an ornamental plant as far north as New York.

In an attempt to develop edible citrus fruits that are hardier than oranges, many **hybrids** have been produced. Some of these, like the citrange, a hybrid between the trifoliate and the sweet orange, are grown to a considerable extent in the Southern states. Others include the tangelo (tangerine × grapefruit), limequat (kumquat × lime), orangequat (kumquat × orange), citrangequat (citrange × kumquat), tangor (orange × tangerine), tangerona, and orangelo.

OTHER TROPICAL FRUITS

Banana

The banana (*Musa paradisiaca* subsp. *sapientum*) is one of the most familiar and important of all tropical fruits. From its original home in the humid tropics of India or Malaya it has spread all over the tropical world, and today there is almost no warm region where it is not grown, except the Sahara Desert, which is too arid. It is likewise a very ancient plant, possibly the world's oldest cultivated crop. We know it was important in Assyria in 1100 B.C., and it was well known to all the other early civilizations. It reached Polynesia, also, at a very early date and was carried to the West Indies in A.D. 1500.

The banana is one of the tallest of the herbaceous plants. Its robust treelike stem is composed of the sheathing spiral leaf bases, which contain fibers of sufficient strength to make possible the erect habit. At the summit of the 10- to 30-ft. stem there is produced a crown of large oval deep-green leaves. These may be up to 12 ft. in length and 2 ft. in width, with a prominent midrib.

Each plant produces a single inflorescence. This consists of clustered flowers which are nearly surrounded by large, fleshy, reddish, spathelike scales, which drop off as the fruits mature. The flower stalk develops from the rootstalk and pushes its way up through the hollow stem,

FIG. 207. Transporting bunches of green bananas (*Musa paradisiaca* subsp. *sapientum*) from a plantation to a railroad. (*Courtesy of the United Fruit Company.*)

emerging in the center of the crown. It soon curves over owing to its own weight.

These drooping inflorescences develop into the familiar "bunches" of bananas. Marketable bunches (Fig. 207) weigh from 80 to 140 lb., and consist of from six to fifteen clusters, known as "hands" or "combs." Normally each hand contains from 10 to 20 individual bananas, or "fingers." Bunches with as many as 22 hands and 300 individual bananas have been produced, but these are unusual. The fruit of the cultivated banana is a modified berry and lacks seeds. Wild species occur, however, which produce normal seeds. As soon as the tree bears, it dies or is cut down, and suckers develop from the rhizome, which give rise to new plants (Fig. 208). A single clump may be productive for several years. Bananas are rapid growers and have a very high yield.

This varies with the locality and may be as low as 125 or as high as 300 to 400 bunches to the acre. It has been estimated that a unit area of land produces 33 lb. of wheat, 98 lb. of potatoes, and 4400 lb. of bananas. The commercial production of bananas for export is an important business. Jamaica, Mexico, and Central America supply 85 per cent of the output.

FIG. 208. A plantation of bananas (*Musa paradisiaca* subsp. *sapientum*) in Costa Rica, showing vegetative reproduction. The plants are nine months old. (*Courtesy of the United Fruit Company.*)

In the various tropics about 300 varieties of bananas are grown. Only a few of these reach the markets of the United States, the principal type being the Gros Michel. Tropical American markets have about 15 kinds, but the vast majority are grown in tropical Asia. There are other species of bananas as well, one of which, the dwarf banana (*Musa nana*), is occasionally seen in our markets. This species is especially important in Southern Asia, Africa, and the Pacific islands. Red bananas are quite common.

Bananas are picked and shipped green. When thoroughly ripe, as indicated by brown blotches on the yellow skin, they constitute one of the most healthful and nourishing foods. They have a high content of carbohydrates with some fats and proteins as well. In fact, their food value is three times that of wheat. Bananas are usually eaten raw, but may be cooked. Banana flour is made from dried green fruits. Banana "figs" are dried slices of ripe fruits. Banana wine has long been known.

The **plantain** (*Musa paradisiaca*), a close relative of the banana, is one of the great food plants of the tropics. A native of Southern Asia, it has furnished food for all tropical peoples for centuries. There are some 75 varieties, all so old that they have never been propagated by seed within recorded time. Plantains are always eaten cooked or made into flour. They are very digestible and are a valuable food for children and invalids.

Custard Apples

The term "custard apple" is rather loosely applied to several fruits of the American tropics belonging to the genus *Annona*. In all these the

Fig. 209. Fruits of the sugar apple (*Annona squamosa*), one shown in section. (*Photo by Walter H. Hodge.*)

fruit is a fleshy syncarp, formed by the fusion of numerous ripened ovaries and the receptacle.

The **cherimoya** (*Annona Cherimolia*) is highly esteemed as a dessert fruit It is very old and originated in the Andes of Peru and Ecuador. It is now grown in Central America, Mexico, the West Indies, Africa, and India. The cherimoya is a shrub or small tree with fragrant flowers. The light-green fruits are globular or conical and from 4 to 10 in. long. The luscious white or yellowish flesh is very aromatic and fragrant, with a soft and custardlike consistency.

The **sweetsop** or sugar apple (*Annona squamosa*) is a native of South America and the West Indies, and is now widely grown in the tropics of both hemispheres. The yellowish-green tuberculate fruit (Fig. 209) is about 2 or 3 in. in diameter. It has a white custardlike pulp and is the best of the group for eating.

The **soursop** or **guanabana** (*Annona muricata*) of the West Indies is a small slender tree with a large ovoid spiny fruit, deep green in color. The fruit may be as much as 12 in. in length and weigh as much as 8 to 10 lb. The white juicy flesh is very aromatic. The soursop is unrivaled for sherbets and drinks.

The true **custard apple** or bullock's-heart (*Annona reticulata*) is a common tropical fruit. The fruit itself is heart-shaped, 4 to 6 in. in length, and brownish or reddish in color. The soft white or cream-colored, sweetly aromatic pulp is slightly granular toward the rind and rather insipid and cloying. It is a native of the West Indies.

The family *Annonaceae* contains about 600 species, most of which have edible fruits, and affords great promise for future development.

Date

The date (*Phoenix dactylifera*) is one of the oldest of crops, going back at least 5000 years. It probably is a native of India or Arabia, but has long been domesticated throughout Southwestern Asia and Northern Africa. It was in Arabia before the dawn of history, was of great importance in Babylonia, and had reached Egypt long before the Christian era.

The date (Fig. 210) is a palm with a slender trunk, 70 to 100 ft. or more in length. It tends to produce offshoots from the base, and so is often found growing in clumps. It has a crown of stiff, pinnate, ascending and descending leaves 10 to 20 ft. in length. The numerous flowers, sometimes 10,000 to an inflorescence, are surrounded by a spathe. Male and female flowers are produced on different trees, and in cultivation 90 per cent of the male trees are removed. The fruit, a nearly round drupe, or one-seeded berry, is hard and green at first, but later turns yellow or red. The flesh is thick and very sweet, and is soft or dry and hard, depending on the variety. The date can grow with less water than any other crop, and so is of the utmost importance to desert peoples. It often serves as a staple food, as well as the main source of fruits and sugar. It is distinctly a plant of hot sunny climates with low humidity. In exceedingly arid areas some irrigation is necessary. A favorite habitat is the oasis. Dates are propagated by seeds or cuttings. They are very long-lived, often reaching an age of 200 years. Over 1000 varieties are grown. The fruits are ripened off the tree and are dried before shipping. Unlike most fruits they have a high food value, with 54 per cent sugar and 7 per cent protein as well as pectins and gums.

Dates are used as a table fruit and in jams, pastes, cooking, and alcoholic beverages. In desert regions it has been estimated that the articles that are made from the date palm and the uses to which it is put number more than 800. Every part of the plant is utilized, and the fruits even serve as money.

Iraq, the site of ancient Mesopotamia, produces 80 per cent of the dates of commerce and has 20,000,000 trees under cultivation. Arabia and Northern Africa are also large producers. Dates have been grown in California since the eighteenth century, but have been commercially

FIG. 210. A date palm (*Phoenix dactylifera*) in fruit. (*Courtesy of S. J. Record.*)

important only since 1890. As a result of scientific study we now know more about the date and its requirements than about any other tropical crop. In 1947 about 10,180 tons were produced in California and Arizona. Dates are also grown in Mexico.

Durian

The durian (*Durio zibethinus*) is one of the most interesting of all tropical fruits. It is a native of Malaya, and occurs locally in that region, and to some extent in Burma and the East Indies. It is rarely cultivated. It is a tall handsome tree, up to 80 ft. in height, with large conical spiny fruits, 6 to 8 in. in diameter. The leaves are densely covered with golden hairs on the underside. The flowers are yellow or creamy white. The custardlike flesh has an exquisite flavor and is at the same time aromatic and sweet with a strange balsamic taste. The odor, however, is extremely offensive and is a serious obstacle to the popularity of the fruit. Durians are much sought after by the native people, and by animals as well. It has been claimed that the fruit possesses great rejuvenating powers.

Fig

The fig (*Ficus Carica*) has been cultivated since earliest time. It originated in Southern Arabia and early spread to the Mediterranean region. It is frequently mentioned in the Bible. Theophrastus was familiar with many varieties and in his "Enquiry into Plants" gives a detailed account of fig cultivation. Today figs are grown in nearly all subtropical countries. The fig (Fig. 211) is a shrub, or small tree, with characteristically lobed leaves. The fruit is a syconium, a fleshy hollow receptacle with a narrow aperture at the tip. The true fruits, which are small achenes, are borne on short stalks on the inside of the syconium.

Several different types of figs occur: common figs, caprifigs, Smyrna figs, and San Pedro figs.

In the **common figs** there are no staminate flowers, and consequently the fruits develop without pollination and have no seeds. Two crops are produced annually. The first crop (brebas) are larger and more juicy and are usually eaten fresh. They are borne on the old wood. The second crop is produced in the axils of the leaves. They are used fresh or are dried. Between 600 and 800 kinds of common figs are known. They are usually propagated by cuttings.

The **caprifigs** are wild figs that grow naturally in the Mediterranean region and Western Asia and probably represent the primitive type. Although of no commercial value as fruit, they are cultivated in most fig-raising countries, for they are essential to the development of the Smyrna fig. The life history of the caprifig is closely connected with that of a small wasp (*Blastophaga psenes*), which brings about cross pollination. Caprifigs produce three crops of fruit a year. The spring crop (profichi) contains staminate flowers and the so-called gall-flowers. These are similar to pistillate flowers but have short-styled ovaries. The fig wasp enters the young figs and lays eggs in the gallflowers. In about two months the new generation of wasps hatches out and emerges from the fig, becoming covered with pollen in so doing. By this time the summer crop of figs (mammoni) have been produced, which contain chiefly gall-flowers. The wasps enter these and deposit eggs in most of them. Although these are pollinated by the wasp, the presence of the larvae inhibits seed development. Any flowers in which eggs were not placed are able to develop fertile seed. There is often a continual crop of these summer figs until cold weather sets in, and figs and wasps can be found in all stages of development. Late in the season the winter crop of figs (mamme) is developed and after visitation by the wasps remains on the tree over winter. The larvae mature in April, when a new crop of profichi figs is ready to receive the wasps, and the annual cycle is resumed.

FIG. 211. Glass model of the leaves and fruit of the fig (*Ficus Carica*). (*Courtesy of the Botanical Museum of Harvard University.*)

In **Smyrna figs** no staminate flowers are produced, and consequently these figs are absolutely dependent on cross pollination from caprifigs. This process is known as caprification and is brought about artificially. Branches of caprifigs of the profichi crop are suspended on the Smyrna tree. The wasps, on emerging, enter the partly developed Smyrna figs and effect pollination. Unlike the caprifigs, the ovaries have styles so long that the wasp is unable to deposit eggs in the proper place, so the ovules are able to develop normally after fertilization. The wasps, thwarted in their attempt to raise another generation, emerge from the fig or die within the cavity. Smyrna figs have a superior nutty flavor due to the presence of the fertile seeds. They are the most important commercial fig, and are extensively grown in Asia Minor, Greece, Algeria, and parts of Portugal and California. The early attempts at cultivation in California and elsewhere met with continued failure until the profichi crop of caprifigs and the fig wasp were introduced.

The **San Pedro figs,** which are grown to some extent in California, have two crops annually. The first develops without pollination, while the second fails to mature and falls from the tree, unless it is caprified.

Figs are used fresh, dried, preserved, or canned. A considerable amount is used in baking, and ground up for fig coffee. In addition to their food value, they have definite laxative properties and are of importance in medicine.

In 1947 the United States produced 30,500 tons of dried figs in California and Texas, and also imported a considerable amount, chiefly from Turkey, Greece, and Italy.

Granadilla

The granadillas are the edible fruits of various species of passion flower. These are woody tendril-bearing vines with solitary showy flowers and a many-seeded berry. They are natives of tropical America.

The **purple granadilla** (*Passiflora edulis*) is a native of Brazil, but is cultivated all over the world. In Australia it is of considerable economic importance. It is also grown in Ceylon, the Mediterranean area, and the southern United States. The flowers are white, with a white and purple crown. The deep-purple fruit, about 3 in. in length, is used as a table fruit and in sherbets, candy, and beverages.

Other common species include the **giant granadilla** (*Passiflora quadrangularis*), with greenish-yellow fruits reaching 10 in. in length; and the **sweet granadilla** (*Passiflora ligularis*).

Guava

The guava (*Psidium Guajava*) is another tropical American fruit that has been cultivated for centuries. It was known to the Incas, and had spread all over tropical America before the time of the early navigators. Today it is common in the tropics everywhere and is of increasing importance in Florida and California. The plant is a shrub or small tree with large white flowers. The yellow berrylike fruit is about 4 in. long and has a variously colored flesh. The guava is a very aromatic, sweet, juicy, and highly flavored fruit with a fine balance between the content of acid, sugar, and pectin. It is one of the richest sources of vitamins A, B, and C and of ascorbic acid. It is usually used for jellies, preserves, and pastes, but is equally good as a fresh fruit. The powder from dehydrated fruits is used to fortify other jellies and jams.

The **strawberry guava** (*Psidium littorale*), a native of Brazil and cultivated elsewhere, has small red fruits with a very sweet juicy pulp. They are used fresh or for beverages.

The family *Myrtaceae*, to which the guavas belong, probably has as many, if not more, species with edible fruits than any other family. In addition to the guavas, of which there are about 150 species, there are the eugenias and syzygiums with nearly 700 species. Some of these are of considerable importance. The **rose apple** (*Syzygium Jambos*) has spread from tropical Asia and is extensively cultivated in other areas. It is grown in Florida for its greenish-yellow fruits, which are used in preserves and candy. The **pitanga** or **Surinam cherry** (*Eugenia uniflora*), considered to be one of the best of the genus, is grown in Florida and California for use as a fresh fruit and in jellies and sherbets. The Brazilian **grumichama** (*E. Dombeyi*), the **jambolan** or **Java plum** (*S. Cumini*), and the **ohia** or **mountain apple** (*Syzygium malaccensis*), species native to Southeastern Asia, are of lesser importance. Related species of note include the **feijoa** (*Feijoa Sellowiana*) and the **jaboticaba** (*Myrciaria cauliflora*). The latter, a strikingly beautiful tree native to Brazil, is often planted as an ornamental. The grapelike fruits, borne on the branches and trunk, are used for fresh fruit, jellies, wines, and cordials.

Jujube

The jujube (*Zizyphus Jujuba*) is a native of China and has been cultivated in that country for at least 4000 years. It is still one of the five chief fruits of China, and is also grown elsewhere in Southeastern Asia and in New Zealand. It can be grown in all tropical and subtropical countries, and since 1900 has been grown to an increasing extent in California, Texas, and Mexico. It promises well as a fruit tree for the Southwestern states as it is remarkably free from pests. The jujube is a large bush or small spiny tree with a small dark-brown fleshy drupe (Fig. 212), which has a white, crisp, rich flesh. It is used fresh, dried, or preserved, and is useful in cooking and candymaking.

Litchi

The litchi (*Litchi chinensis*), a native of Cochin China and Siam, has been an important fruit in Southeastern Asia for over 2000 years. It is now widely grown in the tropics, and has been introduced into California and southern Florida. The tree is valuable for ornamental purposes. It reaches a height of 35 or 40 ft. and has a broad round-topped crown and leathery shiny leaves. The fruits are very distinctive. They are round, 1 or 2 in. in diameter, and are borne in loose clusters. The pericarp is bright red and leathery, becoming brown and brittle on drying. The translucent white flesh surrounds a single large seed. In the dried fruit, the litchi "nuts" of the Chinese restaurants, the flesh has a raisinlike consistency. The fresh fruit is a great delicacy in China. In Kwantung alone 30,000,000 lb. are produced. Litchis are canned for the export trade.

FIG. 212. A fruiting branch of the jujube (*Zizyphus Jujuba*).

Loquat

The loquat (*Eriobotrya japonica*) is one of the few tropical fruits belonging to the *Rosaceae*, a family which furnishes such a large number of edible fruits in temperate regions. It is a native of China, but is now grown in most tropical and subtropical countries. It has been introduced into California, Florida, and the Gulf States. The loquat is a small evergreen tree with broad leaves and fragrant white flowers, which appear in the fall. The small, round, downy, yellow-orange fruits (Fig. 213) are produced in the spring. The flesh is slightly acid and not so sweet and rich as most tropical fruits. It is highly esteemed in the Orient and has been grown since antiquity. Japan produces 20,000,000 lb. The fruit is used fresh, and is made into jellies, pies, and sauces.

Another rosaceous species, the **coco plum** (*Chrysobalanus Icaco*), found on sandy shores from southern Florida to the West Indies and Brazil, has plumlike fruits which make excellent conserves but are too acid for use as a fresh fruit.

FIG. 213. The fruit of the loquat (*Eriobotrya japonica*).

Mango

The mango (*Mangifera indica*) is one of the oldest and most important of tropical fruits. It has been cultivated for nearly 6000 years. Few fruits have the same historical background or are so closely connected with folklore and religion. It is a sacred tree in India. A native of Southern Asia, it is now widely grown in Malaya, Polynesia, Africa, and tropical America, including southern Florida and California. The mango is one of the few tropical plants that have been improved under cultivation, and upward of 500 horticultural varieties are now grown. The tree is a beautiful evergreen, growing to 90 ft. in height, with small pink flowers in large panicles. The fruit (Fig. 214) is a fleshy drupe with a thick yellowish-red skin and a large seed. The size, shape, and quality of the mango vary greatly. The length is 3 to 5 in. The pulp is orange, yellow, or red in color, and when ripe has a rich, luscious, aromatic flavor with a perfect blending of sweetness and acidity. Young and inferior fruits are often fibrous and unpleasantly acid and sometimes prejudice the consumer against all mangos.

FIG. 214. Mango fruits. The mango (*Mangifera indica*) is one of the oldest and most important of tropical fruits. (*Reproduced from the National Geographic Magazine, October, 1911.*)

Mangos occupy a more important position among tropical fruits than apples among temperate fruits, and they furnish food for at least one-fifth of the world's inhabitants. Ninety-nine per cent of them are eaten fresh. They are used in preserves, salads, and sauces, such as chutney. They are also cooked, dried, and canned. The total production is estimated to be over 100,000 tons.

The related genus *Spondias* contains three species which furnish fruits frequently encountered in tropical markets. The **golden apple, Otaheite apple,** or **ambarella** (*S. cytherea*), native to the Society Islands, is cultivated in both hemispheres (Fig. 215). Though inferior to the mango, it is eaten fresh, cooked, and in sherbets and beverages. The **yellow mombin** or **hog plum** (*S. Mombin*) and the **red mombin** or **Spanish plum** (*S. purpurea*), natives of tropical America, are widely distributed in this area, as both wild and cultivated trees. The fruits are eaten raw, cooked, or in jams and jellies.

Mangosteen

The mangosteen (*Garcinia Mangostana*) is generally considered to be the world's best flavored fruit, and is highly prized in regions where it can be grown. It is a native of the Malayan region and is common in the East Indies, Cochin China, and Ceylon. A few plants have been introduced into the West Indies, but the fruit is as yet unavailable in America. The tree is small, rarely over 30 ft. in height, with deep-green foliage. The fruit

FIG. 215. A fruiting branch of the Otaheite apple (*Spondias cytherea*).

FIG. 216. Fruits of the mangosteen (*Garcinia Mangostana*) tied in preparation for street vending in Java. Several fruits were cut later to show the contents. (*Photo by Oakes Ames.*)

(Fig. 216) is a dark-purple berry, 2 or 3 in. in length, with adherent sepals at the base. The rind is ½ in. in thickness and the flesh is so delicate that it melts in the mouth like ice cream. The pulp is white or yellowish, with crimson veins, and exudes a yellow juice of an exquisite flavor.

There are over 200 species of *Garcinia*, two-thirds of which have edible fruit, and many of these are being grown experimentally in this country. The **mamey** or **mammee apple** (*Mammea americana*), a close relative of the garcinias, is an important edible fruit in the West Indies and tropical America and should be cultivated in the Old World.

Olive

The olive (*Olea europaea*) is one of the oldest of fruits and has been grown from prehistoric time. It was known in Egypt in the seventeenth

FIG. 217. Leaves and fruit of the olive (*Olea europaea*). (*Courtesy of the Arnold Arboretum.*)

century B.C. and is frequently mentioned in the Bible and in Greek and Roman writings. At the present time it is cultivated everywhere in the Mediterranean region and has been widely introduced throughout the tropics and subtropics. It has been grown in California since 1769 for ornamental purposes. The fruit has been of importance commercially in that state since 1890.

The tree is a small evergreen 25 to 40 ft. in height with leathery entire leaves. It bears whitish flowers and a one-seeded drupe (Fig. 217). The fruit is a shiny purplish black when ripe. Although living to a great age under favorable conditions, olives require careful cultivation. A deep fertile soil and a temperature averaging 57°F., and never going below 14°F., are desirable. Irrigation is often necessary. The tree is propagated by cuttings.

Olives contain a bitter glucoside and have to be processed before they are palatable. This is accomplished by pickling and heating with sodium hydroxide. Ripe olives have a high food content for they are one of the few fruits rich in oil. They are cultivated for eating, but more especially as a source of olive oil, which has already been discussed. Green olives are also a favorite food. These are picked by hand when fully grown, but still unripe. They are cleaned, heated with lye, which softens them and removes the bitter principle that is present, and pickled in brine. Stuffed olives, with the stone removed and replaced by a pimiento or nut, are a familiar product. The United States has produced as much as 69,000 tons of olives but has always imported large quantities as well. Spain, Italy, Portugal, Tunisia, Turkey, and Greece are the chief producing countries.

Papaya

The papaya or papaw (*Carica Papaya*), a native of the West Indies or Mexico, is now widely dispersed. It is a valuable food and drug plant in the West Indies and Ceylon, and is also grown in India, Malaya, California and Florida, and the Hawaiian Islands, where it is of great importance. The papaya tree, which is really a giant herb 25 ft. in height, is dioecious. The straight stem is rather succulent, with a crown of large, deeply seven-lobed leaves and yellow flowers. The fruits are fleshy berries, resembling melons in appearance. They are yellow-orange in color, weigh up to 20 lb., and are borne on long stalks just below the crown of leaves (Fig. 218). Few trees grow as rapidly or yield more heavily. The papaya is an excellent breakfast fruit, the orange flesh having a sweet musky taste. It is also used for salads, pies, sherbets, and confections. Unripe fruits are cooked or preserved. The fruit and other parts of the plant contain a latex that is used in chewing-gum manufacturing. One of the constituents of the latex is a digestive ferment, papain, which acts on proteins in a manner similar to pepsin. This ferment is important in medicine and is also used for tenderizing meat. Considerable papaya is imported, chiefly from Ceylon.

This true papaya should not be confused with the **papaw** (*Asimina triloba*) of temperate North America. This deciduous tree, with drooping leaves, axillary purple flowers that appear before the leaves, and edible fruits, grows from New York to Florida and Texas.

Fig. 218. A papaya tree (*Carica Papaya*) in fruit in Florida.

Persimmon

The **Japanese persimmon** or kaki (*Diospyros Kaki*) is a native of China and has spread from there around the world. Over 800 varieties are grown in Japan. It is cultivated in France and other Mediterranean countries, and is common in California, Texas, Florida, and the Gulf States. It is a large tree, 40 ft. or more in height, with orange-red fruits 3 in. in diameter. These are edible berries, with an enlarged calyx at the base. They are eaten fresh or dried.

The **native persimmon** (*Diospyros virginiana*) of the eastern United States is a hardier and smaller temperate relative. The ripe fruits are of high quality and delicious flavor and should be more widely used. The unripe fruits, which often reach our markets, are, however, very acid, while ripe fruits are somewhat pulpy and are hard to transport. These characteristics have tended to offset the more favorable features, and the Japanese persimmon is more popular.

Pineapple

The pineapple (*Ananas comosus*) is one of the few tropical fruits that has been exploited commercially, and it is known the world over. No other tropical crop except rubber has had a more rapid rise in international trade, owing to good luck and good management, as well as to its own intrinsic qualities.

The pineapple is a native of Northern South America. It had reached the West Indies before the coming of the white man and the native

FIG. 219. A field of pineapples (*Ananas comosus*) in Florida.

Indians were growing it everywhere. Wild forms are still found in Brazil. The ananas, as they were called, were carried by the Spaniards and Portuguese to the Old World, and spread all over tropical Asia, Africa, the East Indies, and Polynesia.

The plant (Fig. 219) is a biennial, with a short stem and rosette of stiff leaves, 3 ft. in length, with spiny tips and prickly margins. The flowers are borne in dense heads and are crowned by a tuft of leaves. The large fruits, which weigh from 1 to 20 lb., are syncarps. These are multiple accessory fruits formed from the whole inflorescence. The individual ripened ovaries are embedded in a fleshy mass formed from the bracts, sepals, petals, and axis of the inflorescence. The cultivated varieties are usually seedless. The pineapple is a very dependable crop.

The plants are propagated by suckers, slips, or by planting the crown. They can be grown in a poor dry sandy soil. A large number of varieties are known.

There are few larger, better flavored, or more wholesome fruits in the market. In addition to the content of sugar and fruit acids, a valuable digestive ferment, bromelin, is present. Pineapples must be left on the plant until fully ripe in order for the full flavor to develop. Most of the fruits that are available outside the regions where they are grown are picked before maturity, so that the majority of people have never experienced the delicious flavor of a fully ripe "pine." Since 1900 pineapples have been canned and today the canning industry is very important. It has been estimated that the total annual output is 190,000,000 cans, with a value of over $30,000,000. Machines have been developed which do all the work in a very few seconds. The waste products are all utilized in one way or another. Pineapple juice, which is used in beverages or alone, is also canned extensively.

Although grown everywhere in the tropics, the chief commercial area lies outside the true tropics. Hawaii leads in production, with 75 per cent of the world's crop, most of which is canned. Cuba, Puerto Rico, and Central America supply the United States with most of its fresh fruit. At one time Florida was important, but at present pineapple growing is of only local interest. Borneo and the Malay States are large producers, and grow some very fine types.

Piña fiber, which is obtained from the leaves, has already been discussed.

Pomegranate

The pomegranate (*Punica Granatum*) is a native of Iran. It has been cultivated for centuries and early spread to the Mediterranean region and

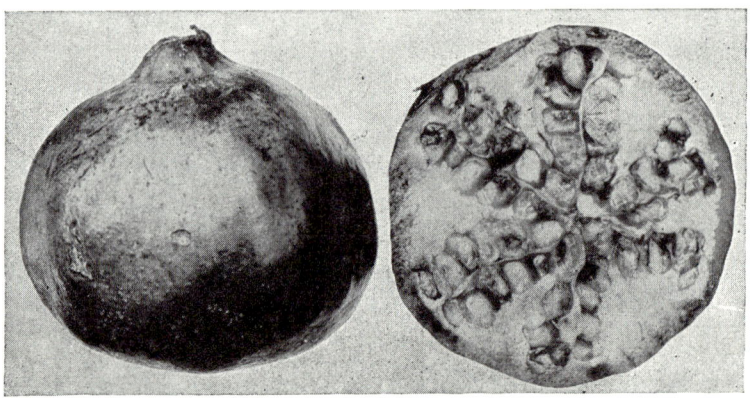

Fig. 220. The fruit of the pomegranate (*Punica Granatum*), showing the whole fruit and a transverse section.

Southern Asia. It was grown in the famous Hanging Gardens of Babylon. It is a bush or low tree with orange-red flowers. The round berry-like brownish-yellow or reddish fruits are 2 to 4 in. in diameter and are crowned with the thick persistent calyx. They have a hard rind, an edible pulp with amethyst-colored juice and many seeds (Fig. 220). Pomegranates are a very refreshing fruit and are used as a table or salad fruit and in beverages. The roots, rind, and seeds are medicinal. Pomegranates have been widely introduced in arid tropical and subtropical regions, and are grown in southern California, Arizona, and New Mexico.

Sapodilla

The sapodilla (*Achras Zapota*) is one of the best dessert fruits of tropical America. The tree is a stately evergreen, 75 ft. in height, with a dense crown and horizontal branches. It has white flowers and a large rough brown fruit 3 to 4 in. in diameter. The yellowish-brown flesh is translucent and very sweet and wholesome. Young fruits contain considerable tannin and are unpalatable. The tree is now grown in Florida and in the tropics and subtropics of the Old World. The chief commercial product of the tree, however, is not the fruit but the milky latex, which is the chief source of chicle.

The family *Sapotaceae*, to which the sapodilla belongs, contains over 400 species, most of which are edible. Of these, the sapote, star apple, and canistel are of considerable importance.

The **sapote** (*Calocarpum Sapota*) is one of the commonest fruit trees in Central America and the West Indies. The russet-brown fruit, 3 to 6 in. in length, has a sweet spicy flesh and is an important adjunct to the diet. It is eaten fresh, in salads, or as a conserve. The **star apple** (*Chrysophyllum Cainito*), native to the West Indies, has a good-flavored fruit the size of an apple which is eaten only fresh. The **canistel** or **egg fruit** (*Lucuma nervosa*), a very ornamental tree, has an orange-yellow fruit with a sweet and aromatic pulp. A native of Northern South America, it is cultivated in Brazil and has been naturalized in Florida and the West Indies. The fruits are used fresh, in salads, or for pies, puddings, and jam.

Tamarind

The tamarind (*Tamarindus indica*) probably originated in tropical Africa or Southern Asia. It is a large tree, 80 ft. in height, with a dense crown, and is frequently grown for shade and ornamental purposes in semiarid regions. The fruits are brown pods, 3 to 8 in. in length. The pulp contains 12 per cent tartaric acid, as well as 30 per cent sugar, and so has a sour taste. The tamarind is extensively used in India and the Orient as a fresh fruit, in beverages, for preserving, and in medicine. The fruits were valued in Europe in the Middle Ages. The tamarind is commonly grown in the West Indies and also in southern Florida.

Tropical Fruits of Minor Importance

Travelers to the tropics are always impressed by the wealth of small fruits which, although of no commercial value, nevertheless are of considerable importance in the economy of the inhabitants. The time may

FIG. 221. The Otaheite gooseberry (*Phyllanthus acidus*), showing flowers and fruit.

come when improved methods of refrigeration and transportation will make it possible to export some of these fruits. Several of them have already been mentioned in this chapter. Others, worthy of comment, include both introduced and native American species.

The **bilimbi** (*Averrhoa Bilimbi*) and the **carambola** (*A. Carambola*), natives of Southeastern Asia, are grown in the tropics of both hemispheres. The small fruits are quite acid, so are usually cooked with sugar. The **Otaheite gooseberry** (*Phyllanthus acidus*), with yellow cherrylike fruits (Fig. 221), borne in clusters on the trunk and branches, is likewise too acid to eat raw but is excellent with sugar. The small tree is ornamental and has run wild in Florida and the West Indies. The **governor's plum** (*Flacourtia indica*), another Asiatic species, has excellent fruits. The

Natal plum (*Carissa grandiflora*), from South Africa, is a spiny ornamental and has been introduced into Florida as a hedge plant. The small scarlet fruits are eaten raw or cooked or used for jellies and preserves.

Native species of some importance include the following: The **Barbados cherry** (*Malpighia glabra*), found from Texas to Northern South America, has juicy red cherrylike fruits which are usually cooked or used for beverages. The **ceriman** (*Monstera deliciosa*), an ornamental aroid often planted in greenhouses, has long conelike fruits with a pleasant pineapple-like flavor when fully ripe. The **white sapote** (*Casimiroa edulis*), native to the highlands of Mexico and Central America, has been introduced

FIG. 222. Naranjillas (*Solanum quitoense*), a popular Colombian fruit, shown entire and in section. (*Photo by Walter H. Hodge.*)

into California, the southern United States, and the West Indies. The applelike fruits have a soft, yellow, sweet, custardlike pulp.

Two common fruits of the Andean region of South America have so much promise for the future that they should be referred to, even though they are at present but little known outside their native home. The **tree tomato** (*Cyphomandra betacea*), a native of Peru, is extensively cultivated in the whole Andean region. It has been introduced into Puerto Rico and Southeastern Asia. The oval, reddish-orange fruits are eaten raw or cooked. The **naranjilla** or **lulo** (*Solanum quitoense*) is a robust herb with gigantic leaves and orange fruits (Fig. 222) produced in great abundance throughout the year. It is common in the high Andes from Peru to Colombia. The fruit has a particularly delicious and refreshing juice, rich in proteins and minerals, and is worthy of exploitation.

CHAPTER XX

SPICES AND OTHER FLAVORING MATERIALS

The story of spices, condiments, and the other flavoring materials is one of the most romantic chapters in the history of vegetable products. From the earliest time spices have been as eagerly sought after as gold. The craving for spices has been one of the great factors in human progress, and has done much to change the course of history and geography and to promote international relations. The discovery of new lands and of shorter trade routes and the colonization of spice-producing countries have resulted, in part, from this interest in aromatic plants. The quest for spices created a furor comparable only to the Crusades, and was one of the dominant factors in European history during the Middle Ages and as late as the sixteenth century. The use and cultivation of spices, however, go back to the beginnings of history. They have played a prominent part in all the civilizations of antiquity, in ancient China and India, in Babylon and Egypt, and in Greece and Rome. The majority of spices originated in the Asiatic tropics and were among the first objects of commerce between the East and the West. The first traders were the Arabs, who brought the products of southern India and the Spice Islands by caravan to Arabia, and thence to Europe. Later other countries took over the spice trade. For many years Venice was the leader. In the sixteenth century the Portuguese assumed control and held a virtual monopoly for 200 years. They were supplanted by the Dutch, who were supreme for many years. Later the British Empire shared with Holland most of the spice trade of the world.

In the olden days spices were put to many uses. They not only served to season insipid foods and give zest to an otherwise monotonous diet, but acted as preservatives as well. Their aromatic qualities were useful in overcoming the odors of bad food and unwashed humanity. They were used in beverages, in medicine, and even in lieu of money. Sought after by rich and poor alike, and expensive because of the demand and the difficulty of obtaining them, they were the basis of many great fortunes made between A.D. 1300 and 1700.

The use of spices is not so widespread at the present time, but the United States still pays from $10,000,000 to $20,000,000 annually for crude spices, which are worth twice as much in the retail trade. The practice of importing the various aromatic substances in a crude state

(Fig. 223) and converting them into a powdered form is still followed in an attempt to prevent adulteration and to ensure the quality of the final product. Essential oils, obtained from these aromatics, are also imported in large amounts.

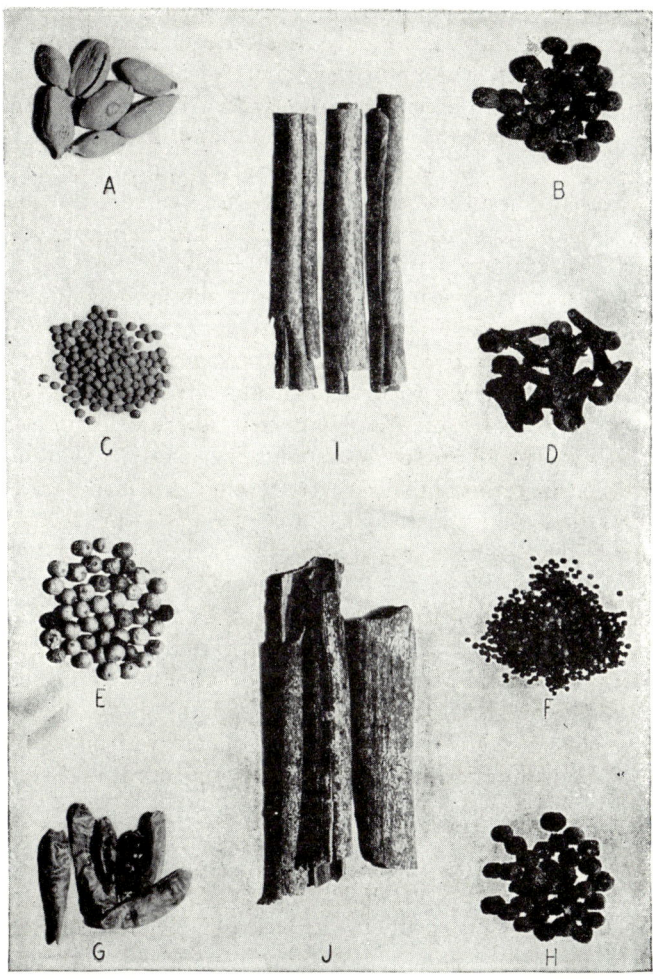

Fig. 223. Crude spices. *A*, cardamom (*Elettaria Cardamomum*); *B*, allspice (*Pimenta dioica*); *C*, white mustard (*Brassica hirta*); *D*, cloves (*Syzygium aromaticum*); *E*, white pepper; *F*, black mustard (*Brassica nigra*); *G*, capsicum (*Capsicum frutescens*); *H*, black pepper (*Piper nigrum*); *I*, cinnamon (*Cinnamomum zeylanicum*); *J*, cassia (*Cinnamomum Cassia*).

Spices cannot be classed as foods, for they contain little of nutritive value. They do, however, give an agreeable flavor and aroma to food, and add greatly to the pleasure of eating. They stimulate the appetite and increase the flow of the gastric juices. For this reason they are often referred to as food accessories or adjuncts. Whatever value they have is due to the presence of the essential oils, and occasionally to other aromatic principles.

The medicinal value of spices is not so great as was thought during the Middle Ages, but a considerable number of them are still official drugs in both Europe and America. They are used as carminatives and antiseptics and to disguise the unpleasant taste of other drugs. They also play an important part in many of the industries and are used in perfumery, soaps, incense, as dyes, in histology, and in various arts.

The majority of spices are still obtained from the tropics, chiefly from Asia. Africa supplies the grains of paradise, while tropical America furnishes vanilla, red pepper, and allspice. A small number occur in the cooler temperate regions of the Old World.

The classification of spices, as in the case of all plant products that contain essential oils, is very difficult and there are no sharp boundaries between the various groups. Usually all aromatic vegetable products that are used for flavoring foods and drinks are included under spices. In other cases the term "spice" is restricted to hard or hardened parts of plants, which are usually used in a pulverized state. Condiments are spices or other flavoring substances that have a sharp taste, and are usually added to food after it has been cooked. Savory seeds are small fruits or seeds that are used whole. In the sweet or savory herbs, fresh or dried leaves are used for flavoring or garnishing. Essences are aqueous or alcoholic extractions of the essential oils. In view of the difficulty of distinguishing between spices, condiments, and the other flavoring materials, it seems best to consider this group on a morphological basis—the nature of the plant part utilized. Of the hundreds of spices that are used today, only a few can be discussed. These will be treated under roots, barks, buds and flowers, fruits, seeds, and leaves and stems.

SPICES OBTAINED FROM ROOTS AND ROOTSTALKS

Angelica

The angelica plant (*Angelica Archangelica*) is a stout perennial herb with large pinnately compound leaves and small greenish-white flowers in terminal compound umbels. It is a native of Syria and now occurs in many parts of Europe and Western Asia in low ground. It even reaches boreal regions in Lapland and the Alps.

All parts of the plant are aromatic. The roots and fruits are dried and used for flavoring cakes, candy, and beverages, such as vermouth and the various bitters and liqueurs. The young stems and leafstalks are candied by steeping them in syrups of increasing strength. This candied angelica is much used for decorating and flavoring candy and cakes because of its attractive bright green color and aromatic taste. The oil, usually distilled from the fruits, is used in flavoring, perfumery, and medicine. Once grown in every garden, it is now cultivated only in Germany. Its use dates back to A.D. 1500.

Galangal

The **lesser galangal** (*Alpinia officinarum*) is a native of southern China, and was used at an early date in that country. It is a perennial herb with a raceme of showy flowers and ornamental foliage. The reddish-brown rhizomes have an aromatic, spicy odor and a pungent taste, like a mixture of pepper and ginger. Galangal was much more important formerly than it is today, but it is still used to some extent in cooking, in medicine, and for flavoring liqueurs and bitters.

The **greater galangal** (*Alpinia Galanga*), a larger plant of Java and Malaya, is also used somewhat for flavoring purposes.

Ginger

Ginger is the most important of the spices obtained from roots. It has had a long and interesting history. A native of Southeastern Asia, it was early used in China and India, and was brought by caravans to Asia Minor before the time of Rome. It was among the first of the oriental spices to be known in Europe, where it was prominent early in the Middle Ages. For many years it was an important drug. It was the principal ingredient of a remedy for the plague, which was much used in England during the reign of Henry VIII. Today ginger is cultivated over a wider area than most spices, owing probably to the ease which the roots can be transported. It was one of the first Asiatic spices to be grown in the Western Hemisphere.

The ginger plant (*Zingiber officinale*) is an erect perennial herb (Fig. 224) with thick scaly rhizomes that branch digitately and are known as "hands." The stem reaches a height of 3 ft. and is surrounded by the sheathing bases of the leaves. The flowers are borne in a spike with greenish-yellow bracts subtending the yellowish flowers, which have a purple lip. Ginger is cultivated for the most part in small home gardens. A rich moist soil, partial shade, and a strictly tropical climate are desirable. The plant is propagated by the rhizomes.

The rhizomes are pale yellow in color externally and a greenish yellow inside. They contain starch, gums, an oleoresin, and an essential oil as well. The several varieties differ in the content of the latter two principles. The rhizomes are dug after the aerial parts of the plant have withered.

Ginger is prepared in two different ways. *Preserved* or *green ginger* is a product of southern China. Young juicy rhizomes are dried, cleaned, and boiled in water until tender. They are then peeled, scraped, and boiled several times in a sugar solution, and finally packed in a similar solution. Occasionally preserved ginger is prepared in a dry state by dusting the drying rhizomes with powdered sugar.

Dried or *cured ginger* is the product of the other ginger-growing countries. The rhizomes are cleaned, carefully peeled, and dried in the sun. They are sometimes parboiled in water or lime juice before peeling. This is the black ginger of commerce. White ginger is made by bleaching the rhizomes.

The aromatic odor of ginger is due to the essential oil, while the pungent taste is due to the presence of the nonvolatile oleoresin, gingerin.

FIG. 224. Two bunches of fresh ginger (*Zingiber officinale*). (*Photographed in Paoki, Shensi, China by F. N. Meyer. Courtesy of the Arnold Arboretum.*)

Ginger is used more as a condiment than as a spice. It dilates the blood vessels in the skin, causing a feeling of warmth, and increases perspiration, with an accompanying drop in temperature. For this reason it is much used in warm countries.

In medicine, ginger is used as a carminative and a digestive stimulant. It is extensively used in culinary preparations, such as soups, puddings, pickles, gingerbread, and cookies, and is an ingredient of all curries, except those used with fish. Ginger is exceedingly popular for flavoring beverages, such as ginger ale and ginger beer. It was formerly used for spicing wine and porter. The oleoresin is extracted and used in medicine and flavoring. The essential oil is also extracted. As this lacks pungency, cayenne pepper is usually added when it is used for flavoring purposes.

Ginger is grown chiefly in China, Japan, Sierra Leone, Queensland, Indonesia, and Jamaica and other Caribbean islands, where the soil and climate are unsurpassed. The United States imports crystallized ginger from China and dried ginger from Jamaica, India, and Sierra Leone.

Horse-radish

The horse-radish (*Armoracia lapathifolia*) is a native of Southeastern Europe. The plant is extensively grown in both Europe and America and frequently escapes from cultivation and becomes established. In many places it is a troublesome weed.

It is a tall hardy plant with glossy green toothed leaves and masses of small white flowers. The large, fleshy, white cylindrical roots (Fig. 225) are usually dug in the fall. They are scraped or grated, and used as a condiment, either fresh or preserved in vinegar. The pungent taste is due to a glucoside, sinigrin, which is broken down in water by enzyme action. It is similar to mustard oil in its properties. Horse-radish is a valuable condiment and has been used for many years. It aids digestion and prevents scurvy.

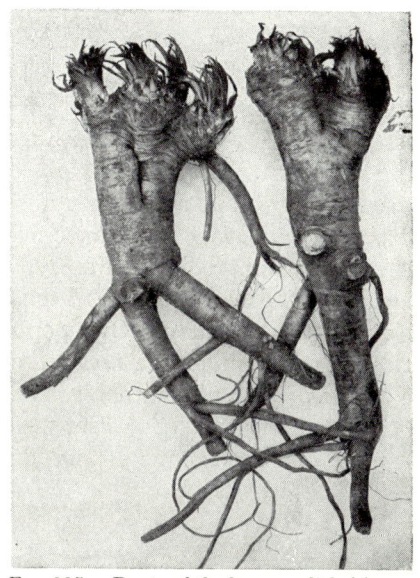

FIG. 225. Roots of the horse-radish (*Armoracia lapathifolia*).

Sarsaparilla

Sarsaparilla is obtained from the dried roots of several tropical species of *Smilax*, among them *S. aristolochiaefolia* of Mexico, *S. officinalis* of Honduras, and *S. Regelii* of Jamaica. These plants are climbing or trailing vines with prickly stems and are found in dense moist jungles. They may be propagated by seed, layering, cuttings, or suckers. They have a short thick rhizome and very long thin roots up to 10 ft. in length. This makes the collection of the roots a very arduous task. They are harvested when two or three years of age. The roots contain a bitter principle much used for flavoring purposes. Sarsaparilla is rarely used alone, but usually in combination with wintergreen and other aromatics. At one time it was used in medicine.

Turmeric

Turmeric (*Curcuma longa*) combines the properties of a dyestuff and a spice, and has already been discussed in the former connection. It is native to Cochin China and the East Indies and is widely cultivated in the tropics of both hemispheres. Turmeric is especially popular in India, where 60,000 acres are devoted to it, and enormous quantities are used, as they have been for centuries. The plant is a robust perennial (Fig. 64) with a short stem and tufted leaves. The pale-yellow flowers are borne in dense spikes, topped by a tuft of pinkish bracts. The rhizomes, which are the source of the colorful condiment, are short and thick, with blunt tubers. They are cleaned, washed, and dried in the sun. Turmeric is very aromatic with a musky odor, and has a pungent bitter taste. It is used to flavor, and at the same time color, butter, cheese, pickles, mustard, and other foodstuffs. A considerable quantity is exported to Europe and America for this purpose. Turmeric is one of the principal ingredients of curry. Curry is not a single substance, but a compound of many spices. Each type of meat or other food requires its own particular curry. One popular recipe for a meat curry includes turmeric, coriander, cinnamon, cumin, ginger, cardamom, fenugreek, cayenne pepper, pimiento, black pepper, long pepper, cloves, and nutmeg. Another curry, used for fish, is made from turmeric, coriander, black pepper, cumin, cayenne pepper, and fenugreek.

Zedoary

Zedoary (*Curcuma zedoaria*) is a plant of similar habit with pale-yellowish or white flowers and showy crimson or purple bracts. It is much grown in India for the large tuberous rhizomes, which are sliced and dried. It was formerly an important spice, and is still used for flavoring liqueurs and curries. Its chief use today is in medicine, perfumery, and cosmetics.

SPICES OBTAINED FROM BARKS

Cassia

Cassia or Chinese cinnamon is one of the oldest of spices. It was known in China as long ago as 2500 B.C., in Egypt in the seventeenth century B.C., and was familiar to all the people of the Mediterranean area at an early date. In the earlier records it is likely to be confused with cinnamon. Cassia is obtained from *Cinnamomum Cassia*, a native tree of Burma. It is an evergreen, 40 ft. in height, with smooth pale bark, small pale-yellow flowers, and a fleshy drupelike fruit. The tree is grown in southern China from seed, usually on terraced hillsides. Young trees from 6 to 10 years of age are cut down and cut up into short lengths. The bark is loosened, stripped off, and dried. Cassia bark (Fig. 223, *J*) reaches the markets in the form of dark-reddish-brown "quills," usually with some patches of grayish cork on the outside. It varies in quality, but is always very aromatic, though not so delicate as cinnamon. Among the several ingredients are tannin, sugar, starch, a dyestuff, a fixed oil, and the essential oil, which is distilled and used in medicine and flavoring.

Cassia buds are the dried unripe fruits. These contain the same essential oil, and should be used more. They are picked when only one-fourth grown and resemble little cloves.

There are several other sources of cassia of less importance. Indian cassia comes from *Cinnamomum Tamala;* Padang cassia, with smooth bark and no cork, from *C. Burmannii*, a tree of Indonesia. Large amounts of this cassia are imported in the United States. Oliver's bark (*C. Oliveri*) of Australia and Massoia bark (*C. Massoia*) of New Guinea are inferior substitutes.

Cassia bark and cassia oil are used in medicine, for flavoring purposes, and in soap, candy, and perfumery. The United States imports cassia chiefly from China and Indonesia.

Cinnamon

Cinnamon was discovered much later than cassia and immediately superseded the older spice. It was used by the natives long before it attracted the attention of the white man. It is a native of Ceylon, and is often called Ceylon cinnamon. For many years it was grown only in Ceylon and was a monopoly of the Portuguese, Dutch, and English in succession. Now the tree is grown in southern India, Burma, parts of Malaya, and to some extent in the West Indies and South America.

Cinnamomum zeylanicum, the source of cinnamon, is an evergreen shrub or small tree, with beautiful dark coriaceous aromatic leaves, numerous inconspicuous yellow flowers, and blackish berries. In cultivation young trees are cut back, and sucker shoots develop from the roots. These are long and slender and furnish the commercial product. They are cut twice a year (Fig. 226), the bark is removed, and the outer and inner portions are scraped off. After drying, compound quills are tied up, ready for shipment (Fig. 223, *I*). The waste is used as a source of oil of cinnamon. The leaves and roots of the plant are also aromatic, but the essential oil differs from that in the bark and is of little value. Cinnamon is one of the most popular spices used for flavoring foods. It is also used in candy, gum, incense, dentifrices, and perfumes. The oil is used in medicine as a carminative, antiseptic, and astringent, and as a source of cinnamon extract.

Saigon cinnamon or Saigon cassia (*Cinnamomum Loureirii*) is grown in French Indo-China. Its coarse bark is highly esteemed in China and Japan, and is also used in the United States, where it is recognized as an official cinnamon in the U.S. Pharmacopoeia.

FIG. 226. Cutting cinnamon (*Cinnamomum zeylanicum*) in Ceylon. (*Reproduced by permission of the Philadelphia Commercial Museum.*)

Sassafras

Sassafras is not really a spice, but is a flavoring material of considerable importance. It is obtained from the bark on the roots of *Sassafras albidum* of eastern North America. The sassafras is a tree from 60 to 100 ft. in height, with characteristically lobed leaves and greenish-yellow dioecious flowers, produced before the leaves, and dark-blue drupes with red stalks. The spicy root bark was used by the Indians and early settlers. All parts of the plant are aromatic. The bark is gathered in the spring or fall, deprived of the outer corky layers, and dried. The supply comes chiefly from Virginia, Tennessee, North Carolina, and Kentucky. Sassafras is used for flavoring tobacco, patent medicines, root beer and other beverages, soaps, perfumes, dentifrices, and gum. Both sassafras bark and sassafras pith are used in medicine. The oil, obtained by distillation, is used for flavoring and is one of the sources of artificial heliotrope. It is also an ingredient of many soaps and of floor and polishing oils.

SPICES OBTAINED FROM FLOWERS OR FLOWER BUDS

Capers

The caper bush (*Capparis spinosa*) is a trailing spiny shrub, a few feet in height. It is a native of the Mediterranean region and is cultivated in Southern Europe and the southern United States. The solitary berry-like fruits are borne on thick stalks. The unopened flower buds are gathered every morning and pickled in salt and strong vinegar. These capers are roughly spherical and round angled, and dark green in color. They have a very pungent taste and are used as condiments with meat and in sauces and pickles.

Cloves

Cloves are one of the most important and useful of the spices. They were in use as early as the third century B.C. in China, were well known to the Romans, and reached Europe during the Middle Ages. Their source and place of origin were unknown until the Portuguese discovered the Molucca Islands in the sixteenth century. For a time cloves were a Portuguese and later a Dutch monopoly. Today they are grown in many tropical countries in both the Old and New Worlds.

Cloves are the unopened flower buds of *Syzygium aromaticum*, often called *Eugenia caryophyllata*, a small, conical, and very symmetrical evergreen tree. In the wild state it produces clusters of crimson flowers, but in cultivation (Fig. 227) it never reaches the flowering state. The flower buds are greenish or reddish when fresh and become brown and brittle on drying. Their shape is nail-like, and the name "clove" is derived from the French word for nail, *clou*. They have a slightly cylindrical base, surmounted by the plump, ball-like, unopened corolla, which is surrounded by the four-toothed calyx. The cloves (Fig. 223, *D*) are picked by hand, "stemmed," and dried in the sun or in kilns. The crop is an uncertain one and is hard to grow. Cuttings are useless and the seeds germinate and grow slowly; so nursery seedlings are usually utilized for large-scale propagation. The yield is rather low until the trees are at least 20 years old. Considerable moisture in the soil is necessary, and there is an old saying that clove trees must be able "to see the sea."

Cloves are very aromatic and fine flavored and impart warming qualities. They have almost endless uses, both whole and in the ground state, as a culinary spice, for the flavor blends well with both sweet and savory dishes. They are used for flavoring pickles, curries, ketchup, and sauces; in medicine; and for perfuming the breath and the air in rooms. Cloves have stimulating properties and are one of the ingredients of betel-nut chew. Clove cigarettes are smoked in Java.

The essential oil, which is obtained by distilling cloves with water or steam, has even more uses. It is used in medicine as an aid to digestion and for its antiseptic and antispasmodic action. It is often used as a local antiseptic in toothache. Externally it has a counterirritant action. It is an ingredient of many toothpastes and mouthwashes. The oil has

Fig. 227. An orchard of cloves (*Syzygium aromaticum*) in Zanzibar. (*Reproduced by permission from Allen, Africa, Australia and the Islands of the Pacific, Ginn and Company.*)

many industrial applications and is extensively employed in perfumes, in scenting soap, and as a clearing agent in histological work. The chief constituent of the oil, eugenol, is extracted and used as an imitation carnation in perfumes and for the formation of artificial vanilla.

Clove stems are a commercial product, with a smaller content of the essential oil, and they are often used to adulterate cloves. The dried fruits, known as mother cloves, are also of some value.

The chief clove-producing countries are Zanzibar, which grows 90 per cent of the total output, Indonesia, Mauritius, and the West Indies.

Saffron

The cultivation of the saffron crocus (*Crocus sativus*) dates back to the time of the Greeks and Hebrews and is still carried on in many parts of

Europe and the Orient. The dried stigmas and tops of the styles are used as a spice and as a dyestuff. Saffron was of great importance during the Middle Ages because of both its real and its fancied value in medicine. Today it is used as a flavoring material to some extent. Saffron cakes are popular in some parts of England. Saffron is an ingredient of many Continental dishes, particularly the famous French bouillabaisse.

Flavoring Materials from Flowers

Certain flowers contain essential oils that are frequently used for flavoring candy, cakes, and similar products, although, as in the case of perfumes, synthetic substances have almost replaced the natural ones. Otto of roses and the oil from sweet violets, however, are still used. Floral syrups are also prepared and used for flavoring ices and beverages. Crystallized flowers are being used more and more. These are prepared by placing fresh flowers in baskets and allowing a sugar syrup to trickle over them until no more can be absorbed. They are then dried in the sun or with artificial heat. These confections have the flavor imparted by the respective essential oils. The industry centers in Grasse, France. The flowers utilized include violets, rose petals, lavender, carnations, lilac, and orange.

SPICES OBTAINED FROM FRUITS

Allspice

The dried unripe fruits (Fig. 223, B) of *Pimenta dioica*, a small tree native to the West Indies and parts of Central and South America, constitute the spice known as allspice, pimento, or Jamaica pepper. The name "allspice" is due to the fact that its flavor resembles a combination of cinnamon, clove, and nutmeg.

The allspice tree is an evergreen, 20 to 30 ft. in height, with greenish-white flowers and purple fruits. Since the ripe fruits lose most of their aromatic qualities, the commercial product is collected when the berries are mature but still green. Branches are broken off and the fruits removed by hand or flails. The ripe and undersized berries are discarded, and the desirable ones are dried for several days (Fig. 228). They become wrinkled and turn a dull reddish brown, while the aroma becomes more pronounced. The allspice tree is so common in Jamaica that it does not have to be cultivated. It grows slowly and begins to bear when about seven years of age and continues for 12 years, with an average yield of 75 to 100 lb. per tree.

Allspice is used as a culinary spice alone, or in mixture. It is much favored for sauces, pickles, sausages, and soups. The extracted oil is used for flavoring and perfumery. The leaves contain an inferior oil of bay, which is sometimes used to adulterate bay rum. The wood is used for canes and umbrella handles.

Although grown to some extent in all the American tropics, allspice is so abundant in Jamaica that the island has a virtual monopoly, exporting

FIG. 228. Drying allspice in Jamaica. The crude spice consists of the unripe fruits of *Pimenta dioica*. (*Reproduced by permission of the Philadelphia Commercial Museum.*)

some 5,000,000 lb. annually. Mexico and Guatemala produce a much smaller amount.

Capsicum

America's most important contribution to the spices is capsicum or red pepper. This familiar condiment is obtained from the fruits of several different plants, all belonging to the genus *Capsicum*. This genus is native to tropical America and the West Indies. The capsicums are very old, extending back to pre-Inca days. Shortly after the voyages of Columbus, who found all the West Indian natives using red pepper, the spice reached Europe, and by 1600 it was widespread in the Eastern tropics. Today capsicums are grown all over the world, except in the colder parts, and in many countries they are the most important spice.

The long period of cultivation has resulted in many varieties, differing in habit and in the size, shape, color, and pungency of the fruit. Among these are the bell peppers, chilis, paprikas, pimientos, tabascos, and others. By some authorities these are considered to belong to two or three distinct species, while others interpret them as derivatives of a single species, known variously as *Capsicum frutescens* or *C. annuum.*

The **sweet** or **bell peppers** (var. *grossum*) are herbs or slightly woody plants, 2 or 3 ft. in height, with ovate leaves, white flowers with a rotate corolla, and many-seeded fleshy fruits, which are technically berries. The fruits are large and puffy with a depression at the base, and are yellow or red in color when ripe. These peppers are the mildest of all the capsicums, as the pungent principle is restricted to the seeds. They are more favored in the northern part of the United States, where they are used as a fruit vegetable rather than as a spice. Both green and ripe peppers are eaten raw in salads, or are cooked in various ways, stuffed peppers being especially popular. They are also used in pickles. The plants are grown as annuals or biennials, depending on the climate. They require a long season in which to develop, but even so are well adapted to cooler areas, for they will withstand a little frost.

The **paprikas** are European varieties with large mild fruits. Spanish paprika, which is better known as pimiento, produces attractive fruits with a characteristic flavor, but entirely lacking in pungency. These are preserved, and are used in cheese preparations and stuffed olives. They are also grown in South America, California, and Georgia. Hungarian paprika has long pointed fruits which are more pungent. These are dried and used for powdered paprika, which is a familiar condiment. Only the pericarps and seeds are utilized. The uses of paprika as a condiment and in cooking are too well known to need mention. It has a high vitamin content. Perhaps the best known dish, in which paprika is an essential element, is Hungarian goulash.

The **chilis** (var. *longum*) or capsicums, as they are known in medicine, are strictly tropical and subtropical plants. They are more woody and taller (Fig. 229), with small podlike berries and innumerable small flat seeds. The crimson or orange-red fruits (Fig. 223, *G*) are elongate, conical, somewhat flattened, and very pungent. The pungent principles are present in the flesh and rind as well as the seeds. These peppers are cultivated everywhere in the tropics. The African varieties are the hottest, but Japanese chilis are more favored for culinary purposes. The ripe fruits are dried in the sun and used whole or powdered. The ground fruits constitute the cayenne pepper or red pepper of commerce. Capsicum is used in medicine internally as a powerful stimulant and carminative and to prevent fever; it is used externally as a counterirritant. It is extensively used in such beverages as ginger ale because of its pungency. The culinary uses are too numerous to mention. These small peppers are especially favored in the American tropics, where they are used in chili con carne, tamales, and other local dishes. Pepper sauce is made by extracting the pulp by pressure and pickling in brine or strong vinegar.

Tabasco sauce is pepper sauce made from a small variety grown in Louisiana.

The United States imports paprika from Hungary, and capsicum chiefly from Japan, British East Africa, and Mexico.

FIG. 229. Chili peppers (*Capsicum frutescens* var. *longum*) showing the habit. (*Photo by Walter H. Hodge.*)

Juniper

The "berries" of the common juniper (*Juniperus communis*) are used as a flavoring material. The juniper is a small tree or prostrate shrub with evergreen needlelike leaves and a berrylike cone, formed by the fleshy coalesced scales. The tree is a native of the cooler parts of Asia, Europe, and North America. The berries have a sweetish pulp with a characteristic ginlike aroma. They are purple in color, with a greenish bloom. They are dried and used in flavoring game and various meats, but more particularly for gin. The volatile oil that is extracted from crushed berries by steam distillation is also used for flavoring gin, and to some extent in medicine.

Pepper

Pepper has always been one of the most important of spices, and it is one of the most ancient. It has been highly esteemed in the East from time immemorial. It was an important commodity in Greece and Rome, and was the chief spice during the Middle Ages, when tributes were often levied in pepper. As early as 1180 the Guild of Pepperers was one of the leading trade guilds in England. It is interesting to note that London is still the center of the pepper trade. The high price of pepper was one of the chief incentives for the search for a sea route to India. Today no other spice is better known or more widely used.

Black pepper is the dried unripe fruit of *Piper nigrum*, a vine indigenous to India or the Indo-Malayan region. It is now cultivated everywhere in

FIG. 230. Picking pepper in Singapore. The pepper plant (*Piper nigrum*) is a weak climbing shrub, and is usually supported on posts when under cultivation. (*Reproduced by permission of the Philadelphia Commercial Museum.*)

the Eastern tropics from Africa to India, Siam, the Philippine Islands, the East Indies, and the South Sea Islands. The pepper plant is a weak climbing or trailing shrub with adventitious roots, reaching a length of 30 ft. in the wild state. It has coriaceous evergreen leaves and very small flowers in catkins. The fruits are small one-seeded berry-like drupes, about 50 to a catkin. In ripening they change in color from green to bright red and then to yellow. Pepper requires a hot humid climate and at least partial shade. Various soils can be utilized. The plants are supported on posts or living trees (Fig. 230). When they are about 2 ft. in height, the tip is removed to promote the development of lateral buds. The crop begins to yield in two or three years and reaches full bearing in seven years. Propagation is by seed or cuttings from the tips of the vines.

For the preparation of the black pepper of commerce the fruits are gathered when at least a few of the berries in each spike are red. They are picked by hand. The spikes are dried in the sun or in smoke and are sometimes treated with boiling water preparatory to drying. When dry, the berries or peppercorns (Fig. 223, *H*) are rubbed off, winnowed, and packed for shipment. They are reddish brown or black with a wrinkled surface and measure 3 to 5 mm. in diameter.

White pepper is prepared from berries that are nearly ripe. They are picked and piled in heaps to ferment or are soaked in water. The pulp and outer coating of the seed are then removed. White pepper is a yellowish-grey color, and the surface is smooth (Fig. 223, *E*). Frequently white pepper is prepared from black peppercorns by grinding off the outer parts by machinery. Although not so pungent as black pepper, white pepper is preferred in the trade. Commercial ground pepper is often a blend.

The aromatic odor of pepper is due to a volatile oil, while the pungent taste is caused by an oleoresin. An alkaloid is also present. Pepper stimulates the flow of saliva and the gastric juices and has a cooling effect. The culinary uses are numerous, and it is especially valuable as a condiment. Pepper itself as well as the oleoresin and alkaloid are used in medicine. The alkaloid is used as a source of synthetic heliotrope. The United States imports more pepper than any other spice.

Long Pepper

Long pepper is obtained from *Piper retrofractum* of Java and *Piper longum* of India. The former species, a climbing woody plant native to Malaya, is cultivated in Java, Bali, and adjacent islands. The latter is more shrubby. It is a native of India, Ceylon, and the Philippine Islands, and is grown chiefly in Bengal.

Long pepper was more highly esteemed by the Romans than black pepper and was also important in the Middle Ages. The tiny fruits are fused into cylindrical spikelike cones. These are collected when unripe and are dried quickly in the sun or over fires. Long pepper contains the same principles as black pepper, but is very aromatic and somewhat sweeter. It is grown in the same way as the ordinary pepper. It is almost a forgotten spice, except in the tropics, where it is extensively used in pickles, preserves, and curries.

Star Anise

Star anise is the fruit of a small evergreen tree (*Illicium verum*) which is probably a native of China. The star-shaped reddish-brown fruits consist of eight carpels, each with a hard shiny seed. Both the seeds and the fruit are highly aromatic with a flavor of anise. The plant is cultivated from seed only in southern China and Indo-China. It is not grown more frequently because of its special climatic requirements and its slow and hazardous development. The tree yields at from 6 to 100 years of age, often producing two crops a year. The fruits are collected before they are ripe and are dried, or are immediately distilled for the oil. Star anise is used as a culinary spice only in the East. It is often chewed to sweeten the breath and aid digestion. The oil is used in medicine as a carminative, expectorant, and flavoring material, and also in liqueurs, aperitifs, and perfumery.

Vanilla

A climbing orchid, *Vanilla planifolia*, native to the hot moist forests of tropical America, is the chief source of vanilla. This favorite flavoring

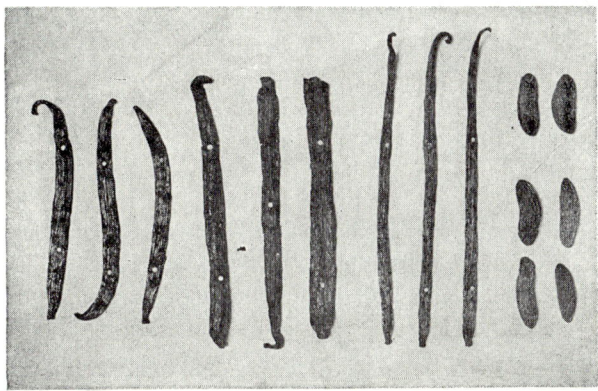

FIG. 231. Vanilla and tonka beans. From left to right: Tahiti vanilla; Pompona vanilla; Mexican vanilla; tonka beans, a substitute for vanilla. Tahiti and Mexican vanilla come from *Vanilla planifolia*, while Pompona vanilla is the fruit of *V. Pompona*. Tonka beans are the seeds of *Dipteryx odorata* and *D. oppositifolia*. (*Reproduced by permission from Youngken, Textbook of Pharmacognosy, P. Blakiston's Son & Company.*)

material is obtained from the cured, fully grown, but unripe fruits. Its use antedates the discovery of America. The Spaniards, who found the Aztecs using it to flavor chocolate, carried vanilla to Europe. It soon reached the Eastern tropics and was cultivated in many places.

The plant is a climbing vine with fleshy adventitious roots, large succulent leaves, and greenish-yellow flowers. The fruits are long, thin, yellow, podlike capsules, known as vanilla beans. Vanilla is a strictly tropical species and requires a hot climate with frequent rains. In cultivation it is grown from cuttings, and is trained on posts or living trees. The flowers are artificially pollinated by hand.

The flavor and aroma are not present in the pods until they have been cured. The unripe fruits are picked at just the right time and submitted to a sweating process. They are exposed to the sun during the morning, and are then protected by blankets during the afternoon, while at night they are placed in airtight boxes. During this curing process a glucoside is changed by enzyme action into a crystalline substance, vanillin, which possesses the characteristic odor and flavor. The pods (Fig. 231) become tough and pliable and very fragrant, and turn dark brown in color. Often crystals of vanillin appear on the surface. Vanilla is cultivated in many tropical countries, regions with an island climate being particularly favorable. Mexico, the Seychelles Islands, Madagascar, the Comoro Islands, Reunion, Tahiti, Dominica, Puerto Rico, and Guadeloupe are the chief producing areas. West Indian or Pompona vanilla is obtained from *Vanilla Pompona*, a species with shorter, thicker pods.

Vanilla is used in flavoring chocolate, ice cream, candy, puddings, cakes, beverages, etc. Occasionally the beans are used, but more often an extract is prepared by extracting crushed beans with alcohol. The manufacture of synthetic vanillin from eugenol, which occurs in clove oil, has seriously threatened the vanilla industry, but the demand for the natural product is now increasing. Several other plants have been used as substitutes for the true vanilla, but they are of little value.

Savory "Seeds"

The great family of the *Umbelliferae* is characterized, among other things, by the possession of aromatic fruits. These fruits (Fig. 232) consist of two one-seeded carpels, or mericarps, with numerous oil ducts containing essential oils. The mericarps separate readily and are so seedlike in appearance that they are commonly called seeds. These savory "seeds" are usually used whole for flavoring purposes. The commonest commercial species are anise, caraway, celery, coriander, cumin, dill, and fennel.

Anise. Anise (*Pimpinella Anisum*) is one of the earliest aromatics to be mentioned in literature. It was well known to the Hebrews, Greeks, and Romans and was highly valued during the Middle Ages for its real or reputed medicinal value. The plant (Fig. 233) is an annual, about 2 ft. in height, with simple or ternate basal leaves and once- or twice-pinnate stem leaves. The small fruits (Fig. 232, D) are grayish-brown and covered with short hairs. Anise is extensively cultivated in Europe, Asia Minor, India, Mexico, and parts of South America. It is a native of the Mediterranean region. Anise is used for flavoring cakes, curries, pastry, and candy. The oil is distilled and used in medicine, perfumery, soaps and other toilet articles, and beverages. The liqueur anisette is well known.

Caraway. Caraway (*Carum Carvi*) is the most important of the umbelliferous fruits. The plant is a native of Europe and Western Asia, but is now widely distributed in temperate regions of both hemispheres, often

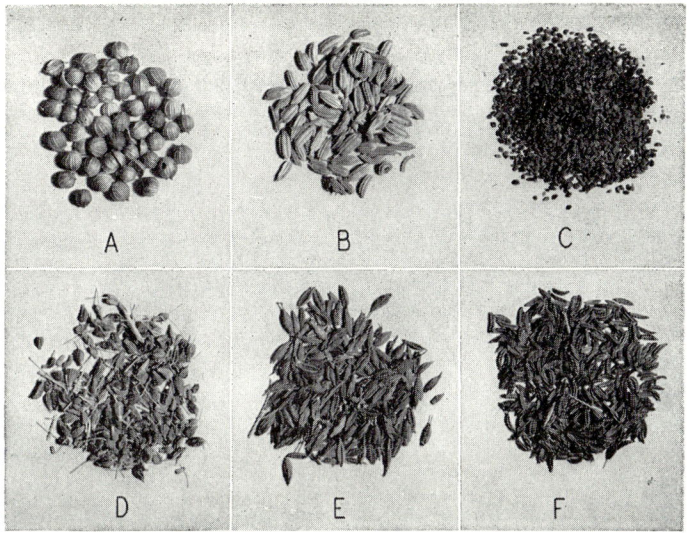

FIG. 232. Savory "seeds," the fruit of various members of the *Umbelliferae*. *A*, coriander (*Coriander sativum*); *B*, fennel (*Foeniculum vulgare*); *C*, celery (*Apium graveolens* var. *dulce*); *D*, anise (*Pimpinella Anisum*); *E*, cumin (*Cuminum Cyminum*); *F*, caraway (*Carum Carvi*).

FIG. 233. An anise plant (*Pimpinella Anisum*) showing flowers and fruit.

occurring as a weed. It has been cultivated since before the time of the Lake Dwellers in Europe. Caraway is a perennial with thick roots, compound leaves with linear segments, and small white flowers. The brown fruits (Fig. 232, *F*) are slightly curved and tapering. These "seeds" are used by the baking industry, in perfumery, medicine, and beverages, such as the liqueur kümmel. Caraway is grown commercially throughout Northern Europe and to some extent in the United States and Canada.

Celery. The "seeds" of the celery (*Apium graveolens* var. *dulce*), already discussed under vegetables, are much used for flavoring. These fruits (Fig. 232, *C*) are small and dark brown with a pronounced celery flavor. The oil has some medicinal value, but is used chiefly for flavoring in the form of an extract. Salt, flavored with celery-seed oil or the ground seeds, is in great demand for culinary purposes.

Coriander. Coriander is another very old flavoring substance. It is mentioned in Egyptian, Sanskrit, Hebrew, and Roman literature. During the Middle Ages it had many curious uses, such as love potions, incense, etc. The plant (*Coriandrum sativum*) is a native of the Mediterranean region, and is extensively grown in Europe, Morocco, India, and South America. It is a rank-smelling perennial, 3 ft. in height, with small white or pinkish flowers. The lower leaves have broad segments, while the upper are very narrow. The globular yellow-brown fruits (Fig. 232, *A*) have an unpleasant odor when fresh. The dried fruits, however, are pleasantly aromatic and serve as a common flavoring substance for both sweet and savory dishes, especially in Europe and India. The fruits are often candied in a sugar solution and sold as "sugar plums." Oil of coriander is used in medicine and in flavoring beverages, such as gin, whisky, and various liqueurs. The extract or essence is a better flavoring substance than either the dried fruit or the oil.

Cumin. Cumin (*Cuminum Cyminum*) has been cultivated for so long that it is difficult to say where it is native. Like so many others of this group, it probably originated in the Mediterranean area. The plant is an attractive little annual with small pinkish flowers. The elongated oval fruits (Fig. 232, *E*) are light brown in color and hot and aromatic. Cumin was highly prized by the ancients and is frequently mentioned in the Bible. Today it is extensively grown in Southern Europe and India. The fruits are used in soup, curries, cake, bread, cheese, and pickles, and are often candied. The oil is used in perfumery and for flavoring beverages.

Dill. Dill (*Anethum graveolens*), a native of Eurasia, still occurs spontaneously in many places. It was grown in Greece, Rome, and ancient Palestine, where it was held in high repute. It is now cultivated in Europe, India, and the United States. Dill is a small annual or biennial with light-green leaves and yellow flowers. The "seeds" are oval, light brown, and much compressed. In the United States dill is used chiefly for flavoring pickles. In France, India, and other countries it is much used in soups, sauces, and stews and for other culinary purposes. Dill oil is often used as a substitute for the seeds. Both the seeds and the oil are used in medicine. In India and the U.S.S.R. the leaves are also utilized.

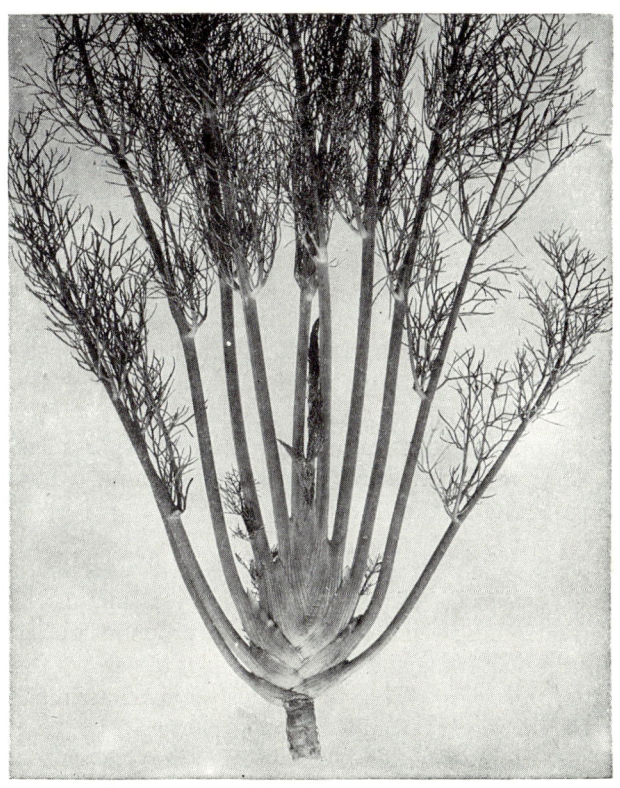

FIG. 234. Sweet or Florence fennel (*Foeniculum vulgare* var. *dulce*). The thickened and blanched leafstalks of this variety are used as a vegetable. (*Courtesy of Breck and Company*.)

Fennel. Fennel (*Foeniculum vulgare*) has had a long and interesting history. A native of the Mediterranean region, it has spread all over the world and often occurs as an escape from cultivation. It was known to the ancient Chinese, Hindus, and Egyptians as a culinary spice. The Romans cultivated it for its aromatic fruits and edible shoots. It is indispensable in modern French and Italian cooking. All parts of the plant are aromatic and can be utilized in various ways. Fennel is a tall perennial with finely divided leaves and yellow flowers. The "seeds" are oval and greenish or yellowish brown (Fig. 232, *B*). They are used in cooking and for candy and liqueurs. The oil is used in perfumes, soaps, and medicine. The thickened leafstalks (Fig. 234) of one variety, finochio or Florence fennel (var. *dulce*), are blanched and used as a vegetable.

SPICES OBTAINED FROM SEEDS

Cardamom

For centuries the highly aromatic seeds of the cardamom (*Elettaria Cardamomum*) have been an important spice in the Orient. The plant is a native of India, and is cultivated chiefly in that country and Ceylon, although it has been introduced into other tropical countries. Large quantities are grown in Central America, particularly in Guatemala. It is a perennial herb, 6 to 12 ft. in height, with long lanceolate leaves with sheathing bases. The white flowers, with a blue and yellow lip, are borne on a separate elongated stalk. The fruits, which are triangular paper-thin capsules, are borne the year round. The small seeds (Fig. 223, *A*) are light colored and have a delicate flavor. They are usually kept in the fruit until needed, for the flavor is better. In other cases, seeds of either wild or cultivated plants are gathered when dead ripe and dried in the sun. Few spices are handled with greater care. Cardamoms are used in curries, cakes, pickles, and for other culinary purposes, as well as in medicine. They are a favorite masticatory in India. The oil is used somewhat in cooking and in flavoring beverages.

Fenugreek

Fenugreek (*Trigonella Foenum-graecum*) is an annual legume with white flowers and long slender pods with a pronounced beak. It is a native of Southern Europe and Asia. The plant is grown for forage and for ornamental purposes. The small seeds are used in India for curries, in dyeing, and in medicine. The extract is used with other aromatic substances in making an artificial maple flavoring.

Grains of Paradise

Aframomum Melegueta, a perennial herb of West Africa, is the source of the aromatic seeds known as grains of paradise. This plant has large rootstalks which send up an erect stem, 8 ft. or so in height, with long fragrant leaves and showy yellow orchidlike flowers in dense spikes. The fruits are orange pear-shaped capsules, and contain the golden-brown seeds with a distinctive aroma. These seeds are very pungent and during the Middle Ages rivaled pepper in popularity. They are still used somewhat in medicine and for flavoring beverages, and deserve wider recognition. The seeds of various species of *Amomum* are sometimes utilized as substitutes.

Mustard

Mustard was well-known to the ancients. It is frequently mentioned in the Bible and in Greek and Roman writings. During its long history it has had many curious uses. Today mustard is grown as a field crop in most temperate regions, especially in North America, Europe, China, and Japan. Although cultivated chiefly for the seeds, the tops are used somewhat as pot herbs and salad plants. Two species are utilized.

White Mustard (*Brassica hirta*) is a freely branching annual, 2 to 6 ft. in height, with yellow flowers, hairy lobed leaves, and a bristly pod with a long beak. The small round seeds (Fig. 223, *C*) are yellow on the outside and white within. They contain, among other substances, mucilage, proteins, a fixed oil, and a glucoside, sinalbin. When ground seeds are treated with water, this glucoside is broken down through enzyme activity and yields a nonvolatile sulphur compound with a characteristic sharp taste and pungency. White mustard is used in medicine and as a condiment. The fixed oil is expressed and used externally as a counterirritant. It can also be utilized as a lubricant and illuminant.

Black mustard (*Brassica nigra*), also a native of Eurasia, is grown more commonly in almost all civilized countries. It is a common weed in the United States, and is cultivated commercially in many places, especially in California, Montana, and Kentucky. The plant is smaller than the white mustard, and has smooth pods with dark-brown seeds (Fig. 223, *F*), which are yellow inside. Black-mustard seed has the same general constituents as white-mustard seed. The glucoside, sinigrin, however, yields on decomposition a volatile oil containing sulphur, which is responsible for the pungent, aromatic odor and flavor. This essential oil is very powerful and dangerous to handle as it readily blisters the skin. It also attacks the membranes of the nose and eyes. When well diluted it is used in medicine as a counterirritant, and to some extent in condiments. The expressed fixed oil has a mild taste. It is used in soapmaking and medicine.

Ground mustard is much used as a condiment and in preparing pickles, sardines, salad dressing, etc. It has a stimulating effect on the salivary glands and also increases the peristaltic movements of the stomach. Mustard and warm water form an efficient emetic. The more pungent black mustard is preferred on the continent of Europe, while white mustard is more popular in England. Ground mustard, however, is usually a combination of the two kinds. The familiar mustard paste is prepared by treating ground mustard with salt, vinegar, and various aromatics.

Indian mustard (*Brassica juncea*) is used in India and parts of Europe as a spice and in cooking. Its properties are similar to those of black mustard. The fixed oil is expressed and is used in cooking and to anoint the body.

Nutmeg and Mace

Nutmeg and mace are both obtained from *Myristica fragrans*, a native of the Moluccas or Spice Islands, and now grown in the tropics of both hemispheres, particularly in the East Indies and the British West Indies. It is doubtful whether these spices, which are now so popular, were

FIG. 235. A fruiting branch of the nutmeg (*Myristica fragrans*), from a plant growing in the Peradeniya Botanical Garden, Ceylon. Both nutmeg and mace are obtained from the seeds of this species. (*Courtesy of the Arnold Arboretum.*)

known to the ancients. They had reached Europe, however, by the twelfth century. Upon the discovery of the Spice Islands in 1512, the Portuguese obtained a monopoly of nutmeg and mace, which was later wrested from them by the Dutch. At a later date trees were smuggled into French and British possessions and the monopoly was broken.

The nutmeg tree is a handsome evergreen with dark leaves and reaches a height of 30 to 60 ft. It is normally dioecious, with small pale-yellow flowers that are fleshy and aromatic. The ripe fruits (Fig. 235) are golden-yellow and resemble apricots or plums. They gradually dry out and, when dead ripe, the husk splits open revealing the shiny brown seed covered with a bright-red branching aril (Fig. 236). Inside the seed is the kernel, which is the nutmeg of commerce. The aril is the source of mace.

The nutmeg is propagated from seed in nurseries and later transplanted. It requires a hot moist climate and thrives best when near the sea, so that islands are very favorable for its growth. The trees come into full bearing when about 20 years of age and continue for 30 or 40 years. The yield is very high, a large tree furnishing about 1000 nutmegs annually. Fruits are produced all the year round. After the husks split open, the fruits are picked, the pericarp is removed, and the mace is stripped from the shell, flattened, and dried. It turns a yellowish brown. The seeds

Fig. 236. Fruit of the nutmeg (*Myristica fragrans*), showing the whole fruit, a sectioned fruit with the aril-covered seed, the aril (the source of mace) removed, and the kernel (the nutmeg of commerce). (*Photo by Walter H. Hodge.*)

are dried and the shell cracked off. The kernels are removed, sorted, and often treated with lime to prevent insect attack.

Mace is one of the most delicately flavored of spices and is used with savory dishes and in making pickles, ketchup, and sauces.

Nutmegs have been used medicinally and as a culinary spice for a long time. Grated nutmeg is used with puddings, custards, and other sweet dishes, and with various beverages. A jelly is made from the fresh husks of the ripe fruit. An essential oil is extracted for use in medicine and as a flavoring agent. This oil contains a highly toxic substance, myristicin, and can be used only in small amounts. Caution must also be exercised in the use of nutmeg and mace. Nutmeg oil is also used in the perfume and tobacco industries and in dentifrices. Nutmegs contain a fixed oil, known as nutmeg butter, the uses of which have already been discussed.

Tonka Beans

Two species of tropical South American trees (*Dipteryx odorata* and *D. oppositifolia*) are the source of tonka beans, which are becoming increasingly important as substitutes for vanilla. Most of the commercial supply comes from Brazil, Venezuela, and Colombia. The large trees, up to 110 ft. in height, have curious egg-shaped fruits with a hard shell and pulpy flesh surrounding a single seed. The natives collect the fallen fruits, break them open, and dry the seeds (Fig. 231). These resemble Jordan almonds and have a black wrinkled surface. They contain a crystalline substance, coumarin, which is of considerable importance in the manufacture of perfumes. The odor is that of new-mown hay, and closely suggests vanilla. The beans, or an alcoholic extract, are used for flavoring tobacco, cosmetics, soap, perfumes, liqueurs; as a substitute for vanilla in cocoa, candy, and ice cream; and as a fixative in making dyestuffs.

SPICES OBTAINED FROM LEAVES

The aromatic and sweet-smelling leaves of many plants have long been used for flavoring materials and for their medicinal value. The old-fashioned herb garden with its fragrant plants is returning to popularity and the savories or sweet herbs may again be as familiar as they were a century ago. Many of these plants belong to the mint family, which is characterized among other things by its aromatic odor, square stems, and small bilabiate corollas. Among the more important mints that are used as flavoring materials may be mentioned balm, basil, marjoram, peppermint, sage, savory, spearmint, and thyme.

Balm

Balm (*Melissa officinalis*) is a perennial herb of Southern Europe, which has been introduced into all temperate climates. It has been cultivated for over 2000 years and was well known to the Arabs, Greeks, and Romans. The leaves are used in soups, stews, sauces, dressings, and salads. The essential oil has a lemonlike taste and is used in beverages. The flowers have long been an important source of honey.

Basil

Sweet basil (*Ocimum Basilicum*) is probably a native of India and Africa. It has been used in the former country for centuries as a condiment, and has long been popular in England because of its aromatic qualities. The leaves are used in stews and dressings and as an ingredient of mock turtle soup and the famous Fetter Lane sausages. Basil is also very popular in French cookery. The golden-yellow essential oil, used in perfumery and various beverages, is produced commercially on a small scale in California. It is obtained from several varieties of basil.

Marjoram

Sweet marjoram (*Majorana hortensis*), a native of the Mediterranean region, is another savory herb of great antiquity. It is a sacred plant in India, and is popular in both Europe and the United States. The leaves, flowers, and tender stems are used for flavoring syrups, stews, dressings, and sauces. The essential oil is used for soap and perfumes. Pot marjoram (*Origanum vulgare*) is also used to some extent.

Peppermint

Peppermint (*Mentha piperita*) is one of the most important of the aromatic herbs. It is a perennial plant found wild in moist ground in the temperate parts of Europe, Asia, and America. It is cultivated in Europe and has been an important crop plant in America for over 100 years. Peppermint was first grown in New York, but now Michigan, Indiana, and the Pacific Northwest are the leading states. Mucky soils unsuited to other types of agriculture are utilized. The crop is harvested with mowing machines when in blossom and after drying is hauled to distilleries. Peppermint has a refreshing odor and a persistent cooling taste. The leaves are used to some extent for flavoring purposes, but the oil, obtained by steam distillation, is of much greater importance. This is widely used to flavor gum, candy, dentifrices, and various pharmaceutical preparations. It is valuable in both internal and external medicine and in the perfume and soap industries. Because of its penetrating odor it is often used to detect leaks in pipes. Over 1,000,000 lb. of peppermint oil are produced in the United States annually. Peppermint camphor or menthol, a derivative of the oil extracted by freezing, is a valuable antiseptic and is much used in the treatment of colds.

Japanese peppermint (*Mentha arvensis* var. *piperascens*) is extensively cultivated in Japan, Brazil, and the United States as the chief commercial source of menthol. Although the menthol content is much higher than in peppermint, both the oil and the camphor are very bitter and so are less valuable.

Sage

Sage (*Salvia officinalis*) has long been esteemed as a spice for use in making stuffing for fowl, meats, and sausage and is one of the most important culinary herbs at the present time. It is a shrublike herb (Fig. 237) of the Mediterranean region and is widely cultivated. The grayish-green hairy leaves are very aromatic. It has been used for its reputed health-giving qualities since the time of the Romans. Oil of sage is used in perfumery.

FIG. 237. Leaves and inflorescences of sage (*Salvia officinalis*). (*Photo by H. W. Youngken.*)

Savory

Summer savory (*Satureja hortensis*), a native of the Mediterranean countries, is now grown all over the world. It is cultivated in Ohio, Illinois, Michigan, Indiana, and several of the Western states. The leaves are strongly aromatic, with a warm bitter taste. Formerly savory was used for flavoring cakes, candy, and puddings, but now it is used in dressings, sauces, and gravies and similar culinary products. The Romans used savory as a pot herb as well as flavoring material.

Winter savory (*S. montana*) was formerly a popular flavoring herb in Europe.

Spearmint

Spearmint (*Mentha spicata*), a native of temperate Europe and Asia, has spread all over the world. It is very common in the United States in wet places. Spearmint has been known since Biblical times. Both fresh and dried leaves are used for mint sauce and jelly and to flavor soups, stews, sauces, and beverages, such as mint juleps. It is also used in chewing gum, dentifrices, medicine, and candy. The plant (Fig. 238) resembles peppermint, but has longer and lighter colored leaves and more pointed spikes. It also has a milder flavor. Spearmint is cultivated in Michigan and Indiana.

Fig. 238. The spearmint (*Mentha spicata*) under cultivation.

Thyme

Thyme (*Thymus vulgaris*) is a native of the Mediterranean region, where it is still very common as a wild plant. It is cultivated in most countries and often escapes. Thyme was used by the Greeks as an incense in their temples and by the Romans in cooking and as a source of honey. Today the fresh or dried green parts of the low shrubby plant are used in soups, sauces, dressings, and gravies. The oil is used in perfumery. Thymol, a derivative of the oil, is antiseptic and is used in mouthwashes, tooth pastes, as a fungicide, and as an internal medicine, where it is effective against hookworm. It is also useful in industry.

Bay

The sweet bay (*Laurus nobilis*) is a small tree native to Asia Minor. It is very ornamental and is often cultivated. The leaves constituted the laurel of antiquity, the symbol of victory. They are bitter and aromatic and are much used in cooking. Bay is extensively grown in Europe, where the leaves are used in soups, puddings, and other culinary products. It is an ingredient of the "bouquet," the small bunch of sweet herbs used extensively by the French. The essential oil was formerly used in medicine. Bay leaves also contain a fixed oil.

Parsley

Parsley (*Petroselinum crispum*) is one of the most familiar and widely cultivated of the garden herbs. It is a native of the rocky shores of the Mediterranean, but is found escaped from cultivation in all moist cool climates. The plant is a biennial or short-lived perennial, which produces during the first year a dense tuft of dark-green, finely divided leaves (Fig. 239). The leaves are used as a garnish and for flavoring soups, omelets, and stuffing. They are an excellent source of vitamin C.

FIG. 239. Parsley (*Petroselinum crispum*), one of the most familiar garden herbs. (*Courtesy of the Massachusetts Horticultural Society.*)

In some parts of Europe the tops are used for pot herbs and the roots as boiled vegetables.

Tarragon

Tarragon (*Artemisia Dracunculus*), a small herbaceous perennial of Western Asia, is widely grown in Europe for its pungent, aromatic leaves, which are extensively used in making vinegar and pickles. It is also used for seasoning soups, salads, and various meat preparations. The tender shoots can also be utilized. The essential oil is used to perfume toilet articles.

Wintergreen

Wintergreen or checkerberry is an important flavoring material in the United States. The original source of this material was *Gaultheria procumbens*, a low creeping evergreen plant of eastern North America. The leaves contain a glucoside which breaks down in water to form methyl salicylate or oil of wintergreen. The oil is distilled from the leaves in copper stills. It was formerly an important industry in New England, and is now carried on chiefly in Pennsylvania. The sweet birch (*Betula lenta*) contains the same glucoside in its bark, and the young twigs and bark of this plant have almost entirely displaced the checkerberry as the source of oil of wintergreen. The oil is used in medicine and in flavoring candy, soft drinks, chewing gum, and dentifrices.

Minor Savory Leaves

The following species contain aromatic oils and are used to some extent in medicine and for flavoring purposes. Among the mints may be mentioned: catnip (*Nepeta Cataria*), clary sage (*Salvia Sclarea*), hyssop (*Hyssopus officinalis*), and European pennyroyal (*Mentha Pulegium*). Species belonging to other families include: chervil (*Anthriscus Cerefolium*), lovage (*Levisticum officinale*), rue (*Ruta graveolens*), and tansy (*Tanacetum vulgare*).

OTHER SPICES AND FLAVORING MATERIALS

Several plants that furnish spices or flavoring materials have already been discussed in other connections. These include almonds, calamus root, chives, cubebs, garlic, hoarhound, lavender, lemon, lime, orange, orris root, pistachio, poppy seeds, rosemary and sesame.

CHAPTER XXI

BEVERAGE PLANTS AND BEVERAGES

Beverages of some sort are an essential part of human diet because of their liquid content. From earliest time man has sought for drinks which are palatable and refreshing. He has utilized thousands of species, surprisingly few of which have become of commercial importance. Two categories may readily be recognized: nonalcoholic and alcoholic.

NONALCOHOLIC BEVERAGES CONTAINING CAFFEINE

Beverages that contain caffeine are used the world over for their stimulating and refreshing qualities. As in the case of the cereals, each of the ancient centers of agriculture and civilization had its own beverage plant. Coffee, which originated in regions adjacent to Southwestern Asia, is now used by one-third of the world's population. Tea, which is associated with Southeastern Asia, is used by fully one-half the population of the world. Cocoa, a product of tropical America, today serves as both food and drink for over 300,000,000 people. In addition to these familiar beverages, there are others that are less widely known, but equally important. These include maté, the principal drink of 15,000,000 South Americans; cola, a favorite beverage and masticatory with millions of Africans; khat, used by the Arabs; and guarana, another South American drink, which has a higher caffeine content than any other beverage.

Caffeine is an alkaloid and, like others of this group of plant products, has definite medicinal values, acting as a diuretic and nerve stimulant. Although, as in the case of other drugs, caffeine is harmful in large quantities, it is present in these beverages in such small amounts, rarely over 2 per cent, that the average adult experiences no ill effects from their moderate use. Excessive indulgence should be avoided, and, in the case of sufferers from nerve disorders and of children, caffeine-containing beverages should be used sparingly, if at all.

Coffee

Coffee is the most important beverage plant from a commercial standpoint, in spite of the fact that more people use tea. The world output of coffee has been as high as 3,000,000,000 lb. with a value of over $500,000,000.

The coffee plant is considered to be a native of Abyssinia, and coffee must have been used in that country from very early times. It was carried to Arabia about 500 years ago, and for two centuries Arabia supplied the world. The plant was gradually introduced elsewhere in the tropics, and reached Ceylon and Java by 1700, the West Indies in 1720, and Brazil in 1770. Coffee has been in general use as a beverage for only about 250 years. From Arabia it spread to Egypt and Palestine, and thence to Constantinople. It reached Venice in 1615, Paris in 1645, and London in 1650. In both France and England coffee enjoyed widespread popularity for a time and led to the establishment of the famous coffee-houses, the gathering places of the literary men of the day. In spite of the fact that today the United States is the greatest coffee-consuming country, the beverage was slow in gaining a foothold. The first coffee mill was not built until 1833.

Kinds of Coffee. Coffee belongs to the genus *Coffea*, which contains some 25 species, only three of which are of commercial importance.

Arabian coffee (*Coffea arabica*) is the source of 90 per cent of the world supply. The plant, a native of Abyssinia, is a beautiful shrub or small tree from 15 to 30 ft. in height. The smooth evergreen leaves are borne in pairs. The white, fragrant, starlike flowers (Fig. 240) are clustered in the leaf axils. The fruits, sometimes known as "cherries," are small fleshy berries, changing in color from green through yellow to red or crimson. The two greenish-gray seeds are covered with a thin membrane, the silver skin, and are enclosed in a dry husklike parchment. When only one seed develops, the fruit is known as a "pea berry," and commands a higher price. Coffee is distinctly a tropical crop and requires a hot moist climate. It is restricted to regions lying between 25°N.L. and 25°S.L. It needs at least 50 in. of rainfall, and prefers 75 to 120 in. A high humus content is also necessary. The plant is very susceptible to diseases. There are about 15 kinds of Arabian coffee under cultivation. One of them, Mocha coffee, a small-seeded variety grown in the Red Sea region, is highly esteemed.

Congo coffee (*Coffea robusta*) is a larger and more vigorous plant, with thick leaves. It bears heavily and is more hardy, and so is adapted to a wider range of climate. It is a native of the Congo region of Africa and is cultivated elsewhere. It constitutes the greater part of the coffee acreage of Indonesia. The quality of Congo coffee is not so good as that of Arabian coffee.

Liberian coffee (*Coffea liberica*), a native of the west coast of Africa, is a still larger species, reaching a height of 40 to 50 ft., and with fruits 1 in. in diameter. The plant is more vigorous and less susceptible to disease. This coffee is used chiefly in blends, for the flavor and aroma are inferior.

An extensive breeding program has greatly increased the yield in recent years.

Cultivation and Preparation of Coffee. Coffee can be grown from sea level to an altitude of 6000 ft. and thrives best at the higher elevations, with 4500 ft. the optimum. Under cultivation the plants are grown directly from seed, or seedlings are transplanted at 6-ft. intervals. Shading and constant weeding are essential, and catch crops are often grown.

Fig. 240. Flowers and fruit of the coffee (*Coffea arabica*).

The plants begin to bear in the third year. The best yield is obtained from the fifth year, and continues for about 30 years (Fig. 241).

The coffee berries are usually picked individually by hand when fully ripe, although in Arabia and parts of Brazil they are stripped off or allowed to fall to the ground. After picking and sifting or winnowing to remove the debris, coffee is prepared for the market by either the dry or the wet method. In the former the berries are spread out on drying floors and exposed to the sun, care being taken to protect them from the rain. The berries are constantly stirred so they will be dried uniformly. Eventually the dried skin and pulp are cleaned off by machines and the parchment is removed by pounding in a mortar or by mechanical means. In the wet method the berries are run through a pulping machine, which removes the skin and part of the pulp. They are then placed in vats, where the remainder of the pulp ferments and can be washed off. They are finally dried by the sun or artificial heat. The color of the finished product depends on the amount of moisture. After drying, the brittle parchment is cracked and removed by hulling machines, and the silver

FIG. 241. A coffee tree in full bearing. (*Courtesy of the Arnold Arboretum.*)

skin is rubbed off in polishing machines. The seeds or "coffee beans" are then graded and packed in burlap bags for shipment. Occasionally coffee is exported with the parchment still in place. Eventually the beans are roasted, a process which results in a loss in weight but a gain in bulk, and which is accompanied by many physiological changes. The aroma, flavor, and color develop during this process. No two varieties require the same amount of roasting, and there are many differences in the temperature used and the duration of the process. Before coffee is sold to the consumer it is usually ground. Trade coffee is often made up of different blends. The roasted coffee beans contain from 0.75 to 1.5 per cent caffeine, the stimulating principle, and a volatile oil, caffeol, which is responsible for the aroma and flavor. Glucose, dextrin, proteins, and a fatty oil are also present. The last tends to become rancid if coffee is kept too long.

Production and Consumption of Coffee. The chief areas of coffee production have changed during the years. At first Arabia led, but was replaced in turn by the West Indies, Java, and Brazil. Ceylon was an important producer from 1830–1875 when the industry was destroyed by a blight. Today Brazil stands preeminent and produces over 50 per cent of the world's supply. Coffee is the principal crop and the chief source of revenue, and the economic structure of the country is dependent on the coffee trade. Brazil has suffered in recent years from overproduction and low prices, and has attempted to restrict the industry in all its phases, even using surplus beans as fuel. The Western Hemisphere grows five-sixths of the world's coffee. Colombia, Venezuela, Costa Rica, Guatemala, Salvador, and Haiti are next to Brazil in importance. Central American coffees are milder in flavor and more rich bodied. Indonesia and the British possessions in Africa also grow a large amount of coffee.

The United States leads in coffee consumption, using over half the world's supply. In 1945 the imports amounted to 2,716,479,896 lb., coming chiefly from Brazil and Colombia. The per capita consumption is estimated at from 11 to 13 lb. Other important coffee-using countries are Sweden with 15 lb. per person, Cuba and Denmark with 13 lb., Belgium with 11 lb., and Norway with 10 lb. France uses about 7 lb. per person, Germany 4 lb., and the United Kingdom only 0.07 lb.

There are many ways of using coffee other than the familiar method in use in this country. In Turkey coffee grounds mixed with sugar are eaten, and Turkish coffee is a thick and syrupy concoction. In Sumatra coffee leaves are steeped and yield a wholesome and good-flavored beverage. Coffee extract and soluble coffee, as well as decaffeinized coffee, are also available. Coffee is often adulterated, usually with chicory, the roasted and ground roots of the chicory plant, already discussed. In Europe coffee containing chicory is preferred to the pure product. Substitutes for coffee are also in use, such as Postum and other cereal beverages, which are made from roasted barley or wheat, and which of course lack caffeine.

The waste products, pulp and parchment, are used for fertilizer, fuel, and in the manufacture of cafelite, a plastic material with good insulating properties.

Tea

Tea is the most popular of the caffeine beverages, and it is used by fully one-half of the population of the world. It is prepared from the dried leaves of *Camellia sinensis*, a native of Assam in India or of China. Tea has been associated with the latter country since early times. At first it was valued only for its medicinal properties, but since the fifth century it has served as the principal beverage. The word "tea" comes from "te," which is used in one of the Chinese dialects in place of the more universal "cha." Tea was introduced into Japan about A.D. 1000. It was known in Europe in the sixteenth century, but did not become important until the seventeenth. Great Britain today is the chief consumer of tea, and London is the great tea market of the world. The prominence of the tea-drinking habit in England, however, has come about during the last 100 years.

FIG. 242. The tea plant (*Camellia sinensis*) under cultivation in Japan. The small shrub is often planted in hedges.

Cultivation of Tea. The tea plant in nature is a small tree, but it is grown under cultivation as a shrub, 3 or 4 ft. in height (Fig. 242). The leathery lanceolate leaves have a serrated margin and numerous oil glands. The white or pinkish flowers (Fig. 243) are produced in the axils of the leaves, and are followed by capsular fruits. Constant pruning stimulates the vigorous development of new shoots, and these "flushes," as they are known, are the source of the commercial product.

Tea is a crop of tropical and hot temperate regions. The nature of the plant and the methods of cultivation vary in different localities. Some 1000 varieties are known. The tea plant is propagated from seed or seedlings. The yield may be anywhere from 200 to 1000 lb. per acre, and continues for 50 years or more. There are records in Japan of a single plant living for two centuries. Tea can be grown from sea level to an altitude of 5000 ft. Often steep slopes and soil that is too poor for other types of agriculture are used. In China the tea is grown on small farms, and is prepared for the market by primitive methods. In Ceylon, on the other hand, tea is cultivated on large plantations, and the most modern mechanical methods are used in its preparation.

The tea leaves are picked by hand or with scissors, and an expert can pick from 25 to 75 lb. a day. In China, where growth stops during the

FIG. 243. Glass model of the tea plant, showing the habit of growth, flowers, and flushes. (*Courtesy of the Botanical Museum of Harvard University.*)

winter months, only three or four pickings a year are possible. In the hotter areas, such as Ceylon, where growth continues throughout the year, as many as 25 or 30 pickings can be made. The first picking is usually made when the plants are five years old. The grade of tea depends on the age of the leaves. In *golden tips* the youngest bud only is used; in *orange pekoe* the smallest leaf; in *pekoe* the second leaf; in *pekoe-souchong* the third leaf; in *souchong* the fourth leaf; and in *congou* the fifth and largest leaf to be gathered. The flavor and quality vary with the soil, climate, age of the leaf, time of picking, and method of preparation.

Preparation of Tea. The process of preparing tea from the fresh leaves in general is as follows: The leaves are first exposed to the sun or heated in shallow trays, until they become soft and pliable. They are then rolled, by hand or by machine. This curls the leaves and removes some of the sap. Finally the curled and twisted leaves are completely dried in the sun, over fires, or in a current of hot air. In the final product, known as *green* tea, the dried leaves are dull green with an even texture and quality. In making *black* tea, the leaves are fermented after rolling by covering them up and keeping them warm. This causes them to lose their green color and changes their flavor. After fermentation, the leaves are dried in the usual manner. China produces both green and black tea, Japan mostly green, and India and Ceylon chiefly black. In Formosa the so-called *oolong* tea is produced. This is only partially fermented and is intermediate between black and green, with the color of the former and the flavor of the latter. The various pekoes, souchongs, and congous are black teas, while *gunpowder* and *hyson* are the most important grades of green tea.

Scented teas are prepared by drying the leaves with fragrant flowers, such as jasmine, and then sifting out the dried flowers. Brick tea is made by steaming the coarser leaves, twigs, and even dust for a few minutes and then pressing them into molds, sometimes with the addition of a little rice paste. Brick tea is exported from China to the U.S.S.R. and Tibet. Where tea is grown on large plantations and prepared in factories, it can be packed at once for export. In China and Japan where enormous quantities are prepared by hand labor on small farms, the tea has to be thoroughly redried by the exporters before shipment. Tea is usually shipped in light boxes lined with lead or zinc to protect the tea from air and moisture, or in small packages lined with metal foil. The tea that finally reaches the consumer is usually a blend of several different varieties. Blending is a very delicate operation and can be done only by an expert.

Tea contains from 2 to 5 per cent theine, an alkaloid identical with caffeine, together with a volatile oil and considerable tannin (13 to 18 per cent). When an infusion is made with hot water, the alkaloid and the oil readily dissolve out and the resulting beverage has a stimulating effect and a characteristic taste and aroma. If the leaves are steeped for a longer period, the tannin dissolves and the liquid becomes bitter and loses its beneficial qualities.

Production and Consumption of Tea. For hundreds of years all the tea came from China. Java began to export in 1826, India in 1830, Formosa in 1860, and Ceylon in 1890. Today China produces nearly one-half (48.9 per cent) of the world's tea, while India and Ceylon supply about 22 per cent and 13 per cent, respectively. The world production is about two billion pounds. India and Ceylon, however, are together responsible for one-half the world's tea exports, followed by Indonesia, China, Japan, and Formosa. Tea growing has been experimented with in the West Indies, and even in the Carolinas and other Southern states. The crop, however, does not thrive in this country, and there is not enough cheap labor available to make the industry a financial success. Some tea is grown in the U.S.S.R., Africa, Guatemala, and Brazil. The last country is becoming increasingly important. Great Britain is the great consumer of tea, importing five times as much as any other country. The United States is second, followed by Australia, U.S.S.R., Canada, and Holland. Chinese and Japanese teas are more important in the United States, while Great Britain is the chief consumer of the Indian and Ceylon teas. Java supplies Europe and Australia. The per capita consumption in England is estimated at 9.8 lb., and in the United States only 0.077 lb. The imports of tea into the United States have sometimes amounted to nearly 100,000,000 lb., two-thirds of which is black tea.

Cocoa and Chocolate

Cocoa and chocolate are prepared from the seeds of the cacao or cocoa tree (*Theobroma Cacao*), a native of the lowlands of tropical America. The cultivation and use of cocoa are so ancient that it is improbable that any wild trees exist today. Cocoa is grown throughout tropical South and Central America, in the West Indies, and in many other parts of the world. The use of cocoa and chocolate by other than the native peoples is of recent origin as compared with the use of tea and coffee. The beverage was unknown to Europeans until the voyage of Cortez in 1519, and was not introduced into Europe until 1526. Chocolate was the chief drink of the Aztecs and other native American peoples. It is the most nutritious of all beverages. Since its introduction to the Northern races, cocoa has steadily increased in popularity; today the original regions of cultivation are insufficient to supply the demand, and new areas have been developed in other countries.

Cultivation of Cocoa. Cocoa is distinctly a tropical crop and is grown within 20° of the equator. It also requires special environmental conditions. It is sensitive to drought and wind, and so needs shelter from the direct rays of the sun and protection from strong winds. Catch crops and permanent shade trees are usually grown with cocoa. A deep rich alluvial soil with abundant moisture and suitable drainage is also necessary. Cocoa cannot be grown satisfactorily at altitudes above 2500 ft., and it is injured by temperatures below 60°F. The crop is raised from seed or transplanted seedlings with the individual plants in rows at 4- or 5-ft. intervals.

The cocoa tree is rather small, from 15 to 25 ft. in height, with numerous branches. The shiny leaves are ovate in outline and often 1 ft. in length. The flowers and fruits are borne on short stalks directly on the trunk and larger branches (Fig. 244). The trees begin to bear when four or five years of age, and reach full-bearing during the ages of 12 to 50 years. The plants produce flowers and fruit throughout the year so that several crops annually are possible. The fruits are podlike capsules 6 to

Fig. 244. A cacao tree (*Theobroma Cacao*) in fruit. Note the pods attached directly to the trunk and larger branches.

9 in. long and 3 or 4 in. thick, with tapering ends. They contain a mucilaginous pulp and from 40 to 60 or more seeds (Fig. 245). The fruits ripen in about four months, the color changing from green to a reddish purple or yellow. Dried specimens are often chestnut brown in color.

Kinds of Cocoa. Numerous varieties of cocoa are cultivated. The most important of these are the Criollo and Forastero. In the Criollo type the fruit is soft and thin skinned, with a rough surface and pointed ends. The seeds are plump, pale in color, and whitish within; they are the finest beans for flavoring purposes. The Forastero varieties are of hybrid origin. They are hardier and more resistant and have hard, thick-shelled pods with seeds of a pale to deep-purple color. Most of the commercial crop comes from this type.

Preparation of Cocoa and Chocolate. In preparing the seeds or cocoa beans for market, mature pods are carefully cut off with special knives and are then split open. The pulp and seeds are scooped out, cured, and usually fermented. Occasionally they are merely dried in the sun, but they are more desirable if fermentation takes place. This process may be carried on by piling the seeds in heaps for several days and then spreading them out to dry. Usually specially constructed vats or houses are used, which afford protection from rain and allow the liquids from the disintegrating pulp to run off. The beans in these "sweating boxes" are constantly stirred. During the fermentation process, which lasts about

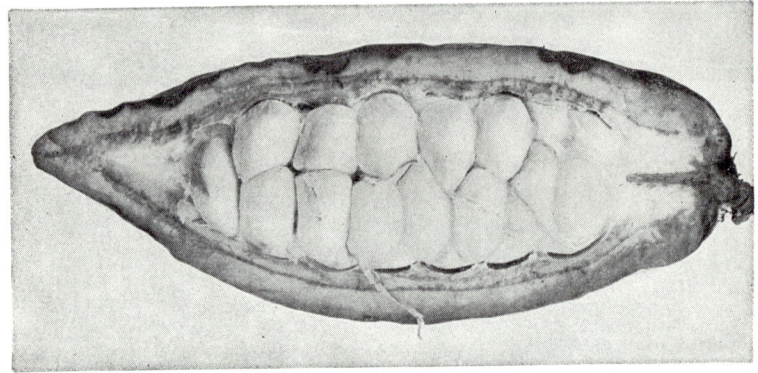

FIG. 245. Section of a cocoa pod showing the seeds or "beans" imbedded in the mucilaginous pulp.

a week, the beans become brownish red in color, lose their bitter taste, and develop an aroma. They are then washed and dried, and polished by machines or the feet of the natives to remove any of the dry pulp.

Commercial cocoa and chocolate are prepared from the beans in European and American factories. The beans are first cleaned to remove any impurities and are then sorted. They are next roasted at a temperature from 257° to 284°F. in iron drums. This develops the flavor, increases the fat and protein content, and decreases the amount of tannin. The shells become dry and brittle and the seeds easier to grind. The beans are now passed between corrugated rollers which break the shells into small fragments. These are removed in a winnowing machine. The seeds or "nibs" are finally ground to an oily paste, constituting the "liquor" or bitter chocolate, which is the starting point for further operations.

When cooled and hardened, this material is the bitter chocolate of commerce. Sweet chocolate is made by adding sugar and various spices or other aromatic materials. Milk chocolate contains milk as well as sugar and spices. Cocoa is prepared by removing about two-thirds of the fatty oil in hydraulic presses and powdering the residue.

The fatty oil present is cocoa butter, the uses of which have already been discussed. The cocoa shells are used for beverage purposes, for adulterating cocoa and chocolate, for fertilizer, and for cattle feed.

Production and Consumption of Cocoa. British West Africa now leads in the production of cocoa with about 64 per cent of the total output. The Gold Coast, Nigeria, and St. Thomas are the leading countries. The Western Hemisphere produces about 32 per cent, chiefly in Brazil. Ecuador, Venezuela, Trinidad, and the Dominican Republic also grow large amounts. Asia and the South Seas produce only a little over 1 per cent.

The United States is the chief consumer of cocoa, with a per capita consumption of over 3 lb. In 1944 imports amounted to 742,312,000 lb. of cocoa beans. Germany, Holland, and Great Britain are other large users of cocoa.

Cocoa is in a class by itself as a beverage, since it is also a food. The seeds contain less than 1 per cent of an alkaloid, theobromine, which, with a few traces of caffeine, is responsible for the stimulating properties. They also contain 30 to 50 per cent of a fatty oil, 15 per cent starch, and 15 per cent protein. A volatile oil develops during the roasting process.

Maté

Maté, yerba maté, or Paraguay tea, as it is variously known, is next to coffee, tea, and cocoa in importance. It is obtained from the leaves of various species of holly, chiefly *Ilex paraguariensis*. These plants grow wild in the mountains of southern Brazil, Paraguay, and Argentina, and they are also cultivated to a considerable extent. The use of maté in South America goes back to antiquity. Originating with the natives, it was adopted by the first white colonists and today is the universal beverage. The plant is an evergreen shrub or small tree (Fig. 246). The oval leaves are from 4 to 5 in. in length, with serrated margins, and resemble tea leaves. Maté can readily be grown from seed and the first crop is ready within a year, although the best yield is obtained from older plants.

Small leafy branches are carefully cut and toasted over fires. They are then beaten with sticks to break off the leaves, which are dried in ovenlike structures. Finally they are threshed and sifted. The leaves contain up to 0.5 per cent of theine, a volatile oil, and some tannin. Maté is greenish in color, has an agreeable aroma and slightly bitter taste, though it is much less astringent than tea. It has valuable restorative and stimulating properties. The beverage is usually prepared in a gourd or cup by pouring boiling water on the leaves, often with sugar and lemon. It is then sucked through a *bombilla*, a hollow tube of silver, brass, or straw with a perforated bowl, which acts as a strainer. Maté is the universal drink of millions of South Americans but has not been too popular in the United States. A process of double toasting produces a brownish beverage with a more tangy flavor, especially for the North American trade.

FIG. 246. A three-year-old plant of the Paraguay tea (*Ilex paraguariensis*). Maté, the universal beverage of millions of South Americans, is obtained from the leaves of this plant. (*Courtesy of the United Fruit Company.*)

Maté is also used somewhat in the preparation of soft drinks. In 1935, 422,149 lb. were imported, but more recently the amount has fallen off.

Guarana

Guarana is one of the most stimulating of all the caffeine beverages, as it has three times as much caffeine as coffee. It is prepared from the seeds of *Paullinia Cupana*, a large woody climber of the Amazon valley. The natives grind up the seeds with water and cassava flour and mold the resulting paste into brown sausage-shaped cylinders or other forms.

These are dried in smoke, becoming hard as stone. They will keep for many years. For use it is grated and added to either hot or cold water. One-half a teaspoonful of this reddish-brown guarana in a cupful of water is equivalent to two or three cups of strong coffee. Guarana contains some tannin and a volatile oil and is bitter and astringent with a bittersweet taste. The beverage is extensively used in Brazil, particularly in Mato Grosso, where the plant is cultivated to some extent (Fig. 247).

FIG. 247. A native collecting the fruits of *Paullinia Cupana* in Brazil. The seeds are the source of guarana, the most stimulating of the caffeine beverages.

Under such conditions it is a small bush. Guarana is used medicinally in the treatment of neuralgia and certain cardiac and intestinal disorders.

Khat

The dark-green leaves of *Catha edulis* are used in Arabia to yield khat, the principal beverage of the natives. This shrub (Fig. 248), which resembles tea, was grown in terraced gardens in Arabia long before coffee was introduced, and may even antedate tea. It grows wild in Abyssinia, and is cultivated in other parts of Northeastern Africa. The leaves and buds contain an alkaloid similar to caffeine, and are used dried or are chewed in the fresh condition for the stimulating effect. Khat is an excellent beverage plant and is worthy of exploitation.

Cola

Cola nuts (Fig. 139), the seeds of *Cola nitida*, already discussed under masticatories, are also used in Africa and elsewhere for beverage purposes. The drink is prepared by powdering the seeds when needed and boiling some of the powder in water for a few minutes. Cola contains 2 per cent caffeine, as well as other ingredients, and so is very invigorating. The seeds are imported into the United States for use in various soft drinks.

Cassine

Cassine is a tealike beverage obtained from a species of holly, *Ilex vomitoria*. The plant is a tall compact shrub or small tree with small oval, evergreen leaves and tough branches. It occurs on the sandy soil,

FIG. 248. The leaves and flowers of *Catha edulis*. Khat, the principal beverage of the Arabians, is obtained from the leaves.

of the coastal plain from Virginia to Mexico and is often found in dense thickets. The Indian inhabitants of this region were the first to use cassine for beverage purposes. They prepared an infusion of fresh or dried leaves which was known as yaupon or black drink. This was used medicinally as a spring tonic and emetic, and also played an important part in religious rites. It was sometimes fermented. Although the use of cassine was first reported from Florida as early as 1562, and was practiced to some extent by the early settlers, it never became very popular. Recently attention has been directed to the beverage as a basis for soft drinks. The leaves and shoots are picked and dried in the sun on trays or are roasted in ovens. Twigs and older leaves are sometimes steamed, dried, and ground. Cassine is prepared by boiling or making an infusion. The beverage is dark colored with a very sharp, bitter taste and tealike odor. It contains caffeine, tannin, and essential oils.

Yoco

The yoco (*Paullinia Yoco*) is an important beverage plant among the Indians of southern Colombia and adjacent Peru and Ecuador. Unlike

FIG. 249. Yoco (*Paullinia Yoco*), showing pieces of the stem in a Kofán Indian hut in Colombia. A stimulating beverage is prepared by rasping the epidermis, cortex, and phloem. (*Photo by Richard E. Schultes.*)

other caffein-containing species, it is the bark (Fig. 249) that is utilized. The caffeine content is usually 3 to 4 per cent but may be as high as 6 per cent. Extractions of the bark are made in cold water. The stimulating and hunger-allaying properties of yoco are very pronounced. So dependent are the Indians on this plant that it controls their whole economy. A scarcity of wild plants often leads to the abandonment of otherwise excellent village sites.

OTHER NONALCOHOLIC BEVERAGES

Ordinary beverages that do not contain alcohol are commonly referred to as *soft drinks*. These include a great variety of preparations, only a few of which can be discussed. They nearly all have a high sugar content and so are good sources of energy. The United States is the outstanding consumer of soft drinks, using an estimated 50,000,000 bottles a day.

Fruit juices are the simplest kind of soft drinks, consisting of the extracted juice alone, or with sugar and water added. Although fresh juice is readily obtainable, synthetic flavors have been all too common in commercial products. The most familiar types of fruit drinks are lemonade, orangeade, etc. Orange juice, grapefruit juice, tomato juice, and pineapple juice are popular. Shrubs and sherbets made from strawberries, raspberries, etc., were more in vogue at an earlier time. Grape juice is made by expressing the fresh fruit and heating the liquid to extract the color and to pasteurize it and thus prevent fermentation. Sweet cider, the expressed juice of apples, and perry, obtained from pears, are well known. These juices contain wild yeasts and will ferment after 24 hours or so unless they are pasteurized, or otherwise treated so as to kill the yeast organisms. Many tropical fruits are used for beverage purposes.

Soda water to the average person in this country means the familiar beverage dispensed at soda fountains. This is one of the most typical soft drinks in the United States, but its use is almost entirely restricted to this country. Today, in spite of its name, the beverage contains no soda, but consists of water charged with carbon dioxide and mixed with a syrup composed of sugar and various natural or artificial flavoring substances. Bottled soda, commonly known as pop, is also important.

An enormous number of bottled soft drinks are available in this country, chief among which are malt beverages, ginger ale, sarsaparilla, root beer, and the cola beverages. The **malt beverages** are made from malted barley, or other grains, before fermentation has started or progressed very far. They include the "near beers," which have an alcoholic content of less than 0.5 per cent. **Ginger ale** consists of acidulated sugar, water, and carbon dioxide, flavored with ginger and capsicum. **Sarsaparilla** and root beer are similar, but the flavor is due to sarsaparilla, wintergreen, and other aromatics. The **cola beverages,** in addition to other materials, contain cola, obtained from cola nuts, which has a high caffein content.

ALCOHOLIC BEVERAGES

The use, and the abuse, of alcoholic beverages have paralleled the entire history of mankind. From earliest time it has been all too easy for man to observe the natural process of fermentation, and to use its products for his own pleasure. In all ages and at all times man has resorted to alcohol, often in connection with religious or other ceremonies. Today the consumption of alcoholic beverages of some sort is world wide, existing in both primitive and civilized countries. Alcohol is a poison, and, when taken to excess, produces very deleterious effects on the human system. The various inebriating beverages bring about cerebral excitation, followed by depression, and may lead to the complete, though temporary, suppression of the functions. The evils of excessive drinking and chronic alcoholism are perfectly obvious to everyone. As regards the moderate use of alcohol, it is not within the province of this book to discuss the relative merits of the question. Strong arguments are produced by believers in prohibition or total abstinence, and on the other hand by those who favor moderate indulgence. It is clear, however, that in any work which purports to treat of useful plants and plant products there should be some discussion of the alcoholic beverages that play so important a part in man's life.

Alcoholic beverages fall naturally into two classes: the fermented beverages, in which the alcohol is formed by the fermentation of sugar present either naturally in the source or produced by the transformation of starch; and the distilled beverages, which are obtained by the distillation of some alcoholic liquor.

Fermented Beverages

Wine

Wine is the most important and also the oldest of the fermented beverages. It was known in remotest time, at least as early as 4000 B.C., and its antiquity is evidenced by the fact that the word for wine is the same in many languages. Wine is produced by the conversion of sugar, which occurs in fruits or other parts of plants, into alcohol and carbon dioxide. This process of alcoholic fermentation is brought about through the agency of wild yeasts which are present on the skins of the fruit. Unless some specific fruit, such as blackberries, is mentioned as the source, wine is always understood to mean the fermented juice of the grape.

The cultivation of grapes for wine making has been carried on for centuries in many parts of the world. Today the industry is most prominent in Southern and Central Europe, although the United States, Australia, and South America have extensive vineyards. The wine grape (*Vitis vinifera*) and its varieties are the principal source. Successful grape growing is not a haphazard business, and a knowledge of the best environmental conditions and many other factors are essential. Wine making is also an art that requires great skill and experience. In general, wines can be classified as beverage wines or fine wines. The former, often called vin ordinaire, comprise about 95 per cent of all wines and are used virtually as a food, chiefly in the regions where they are made. They are inexpensive and constitute the backbone of the wine industry. Fine wines are the more familiar commercial types that enter world trade.

They are very carefully prepared and are more costly. The finest grades are produced in the older vine-growing countries, which have years of experience behind them.

Wines vary considerably in their characteristics. The alcoholic content varies from 7 to 16 per cent. It is impossible to produce a wine naturally with a higher content, for the yeast plant is killed under such conditions and further fermentation is prevented. The sugar content of the grapes is from 12 to 18 per cent. Fermentation of the fruits or the juice is carried on in vats, usually with the aid of selected yeasts. The optimum temperature is 68°F. The agreeable aroma and flavor are due to various aromatic principles present in the fruit. The characteristic bouquet develops only after the wine has been aged for periods varying from four or five years to several decades. Clarification is sometimes necessary.

Red wines are made from grapes with colored skins and derive their own color from the pigments and other substances present in the skins. White wines are made from white grapes, or expressed juice. In the so-called dry wines or sour wines, the sugar is almost completely fermented. In sweet wines, on the other hand, fermentation is stopped before all the sugar is converted, and at least 1 per cent is still present. In sparkling wines, the wine is bottled before fermentation is complete so that carbon dioxide is produced within the bottle. Ninety per cent of the European wines are such natural or table wines. Fortified wines, on the other hand, are always still wines and have a higher alcoholic content, due to the addition of wine, brandy, or alcohol. Two-thirds of the wines produced in the United States are of this type, the dessert or appetizer wines.

PRINCIPAL WINES AND WINE-GROWING COUNTRIES

France. France is the chief wine-consuming and wine-producing country of the world, with over 4,000,000 acres devoted to vineyards. The industry, however, is more or less localized. The region around Bordeaux in the valleys of the Garonne and Gironde produces most of the wine. This district is the most outstanding single wine-growing area in the world and is famous not only for the quantity but for the quality of its output. Here are the most famous vineyards and here are produced the finest wines in great variety. Among these are the Medocs, renowned red wines or clarets; Graves, dry wines with both red and white varieties; and the white sauternes and Barsacs, which are sweeter and richer. The Bordeaux wines include regional wines, which consist of blends from several vineyards, as well as the finest grades that are bottled by individual châteaux and bear their names.

Burgundy wines are produced in the hilly country of the Côte d'Or in east central France. These red and white wines are drier and have more body and flavor than the Bordeaux wines. Both still and sparkling wines are made.

Champagnes are produced in the vicinity of Reims and Epernay. Only wines made in this Champagne region have a right to the name. Black and red grapes are used and the manufacture involves a series of elaborate processes which extend over a period of six or seven years. Because of the popularity of these sparkling wines there are many imitations, made by charging light wines with carbon dioxide.

Other noteworthy French wines are produced in the valleys of the Loire and Rhone and in Alsace and Touraine.

Germany. The Rhine valley has long been famous for its vineyards and every foot of available ground on the hillsides is devoted to grapes. The dry Rhine wines, sometimes called hock, are light colored with a rich flavor and fine bouquet. Other similar wines are produced in the valleys of the Moselle, Neckar, and Main. There are some sweet Rhine wines.

Italy. Chianti, Asti, and other Italian wines have long been known, even prior to the days of Horace who sang the praises of Falernian wine. Although today Italy ranks second to France in wine production, only a few of its products are world famous. The chief wine-producing regions are Piedmont, Tuscany, and the country from Naples southward. Sicily is noted for its Marsala, a sherry-like fortified wine.

Hungary. Hungary is the home of Tokay, a golden-yellow wine with a sweet rich flavor and rare bouquet. It is more of a liqueur than a wine as it has a soft oily taste. Tokay is expensive and the supply is limited. There are many cheap imitations and adulterations.

Spain. Spain is noted for the production of sherry, a dry wine, usually fortified with brandy, and having an alcoholic content of 15 to 24 per cent. Commercial sherries are all blended and several different grades are on the market. Malaga, another Spanish wine, is rich and sweet.

Portugal. The chief wine of Portugal is port, which is heavy and sweet, owing to the presence of considerable unfermented sugar. New port is deep purplish red in color. Wine that has been aged in casks loses some of its color and takes on a tawny hue. Port is often blended and fortified and is frequently adulterated.

Madeira. Madeira is a fortified white wine made from grapes grown on the island of Madeira. Its quality is improved by heat and shaking, and so it is stirred in glass-lined tanks and then heated. Formerly the wine was shipped on long sea voyages, which produced the desired result.

United States. Grapes grown in the United States yield wines of a distinctive nature and general all-round excellence. These domestic wines do not need foreign names, for they are just as good as most European wines. The principal wine-producing states are California, which is responsible for 90 per cent of the commercial output of the United States, New York, Ohio, and Virginia. Wines made from grapes grown in the East and Middle West are known as native or "American" wines; "California" wines, on the other hand, are made from European grapes grown on the West Coast and resemble European wines more closely.

Beer

Whereas normal processes of fermentation are sufficient to transform sugar into alcohol, in making beer and other beverages that utilize starch, special methods are required for first changing the starch into sugar. These involve the use of saliva, malt, or various molds.

The art of brewing alcoholic beverages from cereals is very old. Millet was perhaps the first to be so used, and it is still fermented in India and parts of Africa. Rice, maize, and rye have been used to some extent, but barley has always been the chief source. Barley "wine" was known to the ancient Egyptians and Romans. Beer was popular during the Middle Ages. For a long time the monasteries were the chief source of supply. As early as the thirteenth century beer was a favorite beverage in England. It was home brewed for the most part and was a dark muddy liquid with a high alcoholic content. Early in the nineteenth century the lighter German beers began to replace it in popularity. The commercial manufacture of beer involves two distinct processes, malting and brewing.

Malting. Malting has as its object the conversion of the starch present in the grains into sugar. This is brought about through the agency of an enzyme, diastase, which is produced during the process of germination. Barley is used almost universally for malting. Occasionally some maize is added. Only large, fresh, perfect, light-colored grains are used, which are free from chaff and other impurities. The barley is first steeped in water for from one to four days. During this time the grains absorb their own weight of the water. The grains are then placed in heaps or layers 6 in. deep until germination starts. Next they are spread out on the malting floor in a temperature of 50 to 60°F., and are constantly turned over. When the requisite amount of germination has occurred, the shoots are about one-third the length of the grain. The germinated barley is then kiln dried for 12 hours. This prevents any further germination and consequent loss of sugar. The color of the dried product, which is known as malt, is dependent on the degree of heat.

Brewing. The malt is crushed or coarsely ground in a roller mill and is mixed with water heated to 170°F. In some cases unmalted cereals are added. The sugar dissolves out and the infusion or wort is drawn off. This process of mashing is repeated several times. The residue is fed to cattle. The wort is then boiled with hops for 2 hours. The hops impart the bitter flavor and tonic properties and improve the keeping qualities by preventing bacterial action. The liquid is then cooled rapidly and yeast is added to bring about the fermentation of the sugar. Care must be taken to keep an optimum temperature for enzyme action and to prevent the process from continuing too long, in which case acetic acid might be formed. It is usually stopped before fermentation is complete and the yeast is removed. The beer is then drained off and strained and allowed to cool in casks. A slow fermentation continues, increasing the alcoholic content and forming the carbon dioxide that is responsible for the foaming of the beer. Beer contains from 3 to 8 per cent of alcohol. Its nutritive properties are due to the presence of sugar, dextrin, and various proteins and phosphates.

Kinds of Beer. Differences in temperature during the brewing process are responsible for the heavy and the light beers. *Lager beer* is a term which should be restricted to beer that has been aged for some time. *Bock beer* is a very strong dark beer, usually made in the spring from the first of the new malt and hops. The German *weiz bier* does not contain hops.

Ale originally meant any kind of malt beverage, and this usage continued until hops began to be used. Today the difference between beer and ale is due to differences in the temperature during fermentation. Ale is brewed by "top fermentation" at higher temperatures, around 58°F., while beer is brewed by "bottom fermentation" with temperatures averaging 40°F. The alcoholic content of ale is 4 to 7 per cent, while that of beer is 3 to 5 per cent.

Porter, first brewed in 1722, is a dark-brown beer with a slightly burned taste. It is made from inferior grades of malt. The color is sometimes heightened by the addition of caramel or licorice. Porter is stored for six to eight weeks before it is used. *Stout* is a similar beverage, but much heavier. It is stored for at least a year before it is used.

Beer is usually made from pale or amber-colored malt, ale from brown, and porter and stout from black malt.

Other Fermented Beverages

Only a few of the countless other fermented beverages can be discussed. Most of them are used locally and are of minor importance.

Hard Cider. The fresh juice of apples begins to ferment within 24 hours and gradually increases in alcoholic content until the stage known as hard cider is reached. While the domestic production of hard cider may be used for beverage purposes, a considerable proportion is allowed to undergo acetic acid fermentation and become vinegar. Cider and other fruit vinegars are also made on a commercial scale. Pear juice, or perry, is likewise often fermented.

Root Beer. Root beer consists of an infusion of various roots, barks, and herbs, among them sarsaparilla, ginger, and wintergreen, with the addition of sugar and yeast. Fermentation sets in and the beverage becomes charged with carbon dioxide. Root beer may also consist of an alcoholic extract of various aromatics and bitters. Nonalcoholic root beer is also made. *Spruce beer*, a diffusion of the leaves and twigs of the spruce, and *birch beer*, obtained from the bark of the black birch, are similar in nature.

Mead. Mead is a fermented beverage of great antiquity and it is still used in Africa and to some extent elsewhere. It is fermented from honey

FIG. 250. Maguey or pulque plants (*Agave* sp.) under cultivation in Mexico.

and water and has a winelike flavor. Mead was introduced into England by the Scandinavians, and the beverage played an important part in the nuptial ceremonies of the latter people which lasted for 30 days. Because of this fact, the duration of these ceremonies was commonly known as the honeymoon.

Sake. This important beverage of Japan and China is prepared by fermenting rice. No hops are used. Sake contains more alcohol than beer or wine. It has been used for 2600 years, with records as far back as 90 B.C.

Palm Wine. The fermented juice obtained from the inflorescences of many varieties of palms is another beverage of great antiquity. The sugary exudation has been referred to previously as a source of sugar. Palm wine or toddy was known to Herodotus as early as 420 B.C. It has been prepared by the natives of tropical regions of both hemispheres. The most important species utilized are *Raphia vinifera*, *Elaeis guineensis*, *Borassus flabellifer*, *Arenga pinnata*, *Phoenix dactylifera*, and *Cocos nucifera*. When palm wine is distilled, it yields arrack.

Pulque. The fermented juice of the maguey (*Agave atrovirens*) and other agaves (Fig. 250) was early used in Mexico as a beverage. The juice is obtained by making incisions in the flower stalk or by removing the central cone of leaves and allowing the sap to collect in the cavity. It is often distilled and constitutes mescal.

Chicha. This very common beverage of Peru, Bolivia, and other Andean countries has been known since the days of the Incas. Chicha is prepared from maize by a process of salivation which converts the starch into sugar, followed by a period of fermentation. Similar maize beverages are made in Mexico and Central America.

Other plants whose juices are fermented for use as a beverage include the banana, sugar cane, yucca, sorghum, cassava, algaroba, sweet potato, pineapple, and cactus.

Distilled Beverages

Whisky

Whisky is distilled from a fermented mash of malted or unmalted cereals or potatoes. After several distillations of the mash a product known as "low wines" results. Further distillations yield the "high wines." A mixture of high wines and water constitutes straight whisky. At first this is harsh and unpalatable, owing to the presence of from 20 to 40 volatile principles, such as fusel oil and various ethers and aldehydes. It must be aged to allow these principles to disappear. The finest bonded whiskies are aged in charred oak containers for at least four years and often much longer. Whisky is colorless at first, the color developing during the aging process. Whisky that contains 50 per cent alcohol by volume is known as 100 proof. Similarly that which contains 45 per cent alcohol is 90 proof. A continued distillation of the high wines finally results in the neutralization or elimination of all the volatile substances and yields the so-called neutral spirits. These are used in blended whiskies, in making cordials, and for other purposes. They are made chiefly from maize in the United States and from potatoes in Germany.

American straight whiskies are made from maize or rye, the former constituting the famous "corn" or moonshine of the South. About two-thirds of the commercial whisky is Bourbon, made primarily from maize. Most of the remainder is made from rye. Bourbon was originally made only in Bourbon County, Kentucky. Canadian Club is a blend of various straight whiskies with neutral spirits.

In the preparation of Scotch whisky only barley malt is used. The characteristic flavor is due to the smoke of the peat fires that are used in drying the malt. Irish whisky is made from malt, or a mixture of malt and unmalted grains of barley, oats, and maize. Vodka, the Russian equivalent of whisky, made from fermented wheat mash, is not aged but is bottled immediately after distillation.

Brandy

In a strict sense brandy is distilled only from wine, although the term is also applied to a distillation of the fermented juice of various fruits. The finest brandy is made in France in the Charente district. This product alone has the right to be called cognac. Other French brandies are known as armagnac or *eau de vie*. The best grades are made from white wines. Brandy is a clear colorless liquid and stays so when kept in glass. The brown color develops when it is stored in casks. Brandy is often artificially colored with caramel. The alcoholic content is high, amounting to 65 or 70 per cent.

Among the fruit brandies may be mentioned apricot, peach, cherry, plum, and blackberry brandy. These are often used as cordials. Apple brandy is known commonly as applejack.

Rum

Rum is distilled from various unrefined products of the sugar cane, chiefly the juice and molasses. It is one of the oldest and most widely known of the distilled beverages and is of New World origin. The flavor and aroma, which are due to various aromatic substances, improve with aging. The color is often due to caramel. Rum played an important part in the economic and social life of the American colonies from 1687, particularly in New England, where many distilleries were located. Considerable rum is distilled today in the West Indies, chiefly in Cuba, Jamaica, St. Croix, and Demerara. Rum contains about 40 per cent alcohol.

Gin

Gin is distilled from a fermented mash of malt or raw grain. The best grades are obtained from barley malt and rye. Several distillations are necessary. Gin was invented in Holland and that country still produces the best grade, although England is a close second. The method of production of Holland gin and London gin differs slightly. The flavor of gin and any medicinal value that it may possess are due to oil of juniper. Other aromatic essential oils may be used for flavoring, as in the case of sloe gin and orange gin. Gin has many substitutes and imitations, made chiefly by adding essential oils to grain alcohol.

Liqueurs and Cordials

The various liqueurs and cordials consist of sugar and alcohol flavored with various essential oils. They may be prepared by the addition of the flavoring material to neutral spirits or brandy or by the distillation of fermented fruits. The majority contain no harmful substances other than alcohol. Others, as in the case of absinthe, already discussed, do have deleterious principles and their use is forbidden in most countries.

Liqueurs often consist of various oils and cordials carefully blended according to secret formulas. French monasteries have long been famous for the manufacture of liqueurs. *Benedictine* has been made since A.D. 665 and has been a commercial product since 1792. *Chartreuse* is also made by monks. *Maraschino* is distilled from bruised marasca cherries grown in Dalmatia, sweetened and flavored with cordials. *Curaçao* is distilled from the dried rind of bitter oranges steeped in alcohol and water, with the later addition of sugar and rum. *Kirschwasser*, or black-cherry brandy, is distilled from the fruits, and sugar and alcohol are added.

Essential oils used in the preparation of liqueurs include anise and coriander (anisette), caraway (kümmel), peppermint (créme de menthe), bitter almonds (créme de noyau), and clove.

Aperitifs and Bitters

Many alcoholic preparations are used for their appetizing and tonic effects. *Vermouth*, the best known of the **aperitifs,** is a light bitter wine slightly fortified and sweetened and flavored with an infusion of several bitter and aromatic herbs. It is made chiefly in France and Italy. French Vermouth is dry, while Italian Vermouth is either sweet or dry.

Bitters are prepared by steeping various herbs with bitter principles in water or alcohol. After the principles have dissolved out, the infusion is strained and alcohol is added to prevent decomposition. *Angostura bitters*, which contains quinine and several aromatics, and *orange bitters* are typical of this group.

SYSTEMATIC LIST OF SPECIES DISCUSSED

THALLOPHYTA

Algae

Alaria esculenta (L.) Grev.   Murlins
Chondrus crispus (L.) Stackh.   Irish moss
Eucheuma spinosum (L.) J. Agardh
Gelideum cartilagineum (L.) Gaill.
Gelideum corneum Lam.
Gracilaria confervoides (L.) Grev.
Gracilaria lichenoides (L.) Harv.
Laminaria digitata Lamour.   Kelp
Laminaria saccharina (L.) Lamour.   Kelp
Macrocystis pyrifera (L.) C. Agardh
Nereocystis Luetkeana (Mert.) Post & Rupr.
Porphyra laciniata (Lightf.) Ag.   Pink laver
Rhodymenia palmata (L.) Grev.   Dulse
Ulva lactuca (L.) LeJol.   Sea lettuce.   Green laver

Fungi

Agaricus campestris L.   Meadow mushroom
Amanita muscaria (L.) Pers. ex Fr.   Fly agaric
Bacillus brevis Migula
Bacillus polymixa (Prazmowski) Migula
Bacillus subtilis Cohn emend.   Prazmowsky
Claviceps purpurea (Fr.) Tul.   Ergot
Cortinellus edodes (Berk.) S. Ito & Imai (C. Berkeleyanus S. Ito & Imai.)   Shiitake
Morchella esculenta L.   Morel
Penicillium chrysogenum Thom
Penicillium notatum Westling
Streptomyces aureofaciens Duggar
Streptomyces Fradiae Waksman & Curtis
Streptomyces griseus (Krainsky) Waksman & Henrici
Streptomyces rimosus A. C. Finlay et al.
Streptomyces venezuelae Ehrlich, Gottlieb, Burkholder, Anderson & Pridham
Torulopsis utilis (Henneberg) Lodder.   Torula yeast
Tuber aestivum Vitt.   Truffle
Tuber brumale Gronof.   Truffle
Tuber melanosporum Vitt.   Truffle

Lichens

Evernia furfuracea (L.) Ach.   Oak moss
Evernia prunastri (L.) Ach.   Oak moss
Ramelina calicaris Rönl.
Roccella tinctoria D. C.   Archil.   Cudbear

PTERIDOPHYTA

Dryopteris Filix-Mas (L.) Schott.　Male fern
Dryopteris marginalis (L.) A. Gray.　Marginal shield fern
Lycopodium clavatum L.　Club moss

SPERMATOPHYTA

Gymnospermae

Cycadaceae

Cycas circinalis L.
Zamia floridana A.D.C.

Taxaceae

Dacrydium cupressinum Sol.　New Zealand red pine
Phyllocladus trichomanoides D. Don.　Celery pine
Podocarpus dacrydioides A. Rich.　New Zealand pine
Podocarpus Totara D. Don.　Totara
Taxus baccata L.　Yew

Pinaceae

Abies alba Mill.　(*A. pectinata* D.C.)　Silver fir
Abies balsamea (L.) Mill.　Balsam fir
Abies concolor Lindl. & Gord.　White fir
Abies grandis Lindl.　Lowland white fir
Abies lasiocarpa (Hook.) Nutt.　Alpine fir
Abies magnifica A. Murr.　Red fir
Abies procera Rehd.　(*Abies nobilis* Lindl.)　Noble fir
Agathis alba (Lam.) Foxw.　Amboyna pine
Agathis australis (Lamb.) Steud.　Kauri pine
Araucaria angustifolia (Bert.) O. Ktze.　(*A. brasiliana* Rich.)　Paraná pine
Araucaria araucana (Mol.) K. Koch.　(*A. imbricata* Pav.)　Monkey puzzle
Araucaria Cunninghamii Sweet.　Moreton Bay pine
Cedrus Deodara (Royle ex Lamb.) Loud.　Deodar
Chamaecyparis Lawsoniana (A. Murr.) Parl.　Port Orford cedar
Chamaecyparis nootkatensis (Lamb.) Sudw.　Alaska cedar
Chamaecyparis thyoides (L.) BSP.　Southern white cedar
Juniperus communis L.　Juniper
Juniperus virginiana L.　Red cedar
Larix decidua Mill.　European larch
Larix laricina (Du Roi) Koch.　Tamarack.　Larch
Larix occidentalis Nutt.　Western larch
Libocedrus decurrens Torr.　Incense cedar
Picea Abies (L.) Karst.　(*P. excelsa* (Lam.) Link.)　Norway spruce
Picea Engelmannii Engelm.　Engelmann spruce
Picea glauca (Moench.) Voss.　(*P. canadensis* (Mill.) BSP.)　White spruce
Picea mariana (Mill.) BSP.　Black spruce
Picea rubens Sarg.　(*P. rubra* (Du Roi) Dietr.)　Red spruce
Picea sitchensis (Bong.) Carr.　Sitka spruce.　Tideland spruce
Pinus australis Michx. f. (*P. palustris* Mill.)　Longleaf pine.　Southern yellow pine
Pinus Banksiana Lamb.　Jack pine
Pinus caribaea Morelet.　(*P. cubensis* Gris.) (*P. heterophylla* (Ell.) Sudw.)　Slash pine

Pinus cembroides Zucc., var. edulis (Engelm.) Voss.   (*P. edulis* Engelm.)   Piñon
Pinus cembroides Zucc., var. monophylla (Torr. & Frem.) Voss.   (*P. monophylla* Torr.)   Single-leaf piñon
Pinus contorta Dougl.   Lodgepole pine
Pinus echinata Mill.   Shortleaf pine
Pinus halepensis Mill.   Aleppo pine
Pinus Lambertiana Dougl.   Sugar pine
Pinus monticola Dougl.   Western white pine
Pinus nigra Arnold.   Black pine
Pinus Pinaster Sol.   (*P. maritima* Poir.)   Cluster pine.   Maritime pine
Pinus Pinea L.   Italian stone pine
Pinus ponderosa Doug. in Lawson.   Western yellow pine
Pinus resinosa Ait.   Red pine.   Norway pine
Pinus Sabiniana Dougl.   Digger pine
Pinus Strobus L.   White pine
Pinus succinifera (Göppert) Conw.   Baltic amber
Pinus sylvestris L.   Scotch pine.   Scots pine
Pinus Taeda L.   Loblolly pine
Pinus Torreyana Parry ex Carr.   Torrey pine
Pseudotsuga taxifolia (Poir.) Britt.   (*P. mucronata* (Raf.) Sudw.)   Douglas fir
Sequoia gigantea (Lindl.) Decne.   Big tree
Sequoia sempervirens (Lam.) Endl.   Redwood
Taxodium distichum (L.) Rich.   Bald cypress
Tetraclinis articulata (Vahl) Mast.   (*Callitris quadrivalvis* Vent.)   Sandarac
Thuja occidentalis L.   Northern white cedar.   Arbor vitae
Thuja plicata D. Don.   Western red cedar
Tsuga canadensis (L.) Carr.   Eastern hemlock
Tsuga heterophylla (Raf.) Sarg.   Western hemlock
Tsuga Mertensiana (Bon.) Sarg.   Mountain hemlock

Gnetaceae

Ephedra equisetina Bunge.   Ma-huang
Ephedra sinica Stapf.   Ma-huang

ANGIOSPERMAE

MONOCOTYLEDONAE

Typhaceae

Typha angustifolia L.   Cattail
Typha latifolia L.   Cattail

Pandanaceae

Pandanus tectorius Soland.   Screw pine
Pandanus utilis Bory.   Screw pine

Gramineae

Agrostis alba L.   Redtop
Arundinaria sp.   Bamboo
Avena brevis Roth.   Short oat
Avena byzantina C. Koch.   Red oat
Avena fatua L.   Wild oat
Avena nuda L.   Naked oat

Avena orientalis Schreb. Hungarian oat. Turkish oat
Avena sativa L. Common oat
Bambusa Tulda Roxb. Bamboo
Bambusa sp. Bamboo
Bromus inermis Leyss. Brome
Coix Lachryma-Jobi L. Job's-tears
Cymbopogon citratus (D.C.) Stapf. Lemon grass.
Cymbopogon Martinii (Roxb.) Stapf. Ginger grass. Palmarosa
Cymbopogon Nardus (L.) Rendle. Citronella
Dactylis glomerata L. Orchard grass
Dendrocalamus sp. Bamboo
Echinochloa colona (L.) Link. Shama millet
Echinochloa crus-galli (L.) Beauv. Barnyard millet
Echinochloa frumentacea (Roxb.) Link. Japanese millet
Eleusine coracana (L.) Gaertn. Ragi. African millet
Euchlaena mexicana Schrad. Teosinte
Gigantochloa sp. Bamboo
Guadua angustifolia Kunth. Bamboo
Hierochloë odorata (L.) Beauv. Sweet grass
Hordeum deficiens Steud. Two-rowed barley
Hordeum distichon L. Two-rowed barley
Hordeum intermedium Körnicke. Six-rowed barley
Hordeum spontaneum C. Köch. Wild barley
Hordeum vulgare L. Barley
Lygeum Spartum Loefl.
Muhlenbergia macroura (H.B.K.) Hitchc. (*Epicampes macroura* (H.B.K.) Benth.)
    Broomroot. Zacaton
Oryza sativa L. Rice
Panicum miliaceum L. Proso millet
Pennisetum glaucum (L.) R. Br. Pearl millet
Phleum pratense L. Timothy
Phyllostachys sp. Bamboo
Saccharum officinarum L. Sugar cane
Secale anatolicum Boiss. Wild rye
Secale cereale L. Rye
Secale montanum Guss. Wild rye
Setaria italica (L.) Beauv. Foxtail millet
Setaria viridis (L.) Beauv. Foxtail millet
Sorghum halepense (L.) Pers. Johnson grass
Sorghum virgatum (Hack.) Stapf. Tunis grass
Sorghum vulgare Pers. Sorghum
    var. caffrorum (Retz.) Hubb. & Rehder. Kafir
    var. caudatum (Hack.) A. F. Hill. Feterita
    var. cernuum (Ard.) Fiori & Paoli. White durra
    var. durra (Forsk.) Hubb. & Rehder. Brown durra
    var. nervosum (Hack.) Forbes & Hemsley. Kaoliang
    var. Roxburghii (Hack.) Haines. Shallu
    var. saccharatum (L.) Boerl. Sorgo
    var. subglabrescens (Steud.) A. F. Hill. Milo
    var. sudanense (Piper) Hitchc. Sudan grass
    var. technicum (Koern.) Fiori & Paoli. Broomcorn

Spartina Spartinae (Trin.) Merr.
Stipa tenacissima L.   Esparto
Triticum aestivum L.   (*T. vulgare* Vill.) (*T. sativum* Lam.)   Wheat
Triticum compactum Host.   Club wheat
Triticum dicoccum Schrank.   Emmer
Triticum durum Desf.   Durum wheat
Triticum monococcum L.   Einkorn
Triticum polonicum L.   Polish wheat
Triticum Spelta L.   Spelt
Triticum Timopheevi Zhuk.
Triticum turgidum L.   Poulard wheat
Vetiveria zizanioides (L.) Nash.   Khuskhus.   Vetiver
Zea Mays L.   Maize.   Indian corn
   var. erythrolepis (Bonaf.) Alef.   (var. *amylacea* (Sturt.) Bailey.)   Soft corn
   var. indentata (Sturt.) Bailey.   Dent corn
   var. indurata (Sturt.) Bailey.   Flint corn
   var. praecox Bonaf.   (var. *everta* (Sturt.) Bailey.)   Popcorn
   var. rugosa Bonaf.   (var. *saccharata* (Sturt.) Bailey.)   Sweet corn
   var. tunicata St. Hil.   Pod corn
Zizania aquatica L.   Wild rice
   var. angustifolia Hitchc.   (*Z. palustris* L.)   Wild rice

Cyperaceae

Cyperus Papyrus L.   Papyrus
Cyperus tegetiformis Roxb.   Chinese mat grass

Palmaceae

Areca Catechu L.   Betel-nut palm
Arenga pinnata (Wurmb.) Merr.   (*A. saccharifera* Labill.)   Gomuti palm
Astrocaryum Murumuru Mart.   Murumuru palm
Astrocaryum Tucumu Mart.   Tucuma palm
Astrocaryum vulgare Mart.
Attalea funifera Mart.   Bahia piassava
Borassus flabellifer L.   Palmyra palm
Calamus sp.   Rattan
Caryota urens L.   Toddy palm
Ceroxylon andicola H. & B.   Wax palm
Chamaerops humilis L.   Fan palm
Cocos nucifera L.   Coconut
Copernicia cerifera (Arr.) Mart.   Carnauba wax palm
Daemonorops Draco (Willd.) Bl.   (*Calamus Draco* Willd.)   Sumatra dragon's blood
Elaeis guineensis Jacq.   Oil palm
Leopoldinia Piassaba Wallace.   Para piassava
Metroxylon amicarum (Wendl.) Becc. (*Coelococcus amicarum* (Wendl.) W. F. Wight.)
   Ivory-nut palm
Metroxylon Sagu Rottb.   (*M. Rumphii* Mart.)   Sago palm
Orbignya Cohune (Mart.) Dahlgren.   (*Attalea Cohune* Mart.)   Cohune palm
Orbignya Martiana Barb. Rodr.   Babassu palm
Orbignya oleifera Burret.   Babassu palm
Phoenix dactylifera L.   Date palm
Phoenix sylvestris (L.) Roxb.   Wild date

Phytelephas macrocarpa Ruiz & Pav. Ivory-nut palm. Tagua palm
Raphia pedunculata Beauv. (*R. Ruffia* Mart.) Raffia palm
Raphia vinifera Beauv. West African piassava. Wine palm
Sabal causiarum (Cook) Becc. (*Inodes causiarum* Cook.) Puerto Rico hat palm
Sabal Palmetto (Walt.) Lodd. (*Inodes Palmetto* (Walt.) Cook.) Palmetto. Cabbage palm
Syagrus coronata (Mart.) Becc. Licuri palm. Ouricuri palm

Cyclanthaceae

Carludovica palmata Ruiz & Pav. Panama hat palm.

Araceae

Acorus Calamus L. Sweet flag. Calamus root
Colocasia antiquorum Schott. Taro
Colocasia esculenta (L.) Schott. Dasheen
Monstera deliciosa Liebm. Ceriman
Xanthosoma sagittifolium Schott. Yautia

Bromeliaceae

Aechmea magdalenae (André) André apud Baker. Pita floja
Ananas comosus (L.) Merr. (*A. sativus* Schult. f.) Pineapple
Neoglaziovia variegata Mez. Caroá
Tillandsia usneoides L. Spanish moss

Juncaceae

Juncus effusus L. Rush

Liliaceae

Allium ascalonicum L. Shallot
Allium Cepa L. Onion
Allium Porrum L. Leek
Allium sativum L. Garlic
Allium Schoenoprasum L. Chives
Aloe barbadensis Mill. (*A. vulgaris* Lam.) (*A. vera* L.) Barbados aloes
Aloe ferox Mill. Cape aloes
Aloe Perryi Baker. Socotrine aloes
Asparagus officinalis L. Asparagus
Chlorogalum pomeridianum (Ker-Gawl.) Kunth. California soaproot
Colchicum autumnale L. Meadow saffron. Colchicum root
Convallaria majalis L. Lily of the valley
Dracaena cinnabari Balf. f. Socotra dragon's blood
Hyacinthus orientalis L. Hyacinth
Phormium tenax Forst. New Zealand hemp
Samuela carnerosana Trel. Palma istle
Sansevieria longifolia Sims. Florida bowstring hemp
Sansevieria Roxburghiana Schult. f. Indian bowstring hemp
Sansevieria thyrsiflora Thunb. (*S. guineensis* (Jacq.) Willd.) African bowstring hemp
Sansevieria zeylanica (Jacq.) Willd. Ceylon bowstring hemp
Smilax aristolochiaefolia Mill. Mexican sarsaparilla
Smilax officinalis H.B.K. Honduras sarsaparilla
Smilax Regelii Killip & Morton. Jamaican sarsaparilla
Urginea maritima (L.) Baker. Squills. Sea onion

Xanthorrhoea australis R. Br.   Grass tree
Xanthorrhoea hastilis R. Br.   Grass tree
Xanthorrhoea tateana F. v. Muell.
Yucca sp.   Palma istle

Amaryllidaceae

Agave atrovirens Karw.   Pulque plant
Agave Cantala Roxb.   Manila maguey.   Cantala
Agave fourcroydes Lem.   Henequen.   Mexican sisal
Agave Funkiana C. Koch & Bouché.   Jaumave istle
Agave Lecheguilla Torr.   Tula istle
Agave Letonae F. W. Taylor ex Trel.   Salvador henequen.   Letona
Agave sisalina Perrine.   Sisal
Agave sp.   Mexican maguey
Furcraea Cabuya Trel.   Cabuya
Furcraea gigantea (D. Dietr.) Vent.   Mauritius hemp.   Piteira.   Green aloe
Furcraea hexapetala (Jacq.) Urb.   Pitre.   Cuban hemp
Furcraea macrophylla Baker.   Fique
Narcissus Jonquilla L.   Jonquil
Narcissus Tazetta L.   Narcissus
Polianthes tuberosa L.   Tuberose

Dioscoreaceae

Dioscorea alata L.   Yam

Iridaceae

Crocus sativus L.   Saffron crocus
Iris florentina L.   Orris
Iris pallida L.   Orris

Musaceae

Musa nana Lour.   (*M. Cavendishii* Lamb.)   Dwarf banana
Musa paradisiaca L.   Plantain
  subsp. sapientum (L.) O. Ktze.   (*M. sapientum* L.)   Banana
Musa textilis Née.   Abacá.   Manila hemp

Zingiberaceae

Aframomum Melegueta (Rosc.) K. Schum.   Grains of paradise
Alpinia Galanga (L.) Sw.   Greater galangal
Alpinia officinarum Hance.   Lesser galangal
Amomum sp.   Grains of paradise
Curcuma angustifolia Roxb.   East Indian arrowroot
Curcuma longa L.   Turmeric
Curcuma zedoaria (Berg.)   Zedoary
Elettaria Cardamomum (L.) Maton.   Cardamom
Zingiber officinale Rosc.   Ginger

Cannaceae

Canna edulis Ker-Gawl.   Queensland arrowroot.   Achira

Marantaceae

Calathea lutea G. F. W. Mey.
Maranta arundinacea L.   West Indian arrowroot

Orchidaceae

Vanilla planifolia Andr. (*V. fragrans* (Salisb.) Ames.) Vanilla
Vanilla Pompona Scheide. West Indian vanilla

DICOTYLEDONAE

Archychlamydeae

Piperaceae

Piper Betle L. Betel pepper
Piper Cubeba L. f. Cubeb
Piper longum L. Long pepper
Piper methysticum Forst. Kavakava
Piper nigrum L. Black pepper
Piper retrofractum Vahl. (*P. officinarum* C.D.C.) Long pepper

Salicaceae

Populus balsamifera L. (*P. Tacamahacca* Mill.) Balsam poplar
Populus deltoides Marsh. Cottonwood
Populus grandidentata Michx. Large-toothed aspen
Populus tremuloides Michx. Quaking aspen
Salix alba L. White willow
Salix nigra Marsh. Black willow

Myricaceae

Myrica cerifera L. Wax myrtle
Myrica pensylvanica Lois. (*M. carolinensis* Mill.) Bayberry

Juglandaceae

Carya glabra (Mill.) Sweet. Pignut
Carya illinoensis (Wang.) K. Koch (*C. Pecan* (Marsh) Eng. & Graeb.) Pecan
Carya laciniosa (Michx. f.) Loud. Shellbark hickory
Carya ovata (Mill.) K. Koch. Shagbark hickory
Carya tomentosa (Lam.) Nutt. (*C. alba* (L.) K. Koch.) Mockernut
Juglans cinerea L. Butternut
Juglans nigra L. Black walnut
Juglans regia L. English walnut. Persian walnut

Betulaceae

Alnus glutinosa (L.) Gaertn. European black alder
Alnus rubra Bong. Red alder
Betula lenta L. Black birch. Sweet birch
Betula lutea Michx. f. Yellow birch
Betula papyrifera Marsh. Paper birch. White birch
Betula pendula Roth. (*B. alba* L. in part.) European white birch
Betula pubescens Ehrh. (*B. alba* L. in part.) European white birch
Carpinus Betulus L. European hornbeam
Carpinus caroliniana Walt. Blue beech
Corylus americana Walt. Hazelnut
Corylus Avellana L. Filbert. European hazelnut
Corylus cornuta Marsh. (*C. rostrata* Ait.) Beaked hazelnut
Corylus maxima Mill. Filbert
Ostrya virginiana (Mill.) K. Koch. Hop hornbeam

SYSTEMATIC LIST OF SPECIES DISCUSSED

Fagaceae

Castanea crenata Sieb & Zucc. Japanese chestnut
Castanea dentata (Marsh.) Borkh. Chestnut
Castanea sativa Mill. (*C. vulgaris* Lam.) European chestnut
Fagus grandifolia Ehrh. Beech
Fagus sylvatica L. European beech
Lithocarpus densiflora (Hook. & Arn.) Rehder. California tanbark oak
Quercus alba L. White oak
Quercus bicolor L. Swamp white oak
Quercus borealis Michx. f. (*Q. rubra* DuRoi.) Red oak
Quercus Cerris L. European turkey oak
Quercus coccinea Muench. Scarlet oak
Quercus Frainetto Ten. (*Q. conferta* Ait.) Italian oak
Quercus Garryana Dougl. Oregon white oak
Quercus Ilex L. Holm oak
Quercus imbricaria Michx. Shingle oak
Quercus infectoria Oliv. Aleppo oak
Quercus laevis Walt. (*Q. Catesbaei* Michx.) Turkey oak
Quercus macrocarpa Michx. Bur oak
Quercus macrolepis Klotzsch (*Q. Aegilops* L.) Turkish oak
Quercus montana Willd. (*Q. Prinus* Endl.) Chestnut oak
Quercus palustris Muench. Pin oak
Quercus petraea (Mattuschka) Liebl. (*Q. sessiliflora* Salisb.)
Quercus phellos L. Willow oak
Quercus Prinus L. (*Q. Michauxii* Nutt.) Swamp chestnut oak
Quercus Robur L. English oak
Quercus stellata Wang. Post oak
Quercus Suber L. Cork oak
Quercus texana Buckley. Texas red oak
Quercus velutina Lam. Black oak
Quercus virginiana Mill. Live oak

Ulmaceae

Celtis occidentalis L. Hackberry
Ulmus americana L. White elm
Ulmus procera Salisb. (*U. campestris* Mill. in part.) English elm
Ulmus rubra Muhl. (*U. fulva* Michx.) Slippery elm
Ulmus Thomasi Sarg. (*U. racemosa* Thom.) Rock elm

Moraceae

Antiaris toxicaria (Pers.) Lesch. Upas tree
Artocarpus altilis (Park.) Fosb. (*A. communis* Forst.) Breadfruit
Artocarpus heterophyllus Lam. (*A. integra* (Thunb.) Merr.) Jack fruit
Broussonetia papyrifera (L.) Vent. Paper mulberry
Castilla elastica Cerv. Panama rubber. Castilla rubber
Castilla Ulei Warb. Caucho rubber
Cannabis sativa L. Hemp
Chlorophora tinctoria (L.) Gaud. Old fustic
Ficus Carica L. Fig
Ficus elastica Roxb. India rubber. Assam rubber
Ficus Nekbudu Warb. (*F. utilis* Sim.)

Ficus religiosa L.
Humulus Lupulus L.   Hops
Maclura pomifera (Raf.) Schneid.   Osage orange
Morus alba L.   White mulberry
Morus nigra L.   Black mulberry
Morus rubra L.   Red mulberry
Piratinera guianensis Aubl.   Snakewood.   Letterwood

Urticaceae

Boehmeria nivea (L.) Gaud.   Ramie
   var. tenacissima (Gaud.) Miquel.   Rhea

Proteaceae

Grevillea robusta A. Cunn.   Silky oak
Macadamia ternata F. v. Muell.   Queensland nut.   Macadamia nut

Santalaceae

Santalum album L.   Sandalwood

Polygonaceae

Coccoloba uvifera L.   Sea grape
Fagopyrum sagittatum Gilib.   (*F. esculentum* Moench.)   Buckwheat
Rheum Emodi Wall.   Indian rhubarb
Rheum officinale Baill.   Rhubarb
Rheum palmatum L.   Rhubarb
Rheum Rhaponticum L.   Garden rhubarb
Rumex hymenosepalus Torr.   Canaigre.   Tanner's dock

Chenopodiaceae

Beta maritima L.   Wild beet
Beta vulgaris L.   Garden beet.   Sugar beet.   Mangels
   var. Cicla L.   Chard
Chenopodium ambrosioides L., var. anthelminticum (L.) Gray.   Wormseed
Chenopodium Quinoa Willd.   Quinoa
Spinacia oleracea L.   Spinach

Amaranthaceae

Amaranthus gangeticus L.   Tampala

Aizoaceae

Tetragonia expansa Murr.   New Zealand spinach

Basellaceae

Ullucus tuberosus Calda.   Ullucu.   Melloco

Caryophyllaceae

Dianthus Caryophyllus L.   Carnation
Saponaria officinalis L.   Soapwort.   Bouncing Bet

Ranunculaceae

Aconitum Napellus L.   Aconite.   Monkshood
Hydrastis canadensis L.   Goldenseal

Berberidaceae

Podophyllum Emodi Wall.   Indian podophyllum
Podophyllum peltatum L.   Mandrake.   May apple

Magnoliaceae

Illicium verum Hook. f.   Star anise
Liriodendron Tulipifera L.   Tulip tree
Magnolia acuminata L.   Cucumber tree
Michelia Champaca L.   Champac

Annonaceae

Annona Cherimolia Mill.   Cherimoya
Annona muricata L.   Soursop.   Guanabana
Annona reticulata L.   Bullock's-heart
Annona squamosa L.   Sweetsop
Asimina triloba (L.) Dunal.   Papaw
Cananga odorata (Lam.) Hook. f. & Thoms.   Ylang-ylang
Oxandra lanceolata (Sw.) Baill.   Lancewood

Myristicaceae

Myristica fragrans Houtt.   Nutmeg
Virola spp.   Ucuhuba butter.   Otoba butter

Lauraceae

Aniba panurensis Mez.   Cayenne linaloe.   Bois du rose
Aniba rosaeodora Ducke var. amazonia Ducke.   Brazilian bois du rose
Cinnamomum Burmannii (Nees) Blume.   Padang cassia
Cinnamomum Camphora (L.) T. Nees & Eberm.   Camphor
Cinnamomum Cassia (Nees) Nees ex Blume.   Cassia
Cinnamomum Loureirii Nees.   Saigon cinnamon
Cinnamomum Massoia Schewe.   (*Massoia aromatica* Becc.)   Massoia bark
Cinnamomum Oliveri Bailey.   Oliver's bark
Cinnamomum Tamala (Buch.-Ham.) T. Nees & Eberm.   Indian cassia
Cinnamomum zeylanicum Breyn.   Cinnamon
Laurus nobilis L.   Laurel.   Sweet bay
Ocotea Rodioei (Rob. Schomb.) Mez.   (*Nectandra Rodioei* Schomb.)   Greenheart
Persea americana Mill.   Avocado.   Alligator pear
Sassafras albidum Nees.   (*S. variifolium* (Salisb.) Kuntze.) (*S. officinale* Nees & Eberm.)   Sassafras

Papaveraceae

Argemone mexicana L.   Mexican poppy
Papaver somniferum L.   Opium poppy

Capparidaceae

Capparis spinosa L.   Caper bush

Cruciferae

Armoracia lapathifolia Gilib.   (*A. rusticana* Gaertn.), (*Rorippa Armoracia* (L.) Hitchc.)   Horse-radish
Brassica campestris L.   Field mustard

Brassica chinensis L.   Chinese cabbage
Brassica hirta Moench.   (*B. alba* (L.) Boiss.)   White mustard
Brassica juncea (L.) Cosson.   Indian mustard
Brassica Napobrassica (L). Mill.   Rutabaga
Brassica Napus L.   Rape
Brassica nigra (L) Koch.   Black mustard
Brassica oleracea L.   Wild cabbage
   var. acephala D.C.   Borecole.   Kale
   var. botrytis L.   Cauliflower.   Broccoli
   var. capitata L.   Cabbage
   var. gemmifera Zenk.   Brussels sprouts
   var. gongylodes L.   (*B. caulorapa* Pasq.)   Kohlrabi
Brassica pekinensis (Lour.) Rupr.   Chinese cabbage
Brassica Rapa L.   Turnip
Camelina sativa (L.) Crantz.   False flax
Isatis tinctoria L.   Woad
Nasturtium officinale R. Br.   (*Rorippa Nasturtium-aquaticum* (L.) Britt. & Rendle).
   Water cress
Raphanus sativus L.   Radish

Resedaceae

Reseda Luteola L.   Weld
Reseda odorata L.   Mignonette

Moringaceae

Moringa oleifera Lam.   (*M. pterygosperma* Gaertn.)

Saxifragaceae

Ribes americanum Mill.   Wild currant
Ribes Grossularia L.   Gooseberry
Ribes hirtellum Michx.   Wild gooseberry
Ribes nigrum L.   Black currant
Ribes sativum Syme.   (*R. vulgare* Lam.)   Red currant

Hamamelidaceae

Hamamelis virginiana L.   Witch hazel
Liquidambar orientalis Mill.   Styrax
Liquidambar Styraciflua L.   Red gum.   Sweet gum

Platanaceae

Platanus occidentalis L.   Sycamore
Platanus orientalis L.   European plane tree

Rosaceae

Chrysobalanus Icaco L.   Coco plum
Cydonia oblonga Mill.   (*C. vulgaris* Pers.)   Quince
Eriobotrya japonica (Thunb.) Lindl.   Loquat
Fragaria chiloensis (L.) Duchesne.   Strawberry
Fragaria vesca L.   Strawberry
Fragaria virginiana Duchesne.   Strawberry
Licania rigida Benth.   Oiticica

Mespilus germanica L.   Medlar
Prunus americana Marsh.   Wild plum
Prunus Amygdalus Batsch.   (*P. communis* (L.) Arcang.)   Almond
    var. amara D.C.   Bitter almond
    var. dulcis D.C.   Sweet almond
Prunus Armeniaca L.   Apricot
Prunus avium L.   Sweet cherry
Prunus Cerasus L.   Sour cherry
Prunus domestica L.   European plum
Prunus hortulana Bailey.   Hortulana plum
Prunus insititia L.   Bullace plum
Prunus maritima Marsh.   Beach plum
Prunus nigra Ait.   Wild plum
Prunus Persica (L.) Sieb. & Zucc.   Peach
    var. nectarina (Ait.) Maxim.   (var. *nucipersica* (L.) Schneid.)   Nectarine
Prunus salicina Lindl.   Japanese plum
Prunus serotina Ehrh.   Wild black cherry
Prunus serrulata Lindl.   Japanese flowering cherry
Prunus spinosa L.   Sloe.   Blackthorn
Pyrus baccata L.   (*Malus baccata* (L.) Borkh.)   Siberian crab apple
Pyrus communis L.   Pear
Pyrus Malus L.   (*Malus pumila* Mill.)   Apple
Pyrus pyrifolia (Burm.) Nakai var. culta (Mak.) Nakai.   (*P. serotina* Rehd., var. *culta* Rehd.)   Chinese pear.   Sand pear
Quillaja Saponaria Mol.   Soapbark
Rosa centifolia L.   Cabbage rose
Rosa damascena Mill.   Damask rose
Rubus alleghaniensis Porter.   Blackberry
Rubus argutus Link.   Blackberry
Rubus flagellaris Willd.   Dewberry
Rubus frondosus Bigel.   Blackberry
Rubus Idaeus L.   Red raspberry
    var. strigosus (Michx.) Maxim.   Wild raspberry
Rubus loganobaccus Bailey.   Loganberry
Rubus occidentalis L.   Black raspberry
Rubus trivialis Michx.   Dewberry
Rubus vitifolius C. & S.   California dewberry
Sorbus Aucuparia L.   (*Pyrus Aucuparia* (L.) Ehrh.)   Rowan

Leguminosae

Acacia Catechu (L.f.) Willd.   Black cutch.   Catechu
Acacia decurrens (Wendl.) Willd.   Black wattle
    var. dealbata (Page) F. v. Muell.   (*A. dealbata* Link).   Silver wattle
    var. mollis Lindl.   Green wattle
Acacia Farnesiana (L.) Willd.   Cassie
Acacia melanoxylon R. Br.   Australian blackwood
Acacia nilotica (L.) Delile.   (*A. arabica* (Lam.) Willd.)   Babul
Acacia pycnantha Benth.   Golden wattle
Acacia Senegal (L.) Willd.   Gum arabic
Arachis hypogaea L.   Peanut.   Ground nut
Astragalus gummifer Labill.   Gum tragacanth

Baphia nitida Afzel. ex Lodd.   Camwood
Butea monosperma (Roxb.) Taub.   (*B. frondosa* Roxb.)   Bengal kino
Brya Ebenus (L.) D.C.   Cocus wood.   Granadillo.   American ebony
Caesalpinia brevifolia Baill.   Algarobilla
Caesalpinia coriaria (Jacq.) Willd.   Divi-divi
Caesalpinia echinata Lam.   Brazilwood
Caesalpinia Sappan L.   Sappanwood
Caesalpinia spinosa (Mol.) Ktze.   Tara
Cajanus Cajan (L.) Millsp.   (*C. indicus* Spreng.)   Cajan pea.   Pigeon pea
Canavalia ensiformis (L.) D.C.   Jack bean.   Horse bean
Cassia acutifolia Del.   Alexandrian senna
Cassia angustifolia Vahl.   Indian senna.   Tinnavelly senna
Cassia auriculata L.   Avaram
Ceratonia Siliqua L.   Carob
Cicer arietinum L.   Chick pea
Copaifera conjugata (Bolle) Milne-Redhead.   (*C. Gorskiana* Benth.)   Inhambane copal
Copaifera copallifera (Benn.) Milne-Redhead.   (*C. Guibortiana* Benth.)   Sierra Leone copal
Copaifera Demeusii Harms.   Congo copal
Copaifera mopane J. Kirk.   Congo copal
Copaifera officinalis L.   Copaiba balsam
Copaifera reticulata Ducke.   Copaiba balsam
Copaifera Salikounda Heckel.   Sierra Leone copal
Crotalaria juncea L.   Sunn hemp
Dalbergia latifolia Roxb.   Indian rosewood
Dalbergia nigra Fr. Allem.   Brazilian rosewood
Dalbergia retusa Hemsl.   Cocobolo
Dalbergia Sissoo Roxb.   Sissoo
Daniella Ogea Rolfe.   Accra copal.   Benin copal
Daniella Oliveri (Rolfe) Hutch. & Dalz.   Illurin balsam
Daniella thurifera Benn.   Sierra Leone frankincense
Derris elliptica (Wall.) Benth.   Derris.   Tuba
Derris trifoliata (Lour.) Taub.   (*D. uliginosa* (Roxb.) Benth.)   Derris
Dipteryx odorata (Aubl.) Willd.   Tonka bean
Dipteryx oppositifolia (Aubl.) Willd.   Tonka bean
Dolichos Lablab L.   Hyacinth bean.   Bonavist.   Lablab
Gleditsia triacanthos L.   Honey locust
Glycine Max (L.) Merr.   (*G. Soja* (L.) Sieb. & Zucc.)   Soybean
Glycyrrhiza glabra L.   Licorice
Gymnocladus dioica (L.) K. Koch.   Coffee tree
Haematoxylon Brasiletto Karst.   Brazilette.   Hypernic
Haematoxylon campechianum L.   Logwood
Hymenaea Courbaril L.   South American locust.   West Indian locust
Indigofera suffruticosa Mill.   Indigo
Indigofera tinctoria L.   Indigo
Inga edulis Mart.
Lens culinaris Medik.   (*L. esculenta* Moench.)   Lentil
Lespedeza cuneata G. Don.   (*L. sericea* Miq.)   Lespedeza
Lespedeza stipulacea Maxim.   Korean lespedeza
Lespedeza striata (Thunb.) H. & A.   Lespedeza
Lonchocarpus Nicou (Aubl.) D.C.   Cubé.   Timbo.   Barbasco
Lonchocarpus Urucu Killip & Smith.   Cubé

Lonchocarpus utilis Killip & Smith.   Cubé
Medicago hispida Gaertn.   Bur clover
Medicago lupulina L.   Medic
Medicago sativa L.   Alfalfa
Melilotus alba Desr.   White sweet clover
Melilotus officinalis (L.) Lam.   Yellow sweet clover
Mora excelsa Benth.   (*Dimorphandra Mora* B. & H.f.)   Mora
Myroxylon Balsamum (L.) Harms.   (*M. toluiferum* H.B.K.)   Balsam of Tolu
Myroxylon Pereirae (Royle) Klotzsch.   Balsam of Peru
Pachyrrhizus erosus (L.) Urban.   Yam bean
Parkia biglobosa (Willd.) Benth.   Nitta.   African locust bean
Parkia filicoidea Welw.   African locust bean
Peltogyne paniculata Benth.   Purpleheart
Phaseolus angularis (Willd.) W. F. Wight.   Adzuki bean
Phaseolus aureus Roxb.   Mung bean
Phaseolus calcaratus Roxb.   Rice bean
Phaseolus coccineus L.   (*P. multiflorus* Willd.)   Scarlet runner bean
Phaseolus limensis Macf.   Lima bean
Phaseolus lunatus L.   Sieva bean
Phaseolus vulgaris L.   Common bean
Pisum sativum L.   Pea
Pongamia pinnata (L.) Pierre.   (*P. glabra* (L.) Vent.)   Pongam
Prosopis glandulosa Torr.   Mesquite
Prosopis juliflora (Sw.) D.C.   Algaroba.   Mesquite.   Keawe
Pterocarpus erinaceus Poir.   West African kino.   Barwood
Pterocarpus indicus Willd.   Padouk.   Burmese rosewood
Pterocarpus Marsupium Roxb.   Malabar kino
Pterocarpus santalinus L. f.   Red sanderswood.   Red sandalwood
Pterocarpus Soyauxü Taub.   Barwood
Pueraria lobata (Willd.) Ohwi.   (*P. Thunbergiana* (Sieb. & Zucc.) Benth).   (*P. hirsuta* Schneid.)   Kudzu
Robinia pseudoacacia L.   Black locust
Samanea Saman (Jacq.) Merr.   (*Enterolobium Saman* (Jacq.) Prain.)   Rain tree
Sesbania exaltata (Raf.) Rydb.   (*S. macrocarpa* Muhl.)   Colorado river hemp
Stizolobium Deeringianum Bort.   (*Mucuna Deeringiana* (Bort.) Merr.)   Velvet bean
Tamarindus indica L.   Tamarind
Trachylobium verrucosum (Gaertn.) Oliv.   Zanzibar copal
Trifolium hybridum L.   Alsike clover
Trifolium incarnatum L.   Crimson clover
Trifolium pratense L.   Red clover
Trifolium repens L.   White clover.   Ladino
Trigonella Foenum-graecum L.   Fenugreek
Vicia Faba L.   Broad bean.   Windsor bean
Vicia sativa L.   Vetch
Vicia villosa Roth.   Hairy vetch
Vigna sinensis (L.) Savi.   Cowpea
Xylia xylocarpa (Roxb.) Taub.   (*X. dolabriformis* Benth.)   Acle.   Pyinkado

Oxalidaceae

Averrhoa Bilimbi L.   Bilimbi
Averrhoa Carambola L.   Carambola
Oxalis tuberosa Molina.   Oca

Geraniaceae

Pelargonium graveolens L'Her.   Rose geranium
Pelargonium odoratissimum (L.) Ait.   Rose geranium

Tropaeolaceae

Tropaeolum tuberosum Ruiz & Pavon.   Añu

Linaceae

Linum usitatissimum L.   Flax

Rutaceae

Amyris balsamifera L.   Mexican elemi
Amyris elemifera L.   Mexican elemi
Barosma betulina (Thunb.) Bartl. & Wendl.   Buchu
Barosma crenulata (L.) Hook.   Buchu
Barosma serratifolia (Curt.) Willd.   Buchu
Casimiroa edulis LaLlave & Lex.   White sapote
Chloroxylon Swietenia D.C.   East Indian satinwood
Citrus aurantifolia (Christman) Swingle.   Lime
Citrus Aurantium L.   Sour orange.   Seville orange.   Bitter orange
   subsp. Bergamia (Risso & Poit.) Wight & Arn.   (*C. Bergamia* Risso.)   Bergamot
Citrus grandis (L.) Osbeck.   (*C. maxima* (Burm.) Merr.)   Pomelo.   Shaddock
Citrus Limon (L.) Burm. f.   (*C. Limonia* Osbeck.)   Lemon
Citrus Medica L.   Citron
Citrus paradisi Macfad.   (*C. maxima* (Burm.) Merr. var *uvacarpa* Merr. & Lee.)
   Grapefruit
Citrus reticulata Blanco.   (*C. nobilis* Lour. and vars.)   Mandarin.   Tangerine
Citrus sinensis (L.) Osbeck.   Sweet orange
Feronia Limonia (L). Swingle.   Elephant apple
Fortunella japonica (Thunb.) Swingle.   Kumquat
Fortunella margarita (Lour.) Swingle.   Kumquat
Poncirus trifoliata (L.) Raf.   Trifoliate orange.   Deciduous orange
Ruta graveolens L.   Rue
Zanthoxylum flavum Vahl.   West Indian satinwood

Simarubaceae

Picrasma excelsa (Sw.) Planch.   (*Picraena excelsa* (Sw.) Lindl.)   Jamaica quassia
Quassia amara L.   Quassia

Burseraceae

Boswellia Carteri Birdw.   Olibanum.   Frankincense
Boswellia Frereana Birdw.   African elemi
Boswellia serrata Roxb.   Indian frankincense
Bursera glabrifolia (H.B.K.) Engl.   (*B. aloexylon* (Schlecht.) Engl.)   Mexican linaloe
Bursera gummifera L.   Brazilian elemi
Bursera penicillata (Sesse & Moc. ex D.C.) Engl.   (*B. Delpechianum* Poiss.)   Mexican
   linaloe
Canarium commune L.   Java almond
Canarium luzonicum (Blume) A. Gray.   Manila elemi.   Pili
Canarium ovatum Engl.   Pili nut
Canarium strictum Roxb.   Black damar
Commiphora africana (Arn.) Engl.   African bdellium

Commiphora erythraea (Ehrenb.) Engl. Bisabol myrrh. Sweet myrrh
Commiphora Kataf (Forsk.) Engl. Opopanax
Commiphora Mukul (Hook.) Engl. Indian bdellium
Commiphora Myrrha (Nees) Engl. Herabol myrrh
Commiphora Opobalsamum (L.) Engl. Mecca balsam
Protium heptaphyllum (Aubl.) March. Brazilian elemi

Meliaceae

Carapa guianensis Aubl. Crabwood
Carapa moluccensis Lam.
Cedrela odorata L. Spanish cedar. Cigar-box cedar
Cedrela Toona Roxb. (*Toona ciliata* M. Roem.) Moulmein cedar
Khaya senegalensis (Desr.) A. Juss. African mahogany
Swietenia macrophylla King. Honduras mahogany
Swietenia Mahogani (L.) Jacq. West Indian mahogany

Malpighiaceae

Banisteriosis Caapi (Spruce) Morton. Caapi. Yagé. Ayahuasca
Banisteriopsis inebrians Morton. Caapi
Banisteriopsis quitensis (Niedenzu) Morton. Caapi
Malpighia glabra L. Barbados cherry
Tetrapterys sp. Caapi

Polygalaceae

Polygala Senega L. Senega snakeroot

Euphorbiaceae

Aleurites Fordii Hemsl. Tung-oil tree
Aleurites moluccana (L.) Willd. Candlenut. Lumbang
Aleurites montana (Lour.) Wilson. (*A. cordata* (A. Juss.) Steud.) Mu tree
Cnidoscolus sp. Chilte rubber
Croton Tiglium L. Croton
Euphorbia antisyphilitica Zucc. Candellila wax
Euphorbia Intisy Drake del Castillo. Intisy
Hevea Benthamiana Müll.-Arg.
Hevea brasiliensis (Willd. ex A. Juss.) Müll.-Arg. Hevea rubber. Para rubber
Manihot esculenta Crantz (*M. utilissima* Pohl.) Cassava
Manihot Glaziovii Müll.-Arg. Ceara rubber
Micrandra sp. Caura rubber
Phyllanthus acidus (L.) Skeels. Otaheite gooseberry
Ricinus communis L. Castor bean
Sapium sebiferum (L.) Roxb. Chinese vegetable tallow

Buxaceae

Buxus sempervirens L. Turkish boxwood
Simmondsia chinensis (Link) Schneid. (*S. californica* Nutt.) Jojoba

Anacardiaceae

Anacardium occidentale L. Cashew
Cotinus coggygria Scop. (*Rhus Cotinus* L.) Smoke tree
Mangifera indica L. Mango
Melanorrhoea usitata Wall. Burmese lacquer tree

Pistacia cabulica Stocks.   Bombay mastic
Pistacia lentiscus L.   Chios mastic
Pistacia vera L.   Pistachio.   Green almond
Rhus chinensis Mill.   (*R. semialata* Murr.)   Chinese sumac
Rhus copallina L.   Dwarf sumac
Rhus coriaria L.   Sicilian sumac
Rhus glabra L.   Smooth sumac
Rhus succedanea L. f.   Japanese wax tree
Rhus typhina L.   Staghorn sumac
Rhus verniciflua Stokes.   Lacquer tree
Schinopsis Balansae Engelm.   Quebracho
Schinopsis Lorentzii (Griseb.) Engl.   (*Quebrachia Lorentzii* Griseb.)   Quebracho
Spondias cytherea Sonn.   (*S. dulcis* Forst.)   Golden apple.   Otaheite apple.   Ambarella
Spondias Mombin L.   (*S. lutea* L.)   Yellow mombin.   Hog plum
Spondias purpurea L.   (*S. Mombin* auth.)   Red mombin.   Spanish plum

Aquifoliaceae

Ilex Aquifolium L.   European holly
Ilex opaca Ait.   Holly
Ilex paraguariensis St. Hil.   Maté.   Paraguay tea
Ilex vomitoria Ait.   Yaupon.   Cassine

Celastraceae

Catha edulis Forsk.   Khat

Aceraceae

Acer macrophyllum Pursh.   Oregon maple
Acer nigrum Michx.   Black maple
Acer pseudoplatanus L.   Sycamore maple
Acer rubrum L.   Red maple
Acer saccharinum L.   Silver maple
Acer saccharum Marsh.   Sugar maple

Sapindaceae

Litchi chinensis Sonn.   Litchi
Paullinia Cupana H.B.K.   Guarana
Paullinia Yoco Schultes & Killip.   Yoco
Sapindus Saponaria L.   Soapberry
Schleichera oleosa (Lour.) Merr.   (*S. trijuga* Willd.)   Lac tree

Hippocastanaceae

Aesculus octandra Marsh.   Yellow buckeye

Rhamnaceae

Rhamnus cathartica L.   Buckthorn
Rhamnus globosa Bge.   (*R. chlorophora* Decne.)   Lokao
Rhamnus infectoria L.   Persian berries
Rhamnus Purshiana D.C.   Cascara sagrada
Rhamnus utilis Decne.   Lokao.   Chinese buckthorn
Zizyphus Jujuba Mill.   Jujube.   Chinese date
Zizyphus xylopyrus Willd.   Jujube

Vitaceae

Vitis aestivalis Michx.   Summer grape
Vitis Labrusca L.   Fox grape
Vitis rotundifolia Michx.   Muscadine grape
Vitis vinifera L.   Wine grape
Vitis vulpina L.   Frost grape

Tiliaceae

Corchorus capsularis L.   Jute
Corchorus olitorius L.   Jute
Tilia americana L.   (*T. glabra* Vent.)   Basswood.   Linden
Tilia cordata Mill.   European linden.   Lime

Malvaceae

Abutilon Theophrasti Medic.   (*A. Avicennae* Gaertn.)   China jute.   Indian mallow
Gossypium arboreum L.   Tree cotton
Gossypium barbadense L.   Sea-island cotton.   Egyptian cotton
Gossypium herbaceum L.   Asiatic cotton
Gossypium hirsutum L.   Upland cotton
Hibiscus cannabinus L.   Kenaf.   Gambo hemp.   Deccan hemp.   Ambari hemp
Hibiscus elatus Sw.   Blue mahoe.   Cuba bast
Hibiscus esculentus (L.) Moench.   Okra
Hibiscus Sabdariffa L.   Roselle.   Rama
Hibiscus tiliaceus L.   Mahoe.   Majagua
Sida acuta Burm.
Urena lobata L.   Aramina.   Cadillo

Bombacaceae

Adansonia digitata L.   Baobab
Ceiba aesculifolia (H.B.K.) Britt. & Baker.   Pochote
Ceiba acuminata (S. Wats.) Rose.   Pochote
Ceiba pentandra (L.) Gaertn.   Kapok
Chorisia insignis H.B.K.   Palo borracho
Chorisia speciosa St. Hil.   Samohu
Durio zibethinus L.   Durian
Ochroma pyramidale (Cav.) Urb.   (*O. Lagopus* Sw.)   Balsa
Salmalia malabarica (D.C.) Schott & Endl.   (*Bombax Ceiba* L.)   Red silk cotton

Sterculiaceae

Cola nitida (Vent.) A. Chev.   (*C. acuminata* (Beauv.) Schott & Endl.)   Cola
Sterculia urens Roxb.
Theobroma Cacao L.   Cocoa.   Cacao

Ternstroemiaceae

Camellia Sasanqua Thunb.
Camellia sinensis (L.) O. Ktze.   (*Thea sinensis* L.)   Tea

Guttiferae

Calophyllum inophyllum L.   Indian laurel.   Laurelwood
Garcinia Hanburyi Hook. f.   Gamboge tree
Garcinia Mangostana L.   Mangosteen
Mammea americana L.   Mammee apple

Dipterocarpaceae

Balanocarpus Heimii King.   Damar Penak
Dipterocarpus turbinatus Gaertn. f.   Gurjun balsam
Dryobalanops aromatica Gaertn. f.   Borneo camphor
Hopea micrantha Hook. f.   Damar Mata Kuching
Shorea aptera Burck.   Borneo tallow
Shorea hypochra Hance.   (*S. crassifolia* Ridl.)   Damar Temak
Shorea robusta Gaertn. f.   Sal
Shorea Wiesneri Schiffn.   Batavian damar
Vateria indica L.   White damar

Bixaceae

Bixa Orellana L.   Annatto.   Urucú

Cochlospermaceae

Cochlospermum religiosum (L.) Alston.   (*C. Gossypium* D.C.)   White silk cotton

Violaceae

Viola odorata L.   Violet

Flacourtiaceae

Flacourtia indica (Burm. f.) Merr.   Governor's plum
Gossypiospermum praecox (Griseb.) P. Wils.   (*Casearia praecox* Griseb.)   Venezuelan boxwood.   Zapatero
Hydnocarpus Kurzii (King) Wrbg.   (*Taraktogenos Kurzii* King.)   Chaulmoogra

Passifloraceae

Passiflora edulis Sims.   Purple granadilla
Passiflora ligularis A. Juss.   Sweet granadilla
Passiflora quadrangularis L.   Giant granadilla

Caricaceae

Carica Papaya L.   Papaya.   Papaw

Cactaceae

Lophophora Williamsii (Lem.) Coult.   Peyote.   Mescal buttons

Thymelaeaceae

Daphne cannabina Wall.
Edgeworthia tomentosa (Thunb.) Nakai.   (*E. Gardneri* (Wall.) Meisn.)   (*E. papyrifera* Sieb. & Zucc.)
Lagetta lintearia Lam.   Lacebark
Wickstroemia canescens (Wall.) Meisn.

Lythraceae

Lawsonia inermis L.   Henna

Punicaceae

Punica Granatum L.   Pomegranate

Lecythidaceae

Bertholletia excelsa Humb. & Bonp.   Brazil nut
Couratari Tauari Berg.   Tauary

Lecythis usitata Miers.   Paradise nut.   Sapucaia
Lecythis Zabucajo Aubl.   Paradise nut.   Sapucaia
Lecythis spp.   Paradise nuts

Rhizophoraceae

Rhizophora Mangle L.   Mangrove

Combretaceae

Anogeissus latifolia Wall.   Gum ghatti
Terminalia Bellerica (Gaertn.) Roxb.   Myrobalan
Terminalia chebula Retz.   Myrobalan

Myrtaceae

Eucalyptus camaldulensis Dennhardt.   (*E. rostrata* Schlecht.)   Red gum
Eucalyptus diversicolor F. v. Muell.   Karri
Eucalyptus dives Schau
Eucalyptus globulus Labill.   Blue gum
Eucalyptus marginata Sm.   Jarrah
Eucalyptus occidentalis Endl.   Mallet bark
Eugenia Dombeyi (Spreng.) Skeels.   Grumichama
Eugenia uniflora L.   Surinam cherry.   Pitanga
Feijoa Sellowiana Berg.   Feijoa
Myrciaria cauliflora Berg.   (*Eugenia cauliflora* (Berg.) D.C.)   Jaboticaba
Pimenta dioica (L.) Merr.   (*P. officinalis* Lindl.)   Allspice
Pimenta racemosa (Mill.) J. W. Moore.   (*P. acris* Kostel.)   Bay
Psidium Guajava L.   Guava
Psidium littorale Raddi.   (*P. Cattleianum* Sabine.)   Strawberry guava
Syzygium aromaticum (L.) Merr. & Perry.   (*Eugenia caryophyllata* Thunb.)   Clove
Syzygium Cumini (L.) Skeels.   (*Eugenia Jambolana* Lam.)   Java plum.   Jambolan
Syzygium Jambos (L.) Alston.   (*Eugenia Jambos* L.)   Rose apple
Syzygium malaccensis (L.) Merr. & Perry.   Mountain apple.   Ohia

Araliaceae

Panax Schinseng Nees.   (*P. Ginseng* C. A. Mey.)   Ginseng
Panax quinquefolium L.   American ginseng
Tetrapanax papyriferum (Hook. f.) C. Koch.   (*Fatsia papyrifera* Hook. f.)   Rice paper plant

Umbelliferae

Anethum graveolens L.   Dill
Angelica Archangelica L.   (*Archangelica officinalis* Hoffm.)   Angelica
Anthriscus Cerefolium (L.) Hoffm.   Chervil
Apium graveolens L.   Celery
  var. dulce.   Garden celery
  var. rapaceum D.C.   Celeriac
Arracachia xanthorrhiza Bancroft.   (*A. esculenta* D.C.)   Arracacha
Carum Carvi L.   Caraway
Coriandrum sativum L.   Coriander
Cuminum Cyminum L.   Cumin
Daucus Carota L.   Carrot
Dorema Ammoniacum D. Don.   Ammoniacum

Ferula assafoetida L. Asafetida
Ferula galbaniflua Boiss. & Buhse. Galbanum
Foeniculum vulgare Miller. Fennel
   var. dulce Alef. Florence fennel. Finochio
Levisticum officinale Koch. Lovage
Opopanax Chironium (L.) Koch. Opopanax
Pastinaca sativa L. Parsnip
Petroselinum crispum (Mill.) Mansf. (*P. hortense* Hoffm.) (*Apium Petroselinum* L.)
   Parsley
Pimpinella Anisum L. Anise

Cornaceae

Cornus florida L. Dogwood
Nyssa aquatica L. Tupelo
Nyssa sylvatica Marsh. Sour gum. Black gum. Tupelo

Metachlymydeae

Ericaceae

Erica arborea L. Briar root
Gaultheria procumbens L. Checkerberry. Wintergreen
Gaylussacia baccata (Wang.) C. Koch. Huckleberry
Vaccinium angustifolium Ait. (*V. pennsylvanicum* Lam.) Low-bush blueberry
Vaccinium atrococcum (Gray) Heller. High-bush blueberry
Vaccinium corymbosum L. High-bush blueberry
Vaccinium macrocarpon Ait. Cranberry
Vaccinium membranaceum Torr.
Vaccinium myrtilloides Michx. (*V. canadense* Kalm.) Low-bush blueberry
Vaccinium ovatum. Pursh
Vaccinium Oxycoccus L. Small cranberry
Vaccinium vacillans Kalm. Low-bush blueberry
Vaccinium Vitis-Idaea L. Foxberry
   var. minus Lodd. Mountain cranberry

Sapotaceae

Achras Zapota L. Chicle. Sapodilla. Naseberry
Butyrospermum Parkii (G. Don) Kotschy. Shea butter
Calocarpum Sapota (Jacq.) Merr. Sapote. Marmalade plum
Chrysophyllum Cainito L. Star apple
Ecclinusa Balata Ducke. Abiurana. Coquilana
Lucuma nervosa A.D.C. Canistel. Eggfruit
Madhuca butyracea (Roxb.) Macb. Bassia fat
Madhuca indica J. F. Gmel. Illipe butter
Madhuca longifolia (L.) Macb. Mowra fat
Manilkara bidentata (A.D.C.) A. Chev. (*Mimusops Balata* Pierre.) Balata
Palaquium Gutta (Hook,) Burck. Gutta-percha

Ebenaceae

Diospyros Ebenum Koenig. Macassar ebony
Diospyros Kaki L.f. Japanese persimmon. Kaki
Diospyros virginiana L. Persimmon

Styraceae

Styrax benzoides Craib. Siam benzoin
Styrax Benzoin Dryand. Sumatra benzoin
Styrax tonkinense Craib. Siam benzoin

Oleaceae

Fraxinus americana L. White ash
Fraxinus excelsior L. European ash
Fraxinus nigra Marsh. Black ash
Fraxinus oregona Nutt. Oregon ash
Fraxinus Ornus L. Manna ash
Fraxinus pennsylvanica Marsh. Red ash
  var. lanceolata (Borkh.) Sarg. Green ash
Fraxinus quadrangulata Michx. Blue ash
Jasminum officinale L. var. grandiflorum (L). Kobuski. Jasmine
Olea europaea L. Olive
Syringa vulgaris L. Lilac

Loganiaceae

Strychnos Nux-vomica L. Nux vomica
Strychnos toxifera Schomb. ex Benth. Curare

Gentianaceae

Gentiana lutea L. Gentian

Apocynaceae

Apocynum cannabinum L. Indian hemp
Carissa grandiflora A.D.C. Natal plum
Couma macrocarpa Barb. Rodr. Sorva. Leche caspi
Dyera costulata (Miq.) Hook. f. Jelutong
Funtumia elastica (Preuss.) Stapf. Lagos silk rubber
Hancornia speciosa Gomez. Mangabeira rubber
Landolphia Heudelotii A.D.C. Guinea landolphia rubber
Landolphia Kirkii Dyer. Mozambique landolphia rubber
Landolphia owariensis Beauv. West African landolphia rubber
Strophanthus hispidus P.D.C. Strophanthus
Strophanthus Kombe Oliv. Strophanthus
Strophanthus sarmentosus A.P.D.C.

Asclepiadaceae

Asclepias curassavica L. False ipecac. Bloodflower
Asclepias incarnata L. Swamp milkweed
Asclepias subulata Decaisne. Desert milkweed
Asclepias syriaca L. Common milkweed
Calotropis gigantea (Willd.) Dryand. in Ait. Madar
Calotropis procera (Ait.) Dryand. in Ait. Akund
Cryptostegia grandiflora (Roxb.) R.Br. Rubber vine
Cryptostegia madagascariensis Bojer apud Miq. Rubber vine

Convolvulaceae

Exogonium purga (Hayne) Lindl.   (*Ipomoea purga* Hayne.)   Jalap
Ipomoea Batatas (L.) Poir.   Sweet potato
Rivea corymbosa (L.) Hallier f.   Ololiuqui

Boraginaceae

Alkanna tinctoria (L.) Tausch.   Alkanna
Heliotropium arborescens L. (*H. peruvianum* L.)   Heliotrope

Verbenaceae

Tectona grandis L. f.   Teak

Labiatae

Hedeoma pulegioides (L.) Pers.   Pennyroyal
Hyssopus officinalis L.   Hyssop
Lavandula latifolia Vill.   (*L. Spica* D.C.)   Spike lavender
Lavandula officinalis Chaix.   (*L. Spica* L.) (*L. vera* D.C.)   Lavender
Majorana hortensis Moench.   (*Origanum Majorana* L.)   Sweet marjoram
Marrubium vulgare L.   Hoarhound
Melissa officinalis L.   Balm
Mentha arvensis L., var. piperascens Malinv.   Japanese peppermint
Mentha piperita L.   Peppermint
Mentha Pulegium L.   European pennyroyal
Mentha spicata L.   Spearmint
Nepeta Cataria L.   Catnip
Ocimum Basilicum L.   Basil
Origanum vulgare L.   Pot marjoram
Perilla frutescens (L.) Britton.   (*P. ocimoides* L.)   Perilla
Pogostemon Cablin (Blanco) Benth.   Patchouli
Rosmarinus officinalis L.   Rosemary
Salvia officinalis L.   Sage
Salvia Sclarea L.   Clary sage
Satureja hortensis L.   Summer savory
Satureja montanum L.   Winter savory
Thymus vulgaris L.   Thyme

Solanaceae

Atropa Belladonna L.   Belladonna
Capsicum frutescens L.   (*C. annuum* L.)   Red pepper
  var. grossum (L.) Bailey.   Sweet pepper.   Bell pepper
  var. longum (D.C.) Bailey.   Cayenne pepper.   Chili
Cyphomandra betacea (Cav.) Sendt.   Tree tomato
Datura arborea L.   Maikoa
Datura innoxia Mill.
Datura sanguinea Ruiz & Pav.
Datura Stramonium L.   Jimson weed.   Thorn apple.   Stramonium
Duboisia Hopwoodii F.v.Muell.   Pituri
Hyoscyamus muticus L.   Henbane
Hyoscyamus niger L.   Black henbane
Lycopersicon esculentum Mill.   Tomato
Nicotiana rustica L.   Tobacco

Nicotiana Tabacum L. Tobacco
Solanum Melongena L. Eggplant. Aubergine
Solanum quitoense Lam. Naranjilla. Lulo
Solanum tuberosum L. Potato

Scrophulariaceae

Digitalis purpurea L. Foxglove. Digitalis

Bignoniaceae

Catalpa speciosa Warder. Catalpa
Cybistax Donnell-Smithii (Rose) Seibert. (*Tabebuia Donnell-Smithii* Rose.) Primavera

Pedaliaceae

Sesamum indicum L. (*S. orientale* L.) Sesame

Plantaginaceae

Plantago indica L. French psyllium
Plantago ovata Forsk. Blonde psyllium
Plantago Psyllium L. Spanish psyllium

Rubiaceae

Calycophyllum candidissimum (Vahl.) D.C. Degame. Lemonwood
Cephaelis Ipecacuanha (Brot.) A. Rich. (*Psychotria Ipecacuanha* Stokes.) Ipecac
Cinchona Calisaya Wedd. Quinine
Cinchona Ledgeriana Moens. Quinine
Cinchona officinalis L. Quinine
Cinchona pitayensis Wedd. Quinine
Cinchona succirubra Pav. Quinine
Coffea arabica L. Arabian coffee
Coffea liberica Bull. Liberian coffee
Coffea robusta Linden. Congo coffee
Rubia tinctorum L. Madder
Uncaria Gambir (Hunt.) Roxb. White cutch. Gambier

Valerianaceae

Valeriana officinalis L. Garden heliotrope. Valerian

Cucurbitaceae

Citrullus Colocynthis (L.) Schrad. Colocynth. Bitter apple
Citrullus vulgaris Schrad. Watermelon
  var. citroides Bailey. Citron melon
Cucumis Anguria L. Gherkin
Cucumis Melo L. Melon
Cucumis sativus L. Cucumber
Cucurbita ficifolia Bouché. Malabar gourd
Cucurbita maxima Duchesne. Winter squash
Cucurbita mixta. Pangalo
Cucurbita moschata Duchesne. Squash. Cushaw
Cucurbita Pepo L. Pumpkin. Summer squash
Luffa acutangula (L.) Roxb. Vegetable sponge

Luffa cylindrica (L.) M. Roem.   Vegetable sponge
Sechium edule (Jacq.) Sw.   Chayote

Campanulaceae

Lobelia inflata L.   Indian tobacco

Compositae

Anthemis nobilis L.   Russian chamomile
Artemisia Absinthium L.   Wormwood
Artemisia Cina Berg.   Levant wormseed.   Santonin
Artemisia Dracunculus L.   Tarragon
Carthamus tinctorius L.   Safflower
Chrysanthemum cinerariaefolium (Trev.) Bocc.   Dalmatian insect flowers
Chrysanthemum coccineum Willd.   Persian insect flowers
Chrysanthemum Marschallii Aschers.   Caucasian insect flowers
Chrysothamnus sp.   Rabbit brush.   Chrysil rubber
Cichorium Endivia L.   Endive
Cichorium Intybus L.   Chicory
Cynara Scolymus L.   Globe artichoke
Dahlia pinnata Cav.   Dahlia
Guizotia abyssinica (L. f.) Cass.   Niger-seed
Helianthus annuus L.   Sunflower
Helianthus tuberosus L.   Jerusalem artichoke
Lactuca sativa L.   Lettuce
Lactuca Scariola L.   Wild lettuce
Matricaria Chamomilla L.   German chamomile
Parthenium argentatum A. Gray.   Guayule
Solidago Leavenworthii T. & G.   Goldenrod
Tanacetum vulgare L.   Tansy
Taraxacum kok-saghyz Rodin.   Russian dandelion
Taraxacum officinale Weber.   Dandelion
Tragopogon porrifolius L.   Oyster plant.   Salsify
Trilisa odoratissima (Walt.) Cass.   Deer's tongue.   Wild Vanilla

SUPPLEMENTARY LIST

Menispermaceae

Abuta sp.   Curare
Chondodendron tomentosum Ruiz & Pav.   Curare
Cocculus sp.   Curare

Erythroxylaceae

Erythroxylon Coca Lam.   Coca

Zygophyllaceae

Guaiacum officinale L.   Lignum vitae
Guaiacum sanctum L.   Lignum vitae

BIBLIOGRAPHY

(Books published since 1936)

GENERAL REFERENCES

AMES, O. Economic Annuals and Human Cultures. Botanical Museum of Harvard University. Cambridge. 1939.

BAILEY, L. H. Manual of Cultivated Plants. rev. ed. The Macmillan Company. New York. 1949.

———, and E. Z. BAILEY. Hortus Second. The Macmillan Company. New York. 1941.

BROWN, N. C. Forest Products. 4th ed. John Wiley & Sons, New York. 1950.

CARTER, G. F. Plant Geography and Culture History in the American Southwest. Viking Fund Publications in Anthropology, No. 5. New York. 1945.

CHAPMAN, V. H. Seaweeds and Their Uses. Lange, Maxwell & Springer, Ltd. London. 1950.

Council of Scientific and Industrial Research. The Wealth of India—A Dictionary of Indian Raw Materials and Industrial Products. Vol. I. Parts I and II. New Delhi. 1948–1950.

DARLINGTON, C. D., and E. K. JANAKI AMMAL. Chromosome Atlas of Cultivated Plants. George Allen & Unwin, Ltd. London. 1945.

FERNALD, M. L. Gray's Manual of Botany. 8th ed. American Book Company. New York. 1950.

HAYWARD, H. E. The Structure of Economic Plants. The Macmillan Company. New York. 1938.

HECTOR, J. M. Introduction to the Botany of Field Crops. 2 vols. Central News Agency, Ltd. Johannesburg. 1936.

HOLLAND, J. H. Overseas Plant Products. John Bale, Sons and Company, Ltd. London. 1937.

HUTCHINSON, J., and R. MELVILLE. The Story of Plants and Their Uses to Man. P. R. Gawthorn, Ltd. London. 1948.

HYLANDER, C. J., and O. B. STANLEY. Plants and Man. The Blakiston Company. Philadelphia. 1941.

KELSEY, H. P., and W. A. DAYTON. Standardized Plant Names. 2d ed. J. Horace McFarland Company. Harrisburg. 1942.

KLAGES, K. H. W. Ecological Crop Geography. The Macmillan Company. New York. 1942.

KLOSE, N. America's Crop Heritage. Iowa State College Press. Ames. 1950.

MACMILLAN, H. F. Tropical Planting and Gardening. 5th ed. Macmillan & Co., Ltd. London. 1949.

POOL, R. J. Marching with the Grasses. University of Nebraska Press. Lincoln. 1948.

REHDER, A. Manual of Cultivated Trees and Shrubs. 2d ed. The Macmillan Company. New York. 1940.

SMITH, J. R. Tree Crops. 2d ed. Harcourt, Brace and Company, Inc. New York. 1950.

VERRILL, A. H. Perfumes and Spices. L. C. Page & Company. Boston. 1940.

U.S. Department of Agriculture. Agricultural Statistics, 1937–1949. Washington. 1937–1949.

———. Yearbook of Agriculture, 1937–1949. Washington. 1937–1949.
U.S. Department of Commerce. The Foreign Commerce and Navigation of the United States for the Calendar Years 1936–1945. Washington. 1937–1948.
WILSON, C. M., ed. New Crops for the New World. The Macmillan Company. New York. 1945.

SPECIAL REFERENCES

Industrial Plants

BAILEY, A. E., ed. Cottonseed and Cottonseed Products. Interscience Publishers, Inc. New York. 1948.
BENNETT, H., ed. Commercial Waxes. Chemical Publishing Company, Inc. Brooklyn. 1944.
BROWN, H. P., and A. J. PANSHIN. Commercial Timbers of the United States. McGraw-Hill Book Company, Inc. New York. 1940.
BROWN, N. C. Timber Products and Industries. John Wiley & Sons, Inc. New York. 1937.
DEERR, N. The History of Sugar. 2 vols. Chapman & Hall, Ltd. London. 1950.
DEWEY, L. H. Fiber Production in the Western Hemisphere. *U.S. Dept. Agr. Misc. Pub.* 518. Washington. 1943.
FAUBEL, A. L. Cork and the American Cork Industry. rev. ed. Cork Institute of America. New York. 1941.
GLESINGER, E. The Coming Age of Wood. Simon and Schuster, Inc. New York. 1949.
GRANT, J. Wood Pulp and Allied Products. 2d ed. Leonard Hill, Ltd. London. 1947.
GUENTHER, E. The Essential Oils. 4 vols. D. Van Nostrand Company, Inc. New York. 1948–1950.
HARLAND, S. C. The Genetics of Cotton. Jonathan Cape, Ltd. London. 1939.
HESS, K. P. Textile Fibers and Their Uses. 3d ed. J. B. Lippincott Company. Philadelphia. 1941.
HILDITCH, T. P. The Chemical Constitution of Natural Fats. 2d ed. Chapman & Hall, Ltd. London. 1949.
HOWES, F. N. Vegetable Gums and Resins. *Chronica Botan.* Waltham. 1949.
HUNTER, D. Paper Making. 2d ed. Alfred A. Knopf, Inc. New York. 1947.
HUTCHINSON, J. B., R. A. SILOW, and S. G. STEPHENS. The Evolution of Gossypium. Oxford University Press. New York. 1947.
JAMIESON, G. S. Vegetable Fats and Oils. 2d ed. Reinhold Publishing Corporation. New York. 1943.
KNAGGS, N. S. Adventures in Man's First Plastic. Reinhold Publishing Corporation. New York. 1947.
KNORR, K. E. World Rubber and its Regulation. Stanford University Press. Stanford University, Calif. 1945.
LEEMING, J. Rayon. Chemical Publishing Company, Inc. Brooklyn. 1950.
MANTELL, C. L. The Water Soluble Gums. Reinhold Publishing Corporation. New York. 1947.
MAUERSBERGER, H. R., ed. Matthew's Textile Fibres. 5th ed. John Wiley & Sons, Inc. New York. 1947.
NAVEZ, Y. R., and G. MAZUYER. Natural Perfume Materials. Reinhold Publishing Corporation. New York. 1947.
NEARING, H., and S. NEARING. The Maple Sugar Book. The John Day Company. New York. 1950.

PANSHIN, A. J., E. S. HARRAR, W. J. BAKER, and P. B. PROCTOR. Forest Products. McGraw-Hill Book Company, Inc. New York. 1950.
PEATTIE, D. C. A Natural History of Trees. Houghton Mifflin Company. Boston. 1950.
RECORD, S. J., and R. W. HESS. Timbers of the New World. Yale University Press. New Haven. 1943.
SHERMAN, J. V., and S. L. SHERMAN. The New Fibers. D. Van Nostrand Company, Inc. New York. 1946.
TIEMANN, H. D. Wood Technology. Pitman Publishing Corp. New York. 1942.
TITMUSS, F. H. A Concise Encyclopedia of World Timbers. Philosophical Library, Inc. New York. 1949.
VAN HOOK, A. Sugar. The Ronald Press Company. New York. 1949.
WEINDLING, L. Long Vegetable Fibers. Columbia University Press. New York. 1948.
WILSON, C. M. Trees and Test Tubes. Henry Holt and Company, Inc. New York. 1943.

Drug Plants

ALLPORT, N. L. The Chemistry and Pharmacy of Vegetable Drugs. Chemical Publishing Company, Inc. Brooklyn. 1944.
American Pharmaceutical Association. The National Formulary. 8th ed. Washington. 1946.
DURAN-REYNALS, M. L. The Fever Bark Tree. Doubleday & Company, Inc. New York. 1946.
GARNER, W. W. The Production of Tobacco. The Blakiston Company. Philadelphia. 1946.
GITHENS, T. S. Drug Plants of Africa. University of Pennsylvania Press. Philadelphia. 1948.
IRVING, G. W., and H. T. HERRICK, ed. Antibiotics. Chemical Publishing Company, Inc. Brooklyn. 1949.
MCINTYRE, A. R. Curare. University of Chicago Press. Chicago. 1947.
MUENSCHER, W. C. Poisonous Plants of the United States. The Macmillan Company. New York. 1939.
OSOL, A., and O. E. FARRAR. The Dispensatory of the United States of America. 24th ed. J. B. Lippincott Company. Philadelphia. 1947.
PRATT, R., and J. DUFRENOY. Antibiotics. J. B. Lippincott Company. Philadelphia. 1949.
ROBERT, J. C. The Story of Tobacco in America. Alfred A. Knopf, Inc. New York. 1949.
SPINDEN, H. J. Tobacco Is American: the Story of Tobacco before the Coming of the White Man. New York Public Library. New York. 1950.
TAYLOR, N. Cinchona in Java. Greenberg: Publisher, Inc. New York. 1945.
———. Flight from Reality. Duell, Sloan & Pearce, Inc. New York. 1949.
U.S. Pharmacopoeial Convention. The Pharmacopoeia of the United States of America. 14th decennial rev. Mack Printing Company. Easton. 1950.
YOUNGKEN, H. W. Textbook of Pharmacognosy. 6th ed. The Blakiston Company. Philadelphia. 1948.

Food Plants

AHLGREN, G. H. Forage Crops. McGraw-Hill Book Company, Inc. New York. 1949.

BATCHELOR, L. D., and H. W. WEBBER. The Citrus Industry. Vol. II. The Production of the Crop. University of California Press. Berkeley. 1948.
BEAVEN, E. S. Barley. James Thin. Edinburgh. 1947.
BRAVERMAN, J. B. S. Citrus Products. Interscience Publishing Company. New York. 1949.
BROWN, H. D., and C. T. HUTCHINSON. Vegetable Science. J. B. Lippincott Company. Philadelphia. 1949.
CHILDERS, N. F. Fruit Science. J. B. Lippincott Company. Philadelphia. 1949.
CONDIT, I. J. The Fig. *Chronica Botan.* Waltham. 1948.
CRUESS, W. L. Commercial Fruit and Vegetable Products. 3d ed. McGraw-Hill Book Company, Inc. New York. 1948.
DAHLGREN, B. E. Tropical and Subtropical Fruits. Chicago Museum of Natural History. Chicago. 1947.
FERNALD, M. L., and A. C. KINSEY. Edible Wild Plants of Eastern North America. Idlewild Press. Cornwall-on-Hudson. 1943.
HARDENBAUGH, Z. V. Potato Production. Comstock Publishing Company, Inc. Ithaca. 1949.
HOWES, F. N. Nuts. Faber & Faber, Ltd. London. 1948.
HURT, E. F. Sunflower. Faber & Faber, Ltd. London. 1948.
LAGER, M. The Useful Soybean. McGraw-Hill Book Company, Inc. New York. 1945.
MANGELSDORF, P. C., and R. G. REEVES. The Origin of Indian Corn and Its Relatives. *Texas Agr. Expt. Sta. Bull.* 574. College Station. 1939.
MEDSGER, O. P. Edible Wild Plants. The Macmillan Company. New York. 1939.
MORTON, K. and J. MORTON. Fifty Tropical Fruits of Nassau. Text House, Inc. Coral Gables. 1947.
OLDHAM, C. H. Brassica Crops. Crosby, Lockwood & Sons. London. 1948.
SALAMAN, R. N. The History and Social Significance of the Potato. Cambridge University Press. London. 1949.
SMOCK, R. M. and A. M. NEUBERT. Apples and Apple Products. Interscience Publishers. New York. 1950.
STURROCK, D. Tropical Fruits for Southern Florida and Cuba. Atkins Institution of the Arnold Arboretum of Harvard University. Jamaica Plain. 1940.
THOMPSON, H. C. Vegetable Crops. 4th ed. McGraw-Hill Book Company, Inc. New York. 1949.
VERRILL, A. H. Foods America Gave the World. L. C. Page & Company. Boston, 1937.
VON LOESECKE, H. W. Bananas. Interscience Publishers. New York. 1950.
WEAVER, J. C. American Barley Production. Burgess Publishing Company. Minneapolis. 1950.
WEBBER, H. J., and L. D. BATCHELOR. The Citrus Industry. Vol. I. History, Botany, and Breeding. University of California Press. Berkeley. 1943.
WILSON, H. K. Grain Crops. McGraw-Hill Book Company, Inc. New York. 1948.

Food Adjuncts

PARRY, J. W. The Spice Handbook. Chemical Publishing Company, Inc. Brooklyn. 1945.
TIDBURY, G. E. The Clove Tree. Crosby, Lockwood & Sons. London. 1949.
UKERS, W. H. The Romance of Coffee. *Tea and Coffee Trade J.* New York. 1948.

LIST OF VISUAL MATERIALS

The visual materials listed below and on the following pages are suggested by the publishers to be used to supplement the subject matter of this book. It is recommended, however, that each film be reviewed before using in order to determine its suitability for a particular class or group.

Both motion pictures and filmstrips are included in this list of visual materials, and the character of each one is indicated by the self-explanatory abbreviations "MP" or "FS." Immediately following this identification is the name of the producer; and if the distributor is different from the producer, the name of the distributor follows the name of the producer. Abbreviations are used for the names of producers and distributors, and these abbreviations are identified in the list of sources (with their addresses) at the end of the bibliography.

Unless otherwise indicated, the motion pictures listed in this bibliography are 16-mm. sound films and the filmstrips are 35-mm. silent.

Blue Lupine (MP; USDA; 15 min). How to plant, harvest, clean, store, and care for blue lupine, a nitrogen-producing cover crop used in the South.

Breeding Better Food Crops (MP; Nat Gar Bur; 20 min). Methods used in the production of vegetable seeds.

California's Golden Magic (MP; Mutual Orange; 30 min). California citrus industry; history; care of groves; picking the fruit; washing, drying, waxing, and packing.

Coconut Tree: Source of Wealth (MP; FON; 10 min). Coconut tree—uses and importance to India's economy.

Corn Chemurgy in Action (FS; Creative; 36 fr). The making of both food and non-food products out of corn.

Crops of the Americas (FS; USDA/Photo; 41 fr). Variety of products of South and Central America which are used in trade with the United States.

Crystal of Energy (MP; Sugar; 15 min). Role of sugar in human nutrition and in the economy of many sections of the United States.

Date Culture in the United States (MP; Hoefler; 11 min). Background of date growing in Coachella Valley in California; cycle of date culture from raw land to packaged dates.

Dates and Palms (MP; UWF; 16 min). Shows steps in the growing of date palms in Spain.

European Timber Trees: Selected Types and Their Characteristics (FS; Filmette; 55 fr). Photographs of forests and individual trees with close-ups of foliage, trunks, and some cross sections of typical European timber trees.

From Singapore to Baltimore (FS; McCormick; fr with accompanying record; 15 min). History and uses of spices and extracts.

Gift of Green (MP; Sugar; 20 min). Explains the process of photosynthesis by means of animation; shows how plants built food, particularly sugar, from water and air.

Grain That Built a Hemisphere (MP; IIAA; 11 min). Importance of corn to civilization; historical development of corn; present-day cultivation; and the wide variety of uses of corn. Disney production.

Green Gold (Bananas) (MP color; PAU; 11 min). Comprehensive view of the banana industry in Central America.

Hay Is What You Make It (MP; USDA; 18 min). Hay quality improvement by cutting at the right stage of growth; curing; retaining leaves for proteins.

Life through the Ages (FS; Macmillan; 49 fr). How plants and animals have left a record of their past in the fossils found in the earth's crust which are examples of life in the Paleozoic, Mesozoic, and Cenozoic eras.

New Tobaccoland, U.S.A. (MP; Modern; 30 min). Tobacco-growing areas in the U.S.; tobacco cycle: preparing soil, planting, cultivating, harvesting, curing, and marketing. (Sponsored by Liggett and Myers Tobacco Co.)

Nomads of the Jungle (Malaya) (MP; UWF; 20 min). Creates the illusion of actual participation in the everyday life of a typical nomad family in the Malayan jungle.

One Hundred Million Oranges (MP; Wurtele; 28 min). Orange growing in Florida from the development of the sapling to the canning and marketing of orange juice.

Palmyra (MP; GIIS; 10 min). Explains the wealth of this South Indian tree in the uses made of its various parts.

Photosynthesis (MP; UWF; 20 min). Carbon dioxide entering plant structure of leaf and its conversion in chloroplasts to oxygen; carbon built up into glucose; necessity for chlorophyll and the process of glucose being changed to starch.

Story of Bananas (MP; PAU; 11 min). Tropical agriculture from selection of a plantation site for a banana plantation through the growing and harvesting of bananas.

Story of Coffee (MP; PAU; 11 min). Brazilian coffee from the initial planting of selected seeds to the shipment of coffee for use around the world.

Systematic Botany (FS; Filmette). Series of six filmstrips composed of photographs of different plant types. Identification appears in brief captions in French, German, Latin, and English on each frame.

Blossoming Plants: Part 1 (94 fr).
Blossoming Plants: Part 2 (79 fr).
Needle Trees (67 Fr).
Flowerless Plants: Seaweeds, Mosses, Lichens, etc. (62 fr).
Flowerless Plants: Mushrooms (93 fr).
Botanical Geography (93 fr).

Tea from the Empire (FS; BIS; 44 fr). Development of Britain's tea industries; production of tea.

Today in the Beet Fields of Europe (MP; BSDF; 30 min). Shows various activities of the sugar-beet industry in several European countries.

Tropical Fruits (FS; Schick; 61 fr). Photographs of the following fruits in their native environment—banana, coconut, dates, orange, papaw, lemon, grapefruit, pineapple, and tamarands.

DIRECTORY OF SOURCES

Allis—Allis-Chalmers Company, Milwaukee 1, Wis.
Almanac—Almanac Films, Inc., 516 Fifth Ave., New York 18.
BIS—British Information Services, 30 Rockefeller Plaza, New York 20.
BSDF—Beet Sugar Development Foundation, Box 531, Fort Collins, Colo.
Creative—Creative Arts Studio, Inc., 1200 Eye St. N.W., Washington 5, D.C.
Filmette—Filmette Company, 635 Riverside Dr., New York.
FON—Films of the Nations, Inc., 55 W. 45th St., New York 19.
GIIS—Government of India Information Services, 2111 Massachusetts Ave. N.W., Washington 8, D.C.
Hoefler—Paul Hoefler Productions, Inc., 612½ Ridgeley Dr., Los Angeles 36.

IIAA—Institute of Inter-American Affairs, 499 Pennsylvania Ave. N.W., Washington 25, D.C.
McCormick—McCormick and Company, Inc., McCormick Building, Baltimore 2, Md.
Macmillan—The Macmillan Company, 60 Fifth Ave., New York 11.
Modern—Modern Talking Picture Service, Inc., 45 Rockefeller Plaza, New York 20.
Mutual Orange—Mutual Orange Distributors, Redlands, Calif.
Nat Gar Bur—National Garden Bureau, 407 S. Dearborn St., Chicago 5.
PAU—Pan American Union, Washington 6, D.C.
Photo—Photo Lab, Inc., 3825 Georgia Ave. N.W., Washington 11, D.C.
Schick—Rudolph Schick, 700 Riverside Dr., New York 31.
Sugar—Sugar Information, Inc., 52 Wall St., New York 5.
USDA—U.S. Department of Agriculture, Washington 25, D.C.
UWF—United World Films, Inc., 1445 Park Ave., New York 29.
Wurtele—Wurtele Film Productions, P.O. Box 504, Orlando, Fla.

INDEX

A

Abacá, 20, 34–36, 501
Abies alba, 114, 167, 496
  balsamea, 90, 166, 229, 496
  concolor, 90, 229, 496
  grandis, 90, 496
  lasiocarpa, 496
  magnifica, 90, 496
  nobilis, 496
  pectinata, 496
  procera, 90, 496
Abiurana, 148, 516
Absinthe, 257, 493
Abuta, 248, 520
Abutilon Avicennae, 513
  Theophrasti, 33, 513
Acacia, 182
  arabica, 507
  Catechu, 128, 507
  dealbata, 507
  decurrens, 121, 507
    var. *dealbata*, 121, 507
    var. *mollis*, 121, 507
  Farnesiana, 182, 507
  melanoxylon, 117, 507
  nilotica, 121, 160, 507
  pycnantha, 121, 507
  Senegal, 151, 152, 507
Acaroid resins, 160
Acer macrophyllum, 100, 512
  nigrum, 217, 512
  pseudoplatanus, 114, 512
  rubrum, 100, 512
  saccharinum, 100, 512
  saccharum, 98, 100, 217, 218, 230, 512
Aceraceae, 512
Achira, 368, 501
Achras Zapota, 148, 149, 433, 516
Acle, 116, 509
Aconite, 244–245, 504
Aconitum Napellus, 244, 504
Acorn, 347, 357
Acorus Calamus, 182, 500

Adansonia digitata, 231, 513
Aechmea magdalenae, 40, 500
Aesculus octandra, 106, 512
Aframomum Melegueta, 458, 501
Agar, 244, 263, 294, 295
Agaricus campestris, 292, 495
Agathis alba, 155, 496
  australis, 117, 155, 156, 496
Agave, 38, 490, 501
  atrovirens, 491, 501
  Cantala, 37, 501
  fourcroydes, 36, 37, 501
  Funkiana, 37, 501
  Lecheguilla, 37, 501
  Letonae, 36, 501
  sisalina, 36, 501
Agave fibers, 20, 36–38
Agrostis alba, 344, 497
Aizoaceae, 504
Akund, 49, 517
Alaria esculenta, 295, 495
Alcohol, ethyl, 226, 239, 484
  industrial, 226, 239, 306
  methyl, 226
  wood, 80, 226
Alcoholic beverages, 484–493
Alder, 115
  black, 114, 502
  red, 106, 502
Ale, 489
Aleppo galls, 124, 125
Aleurites cordata, 511
  Fordii, 193, 194, 511
  moluccana, 196, 511
  montana, 193, 511
Aleurone, 12, 297, 307
Alfalfa, 221, 287, 289, 344, 509
Algae, 263–264, 293–295, 495
Algaroba, 345–346, 491, 509
Algarobilla, 124, 508
Algin, 264
Alizarin, 131
Alkaloids, 15
Alkanna, 127, 131, 518

Alkanna tinctoria, 131, 518
Alligator pear (avocado), 201, 287, 379–380, 505
Allium ascalonicum, 372, 500
  Cepa, 371, 372, 500
  Porrum, 372, 500
  sativum, 372, 500
  Schoenoprasum, 372, 500
Allspice, 287, 437, 438, 447–448, 515
Almond, 201, 287, 289, 348, 356–357, 467, 507
  bitter, 177, 356–357, 493, 507
  green, 357, 512
  Java, 354, 510
  sweet, 356, 507
Alnus glutinosa, 114, 502
  rubra, 106, 502
Aloe barbadensis, 252, 500
  ferox, 252, 500
  Perryi, 252, 500
  vera, 500
  vulgaris, 500
Aloes, 174, 242, 252, 500
  green, 38, 501
Alpha cellulose, 234
Alpinia Galanga, 439, 501
  officinarum, 439, 501
Amanita muscaria, 283, 495
Amaranthaceae, 504
Amaranthus gangeticus, 379, 504
Amaryllidaceae, 20, 501
Ambarella, 426, 512
Amber, 153, 158, 497
Amber oil, 158
Ambergris, 178
Ammoniacum, 172–173, 515
Amomum, 458, 501
Amyris balsamifera, 172, 510
  elemifera, 172, 510
Anacardiaceae, 153, 406, 511
Anacardium occidentale, 349, 350, 511
Ananas comosus, 40, 431, 500
  sativus, 500
Andean tubers, 368, 369
Andropogon, 183
Anethum graveolens, 456, 515
Angelica, 438–439, 515
Angelica Archangelica, 438, 515
Aniba panurensis, 187, 505
  rosaeodora var. *amazonia*, 187, 505
Anil, 129
Animé, 154

Anise, 188, 454, 455, 493, 516
  star, 452–453, 505
Annatto, 127, 135, 514
Annona, 417
  Cherimolia, 417, 505
  muricata, 418, 505
  reticulata, 418, 505
  squamosa, 417, 505
Annonaceae, 406, 418, 505
Anogeissus latifolia, 153, 515
Anthemis nobilis, 257, 520
Anthriscus Cerefolium, 467, 515
Antiaris toxicaria, 50, 503
Antibiotics, 261–262
Añu, 368, 369, 510
Aperitifs, 493
Apium graveolens, 376, 515
  var. *dulce*, 376, 377, 455, 456, 515
  var. *rapaceum*, 377, 515
  Petroselinum, 516
Apple, 106, 199, 287, 289, 385, 387–389, 484, 489, 492, 507
  crab, 389, 507
  golden, 426, 512
  mammee, 428, 513
  mountain, 423, 515
  Otaheite, 426, 427, 512
  rose, 423, 515
  star, 433, 516
Applejack, 389, 492
Apocynaceae, 135, 517
Apocynum cannabinum, 34, 146, 517
Apricot, 177, 199, 287, 385, 391, 492, 507
Aquifoliaceae, 512
Araceae, 500
Arachis hypogaea, 200, 341, 342, 507
Araliaceae, 515
Aramina, 34, 513
Araucaria angustifolia, 107, 496
  araucana, 107, 496
  brasiliana, 496
  Cunninghamii, 117, 496
  imbricata, 496
Arbor vitae, 88, 497
Archangelica officinalis, 515
Archil, 134, 495
Areca Catechu, 276, 499
Areca nuts, 276
Arenga pinnata, 218, 491, 499
  saccharifera, 499
Argemone mexicana, 199, 505
Argemone oil, 199

INDEX

Armoracia lapathifolia, 441, 505
  rusticana, 505
Arracacha, 368, 515
Arracacia xanthorrhiza, 368, 515
Arrack, 219, 351, 491
Arrowroot, 224, 501
Arrowroot starch, 221, 223, 224, 368
Artemisia Absinthium, 257, 520
  Cina, 258, 520
  Dracunculus, 466, 520
Artichoke, globe, 287, 289, 373, 520
  Jerusalem, 220, 287, 365–366, 520
Artificial fabrics, 238
Artificial fibers, 51, 234–237
Artocarpus altilis, 380, 503
  communis, 503
  heterophyllus, 381, 503
  integra, 503
Arundinaria, 47, 497
Asafetida, 173, 516
Asclepiadaceae, 517
Asclepias curassavica, 49, 517
  incarnata, 49, 517
  subulata, 146, 517
  syriaca, 34, 49, 517
Ash, 45, 54, 64, 70, 71, 73, 85, 95–96, 114, 115, 517
  black, 96, 517
  blue, 96, 517
  green, 96, 517
  manna, 220, 517
  Oregon, 96, 517
  red, 96, 517
  white, 56, 73, 96, 517
Asimina triloba, 429, 505
Asparagus, 287, 289, 372–374, 500
Asparagus officinalis, 373, 500
Aspens, 85, 102, 229, 232, 502
Aspidium, 264
Assam rubber, 141–143, 503
Astragalus gummifer, 151, 507
Astrocaryum Murumuru, 204, 499
  Tucuma, 204, 499
  vulgare, 204, 499
Atropa Belladonna, 252, 253, 518
Atropine, 253, 257
Attalea Cohune, 499
  funifera, 41, 499
Attar (or otto) of roses, 179, 180, 447
Aubergine (eggplant), 287, 382, 399, 519
Aureomycin, 262
Avaram bark, 121, 508

Avena brevis, 322, 323, 497
  byzantina, 323, 497
  fatua, 322, 497
  nuda, 323, 497
  orientalis, 323, 498
  sativa, 322, 323, 498
Averrhoa Bilimbi, 434, 509
  Carambola, 434, 509
Avocado, 201, 287, 379–380, 505
Ayahusca, 283, 511

B

Babassu oil, 203
Babul, 121, 507
Bacillus brevis, 262, 495
  polymixa, 262, 495
  subtilis, 262, 495
Bacitracin, 262
Bagasse, 212, 213, 231
Balanocarpus Heimii, 157, 514
Balata, 148, 516
Balm, 462, 518
Balsa, 60, 108–109, 513
Balsam, Canada, 166, 167
  copaiba, 167, 170, 508
  gurjun, 171, 514
  illurin, 171, 508
  Mecca, 172, 511
  Oregon, 166
  of Peru, 168–169, 509
  of Tolu, 169, 509
Balsam fir, 62, 85, 90, 166, 229, 232, 496
Balsam poplar, 85, 102, 502
Balsams, 167–170, 178, 252
Bamboo, 44, 46–47, 231, 498
Bambusa, 47, 498
  Tulda, 47, 498
Banana, 231, 287, 289, 386, 414–417, 491, 501
  dwarf, 416, 501
Banisteriopsis Caapi, 283, 511
  inebrians, 283, 511
  quitensis, 283, 511
Baobab, 231, 513
Baphia nitida, 129, 508
Barbadoes aloes, 252, 500
Barbasco, 266, 508
Barcelona nuts, 353
Barley, 44, 231, 287–290, 296, 319–321, 472, 484, 488, 492, 498
Barosma betulina, 254, 510

Barosma crenulata, 254, 510
  serratifolia, 254, 510
Barwood, 127–129, 509
Basellaceae, 504
Basil, 462–463, 518
Baskets, 45
Bassia fat, 205, 516
Basswood, 62, 70, 75, 78, 85, 96, 115, 513
Bast fibers, 20, 26–34
Bay, oil of, 184, 448, 515
  sweet, 465, 505
Bay rum, 184
Bayberry, 207, 208, 502
Bdellium, 174, 510, 511
Beach plum, 394, 395, 507
Bean, adsuki, 339, 509
  bonavist, 343, 508
  broad, 287, 289, 336, 341, 509
  castor, 201, 511
  garden, 338, 509
  horse, 341, 343, 508
  jack, 343, 508
  kidney, 287, 289, 338
  lima, 289, 338–339, 509
  mung, 336, 339, 509
  rice, 339, 509
  scarlet runner, 339, 509
  shell, 338
  sieva, 339, 509
  snap, 338
  soy, 192, 194, 195, 287, 289, 340–341, 508
  velvet, 336, 343, 509
  Windsor, 341, 509
  yam, 368, 509
Beech, 54, 64, 67, 69, 70, 72–75, 79, 85, 96–98, 114, 230, 357, 503
  blue, 106, 502
  European, 357, 503
Beechnut, 357
Beer, 258, 306, 488–489
  birch, 490
  root, 484, 490
Beet, 287, 360, 379
  garden, 214, 360, 504
  sugar, 210, 214–217, 231, 360, 504
Bell pepper, 449, 518
Belladonna, 244, 252–253, 518
Ben, oil of, 201
Benzoin, 169–170, 178, 517
Berberidaceae, 505
Bergamot, 178, 182, 410, 510

Bergamot, oil of, 178, 182, 190, 410
Berries, 399–405
Bertholletia excelsa, 348, 349, 514
Beta maritima, 214, 360, 504
  vulgaris, 214, 360, 504
    var. *Cicla*, 360, 504
Betel, 268, 275–276
Betula alba, 502
  lenta, 97, 467, 502
  lutea, 55, 97, 230, 502
  papyrifera, 97, 502
  pendula, 114, 502
  pubescens, 114, 502
Betulaceae, 502
Beverages, 468–493
  alcoholic, 484–493
  cola, 484
  distilled, 491–492
  fermented, 485–491
  malt, 484
  nonalcoholic, 468–484
Big Tree, 87, 94, 497
Bignoniaceae, 519
Bilimbi, 434, 509
Birch, 54, 64, 68–70, 75, 76, 78, 79, 97, 98, 114, 115, 121, 230, 502
  black, 97, 490, 502
  paper, 85, 97, 502
  sweet, 177, 467, 502
  white, 97, 502
  yellow, 55, 73, 85, 97, 502
Birch beer, 490
Bitter almond, 356–357, 507
  oil of, 177, 357, 493
Bitter apple, 259, 519
Bitterroot, 245
Bitters, 493
Bixa Orellana, 133, 514
Bixaceae, 514
Black gum, 104, 516
Blackberry, 287, 385, 399, 400, 492, 507
Blackthorn, 114, 507
Blackwood, Australian, 117, 507
Blastophaga psenes, 420
Blue gum, 255, 515
Blueberry, 287, 401–402, 516
Boehmeria nivea, 31, 32, 504
  var. *tenacissima*, 32, 504
Bois de rose, 187, 505
Bombacaceae, 20, 513
Bombax Ceiba, 513
Boraginaceae, 518

Borassus flabellifer, 42, 218, 491, 499
Bordeaux turpentine, 167
Borneo tallow, 205, 514
Boswellia Carteri, 174, 510
  Frereana, 172, 510
  serrata, 166, 510
Bouncing Bet, 209, 504
Bowstring hemp, 38, 39, 500
Boxes and crates, 69–70
Boxwood, 109, 114, 511, 514
Brambles, 400–401
Brandy, 492
Brassica alba, 506
  campestris, 199, 362, 505
  caulorapa, 375, 506
  chinensis, 378, 506
  hirta, 437, 459, 506
  juncea, 459, 506
  Napobrassica, 362, 506
  Napus, 192, 199, 506
  nigra, 437, 459, 506
  oleracea, 374, 506
    var. *acephala*, 375, 506
    var. *botrytis*, 375, 506
    var. *capitata*, 375, 506
    var. *gemmifera*, 375, 506
    var. *gongylodes*, 375, 376, 506
  pekinensis, 378, 506
  Rapa, 199, 362, 506
Brazil nut, 347–349, 514
Brazilette, 127, 508
Brazilwood, 126–128, 508
Breadfruit, 287, 289, 380–381, 386, 503
Brewing, 488–489
Briar root, 114, 516
Broad bean, 287, 289, 336, 341, 509
Broccoli, 372, 375–376, 506
Brome grass, 344, 498
Bromeliaceae, 20, 500
Bromelin, 432
Bromus inermis, 344, 498
Broomcorn, 42–43, 325–326, 498
Broomroot, 43–44, 231, 498
Broussonetia papyrifera, 50, 503
Brush fibers, 19, 41–44
Brussels sprouts, 372, 374, 375, 506
Brya Ebenus, 109, 508
Buchu, 254, 510
Buckeye, 78, 106, 512
Buckthorn, 131, 132, 248, 512
Buckwheat, 221, 287, 289, 296, 333–334, 504

Bullet wood, 148
Bullock's heart, 418, **505**
Bur clover, 344, 509
Burgundy pitch, 167
Bursera Aloeoxylon, 510
  Delpechianum, 510
  glabrifolia, 187, 510
  gummifera, 172, 510
  penicillata, 187, 510
Burseraceae, 153, 157, 172, 174, **510**
Butea frondosa, 508
  monosperma, 160, 162, 508
Butternut, 105, 126, 356, 502
Butyrospermum Parkii, 204, 516
Buxaceae, 511
Buxus sempervirens, 114, 511

C

Caapi, 278, 283–284, 511
Cabbage, 287, 289, 372, 374, 375, 506
  Chinese, 378, 506
Cabbage palm, 42, 500
Cabuya, 38, 501
Cacao, 204, 287, 289, 476, 477, 513
Cactaceae, 514
Cactus, 282, 491
Cadillo, 34, 513
Caesalpinia brevifolia, 124, 508
  coriaria, 124, 508
  echinata, 128, 508
  Sappan, 128, 508
  spinosa, 124, 508
Caffeine, 15, 277, 468–484
Cajan pea, 336, 343, 508
Cajanus Cajan, 160, 336, 343, 508
  indicus, 508
Calamus, 46, 499
  Draco, 499
Calamus root, 182, 183, 467, 500
Calathea lutea, 207, 501
Callitris, 161
  quadrivalvis, 161, 497
Calocarpum Sapota, 433, 516
Calophyllum inophyllum, 116, 513
Calotropis gigantea, 49, 517
  procera, 49, 517
Calycophyllum candidissimum, 110, 519
Camelina oil, 199
Camelina sativa, 199, 506
Camellia Sasanqua, 201, 513
  sinensis, 473, 513

Campanulaceae, 520
Camphor, 188–190, 244, 505
  Borneo, 190, 514
Camwood, 128–129, 508
Canada balsam, 166, 167
Canaigre, 124–125, 504
Cananga odorata, 181, 505
Cananga oil, 182
Canarium commune, 354, 510
  luzonicum, 172, 510
  ovatum, 354, 510
  strictum, 157, 510
Canavalia ensiformis, 343, 508
Candelilla wax, 206, 511
Candlenut, 196, 511
Candlenut oil, 196
Canistel, 433, 516
Canna edulis, 224, 368, 501
Cannabis, 242, 268, 278, 281–282
Cannabis sativa, 28, 29, 192, 197, 281, 503
Cannaceae, 501
Cantala, 37, 501
Cantaloupe, 199, 396
Caoutchouc (see Rubber)
Cape aloes, 252, 500
Capers, 445, 505
Capparidaceae, 505
Capparis spinosa, 445, 505
Caprifigs, 420–422
Capsicum, 437, 448–450, 484
Capsicum annuum, 449, 518
  frutescens, 437, 449, 518
    var. *grossum*, 449, 518
    var. *longum*, 449, 450, 518
Carambola, 434, 509
Carapa fat, 204
Carapa guianensis, 110, 204, 511
  moluccensis, 204, 511
Caraway, 188, 454–456, 493, 515
Carbohydrates, 9–11, 290, 297, 357, 359, 386
Cardamom, 175, 276, 437, 458, 501
Carica Papaya, 429, 430, 514
Caricaceae, 514
Carissa grandifolia, 435, 517
Carludovica palmata, 44, 500
Carnation, 185, 447, 504
Carnauba wax, 206, 207
Caroa, 41, 500
Carob, 153, 289, 346, 347, 508
Carpinus Betulus, 114, 502
  caroliniana, 106, 502

Carrageenin, 295
Carrot, 210, 287, 289, 360–361, 515
Carthamus tinctorius, 131, 132, 192, 197, 520
Carum Carvi, 454, 455, 515
Carya alba, 502
  glabra, 99, 502
  illinoensis, 354, 502
  laciniosa, 354, 502
  ovata, 99, 353, 502
  Pecan, 502
  tomentosa, 99, 502
Caryophyllaceae, 504
Caryota urens, 42, 218, 499
Cascara, 244, 246, 248, 512
Casearia praecox, 514
Cashew, 349–350, 511
Casimiroa edulis, 435, 510
Cassava, 364–365, 480, 491, 511
Cassava starch, 224
Cassia, 188, 242, 437, 442–443, 505
Cassia acutifolia, 257, 508
  angustifolia, 257, 508
  auriculata, 121, 508
Cassie, 176, 179, 182, 507
Cassine, 482–483, 512
Castanea crenata, 358, 503
  dentata, 55, 99, 121, 358, 503
  sativa, 114, 121, 358, 503
  vulgaris, 503
Castilla elastica, 140, 141, 503
  Ulei, 140, 503
Castilla rubber, 140, 141
Castor bean, 201, 511
Castor oil, 200–201
Catalpa, 106, 519
Catalpa speciosa, 106, 519
Catechu, 128, 507
Catha edulis, 481, 512
Catnip, 467, 518
Cattail, 49, 497
Cauassú wax, 207
Caucho rubber, 140, 503
Cauliflower, 289, 372, 375–376, 506
Caura rubber, 146, 511
Cayenne pepper, 449, 518
Ceara rubber, 141, 142, 511
Cedar, 62, 71, 72, 74, 88–89, 115, 175, 496, 511
  Alaska, 88, 496
  cigar-box, 109, 511
  eastern red, 88, 190, 496

Cedar, incense, 87, 88, 496, 503
  Moulmein, 116, 511
  northern white, 72, 77, 85, 88, 497
  Port Orford, 75, 88, 496
  southern white, 77, 86, 88, 496
  Spanish, 109, 511
  western red, 72, 77, 78, 87, 89, 497
Cedarwood oil, 177, 190
Cedrela odorata, 109, 511
  Toona, 116, 511
Cedrus deodara, 116, 496
Ceiba acuminata, 49, 513
  aesculifolia, 49, 513
  pentandra, 47, 48, 513
Celastraceae, 512
Celeriac, 377, 515
Celery, 287, 289, 372, 376–377, 455, 456, 515
Cellophane, 239
Celluloid, 238
Cellulose, 7, 8, 10, 226–231, 234–237
  alpha, 234
  regenerated, 237
Cellulose hydrolysis, 239
Cellulose products, 226–239
  acetate, 239
  nitrate, 237–238
Celtis occidentalis, 106, 503
Cephaelis Ipecacuanha, 246, 519
Ceratonia Siliqua, 153, 346, 347, 508
Cereal straw, 44, 45, 49
Cereals, 199, 290, 291, 296–324, 488, 491
Ceriman, 435, 500
Ceroxylon andicola, 206, 499
Chamaecyparis Lawsoniana, 88, 496
  nootkatensis, 88, 496
  thyoides, 88, 496
Chamaerops humilis, 50, 499
Chamomile, 257, 520
Champaca oil, 187, 505
Charcoal, 79–81, 125, 393
Chard, 360, 503
Chaulmoogra, 258–259, 514
Chaulmoogra oil, 206, 258–259
Chayote, 381, 520
Checkerberry, 466, 516
Chenopodiaceae, 504
Chenopodium ambrosioides var. *anthelminticum*, 261, 504
  Quinoa, 334, 504
Cherimoya, 417, 505

Cherry, 97, 114, 199, 287, 289, 385, 392, 492, 493, 507
  Barbados, 435, 511
  Japanese flowering, 392, 507
  sour, 392, 507
  Surinam, 423, 515
  sweet, 392, 507
  wild black, 97, 507
Cherry gum, 153
Chervil, 467, 515
Chestnut, 55, 70–72, 74, 85, 99, 121, 126, 287, 289, 347, 358, 503
  European, 114, 121, 358, 503
  Japanese, 358, 503
Chewing gum, 148–150, 268
Chicha, 491
Chick pea, 336–338, 508
Chicle, 148–150, 516
Chicory, 377, 472, 520
Chilis, 449, 450, 518
Chilte rubber, 146, 511
China grass, 18, 32
Chinawood oil, 193
Chinese green, 131
Chives, 372, 467, 500
Chlorogalum pomeridianum, 209, 500
Chloromycetin, 262
Chlorophora tinctoria, 128, 503
Chlorophyll, 6, 13, 127, 129
Chloroxylon Swietenia, 115, 510
Chocolate, 476, 478, 479
Chondodendron tomentosum, 248, 520
Chondrus crispus, 294, 295, 495
Chorisia insignis, 49, 513
Chrysanthemum cinerariaefolium, 264, 265, 520
  coccineum, 265, 520
  Marschallii, 265, 520
Chrysil rubber, 146, 520
Chrysobalanus Icaco, 425, 506
Chrysophyllum Cainito, 433, 516
Chrysothamnus, 146, 520
Cicer arietinum, 336, 337, 508
Cichorium Endivia, 377, 520
  Intybus, 377, 520
Cider, 389, 484, 489
Cigarettes, 273–274
Cigars, 273
Cinchona, 248–251, 287
Cinchona Calisaya, 248–250, 519
  Ledgeriana, 248, 519
  officinalis, 248, 519

Cinchona pitayensis, 250, 519
  succirubra, 248, 519
Cinnamomum Burmannii, 443, 505
  Camphora, 188, 189, 505
  Cassia, 437, 442, 505
  Loureirii, 444, 505
  Massoia, 443, 505
  Oliveri, 443, 505
  Tamala, 443, 505
  zeylanicum, 437, 443, 444, 505
Cinnamon, 175, 188, 276, 437, 443–444, 505
  Chinese, 442
Circassian walnut, 114
Citron, 406, 413–414, 510
Citron melon, 397, 519
Citronella, oil of, 183, 498
Citrullus Colocynthis, 259, 519
  vulgaris, 397, 519
    var. *citroides*, 397, 519
Citrus aurantifolia, 411, 510
  Aurantium, 182, 409, 510
    subsp. *Bergamia*, 182, 410, **510**
  Bergamia, 510
  grandis, 414, 510
  Limon, 411, 510
  Limonia, 510
  maxima, 510
    var. *uvacarpa*, 510
  Medica, 413, 510
  nobilis, 410, 510
  paradisi, 410, 411, 510
  reticulata, 410, 510
  sinensis, 182, 408, 510
Citrus fruits, 199, 221, 406–414
Citrus hybrids, 414
Civet, 178
Clary sage, 178, 188, 467, 518
Claviceps purpurea, 263, 495
Clove oil, 177, 190, 446, 493
Clover, 221, 344, 509
  alsike, 344, 509
  bur, 344, 509
  crimson, 344, 509
  red, 344, 509
  sweet, 344, 509
  white, 344, 509
Cloves, 177, 188, 276, 437, 445–446, 515
Club moss, 264, 496
Cnidoscolus, 146, 511
Coal, 9, 64, 65
Cob nuts, 353

Coca, 253, 268, 278–279, 333, 520
Cocaine, 253, 254, 278, 279
Coccoloba uvifera, 162, 504
Cocculus, 248, 520
Cochlospermaceae, 514
Cochlospermum Gossypium, 514
  religiosum, 49, 153, 514
Coco plum, 425, 506
Cocoa, 204, 468, 476–479, 513
Cocoa butter, 204, 479
Cocobolo, 110, 508
Coconut, 20, 39, 40, 201, 218, 219, 348, 350–353, 386, 499
Coconut oil, 201–202
Cocos nucifera, 40, 201, 218, 219, 350, 491, 499
Cocus wood, 109, 508
Codeine, 260, 281
Coelococcus amicarum, 499
Coffea, 469
  arabica, 469, 470, 519
  liberica, 469, 519
  robusta, 469, 519
Coffee, 287, 468–472, 519
Coffee tree, 106, 508
Cohune oil, 203, 204
Coir, 39, 40, 230, 351
Coix, 288
Coix Lachryma-Jobi, 332, 498
Coke, 65
Cola, 268, 277, 468, 481, 484, 513
Cola acuminata, 513
  nitida, 277, 481, 513
Cola beverages, 484
Colchicine, 245
Colchicum, 245, 500
Colchicum autumnale, 245, 500
Colewort, 374
Collards, 375
Collodion, 238
Colocasia antiquorum, 370, 500
  esculenta, 370, 500
Colocynth, 259, 519
Colophony, 165
Colza oil, 199
Combretaceae, 515
Commiphora africana, 174, 510
  erythraeae, 173, 511
  Kataf, 174, 511
  Mukul, 174, 511
  Myrrha, 173, 511
  Opobalsamum, 172, 511

Compositae, 175, 520
Condiments, 438
Convallaria majalis, 188, 500
Convolvulaceae, 518
Cooperage, 72–73
Copaiba, 170–171, 508
Copaiba balsam, 167, 170, 508
Copaifera conjungata, 154, 508
  copallifera, 155, 508
  Demeusii, 155, 508
  Gorskiana, 508
  Guibortiana, 508
  mopane, 155, 508
  officinalis, 170, 508
  reticulata, 170, 508
  Salikounda, 155, 508
Copals, 154–156, 508
  kauri, 155, 156
  Manila, 155–156
  South American, 111, 156
Copernicia cerifera, 206, 207, 499
Copra, 201, 351–353
Coquilana, 148, 516
Corchorus capsularis, 30, 513
  olitorius, 30, 513
Cordials, 493
Coriander, 455, 456, 493, 515
Coriandrum sativum, 455, 456, 515
Cork, 81–83
Corn, Indian, 198, 306–314, 499
  (*See also* Maize)
Corn husks, 49
Corn oil, 198
Corn starch, 221, 222, 225, 226
Cornaceae, 516
Cornus florida, 106, 516
Cortinellus Berkeleyanus, 495
  edodes, 293, 495
Cortisone, 261
Corylus americana, 353, 502
  Avellana, 114, 353, 502
  cornuta, 353, 502
  maxima, 353, 502
  rostrata, 502
Cotinus Coggygria, 128, 511
Cotton, 18, 20–26, 44, 228, 230, 235, 287, 289, 513
  Asiatic, 22, 25, 513
  Egyptian, 22–34, 513
  mercerized, 26
  sea-island, 22, 23, 513
  tree, 25, 513

Cotton, upland, 22, 24–25, 513
Cotton industry, 25
Cotton linters, 26, 234, 237, 238
Cotton staples, 22
Cottonseed oil, 26, 197–198
Cottonwood, 75, 78, 102, 502
Couma macrocarpa, 150, 517
Couratari Tauari, 50, 514
Cowpea, 336, 340, 509
Crabwood, 110, 511
Cranberry, 287, 402, 516
Cream nuts, 348
Crin végétal, 50
Crocus, saffron, 131, 446, 501
Crocus sativus, 131, 446, 501
Crotalaria juncea, 32, 508
Croton, 511
Croton oil, 199, 260
Croton Tiglium, 260, 511
Cruciferae, 505–506
Cryptostegia grandiflora, 145, 517
  madagascariensis, 145, 517
Cuba bast, 50, 113, 513
Cubé, 266, 267, 508, 509
Cubebs, 259, 467, 502
Cucumber, 287, 381, 519
Cucumber tree, 106, 505
Cucumis Anguria, 381, 519
  Melo, 396, 519
  sativus, 381, 519
Cucurbita ficifolia, 383, 519
  maxima, 383, 519
  mixta, 382, 519
  moschata, 382, 383, 519
  Pepo, 382, 383, 519
Cucurbitaceae, 519–520
Cudbear, 127, 134, 495
Cumin, 455, 456, 515
Cuminum Cyminum, 455, 456, 515
Curare, 248, 517
Curarine, 248
Curcuma angustifolia, 224, 501
  longa, 130, 131, 442, 501
  zedoaria, 442, 501
Currants, 287, 399, 402–403, 506
Curry, 442
Custard apple, 417–418
Cutch, 127, 128
  black, 128, 507
  white, 123, 124, 128, 519
Cybistax Donnell-Smithii, 113, 519
Cycadaceae, 496

Cycas circinalis, 153, 496
Cycas gum, 153
Cyclanthaceae, 500
Cydonia oblonga, 391, 506
  vulgaris, 506
Cymbopogon, 183
  citratus, 183, 498
  Martinii, 184, 498
  Nardus, 183, 498
Cynara Scolymus, 373, 520
Cyperaceae, 499
Cyperus Papyrus, 231, 499
  tegetiformis, 45, 499
Cyphomandra betacea, 435, 518
Cypress, 62, 70–74, 77, 86, 89, 90, 497

D

Dacrydium cupressinum, 117, 496
Dactylis glomerata, 344, 498
Daemonorops, 161
  Draco, 161, 499
Dahlia, 220, 520
Dahlia pinnata, 220, 520
Dalbergia latifolia, 116, 508
  nigra, 113, 508
  retusa, 110, 508
  Sissoo, 116, 508
Damar, 156–157, 510, 514
Dandelion, 378, 520
  Russian, 144–145, 520
Dandelion rubber, 144–145
Daniella Ogea, 155, 508
  Oliveri, 171, 508
  thurifera, 171, 508
Daphne cannabina, 231, 514
Dasheen, 368, 370, 500
Date, 287, 386, 418–419, 499
  wild, 218, 499
Datura, 284
  arborea, 284, 518
  innoxia, 285, 518
  sanguinea, 284, 518
  Stramonium, 257, 284, 518
Daucus Carota, 360, 515
Deal, white, 114
  yellow, 114
Deer's tongue, 273, 520
Degame, 110, 519
Dendrocalamus, 47, 498
Deodar, 116, 496
Derris, 265–266, 508

Derris elliptica, 265, 508
  trifoliata, 265, 508
  uliginosa, 508
Dewberry, 287, 400, 507
Dextrin, 225
Dextrose, 220
Dianthus Caryophyllus, 185, 504
Digitalis, 244, 254–255, 519
Digitalis purpurea, 254, 519
Dill, 177, 456–457, 515
Dimorphandra Mora, 509
Dioscorea, 364
  alata, 364, 501
Dioscoreaceae, 501
Diospyros Ebenum, 115, 516
  Kaki, 430, 516
  virginiana, 107, 430, 516
Dipterocarpaceae, 115, 153, 157, 514
Dipterocarpus turbinatus, 171, 514
Dipteryx odorata, 162, 453, 462, 508
  oppositifolia, 453, 462, 508
Divi-divi, 124, 508
Dogwood, 106, 516
Dolichos Lablab, 336, 343, 508
Dorema Ammoniacum, 172, 515
Dothidella Ulei, 138
Douglas fir, 64, 66–68, 70, 72–75, 87, 89, 90, 166, 497
Dracaena, 161
  Cinnabari, 161, 500
Dragon's blood, 161, 499, 500
Drug plants, 242–264
Drying oils, 193–197
Dryobalanops aromatica, 190, 514
Dryopteris Filix-Mas, 264, 496
  marginalis, 264, 496
Duboisia Hopwoodii, 285, 518
Dulse, 294, 295, 495
Durian, 419–420, 513
Durio zibethinus, 419, 513
Durra, 326–327, 498
Dyera costulata, 148, 517
Dyes, 13, 126–134

E

Earth vegetables, 359–372
Ebenaceae, 516
Ebony, American, 109, 508
  Macassar, 115, 516
Ecclinusa Balata, 148, 516
Echinochloa colona, 329, 498

Echinochloa crus-galli, 329, 498
  frumentacea, 329, 498
Edgeworthia Gardneri, 514
  papyrifera, 514
  tomentosa, 231, 514
Egg fruit, 433, 516
Eggplant, 287, 382, 399, 519
Einkorn, 297–299, 499
Elaeis guineensis, 202, 491, 499
Elemi, 171–172, 510, 511
Elettaria Cardamomum, 437, 458, 501
Eleusine coracana, 330, 498
Elm, 54, 70, 72, 73, 85, 99, 100, 114, 503
  rock, 99, 503
  slippery, 251, 503
  white, 99, 100, 503
Emmer, 298, 299, 499
Endive, 287, 377, 520
Enfleurage, 176
Enterolobium Saman, 509
Enzymes, 16, 290, 488
Ephedra, 251
Ephedra equisetina, 251, 252, 497
  sinica, 251, 252, 497
Ephedrine, 251, 252
Epicampes macroura, 498
Ergot, 263, 495
Erica arborea, 114, 516
Ericaceae, 516
Eriobotrya japonica, 424, 425, 506
Erythroxylaceae, 520
Erythroxylon Coca, 253, 278, 520
Esparto, 208, 230, 499
Essences, 438
Essential oils, 13, 153, 154, 162, 175–190, 244, 438
Eucalyptus camaldulensis, 162, 515
  diversicolor, 117, 515
  dives, 190, 515
  globulus, 255, 515
  marginata, 117, 515
  occidentalis, 121, 515
  rostrata, 515
Eucalyptus oil, 177, 190, 255–256
Eucheuma spinosum, 263, 495
Euchlaena mexicana, 308, 498
Eugenia caryophyllata, 445, 515
  cauliflora, 515
  Dombeyi, 423, 515
  Jambolana, 515
  Jambos, 515
  uniflora, 423, 515

Eugenol, 446, 454
Euphorbia antisyphilitica, 206, **511**
  Intisy, 145, 511
Euphorbiaceae, 135, 511
Evernia furfuracea, 186, 495
  prunastri, 186, 495
Excelsior, 78
Exogonium purga, 246, 518

F

Fabrics, 19
  artificial, 238
  natural, 50
Fagaceae, 503
Fagopyrum esculentum, 504
  sagittatum, 333, 504
Fagus grandifolia, 96, 97, 230, 357, 503
  sylvatica, 114, 357, 503
Farina, 305, 306
Farinha, 364
Fats, 12, 193, 201–206, 290, 348
Fatsia papyrifera, 515
Fatty oils, 12, 191–206, 244, 259, 260
Feijoa, 423, 515
Feijoa Sellowiana, 423, 515
Fennel, 455, 457–458, 516
  Florence, 457–458, 516
Fenugreek, 458, 509
Fermented beverages, 485–491
Feronia gum, 153
Feronia Limonia, 153, 510
Ferula assafoetida, 173, 516
  galbaniflua, 173, 516
Feterita, 326, 328, 498
Fiber plants, 18–51
Fibers, **7, 8,** 18–51
  agave, 20, 36–38
  artificial, 51, 234–237
  bast, 20, 26–34
  brush, 19, 41–44
  filling, 19, 47–50
  hard, 20, 34–41
  hat, 44
  long, 20
  mixed, 20
  papermaking, 19, 51
  plaiting, 19, 44–47
  protein, 51, 313, 341, 342
  rough weaving, 19, 44–47
  soft, 20, 26–34
  structural, 34–41

Fibers, surface, 20–26
  textile, 18–41
  wood, 6–8, 20, 53, 54, 228–230
Ficus Carica, 420, 421, 503
  elastica, 141, 142, 503
  Nekbudu, 50, 503
  religiosa, 160, 504
  utilis, 503
Fig, 197, 287, 289, 386, 420–422, 503
Filbert, 201, 347, 353, 502
Filling fibers, 19, 47, 50
Finochio, 458, 516
Fique, 38, 501
Fir, 90, 115, 496
  alpine, 87, 496
  balsam, 62, 85, 90, 166, 229, 232, 496
  lowland white, 90, 496
  noble, 90, 496
  red, 90, 496
  silver, 114, 496
  white, 87, 90, 229, 496
Flacourtia indica, 434, 514
Flacourtiaceae, 206, 514
Flag, sweet, 182, 500
Flax, 18, 20, 26–28, 192, 193, 274, 287, 289, 510
  New Zealand, 38
Flaxseed, 192, 193
Fleawort, 260
Floral extracts, 176
Flour, 304–305
Fly agaric, 278, 283, 495
Foeniculum vulgare, 455, 457, 516
  var. *dulce*, 457, 458, 516
Food adjuncts, 436–493
Food plants, 286–435
  history of, 286–288
  nature of, 290–291
  origin and distribution of, 288–290
Forage crops, 343–345
Forage grasses, 344
Forage legumes, 344–345
Forest areas of North America, 85–87
Forest formations of North America, 86
Forest products, 52–83
Forest resources, 84–117
Forests, 84–117
  of Africa, 116–117
  of Asia, 114–116
  of Australia and Oceania, 117
  of Europe, 113–114
  of North America, 84–107

Forests, of South America, 107–113
Fortunella japonica, 413, 510
  margarita, 413, 510
Foxglove, 254, 519
Fragaria chiloensis, 404, 506
  vesca, 404, 506
  virginiana, 404, 506
Frankincense, 166, 171, 174, 187, 508, 510
Fraxinus americana, 56, 96, 517
  excelsior, 114, 517
  nigra, 96, 517
  oregona, 96, 517
  Ornus, 220, 517
  pennsylvanica, 96, 517
    var. *lanceolata*, 96, 517
  quadrangulata, 96, 517
Fructose, 10, 210, 220
Fruit juices, 484
Fruit sugar, 10, 210, 220
Fruit vegetables, 379–384
Fruits, 385–435
  citrus, 199, 221, 406–414
  gourd, 381–383, 396–397
  pome, 387–391
  preservation of, 386–387
  stone, 391–396
  of temperate regions, 385–405
  tropical, 406–435
Fuel, 64–66
Fumitories and masticatories, 268–285
Fungi, 62, 283, 291–293, 495
Funtumia elastica, 143, 517
Furcraea Cabuya, 38, 501
  gigantea, 38, 501
  hexapetala, 38, 501
  macrophylla, 38, 501
Furniture and fixtures, 70
Fustic, 127, 128, 503

G

Galangal, 439, 501
Galbanum, 172, 173, 516
Gambier, 123, 124, 128, 519
Gamboge, 127, 133–134, 174, 513
Garcinia, 428
  Hanburyi, 133, 513
  Mangostana, 427, 513
Garden heliotrope, 247, 519
Garlic, 289, 372, 467, 500
Gaultheria procumbens, 466, 516
Gaylussacia baccata, 401, 516

INDEX 541

Gelidium cartilagineum, 263, 495
   corneum, 263, 495
Gentian, 245, 517
Gentiana lutea, 245, 517
Gentianaceae, 517
Geraniaceae, 510
Geraniol, 179, 183, 184
Geranium, 177, 179
Geranium oil, 179–181
Gherkin, 381, 519
Gigantochloa, 47, 498
Gin, 450, 456, 492
Gingelly oil, 198
Ginger, 175, 439–441, 484, 490, 501
Ginger ale, 484
Ginger-grass oil, 184, 498
Ginseng, 244–246, 515
Gleditsia triacanthos, 100, 347, 508
Glucose, 6, 9, 210, 220, 225–226, 239
Glucosides, 15
Glycine Max, 192, 194, 195, 340, 508
   Soja, 508
Glycyrrhiza glabra, 246, 508
Gnetaceae, 497
Golden apple, 426, 512
Goldenrod, 146, 520
Goldenseal, 244, 245, 504
Gooseberry, 287, 402–403, 506
   Otaheite, 434, 511
Gossypiospermum praecox, 109, 514
Gossypium, 22, 25
   arboreum, 23, 25, 513
   barbadense, 23, 513
   herbaceum, 23, 25, 513
   hirsutum, 23–25, 513
Gourd fruits, 381–383, 396–397
Governor's plum, 434, 514
Gracilaria confervoides, 263, 495
   lichenoides, 263, 495
Grain sorghum, 325–328
Grains of paradise, 438, 458, 501
Gramicidin, 262
Gramineae, 20, 296, 497–499
Granadilla, 422, 514
Granadillo, 109, 508
Grape, 197, 287, 289, 397–399, 484, 513
   fox, 398, 513
   muscadine, 398, 513
   scuppernong, 398
   wine, 397–398, 513
Grape sugar, 6, 9–10, 210, 220, 225
Grapefruit, 287, 410–411, 484, 510

Grapefruit oil, 177
Grass oils, 183–184
Grass sorghums, 325
Grass tree, 501
Grass-tree resins, 160
Grasses, 44, 49, 231, 296
Greenheart, 110, 505
Grevillea robusta, 117, 504
Grossularia, 402
Groundnut, 341, 507
Grumichama, 423, 515
Guadua angustifolia, 46, 47, 498
Guaiacum, 111, 174, 251
Guaiacum officinale, 110, 111, 251, 520
   sanctum, 110, 251, 520
Guanabana, 418, 505
Guarana, 468, 480–481, 512
Guava, 287, 422–423, 515
   strawberry, 423, 515
Guayule, 143–144, 520
Guayule rubber, 143–144
Guizotia abyssinica, 192, 196, 520
Gum arabic, 151, 152, 507
Gum ghatti, 153, 515
Gum guaiac, 251
Gum kino, 161–162
Gum resins, 133, 172–174
Gum tragacanth, 151–152, 507
Gumbo, 328
Gums, 11, 125, 151–153
   British, 225
   chewing, 148, 149, 150, 268
   karaya, 49, 152
   ogea, 155
   spruce, 166–167, 268
Guncotton, 237, 238
Gunny, 30
Gutta-percha, 146–148, 516
Guttiferae, 153, 513
Gymnocladus dioica, 106, 508

H

Hackberry, 73, 106, 503
Haematoxylin, 127
Haematoxylon Brasiletto, 127, 508
   campechianum, 127, 508
Hamamelidaceae, 153, 506
Hamamelis, 256
Hamamelis virginiana, 256, 506
Hancornia speciosa, 146, 517
Hard fibers, 20, 34–41

Hard resins, 154–162
Hashish, 281
Hats, 44, 306
Hazelnut, 114, 347, 353, 502
Hedeoma pulegioides, 256, 518
Hegari, 326, 328
Helianthus annuus, 192, 199, 520
  tuberosus, 220, 365, 520
Heliotrope, 187, 518
Heliotropium arborescens, 187, 518
  peruvianum, 518
Hemicellulose, 10, 240
Hemlock, 67, 68, 70, 72, 74, 85, 90, 91, 118–120, 229, 232, 497
  western, 64, 87, 90, 229, 497
Hemp, 20, 28, 29, 44, 192, 197, 230, 244, 281, 282, 287, 289, 503
  Ambari, 33, 513
  Bowstring, 38, 39, 500
  Colorado River, 34, 509
  Cuban, 38, 501
  Deccan, 33, 513
  Gambo, 33, 513
  Indian, 34, 146, 517
  Manila, 34–36, 230, 501
  Mauritius, 38, 501
  New Zealand, 38, 39, 230, 500
  Sunn, 32, 230, 508
Hempseed oil, 29, 197
Henbane, 244, 256, 285, 518
  black, 285, 518
Henequen, 20, 36, 37, 501
Henna, 127, 129, 514
Herbage vegetables, 372–379
Heroin, 281
Hevea Benthamiana, 136, 511
  brasiliensis, 135–137, 197, 511
Hevea rubber, 136–140
Hibiscus cannabinus, 33, 513
  elatus, 50, 113, 513
  esculentus, 34, 382, 513
  Sabdariffa, 33, 513
  tiliaceus, 34, 513
Hickory, 54, 60, 64, 70, 73, 79, 85, 99, 353–354, 502
  bitternut, 353
  mockernut, 99, 502
  pignut, 99, 502
  shagbark, 99, 353, 502
  shellbark, 354, 502
Hierochloë odorata, 45, 498
Hippocastanaceae, 512

Hoarhound, 256, 467, 518
Holly, 106, 114, 512
Honey, 220–221, 273, 490
Honey locust, 100, 347, 508
Hopea micrantha, 157, 514
Hops, 177, 257–258, 488, 504
Hordeum, 320, 498
  deficiens, 320, 498
  distichon, 320, 498
  intermedium, 320, 498
  spontaneum, 320, 498
  vulgare, 320, 498
Hormones, 17
Hornbeam, 107, 114, 502
Horseradish, 287, 289, 441, 505
Huckleberry, 401, 516
Humulus Lupulus, 257, 258, 504
Hyacinth, 186, 500
Hyacinthus orientalis, 186, 500
Hybrid corn, 313–314
Hydnocarpus Kurzii, 259, 514
Hydrastis canadensis, 245, 504
Hymenaea Courbaril, 111, 156, 508
Hyoscyamine, 253, 256, 257
Hyoscyamus muticus, 285, 518
  niger, 256, 285, 518
Hypernic, 127, 508
Hyssop, 467, 518
Hyssopus officinalis, 467, 518

I

Ilex Aquifolium, 114, 512
  opaca, 106, 512
  paraguariensis, 479, 488, 512
  vomitoria, 482, 512
Illicium verum, 452, 505
Illipe butter, 205, 516
Incense cedar, 87, 88, 496
India rubber, 141, 142, 503
Indian corn, 198, 306–314, 499
  (See also Maize)
Indian mallow, 33, 513
Indian tobacco, 256, 520
Indican, 129
Indigo, 127, 129, 508
Indigofera suffruticosa, 129, 508
  tinctoria, 129, 508
Inga edulis, 347, 508
Ink, 118, 125–126
Inodes causiarum, 500
  Palmetto, 500

Insect flowers, 264, 265, 520
Insecticides, 264–267
Intisy, 145, 511
Inulin, 220, 366
Ionone, 183, 185
Ipecac, 246, 519
Ipomoea Batatas, 362, 363, 518
    purga, 518
Iridaceae, 501
Iris florentina, 182, 501
    pallida, 182, 501
Irish moss, 294–295, 495
Isatis tinctoria, 130, 506
Istle, 37, 500, 501
Ivory, vegetable, 240–241

J

Jaboticaba, 423, 515
Jackfruit, 381, 503
Jaggary, 219
Jalap, 174, 246, 518
Jambolan, 423, 515
Jarrah, 117, 515
Jasmine, 176, 179, 185, 475, 517
Jasminum grandifolium, 517
    officinale var. *grandifolium*, 185, 517
Jaumave istle, 37, 501
Java almond, 354, 510
Java plum, 423, 515
Jelutong, 148, 517
Jerusalem artichoke, 220, 287, 365–366, 520
Jesuit's bark, 249
Jimson weed, 257, 284, 518
Jipijapa, 44
Job's tears, 332, 498
Johnson grass, 325, 344, 498
Jojoba wax, 208, 511
Jonquil, 179, 188, 501
Juglandaceae, 502
Juglans cinerea, 105, 356, 502
    nigra, 104, 105, 355, 502
    regia, 114, 196, 356, 502
Jujube, 423, 424, 512
Juncaceae, 500
Juncus effusus, 45, 500
Juniper, 115, 450, 492, 496
Juniperus communis, 450, 496
    virginiana, 88, 190, 496
Jute, 20, 30, 31, 230, 513
    China, 33, 513

Jute, Java, 33
Jute butts, 31, 230

K

Kafir, 326, 327, 498
Kajú, 350
Kaki, 430, 516
Kale, 372, 374, 375, 506
Kaoliang, 326, 328, 498
Kapok, 47–49, 201, 513
Kapok oil, 201
Karaya gum, 49, 152
Karri, 117, 515
Kauri, 117, 155, 496
Kauri copal, 155, 156
Kavakava, 278, 285, 502
Keawe, 345, 509
Kelp, 263–264, 495
Kenaf, 33, 513
Khat, 468, 481, 482, 512
Khaya senegalensis, 117, 511
Khuskhus, 184, 499
Kino, 161–162, 508, 509
Kittul fiber, 42
Kohlrabi, 375, 376, 506
Kola nuts, 277
    (*See also* Cola)
Kraft paper, 232
Kudzu, 344–345, 509
Kumquat, 412–413, 510

L

Labiatae, 175, 518
Lablab, 336, 343, 508
Lac, 160
Lace bark, 50, 514
Lacewood, 104
Lacquer, 153, 158–159
Lacquer paints, 238
Lacquer tree, 158, 159, 511, 512
Lactuca sativa, 378, 520
    Scariola, 378, 520
Ladino, 344, 509
Lagetta lintearia, 50, 514
Laminaria digitata, 264, 495
    saccharina, 264, 495
Lancewood, 110, 505
Landolphia Heudelottii, 143, 517
    Kirkii, 143, 517
    owariensis, 143, 517
Larch, 68, 72, 85, 91, 115, 496

Larch, European, 114, 121, 167, 496
  western, 55, 64, 87, 91, 496
Larix decidua, 114, 121, 167, 496
  laricina, 91, 229, 496
  occidentalis, 55, 91, 496
Latex, 15, 135–137, 140, 141, 143–145, 429
  (*See also* Rubber)
Lauraceae, 175, 505
Laurelwood, 116, 513
Laurus nobilis, 465, 505
Lavandula latifolia, 184, 518
  officinalis, 184, 185, 518
  Spica, 518
  vera, 518
Lavender, 177, 179, 184, 185, 447, 467, 518
Laver, 295, 495
Lawsonia inermis, 129, 514
Leche caspi, 150, 517
Lecythidaceae, 514–515
Lecythis usitata, 349, 515
  Zabucajo, 349, 515
Leek, 372, 500
Legumes, 291, 335–347
  forage, 344–345
  tree, 345–347
Leguminosae, 20, 115, 153, 265, 335, 507–509
Lemon, 176, 287, 411–412, 467, 484, 510
  oil of, 177, 178, 188, 412
Lemon-grass oil, 183–184, 498
Lemonwood, 110, 519
Lens culinaris, 336, 342, 508
  esculenta, 508
Lentil, 289, 336, 342, 508
Leopoldinia Piassaba, 41, 499
Lespedeza, 345
Lespedeza cuneata, 345, 508
  sericea, 508
  stipulacea, 345, 508
  striata, 345, 508
Letterwood, 110, 504
Lettuce, 287, 289, 372, 378, 520
Levisticum officinale, 467, 516
Levulose, 220, 366
Libocedrus decurrens, 88, 496
Licania rigida, 195, 506
Lichens, 134, 186, 495
Licorice, 231, 246–247, 273, 274, 489, 508
Licuri oil, 203
Lignaloe (*see* Linaloe oil)
Lignite, 64

Lignum vitae, 60, 110–111, 251, 520
Lilac, 447, 517
Liliaceae, 20, 153, 500–501
Lily of the valley, 188, 500
Lima bean, 289, 338, 339, 509
Lime, 287, 412, 467, 510
  oil of, 177, 412
Lime tree, 114, 513
Linaceae, 20, 510
Linaloe oil, 187, 505, 510
Linden, 96, 221, 513
Linen, 18, 26, 230
Linoleum, 83
Linseed oil, 28, 125, 193
Linum usitatissimum, 27, 192, 193, 510
Liqueurs, 493
Liquidambar orientalis, 169, 506
  Styraciflua, 103, 169, 506
Liriodendron Tulipifera, 104, 505
Litchi, 423–424, 512
Litchi chinensis, 423, 512
Lithocarpus densiflora, 119, 503
Litmus, 134
Live oak, 86, 102, 357, 503
Lobelia, 256
Lobelia inflata, 256, 520
Locust, 62, 70, 72, 99, 100
  black, 71, 99, 509
  honey, 100, 347, 508
  South American, 156, 508
  West Indian, 111, 508
Loganberry, 400, 507
Loganiaceae, 517
Logwood, 126, 127, 508
Lokao, 127, 131, 512
Lonchocarpus Nicou, 266, 508
  Urucu, 266, 509
  utilis, 266, 509
Lophophora Williamsii, 282, 514
Loquat, 424, 425, 506
Lovage, 467, 516
Lucuma nervosa, 433, 516
Luffa acutangula, 51, 519
  cylindrica, 51, 520
Lulo, 435, 519
Lumbang oil, 196, 511
Lumber, 66–71
Lycopersicon esculentum, 383–384, 518
Lycopodium, 264
Lycopodium clavatum, 264, 496
Lygeum Spartum, 230, 498
Lythraceae, 514

INDEX 545

M

Ma-huang, 251, 497
Macadamia nuts, 354
Macadamia ternifolia, 354, 504
Macaroni, 305, 306
Macassar oil, 205
Mace, 205, 460, 461
Maclura pomifera, 102, 128, 504
Macrocystis pyrifera, 263, 495
Madar, 49, 517
Madder, 130–131, 519
Madhuca butyracea, 205, 516
    indica, 205, 516
    longifolia, 205, 516
Magnolia, 86, 106
Magnolia acuminata, 106, 505
Magnoliaceae, 505
Maguey, 37, 490, 491
    Manila, 37, 501
    Mexican, 38, 501
Mahoe, 513
    blue, 113, 513
Mahogany, 68, 74, 111–112, 511
    African, 117, 511
    Spanish, 111
    West Indian, 111, 511
Mahua butter, 205
Maikoa, 284, 518
Maize, 198, 210, 222, 287, 289, 290, 296, 306–314, 488, 491, 492, 499
    dent, 309, 310, 499
    flint, 309–310, 499
    flour, 310
    hybrid, 313–314
    pod, 308, 309, 499
    pop, 308–309, 499
    soft, 309, 310, 499
    sweet, 309, 310, 499
    waxy, 310
Majagua, 34, 113, 513
Majorana hortensis, 463, 518
Male fern, 264, 496
Mallet bark, 121, 515
Malpighia glabra, 435, 511
Malpighiaceae, 511
Malt beverages, 484
Malting, 488
Maltose, 220
Malus baccata, 507
    pumila, 507
Malvaceae, 20, 33, 513

Mamey, 428
Mammea americana, 428, 513
Mammee apple, 428, 513
Mandarins, 410, 510
Mandrake, 247, 505
Mangabeira, 146
Mangels, 360, 504
Mangifera indica, 425, 426, 511
Mango, 287, 425–426, 511
Mangosteen, 427–428, 513
Mangrove, 116, 119–121, 515
Manicoba rubber, 141
Manihot esculenta, 364, 365, 511
    Glaziovii, 141, 142, 197, 511
    utilissima, 511
Manila copal, 155–156
Manila elemi, 172, 510
Manila hemp, 34–36, 230, 501
Manilkara bidentata, 148, 516
Manioc, 364
Manna, 220
Mannose, 220, 239
Maple, 54, 64, 68–70, 72–75, 78, 79, 85, 100–101, 114, 115, 512
    black, 217, 512
    hard, 100
    Oregon, 100, 512
    red, 100, 217, 512
    rock, 100
    silver, 100, 217, 512
    sugar, 73, 75, 85, 98, 100, 210, 217, 218, 230, 512
Maple sugar, 217–218
Maple syrup, 217, 218
Maranta arundinacea, 223, 224, 501
Marantaceae, 501
Marihuana, 281
Marjoram, 463
    pot, 463, 518
    sweet, 463, 518
Marron, 358
Marrubium vulgare, 256, 518
Mastic, 153, 161, 512
Mat grass, Chinese, 45, 499
Mat rush, Japanese, 45
Maté, 287, 468, 479–480, 512
Matricaria Chamomilla, 257, 520
Mats and matting, 44–45
May apple, 247, 505
Mead, 221, 490
Meadow saffron, 245, 500
Mechanical pulp, 231–232

Medic, 344, 509
Medicago hispida, 344, 509
  lupulina, 344, 509
  sativa, 344, 509
Medicinal plants, 242–264
Medlar, 387, 390, 507
Melanorrhoea usitata, 159, 511
Meliaceae, 511
Melilotus alba, 344, 509
  officinalis, 344, 509
Melissa officinalis, 462, 518
Melloco, 368, 504
Melon, 396–397, 519
  citron, 397, 519
Menispermaceae, 520
Mentha arvensis var. *piperascens*, 463, 518
  piperita, 463, 518
  Pulegium, 467, 518
  spicata, 464, 465, 518
Menthol, 463
Mescal, 38, 491
Mescal buttons, 282–283, 514
Mespilus germanica, 390, 507
Mesquite, 345, 347, 509
Mesquite gum, 153
Mesta fiber, 33
Metroxylon amicarum, 241, 499
  Rumphii, 499
  Sagu, 224, 499
Michelia Champaca, 187, 188, 505
Micrandra, 146, 511
Mignonette, 188, 506
Milkweed, 34, 49, 517
  desert, 146, 517
Milkwort, Senega, 247
Millet, 287, 289, 296, 328–331, 488
  barnyard, 329, 498
  foxtail, 329, 498
  Japanese, 329, 498
  pearl, 330, 498
  proso, 329–330, 498
  shama, 329, 498
Milo, 326–328, 498
Mimusops Balata, 148, 516
Mine timbers, 71–72
Mineral salts, 264, 290, 291, 294, 359, 386
Mint, 175, 221, 518
Mockernut, 99, 502
Molasses, 213, 273, 492
Mombin, red, 427, 512
  yellow, 427, 512
Monkshood, 244, 504

Monstera deliciosa, 435, 500
Mora, 113, 509
Mora excelsa, 113, 509
Moraceae, 20, 135, 503–504
Morchella esculenta, 293, 294, 495
Mordant, 126
Morel, 293, 294, 495
Moringa oleifera, 201, 506
  pterygosperma, 506
Moringaceae, 506
Morphine, 260, 280, 281
Morus alba, 404, 504
  nigra, 404, 504
  rubra, 404, 504
Mountain apple, 423, 515
Mousse de chêne, 186
Mowra fat, 205, 516
Mu tree, 193, 511
Mucilages, 11, 244, 251, 260, 263
Mucuna Deeringianum, 509
Muhlenbergia macroura, 43, 498
Mulberry, 287, 289, 403–404, 504
  paper, 50, 230, 503
Mung bean, 336, 339, 509
Murlins, 295, 495
Musa Cavendishii, 501
  nana, 416, 501
  paradisiaca, 417, 501
    subsp. *sapientum*, 414–416, 501
  sapientum, 501
  textilis, 34, 35, 501
Musaceae, 20, 501
Mushroom, 283, 291–293, 495
Musk, 178
Muskmelon, 287, 396–397
Mustard, 175, 287, 289, 379, 459–460, 506
  black, 199, 437, 459, 506
  Indian, 459–460, 506
  white, 199, 437, 459, 506
Mutshu cloth, 50
Myrciaria cauliflora, 423, 515
Myrica carolinensis, 502
  cerifera, 207, 502
  pensylvanica, 207, 208, 502
Myricaceae, 502
Myristica fragrans, 205, 460, 461, 505
Myristicaceae, 505
Myrobalans, 123–124, 515
Myroxylon Balsamum, 169, 509
  Pereirae, 168, 509
  toluiferum, 509
Myrrh, 173–174, 187, 242, 511

INDEX 547

Myrtaceae, 175, 406, 423, 515
Myrtle, wax, 207, 502

N

Naranjilla, 435, 519
Narcissus, 188, 501
Narcissus Jonquilla, 188, 501
   Tazetta, 188, 501
Narcotics, 277–285
   solanaceous, 278, 284–285
Naseberry, 148, 516
Nasturtium officinale, 378, 506
Natal plum, 435, 517
Natural fabrics, 19, 50
Naval stores, 162, 165
Nectandra Rodioei, 505
Nectarine, 393, 507
Neoglaziovia variegata, 41, 500
Neomycin, 262
Nepeta Cataria, 467, 518
Nereocystis Luetkeana, 264, 495
Neroli, oil of, 178, 182
Nicotiana rustica, 270, 275, 518
   Tabacum, 197, 269, 519
Nicotine, 264, 269, 275
Niger seed, 192, 196, 520
Niger-seed oil, 196–197
Nitrostarch, 226
Nitta, 347, 509
Noble cane, 212
Nonalcoholic beverages, 468–484
Nondrying oils, 192, 199–201
Nut galls, 124, 125
Nutmeg, 205, 276, 287, 460–461, 505
Nutmeg butter, 205
Nuts, 347–358
   areca, 276
   Barcelona, 353
   Brazil, 347–349, 514
   cob, 353
   cream, 348
   macadamia, 354, 504
   paradise, 349, 515
   pili, 201, 354, 510
   pine, 355
   Queensland, 354, 504
   sapucaia, 349, 515
Nux vomica, 246, 260, 517
Nylon, 51
Nyssa aquatica, 104, 516
   sylvatica, 104, 516

O

Oak, 54, 60, 64, 67–75, 79, 85, 101–102, 114, 115, 118, 119, 357, 503
   Aleppo, 125, 503
   black, 102, 119, 131, 503
   bur, 102, 503
   California tanbark, 119, 503
   chestnut, 102, 119, 503
   cork, 81, 503
   holm, 357, 503
   live, 86, 102, 357, 503
   Oregon, 102, 503
   pin, 102, 503
   post, 102, 503
   red, 73, 75, 102, 119, 503
   scarlet, 102, 503
   shingle, 102, 503
   silky, 117, 504
   swamp chestnut, 102, 503
   swamp white, 102, 503
   Texas red, 102, 503
   turkey, 102, 503
   Turkish, 124, 503
   white, 45, 73, 75, 101, 102, 119, 357, 503
   willow, 102, 503
Oak moss, 178, 186–187, 495
Oakum, 29
Oats, 231, 287, 288, 296, 322–324, 492, 497, 498
Oca, 368, 369, 509
Ochroma Lagopus, 513
   pyramidale, 108, 513
Ocimum Basilicum, 462, 518
Ocotea Rodioei, 110, 505
Ogea gum, 155
Ohia, 423, 515
Oils, drying, 192–197
   essential, 13, 153, 154, 162, 175–190, 244, 438
   fatty, 12, 191–206, 244, 259, 260
   fixed, 191
   grass, 183–184
   nondrying, 192, 199–201
   perfume, 178–188
   semidrying, 192, 197–199
   volatile, 13, 175–190
   (*See also* specific source plants)
Oiticica oil, 195, 506
Okra, 34, 287, 382, 513
Olea europaea, 114, 199, 428, 517
Oleaceae, 517

Oleoresins, 162–172, 178, 186–187, 252, 264
Olibanum, 174, 510
Olive, 114, 199, 287, 289, 428–429, 517
Olive oil, 199–200
Ololiuqui, 283, 518
Onion, 210, 287, 289, 371–372, 500
Opium, 242, 260, 268, 278–281
Opopanax, 174, 511, 516
Opopanax Chironium, 174, 516
Orange, 175, 176, 179, 287, 289, 408–410, 447, 467, 484, 510
  bitter, 182, 409, 493, 510
  deciduous, 414, 510
  King, 410
  mandarin, 410, 510
  Satsuma, 410
  Seville, 409, 510
  sour, 409–410, 510
  sweet, 182, 408–409, 510
Orange hybrids, 414
Orange oil, 177, 182, 409, 410, 492
Orbignya Cohune, 203, 204, 499
  Martiana, 203, 499
  oleifera, 203, 499
Orchard grass, 344, 498
Orchidaceae, 502
Organic acids, 16, 290, 291, 386
Origanum Majorana, 518
  vulgare, 463, 518
Orris, 178, 182, 467, 501
Oryza, 316
  sativa, 315, 498
Osage orange, 60, 62, 71, 102, 127, 128, 504
Osiers, 45
Ostrya virginiana, 107, 502
Otaheite apple, 426–427, 512
Otaheite gooseberry, 434, 511
Otoba butter, 206, 505
Otto (attar or ottar) of roses, 179, 180, 447
Ouricuri, 203, 500
Oxalidaceae, 509
Oxalis tuberosa, 368, 369, 509
Oxandra lanceolata, 110, 505
Oyster plant, 361, 520

P

Pachyrrhizus erosus, 368, 509
Padouk, 115, 509
Palaquium Gutta, 146, 147, 516

Palay rubber, 146
Palm, 210
  betel-nut, 276, 499
  babassu, 203, 499
  cabbage, 42, 500
  carnauba, 206, 207, 499
  cohune, 203, 204, 499
  date, 10, 418–419, 499
  dwarf fan, 50, 499
  gomuti, 218, 499
  hat, 44, 500
  ivory-nut, 10, 240, 499, 500
  licuri, 203, 208, 500
  murumuru, 204, 499
  oil, 202, 203, 499
  ouricuri, 203, 500
  palmyra, 42, 218, 499
  raffia, 45, 208, 500
  sago, 224–225, 499
  tagua, 240, 500
  toddy, 42, 218, 499
  tucum, 204, 499
  wax, 206, 207, 499
  wine, 41, 500
Palm-kernel oil, 203–204
Palm oil, 202–204
Palm sugar, 218–219
Palm wine, 351, 490–491
Palma istle, 37, 500, 501
Palmaceae, 20, 499–500
Palmarosa oil, 179, 184, 498
Palmetto, 42, 125, 500
Palmyra fiber, 42
Palo borracho, 49, 513
Pan, 275
Panama rubber, 140, 141, 503
Panax Ginseng, 515
  quinquefolium, 246, 515
  Schinseng, 245, 515
Pandanaceae, 497
Pandanus tectorius, 45, 497
  utilis, 45, 497
Panicum miliaceum, 329, 330, 498
Papain, 429
Papaver somniferum, 197, 260, 280, 505
Papaveraceae, 505
Papaw, 429, 505, 514
Papaya, 429–430, 514
Paper, 228–234
Papermaking fibers, 19, 51, 228–231
Papier-mâché, 34, 234
Papreg, 234

INDEX

Paprika, 449
Papyrus, 231, 499
Para rubber, 136, 137, 511
Paraguay tea, 479–480, 512
Parchment, vegetable, 234
Parkia biglobosa, 347, 509
  filicoidea, 347, 509
Parsley, 287, 289, 466, 516
Parsnip, 210, 287, 289, 361, 516
Parthenium argentatum, 143, 144, 520
Passiflora edulis, 422, 514
  ligularis, 422, 514
  quadrangularis, 422, 514
Passifloraceae, 514
Pastinaca sativa, 361, 516
Patchouli oil, 178, 187, 518
Paullinia Cupana, 480, 481, 512
  Yoco, 483, 512
Peach, 199, 287, 289, 392–393, 492, 507
Peanut, 200, 287, 341–342, 347, 507
Peanut oil, 200
Pear, 199, 287, 385, 389–390, 484, 507
Peas, 335–337, 509
  cajan, 336, 343, 508
  chick, 336–338, 508
  field, 336
  garden, 287, 289, 336, 337
  pigeon, 343, 508
Peat, 64, 65, 231
Pecan, 201, 287, 348, 354, 502
Pectin, 10, 11, 386
Pedaliaceae, 519
Pelargonium, 179
  graveolens, 181, 510
  odoratissimum, 180, 181, 510
Peltogyne paniculata, 113, 509
Penicillin, 261–262
Penicillium chrysogonum, 261, 495
  notatum, 261, 495
Pennisetum glaucum, 330, 498
Pennyroyal, 177, 256, 518
  European, 467, 518
Pentose, 239
Pepitos, 382
Pepper, 287, 450–452
  bell, 449, 518
  betel, 276, 502
  black, 437, 451–452, 502
  cayenne, 449, 518
  long, 452, 502
  red, 287, 438, 449, 518
  sweet, 449, 518

Pepper white, 437, 452
Peppermint, 177, 188, 463, 493, 518
  Japanese, 177, 463, 518
Perfume oils, 179–188
Perfumes, 178–188
Perilla, 518
Perilla frutescens, 195, 196, 518
  ocimoides, 518
Perilla oil, 195–196
Perry, 390, 484
Persea americana, 379, 505
Persian berries, 127, 132, 512
Persimmon, 60, 107, 287, 289, 430, 516
  Japanese, 430, 516
Peruvian bark, 249
Petitgrain oil, 179, 182
Petroleum, 65
Petroselinum crispum, 466, 516
  hortense, 516
Peyote, 278, 282–283, 514
Pharmacognosy, 243
Pharmacology, 243
Phaseolus angularis, 339, 509
  aureus, 336, 339, 509
  calcaratus, 339, 509
  coccineus, 339, 509
  limensis, 338, 339, 509
  lunatus, 339, 509
  multiflorus, 509
  vulgaris, 338, 509
Phleum pratense, 344, 498
Phoenix dactylifera, 418, 419, 491, 499
  sylvestris, 218, 499
Phormium tenax, 38, 39, 500
Photosynthesis, 6, 7
Phulwara butter, 205
Phyllanthus acidus, 434, 511
Phyllocladus trichomanoides, 121, 496
Phyllostachys, 47, 498
Phylloxera, 398
Phytelephas macrocarpa, 240, 500
Piassaba, 41
Piassava, 41, 42, 499, 500
Picea Abies, 114, 121, 167, 496
  canadensis, 496
  Engelmannii, 95, 496
  excelsa, 496
  glauca, 95, 229, 496
  mariana, 95, 496
  rubens, 95, 166, 167, 229, 496
  rubra, 496
  sitchensis, 95, 229, 496

Picraena excelsa, 510
Picrasma excelsa, 251, 510
Pigeon pea, 343, 508
Pigments, 13, 126–134
Pignolia, 355
Pignut, 99, 502
Pili nut, 201, 354, 510
Pili tree, 172
Pimenta acris, 515
  dioica, 437, 447, 448, 515
  officinalis, 515
  racemosa, 184, 515
Pimiento, 449
Pimpinella Anisum, 454, 455, 516
Piña cloth, 40, 432
Pinaceae, 153, 496–497
Pine, 72, 73, 91–94, 115
  cluster, 114, 497
  Cuban, 163
  digger, 355, 497
  jack, 85, 229, 232, 496
  kauri, 117, 155, 156, 496
  loblolly, 86, 94, 497
  lodgepole, 72, 87, 93, 497
  longleaf, 64, 86, 93, 162–164, 496
  maritime, 165, 497
  Norway, 94, 497
  nut, 355
  Parana, 107, 496
  red, 85, 94, 497
    New Zealand, 117, 496
  Scotch, 114, 497
  screw, 45, 497
  shortleaf, 86, 94, 497
  slash, 86, 94, 496
  southern, 66, 70, 72, 74
  stone, 114, 355, 497
  sugar, 78, 87, 93, 497
  Torrey, 355, 497
  white, 66–68, 70, 78, 85, 91–93, 497
    New Zealand, 117, 496
    western, 87, 93, 497
  yellow, 68–70, 78, 91–94
    southern, 67, 68, 81, 93, 229, 496
    western, 64, 67, 68, 70, 75, 87, 94, 163, 497
Pine nuts, 355
Pineapple, 20, 40, 287, 431–432, 484, 491, 500
Piñons, 355, 497
Pinus australis, 93, 162, 163, 229, 496
  Banksiana, 229, 496

Pinus, caribaea, 94, 163, 496
  cembroides var. *edulis*, 355, 497
    var. *monophylla*, 355, 497
  contorta, 93, 497
  cubensis, 496
  echinata, 94, 497
  edulis, 497
  halepensis, 166, 497
  heterophylla, 496
  Lambertiana, 93, 497
  maritima, 165, 497
  monophylla, 497
  monticola, 93, 497
  nigra, 166, 497
  palustris, 496
  Pinaster, 114, 165–167, 497
  Pinea, 114, 166, 355, 497
  ponderosa, 94, 163, 497
  resinosa, 94, 497
  Sabiniana, 355, 497
  Strobus, 92, 497
  succinifera, 158, 497
  sylvestris, 114, 166, 497
  Taeda, 94, 497
  Torreyana, 355, 497
Piper Betle, 276, 502
  Cubeba, 259, 502
  longum, 452, 502
  methysticum, 285, 502
  nigrum, 437, 451, 502
  officinarum, 502
  retrofractum, 452, 502
Piperaceae, 502
Piratinera guianensis, 57, 110, 504
Pistachio, 357, 467, 512
Pistachio-nut oil, 201
Pistacia cabulica, 161, 512
  lentiscus, 161, 512
  vera, 357, 512
Pisum sativum, 335, 337, 509
Pita floja, 40–41, 500
Pitanga, 423, 515
Piteira, 38, 501
Pitre, 38, 501
Pituri, 285, 518
Plaiting fibers, 19, 44–47
Plane tree, 114, 506
Planing-mill products, 68–69
Plant products, importance of, to man 1–5
  nature of, 5–17
Plant skeleton, 7–9

Plantaginaceae, 519
Plantago indica, 260, 519
    ovata, 260, 519
    Psyllium, 260, 519
Plantain, 386, 417, 501
Platanaceae, 506
Platanus occidentalis, 103, 506
    orientalis, 114, 506
Plum, 199, 287, 289, 385, 393–396, 492, 507
    coco, 425, 506
    hog, 427, 512
    governor's, 434, 514
    Java, 423, 515
    marmalade, 513
    Natal, 435, 517
    Spanish, 427, 512
Plywood, 76
Pochote, 49, 513
Podocarpus dacrydioides, 117, 496
    Totara, 117, 496
Podophyllum, 174, 247, 505
Podophyllum Emodi, 247, 505
    peltatum, 247, 505
Pogostemon Cablin, 187, 518
Poles and piling, 72
Polianthes tuberosa, 188, 501
Polygala Senega, 247, 511
Polygalaceae, 511
Polygonaceae, 504
Polymixin, 262
Pomades, 176
Pome fruits, 387–391
Pomegranate, 289, 432–433, 514
Pomelo, 410, 414, 510
Poncirus trifoliata, 414, 510
Pongam oil, 205
Pongamia glabra, 509
    pinnata, 205, 509
Poplar, 76, 78, 102, 115
    balsam, 85, 102, 502
Poppy, 287, 467
    opium, 197, 260, 279–280, 505
Poppy oil, 197
Populus balsamifera, 102, 502
    deltoides, 102, 502
    grandidentata, 102, 229, 502
    Tacamahacca, 502
    tremuloides, 102, 229, 502
Porcupine wood, 351
Porphyra laciniata, 295, 495
Porter, 489

Posts, 71
Potato, sweet, 287, 362–363, 491, 518
    white, 287, 289, 366–368, 491, 519
Potato starch, 221–223, 225
Prima vera, 113, 519
Proso millet, 329–330, 498
Prosopis glandulosa, 153, 347, 509
    juliflora, 153, 345, 346, 509
Proteaceae, 504
Protein fibers, 51, 313, 340, 342
Proteins, 12, 13, 290, 335, 356, 357
Protium heptaphyllum, 172, 511
Protoplasm, 5–6, 9
Prunes, 395, 396
Prunus, 153, 391
    americana, 395, 507
    Amygdalus, 356, 507
        var. *amara*, 356, 507
        var. *dulcis*, 356, 507
    Armeniaca, 391, 507
    avium, 392, 507
    Cerasus, 114, 392, 507
    communis, 507
    domestica, 394, 395, 507
    hortulana, 395, 507
    insititia, 394, 507
    maritima, 394, 395, 507
    nigra, 395, 507
    Persica, 392, 393, 507
        var. *nectarina*, 393, 507
        var. *nucipersica*, 507
    salicina, 395, 507
    serotina, 97, 507
    serrulata, 392, 507
    spinosa, 114, 394, 507
Pseudotsuga mucronata, 497
    taxifolia, 89, 90, 497
Psidium Cattleianum, 515
    Guajava, 422, 515
    littorale, 423, 515
Psychotria Ipecacuanha, 519
Psyllium, 260–261, 519
Pteridophyta, 496
Pterocarpas erinaceus, 129, 161, 509
    indicus, 115, 509
    Marsupium, 161, 509
    santalinus, 129, 509
    Soyauxii, 129, 509
Pueraria hirsuta, 509
    lobata, 344, 509
    Thunbergiana, 509
Pulpwood, 228–230

Pulque, 38, 490, 491, 501
Pulses, 335
Pumpkin, 199, 287, 382–383, 519
Punica Granatum, 432, 514
Punicaceae, 514
Purpleheart, 113, 509
Pyinkado, 116, 509
Pyrethrum, 244, 264–265
Pyroxylin, 237–238
Pyrus Aucuparia, 507
   baccata, 389, 507
   communis, 389, 390, 507
   Malus, 106, 387, 507
   pyrifolia var. *culta*, 390, 507
   serotina var. *culta*, 507

Q

Quassia, 251–252, 510
Quassia amara, 251, 510
Quebrachia Lorentzii, 512
Quebracho, 107, 122, 512
Quercitron, 119, 127, 131
Quercus, 357
   Aegilops, 503
   alba, 101, 102, 119, 357, 503
   bicolor, 102, 503
   borealis, 102, 119, 503
   Catesbaei, 503
   Cerris, 114, 503
   coccinea, 102, 503
   Garryana, 102, 503
   Ilex, 357, 503
   imbricaria, 102, 503
   infectoria, 124, 125, 503
   laevis, 102, 503
   macrocarpa, 102, 503
   macrolepis, 124, 503
   Michauxii, 503
   montana, 102, 119, 503
   palustris, 102, 503
   petraea, 114, 503
   phellos, 102, 503
   Prinus, 102, 503
   Robur, 114, 503
   rubra, 503
   sessiliflora, 503
   stellata, 102, 503
   Suber, 81, 503
   texana, 102, 503
   velutina, 102, 119, 131, **503**
   virginiana, 102, 357, 503

Quillaja Saponaria, 209, 507
Quince, 287, 289, 391, 506
Quinine, 244, 248–251, 493, 519
Quinoa, 279, 287, 334, 504

R

Rabbit brush, 146, 520
Radish, 287, 289, 361, 506
Raffia, 45
Ragi, 330–331, 498
Railroad cars, 70
Railroad ties, 73–74
Rain tree, 347, 509
Raisins, 197, 398–399
Rama, 33, 513
Ramelina calicaris, 186, 495
Ramie, 20, 31–32, 230, 504
Ranunculaceae, 504
Rape, 192, 199, 506
Rape oil, 199
Raphanus sativus, 361, 506
Raphia pedunculata, 45, 500
   Ruffia, 500
   vinifera, 41, 491, 500
Raspberry, 287, 385, 399–401, 484, 507
Raticides, 264, 267
Rattan, 44, 46, 499
Rayon, 234–237
Reconstructed wood, 76
Red gum, 67, 68, 70, 73–75, 78, 86, 103, 506
   Australian, 162, 515
Redtop, 344, 497
Redwood, 62, 68, 71–73, 77, 78, 87, 94–95, 497
Reeds, 44
Regenerated cellulose, 235
Reseda Luteola, 130, 506
   odorata, 188, 506
Resedaceae, 506
Reserve cellulose, 10, 11
Reserve food, 9–13
Resins, 15, 153–174, 244, 251, 252, 258
   acaroid, 160
   East India, **157**
   grass-tree, 160
   hard, 154–162
Rhamnaceae, 512
Rhamnus cathartica, **132**, 512
   chlorophora, 512
   globosa, 131, 512

INDEX

Rhamnus infectoria, 132, 512
  Purshiana, 246, 248, 512
  utilis, 131, 512
Rhea, 32, 504
Rheum Emodi, 247, 504
  officinale, 247, 504
  palmatum, 247, 504
  Rhaponticum, 378, 504
Rhizophora Mangle, 120, 515
Rhizophoraceae, 515
Rhodymenia palmata, 295, 495
Rhubarb, 247, 504
  garden, 287, 289, 372, 378, 504
Rhus chinensis, 125, 512
  copallina, 122, 512
  coriaria, 123, 512
  Cotinus, 511
  glabra, 122, 512
  semialata, 512
  succedanea, 159, 208, 512
  typhina, 122, 512
  verniciflua, 158, 512
Ribes, 402
  americanum, 403, 506
  Grossularia, 403, 506
  hirtellum, 403, 506
  nigrum, 403, 506
  sativum, 402, 403, 506
  vulgare, 506
Rice, 44, 231, 287–290, 296, 314–318, 488, 490, 498
  wild, 331–332, 499
Rice paper, 231
Rice starch, 221, 223
Ricinus communis, 200, 201, 511
Rivea corymbosa, 283, 518
Robinia pseudoacacia, 99, 509
Roccella tinctoria, 134, 495
Rolled oats, 324
Root beer, 484, 490
Rorippa Armoracia, 505
  Nasturtium-aquaticum, 506
Rosa centifolia, 179, 507
  damascena, 179, 180, 507
Rosaceae, 424, 506–507
Rose, 175, 176, 179, 447, 507
  cabbage, 179, 507
  damask, 179, 180, 507
Rose apple, 423, 515
Rose geranium, 180, 510
Rose oil, 179
Roselle, 33, 513

Rosemary, 178, 179, 185–186, 467, 518
Rosewood, 116, 508, 509
  Brazilian, 113, 508
  Burmese, 115, 509
Rosin, 125, 162–165
Rosmarinus officinalis, 185, 518
Rotenone, 265–267
Rough-weaving fibers, 19, 44–47
Rowan, 114, 507
Rubber, 135–146, 287
  Assam, 141–143, 503
  Castilla, 140, 141, 503
  caucho, 140, 503
  caura, 146, 511
  Ceara, 141, 142, 511
  chilte, 146, 511
  chrysil, 146, 520
  dandelion, 144–145
  guayule, 143–144
  Hevea, 136–140, 511
  India, 141, 142, 503
  Lagos silk, 143, 517
  Landolphia, 143, 517
  manicoba, 141
  palay, 146
  Panama, 140, 503
  Para, 136, 137, 511
Rubber vine, 517
Rubia tinctorum, 130, 519
Rubiaceae, 519
Rubus, 400
  alleghaniensis, 400, 507
  argutus, 400, 507
  flagellaris, 400, 507
  frondosus, 400, 507
  Idaeus, 401, 507
    var. *strigosus*, 401, 507
  loganobaccus, 400, 507
  occidentalis, 401, 507
  trivialis, 400, 507
  ursinus, 400
  vitifolius, 400, 507
Rue, 467, 510
Rum, 273, 492
Rumex hymenosepalus, 124, 504
Rushes, 44, 45, 231, 500
Ruta graveolens, 467, 510
Rutabaga, 362, 506
Rutaceae, 406, 510
Rutin, 275, 334
Rye, 44, 231, 263, 287–289, 296, 321–322, 488, 491, 492, 498

S

Sabal causiarum, 44, 500
  Palmetto, 42, 125, 500
Saccharum officinarum, 210–212, 498
Safflower, 127, 131, 132, 192, 197, 520
Safflower oil, 197
Saffron, 127, 131–132, 446–447
  meadow, 245, 500
Saffron crocus, 131, 446, 501
Sage, 178, 463–464, 518
  clary, 178, 467, 518
Sago, 221, 224–225
Sake, 490
Sal, 116, 514
Sal damar, 157
Salicaceae, 502
Salix alba, 114, 502
  nigra, 107, 502
Salmalia malabarica, 49, 513
Salsify, 361, 520
Salvia officinalis, 463, 464, 518
  Sclarea, 188, 467, 518
Samanea Saman, 347, 509
Samohu, 49, 513
Samuela carnerosana, 37, 500
Sandalwood, 116, 127, 187, 504
  red, 129, 509
Sandalwood oil, 178, 187
Sandarac, 153, 161, 497
Sanderswood, red, 129, 509
Sansevieria guineensis, 500
  longifolia, 39, 500
  Roxburghiana, 39, 500
  thyrsiflora, 39, 500
  zeylanica, 39, 500
Santalaceae, 504
Santalum album, 116, 187, 504
Santonin, 258, 520
Sap green, 132
Sapindaceae, 406, 512
Sapindus Saponaria, 209, 512
Sapium, 146
  sebiferum, 205, 511
Sapodilla, 148, 149, 433, 516
Saponaria officinalis, 209, 504
Saponins, 208
Sapotaceae, 146, 406, 433, 516
Sapote, 433, 516
  white, 435, 510
Sappanwood, 127, 128, 508
Sapucaia nuts, 349, 515

Sarsaparilla, 441, 484, 490, 500
Sassafras, 107, 177, 444–445, 505
Sassafras albidum, 107, 444, 505
  officinale, 505
  variifolium, 505
Satin walnut, 103
Satinwood, East Indian, 115, 510
  West Indian, 113, 510
Satureja hortensis, 464, 518
  montana, 464, 518
Sauerkraut, 375
Savory, summer, 464, 518
  winter, 464, 518
Savory herbs, 438, 462–467
Savory seeds, 438, 454–458
Sawdust, 78
Saxifragaceae, 506
Schinopsis Balansae, 107, 122, 512
  Lorentzii, 107, 122, 512
Schleichera oleosa, 160, 205, 512
  trijuga, 512
Scopolamine, 256, 257, 285
Screw pine, 45, 497
Scrophulariaceae, 519
Sea onion, 247, 500
Secale anatolicum, 321, 498
  cereale, 321, 322, 498
  montanum, 321, 498
Sechium edule, 381, 520
Secretions and excretions, 13–17
Secretory tissues, 14
Sedges, 44
Semidrying oils, 192, 197–199
Semolina, 305, 306
Senega, 247
Senna, 257, 508
Sennit, 39, 40
Sequoia gigantea, 94, 497
  sempervirens, 94, 497
Sesame, 192, 198, 467, 519
Sesame oil, 198
Sesamum indicum, 192, 198, **519**
  orientale, 519
Sesbania exaltata, 34, 509
  macrocarpa, 509
Setaria italica, 329, 498
  viridis, 329, 498
Shaddock, 410, 414, 510
Shagbark hickory, 99, 353, **502**
Shakes, 78
Shallots, 372, 500
Shallu, 326, 328, 498

Shavings, 78
Shea butter, 204, 516
Shellac, 159–160
Shield fern, marginal, 264, 496
Shiitake, 293, 495
Shingles, 77–78
Shorea aptera, 205, 514
  crassifolia, 514
  hypochra, 157, 514
  robusta, 116, 157, 514
  Wiesneri, 157, 514
Sida acuta, 34, 513
Silage, 312
Silk cotton, 47–49
  red, 49, 513
  white, 49, 514
Silky oak, 117, 504
Simal, 49
Simarubaceae, 510
Simmondsia californica, 511
  chinensis, 208, 511
Sinamay, 34
Sisal, 20, 36, 37, 230, 501
  Mexican, 36, 501
Sissoo, 116, 508
Slippery elm, 251, 503
Sloe, 114, 394, 492, 507
Small grains, 291, 324–331
Smilax aristolochiaefolia, 441, 500
  officinalis, 441, 500
  Regelii, 441, 500
Snakeroot, Senega, 247, 511
Snakewood, 57, 110, 504
Snuff, 273
Soap, 191
Soap substitutes, 208, 209
Soapbark, 209, 507
Soapberries, 209, 512
Soaproot, 209, 500
Soapwort, 209, 504
Socotrine aloes, 252, 500
Soda pulp, 232
Soda water, 484
Soft drinks, 483–484
Soft fibers, 20, 26–34
Solanaceae, 284, 518–519
Solanaceous narcotics, 278, 284–285
Solanum Melongena, 382, 519
  quitoense, 435, 519
  tuberosum, 366, 519
Solidago Leavenworthii, 146, 520
Soluble starch, 222, 225

Sorbus Aucuparia, 114, 507
Sorghum, 210, 287, 296, 324–328, 491, 498
  grain, 326–328
  grass, 325
  sweet, 219, 326
Sorghum halepensis, 325, 344, 498
  virgatum, 325, 344, 498
  vulgare, 325, 498
    var. *caffrorum*, 327, 498
    var. *caudatum*, 328, 498
    var. *cernuum*, 326, 498
    var. *durra*, 327, 498
    var. *nervosum*, 328, 498
    var. *Roxburghii*, 328, 498
    var. *saccharatum*, 219, 326, 498
    var. *subglabrescens*, 327, 498
    var. *sudanensis*, 325, 344, 498
    var. *technicum*, 42, 325, 498
Sorghum syrup, 219–220
Sorgo, 219, 326, 498
Sorva, 150, 517
Sour gum, 104, 516
Soursop, 418, 505
South American tubers, 368
Soybean, 192, 194, 195, 287, 289, 340–341, 508
Soybean oil, 194–195
Spanish moss, 50, 500
Spartina, 43
Spartina Spartinae, 43, 499
Spearmint, 177, 464–465, 518
Spelt, 298, 299, 499
Spices, 436–467
  crude, 436–437
Spike lavender, 184, 518
Spinach, 289, 372, 378, 504
  New Zealand, 378, 504
Spinacia oleracea, 378, 504
Spirits of turpentine, 162, 165
Spondias, 426
  cytherea, 426, 427, 512
  dulcis, 512
  lutea, 512
  Mombin, 427, 512
  purpurea, 427, 512
Spruce, 68, 70, 78, 95, 115, 229, 232, 490, 496
  black, 85, 95, 496
  Engelmann, 87, 95, 496
  Norway, 114, 121, 496
  red, 85, 95, 167, 229, 496

Spruce, Sitka, 75, 87, 95, 229, 496
   tideland, 95, 496
   white, 85, 95, 229, 496
Spruce beer, 490
Spruce gum, 166–167, 268
Squash, 287, 289, 382–383, 519
Squills, 247, 500
   red, 264, 267
Staple fiber, 237
Star anise, 452–453, 505
Star apple, 433, 516
Starch, 9, 221–226
   arrowroot, 221, 223, 224, 368
   bean, 221
   cassava, 224
   corn, 221, 222, 225, 226
   potato, 221–223, 225, 226
   rice, 221, 223
   sago, 221, 224–225
   soluble, 222, 225
   wheat, 221, 223
Starch grains, 11, 12, 221
Starch products, 225–226
Sterculia urens, 152, 513
Sterculiaceae, 513
Stipa tenacissima, 230, 499
Stizolobium Deeringianum, 336, 343, 509
Stone fruits, 391–396
Storax, 103, 169
Stout, 489
Stover, 312
Stramonium, 244, 257, 518
Strasbourg turpentine, 167
Strawberry, 287, 385, 399, 404–405, 484, 506
Streptomyces aureofaciens, 262, 495
   Fradiae, 262, 495
   griseus, 262, 495
   rimosus, 262, 495
   venezuelae, 262, 495
Streptomycin, 262
Strophanthus, 261, 517
Strophanthus hispidus, 261, 517
   Kombe, 261, 517
   sarmentosus, 261, 517
Structural fibers, 34–41
Structural timbers, 67–68
Strychnine, 260
Strychnos Nux-vomica, 246, 260, 517
   toxifera, 248, 517
Styracaceae, 153, 517
Styrax, 169, 178, 506

Styrax benzoides, 169, 517
   Benzoin, 170, 517
   tonkinense, 169, 517
Subtilin, 262
Succinite, 158
Sucrose, 10, 210, 212, 220
Sudan grass, 325, 344, 498
Sugar, 9, 10, 210–221, 273, 274
   beet, 214–217
   cane, 10, 210, 213, 214, 225
   corn, 225
   fruit, 10, 210, 220
   grape, 6, 9, 10, 210, 220, 225
   invert, 220
   maple, 217–218
   palm, 218–219
   wood, 239
Sugar apple, 417
Sugar beet, 210, 214–217, 231, 360, 504
Sugar cane, 208, 210–214, 220, 287, 289, 491, 492, 498
Sugar maple, 73, 75, 85, 98, 100, 210, 217–218, 230, 512
Sugar refining, 214
Sulphate pulp, 232–233
Sulphite pulp, 232
Sumac, 122, 512
   Sicilian, 123, 512
Sumac galls, 125
Summer savory, 464, 518
Sunflower, 192, 199, 520
Sunflower oil, 199
Sunn hemp, 32–33, 230, 508
Surface fibers, 20–26
Surinam cherry, 423
Sweet clover, 344, 509
Sweet flag, 182, 500
Sweet grass, 45, 498
Sweet gum, 103, 169, 506
Sweet potato, 287, 362–363, 491, 518
Sweet sorghums, 325, 326
Sweetsop, 417, 505
Swietenia macrophylla, 112, 511
   Mahogani, 111, 112, 511
Syagrus coronata, 203, 500
Sycamore, 67, 75, 103–104, 506
Syringa vulgaris, 517
Syrup, corn, 225
   maple, 217, 218
   sorghum, 219–220
Syzygium aromaticum, 437, 445, 446, 515
   Cumini, 423, 515

Syzygium Jambos, 423, 515
   malaccensis, 423, 515

T

Tabebuia Donnell-Smithii, 519
Tachardia lacca, 160
Tall oil, 165, 197
Tallows, 192
Tamarack, 71, 72, 74, 91, 229, 232, 496
Tamarind, 433, 434, 509
Tamarindus indica, 433, 509
Tampala, 379, 504
Tanacetum vulgare, 467, 520
Tanbark oak, 119, 503
Tanekaha bark, 121
Tangerine, 410, 510
Tanner's dock, 124–125, 504
Tanning industry, 118–119
Tanning materials, 118–126
Tannins, 11, 13, 14, 118, 125, 244, 475
Tansy, 177, 467, 520
Tapa cloth, 50
Tapioca, 224, 365
Tapioca starch, 226
Tara, 124, 508
Taraktogenos Kurzii, 514
Taraxacum kok-saghyz, 144, 520
   officinale, 378, 520
Taro, 368, 370, 500
Tarragon, 466, 520
Tauary, 50, 514
Taxaceae, 496
Taxodium distichum, 89, 497
Taxus baccata, 114, 496
Tea, 287, 468, 472–476, 513
Tea-seed oil, 201
Teak, 115, 116, 518
Tectona grandis, 115, 116, 518
Tempe, 340
Temperate fruits, 385–405
Teosinte, 308, 498
Terminalia Bellerica, 124, 515
   chebula, 124, 515
Ternstroemiaceae, 513
Terramycin, 262
Tetraclinis articulata, 161, 497
Tetragona expansa, 378, 504
Tetrapanax papyriferum, 231, 515
Tetrapterys, 283, 284, 511
Textile fibers, 18–41
Thea sinensis, 513

Theine, 475
Theobroma Cacao, 204, 476, 477, 513
Theobromine, 15, 479
Thorn apple, 257, 518
Thuja occidentalis, 88, 497
   plicata, 89, 497
Thyme, 179, 188, 465, 518
Thymelaeaceae, 514
Thymol, 465
Thymus vulgaris, 465, 518
Tilia americana, 96, 513
   cordata, 114, 513
   glabra, 513
Tiliaceae, 20, 513
Tillandsia usneoides, 50, 500
Timber, 66
Timbo, 266, 508
Timothy, 344, 498
Toadstools, 292
Tobacco, 197, 231, 268–275, 287, 289, 518, 519
Toddy, 219, 490
Tomato, 199, 287, 289, 383–384, 399, 484, 518
   tree, 435, 518
Tonka beans, 273, 453, 462, 508
Toona ciliata, 511
Toquilla, 44
Torula yeast, 239, 495
Torulopsis utilis, 239, 495
Totaquina, 251
Totara, 117, 496
Trachylobium verrucosum, 154, 509
Tragasol, 153, 346
Tragopogon porrifolius, 361, 520
Tree cotton, 25, 513
Tree legumes, 345–347
Tree tomato, 435, 518
Trifolium hybridum, 344, 509
   incarnatum, 344, 509
   pratense, 344, 509
   repens, 344, 509
Trigonella foenum-graecum, 458, 509
Trilisa odoratissima, 273, 520
Tripsacum, 308
Triticum aestivum, 297, 499
   compactum, 298, 499
   dicoccum, 298, 499
   durum, 298, 499
   monococcum, 298, 499
   polonicum, 298, 499
   sativum, 298, 499

Triticum Spelta, 298, 499
  tenax, 298
  Timopheevi, 298, 300, 499
  turgidum, 298, 499
  vulgare, 298, 499
Tropaeolaceae, 510
Tropaeolum tuberosum, 368, 369, 510
Tropical fruits, 406–435
Truffles, 291, 293, 495
Tsuga canadensis, 90, 91, 119, 229, 497
  heterophylla, 90, 119, 229, 497
  Mertensiana, 90, 497
Tuba, 265, 508
Tuber aestivum, 293, 495
  brumale, 293, 495
  melanosporum, 293, 495
Tuberose, 176, 179, 188, 501
Tubers, 359
  Andean, 369
  South American, 368
Tula istle, 37, 501
Tulip tree, 70, 75, 78, 85, 104, 505
Tung oil, 193–194
Tung-oil tree, 193, 194, 511
Tunis grass, 325, 344, 498
Tupelo, 67, 69, 75, 86, 104, 516
Turmeric, 127, 130, 131, 442, 501
Turnip, 287, 289, 362, 379, 506
Turpentine, 81, 162–167
  Bordeaux, 167
  Jura, 167
  oil of, 162, 164, 165, 177, 190
  spirits of, 162, 165
  Strasbourg, 167
  Venetian, 167
Turpentine industry, 162–166
Turpentines, 162–167
Typha angustifolia, 49, 497
  latifolia, 49, 497
Typhaceae, 497
Tyrothricin, 262

U

Ucuhuba butter, 206, 505
Ullucu, 368, 369, 504
Ullucus tuberosus, 368, 369, 504
Ulmaceae, 503
Ulmus americana, 99, 100, 503
  campestris, 503
  fulva, 503
  procera, 114, 503

Ulmus racemosa, 503
  rubra, 251, 503
  Thomasi, 99, 503
Ulva lactuca, 295, 495
Umbelliferae, 153, 172, 174, 175, 454, 455, 515–516
Uncaria Gambir, 123, 124, 519
Upas tree, 50, 503
Urena lobata, 34, 513
Urginea maritima, 247, 267, 500
Urticaceae, 20, 504
Urucú, 133, 514

V

Vaccinium angustifolium, 402, 516
  atrococcum, 402, 516
  canadense, 516
  corymbosum, 401, 402, 516
  macrocarpon, 402, 516
  membranaceum, 402, 516
  myrtilloides, 402, 516
  ovatum, 402, 516
  Oxycoccus, 402, 516
  pennsylvanicum, 516
  vacillans, 402, 516
  Vitis-Idaea, 402, 516
    var. *minus,* 402, 516
Valerian, 247–248, 519
Valeriana officinalis, 247, 519
Valerianaceae, 519
Valonia, 124
Vanilla, 287, 438, 453–454, 502
  wild, 273, 520
Vanilla fragrans, 502
  planifolia, 453, 502
  Pompona, 453, 454, 502
Vateria indica, 157, 514
Vegetable fats, 192, 193, 201–206
Vegetable ivory, 240–241
Vegetable parchment, 234
Vegetable sponge, 51, 519, 520
Vegetable tallow, Chinese, 205, 511
Vegetables, 359–384
  earth, 359–372
  fruit, 379–384
  herbage, 372–379
Vehicles, 70
Velvet bean, 336, 343, 509
Veneers, 74–75
Venetian turpentine, 167
Verbenaceae, 518

INDEX 559

Vermouth, 493
Vetch, 345, 509
Vetiver, oil of, 184, 499
Vetiveria zizanioides, 184, 499
Vicia Faba, 336, 341, 509
  sativa, 345, 509
  villosa, 345, 509
Vigna sinensis, 336, 340, 509
Vinegar, 351, 389, 489
Viola odorata, 185, 514
Violaceae, 514
Violet, 176, 179, 185, 447, 514
Virola, 206, 505
Viscose products, 239
Vitaceae, 513
Vitamins, 16, 264, 290, 291, 294, 359, 373, 407, 412, 422
Vitis aestivalis, 398, 513
  Labrusca, 398, 513
  rotundifolia, 398, 513
  vinifera, 397, 398, 485, 513
  vulpina, 398, 513
Volatile oils, 13, 175–190

W

Walnut, 54, 68, 70, 104, 115, 287, 289' 348, 355–356, 502
  black, 75, 85, 104–105, 287, 355, 502
  English, 114, 196, 356, 502
Walnut oil, 196
Water cress, 378, 506
Watermelon, 199, 287, 397, 519
Wattle, 121, 507
Wax, 15, 206–208
  candelilla, 206, 511
  carnauba, 206, 499
  cauassú, 207
  jojoba, 208
  myrtle, 207
Wax tree, Japanese, 208, 512
Weld, 127, 130, 506
Wheat, 12, 44, 231, 287–290, 296–306, 472, 492, 499
  club, 298–301, 499
  common, 298, 300, 301
  durum, 297, 298, 300, 301, 306, 499
  hard, 288, 300, 306
  Polish, 298, 299, 499
  poulard, 298, 299, 499
  soft, 288, 297, 300, 306
  spring, 300, 301
  winter, 300, 301

Wheat starch, 221, 223
Whisky, 491–492
White potato, 287, 289, 366–368, 491, 519
Whitewood, 104
Wickerwork, 46–47
Wickstroemia canescens, 231, 514
Wild rice, 331–332, 499
Willow, 44, 45, 79, 107, 114, 121, 502
Wine grape, 397–398, 405, 513
Wine-producing countries, 486–488
Wines, 485–488, 492
Wintergreen, 175, 177, 188, 466–467, 484, 490, 516
Witch hazel, 177, 256, 506
Woad, 127, 130, 506
Wood, 1, 7, 52–81
  decay in, 62
  defects in, 61–63
  diagnostic features of, 54–57
  figure in, 56–57
  grain in, 56–57
  importance of, 52–53
  insect damage to, 61–62
  mechanical properties of, 57–59
    factors influencing, 59–63
  porous and nonporous, 54, 55
  preservation of, 62–63
  seasoning of, 61
  structure of, 53–57
  uses of, 63–81
Wood alcohol, 80–81, 226
Wood alloys, 76
Wood distillation, 80, 81
Wood fibers, 8, 20, 53, 54, 228–230
Wood flour, 79
Wood gas, 80, 81
Wood oil, 80, 171
Wood pulp, 231–233, 234
Wood sugar, 239
Wood tar, 80, 81
Wood turpentine, 81
Wood wool, 78
Wood-working industries, 68–71
Woods, of temperate North America, 88–107
  of tropical America, 108–113
Wormseed, 177, 244, 258, 261, 504, 520
Wormwood, 177, 257, 520

X

Xanthorrhoea australis, 160, 501
  hastilis, 160, 501

Xanthorrhoea tateana, 160, 501
Xanthosoma sagittifolium, 371, 500
Xylia dolabriformis, 509
  xylocarpa, 116, 509

Y

Yajé, 283, 511
Yam, 287, 289, 363, 364, 501
Yam bean, 368, 509
Yaupon, 482, 512
Yautia, 371, 500
Yellow poplar, 104
Yew, 114, 115, 496
Ylang-ylang, 181–182, 505
Yoco, 483, 512
Yuca, 364
Yucca, 37, 491, 501

Z

Zacaton, 43, 498
Zamia floridana, 224, 496
Zanthoxylum flavum, 113, 510
Zapatero, 109, 514
Zea, 307
  Mays, 306, 308, 499
    var. *amylacea*, 499
    var. *erythrolepis*, 499
    var. *everta*, 499
    var. *indentata*, 499
    var. *indurata*, 499
    var. *praecox*, 499
    var. *rugosa*, 499
    var. *saccharata*, 499
    var. *tunicata*, 499
Zedoary, 188, 442, 501
Zein, 313
Zingiber officinale, 439, 440, 501
Zingiberaceae, 501
Zizania aquatica, 331, 499
    var. *angustifolia*, 331, 499
  palustris, 499
Zizyphus Jujuba, 160, 423, 424, 512
  xylopyrus, 160, 512
Zygophyllaceae, 520